Biogeography

Biogeography

An ecological and evolutionary approach

C. Barry Cox MA, PhD, DSc

Former Head of Biological Sciences at King's College London

Peter D. Moore PhD

Emeritus Reader in Ecology at King's College London

EIGHTH EDITION

John Wiley & Sons, Inc.

VICE PRESIDENT AND	
EXECUTIVE PUBLISHER	Jay O'Callaghan
EXECUTIVE EDITOR	Ryan Flahive
MARKETING MANAGER	Margaret Barrett
SENIOR PRODUCTION EDITOR	William A. Murray
ASSISTANT EDITOR	Veronica Armour
EDITORIAL ASSISTANT	Meredith Leo
CREATIVE DIRECTOR	Harry Nolan
COVER DESIGN	Jeof Vita
PHOTO RESEARCHER	Sheena Goldstein
COVER PHOTO	Chad Ehlers/Photolibrary Group Limited

This book was set in 9/11.5 Trump Mediaeval by Aptara and printed and bound by R.R. Donnelley.

This book is printed on acid free paper. ∞

ISBN: 978-0-470-63794-4

Printed in the United States of America

10 9 8 7 6 5 4 3 2

Contents

THE ENGINES OF THE PLANET

ISLANDS & OCEANS

HISTORICAL BIOGEOGRAPHY

Preface

To interpret the phenomena of biogeography, we need to understand many different areas of science—for example, evolution, taxonomy, ecology, geology, palaeontology, and climatology. Though each area makes its own individual contribution, a textbook such as this therefore has to cover a similar scope and must be suitable for students with a variety of different backgrounds. This is particularly necessary today, when the final coming-together of molecular methods of demonstrating relationship, and the cladistic technique of imposing pattern upon the resulting data, are now revolutionizing our understanding of biogeography.

Many changes have taken place in the study of biogeography over the 36 years and eight editions that have passed in the life of this textbook. Back in 1973, the depth of the problem that our species poses for our planet's biota and climate was hardly appreciated, and the "greenhouse effect" was still mainly a concern for horticulturalists rather than a worry for the whole planet.

It was not until the 1980s that the evidence that the Earth's climate is changing, and that this is increasingly the result of human activities, became steadily greater. This eventually led to greater public debate and scientific involvement. In interpreting the interaction between the physical world and the living world, and of the human impact on each of these, biogeography clearly has a major role to play in assessing the likely results of climatic change and in suggesting how best to counter them. As climatic change renders old crop areas less fertile, will it be possible to find new areas to replace them—and, if so, where? Or shall we have to find new varieties of plant, adapted to these new conditions—and, if so, where are we likely to find them? This was probably the reason for the great increase in the amount of research on biogeography that took place during the 1990s.

But it is not only our foodstuffs that are threatened by climate change, it is the diversity of life that inhabits environments that are shrinking and disappearing. This is not just a concern for the curators of museums and herbaria, for we are also becoming aware of the extent to which we rely on this diversity for new drugs as well as for new food-plants. So we became aware of the need to census this diversity, in order to appreciate where it is greatest and where it is under particular threat. Which habitats are threatened, where and how should we attempt to preserve them? One response of biogeographers to these new imperatives was to examine whether it was possible to use the integrative Theory of Island Biogeography, put forward in 1963, as a guide to the design of nature reserves. Though some of the claims for the predictive power of this theory now seem to have been overplayed, it remains an important concept in this area of human concern.

Biogeography textbooks have traditionally not treated marine biogeography at similar depth and detail as continental biogeography and island biogeography. Perhaps that was partly because marine biogeography was comparatively not well understood and, maybe, less important. However, we are now realizing that the seemingly limitless harvest of the seas is faltering under the impact of our soaring demands for food. Fortunately, the efforts of marine biologists to understand the dynamics of the life of the seas has recently been greatly aided by new techniques of evaluation, made possible by the use of satellite-based photographic monitoring. That is why this textbook includes a chapter on marine biogeography, so that students can compare and evaluate the data, patterns, and problems arising from continental, marine, and island biogeography within this single book.

The book begins with a chapter on the history of biogeography, which has a particular role in the teaching of the subject. It is very tempting for students to believe that what they are taught is the simple, unalloyed truth. History not only shows us how the subject has developed, but also how new theories are tested and, eventually, modified or rejected. It also, by implication, tells us that the science that we are taught today is not some semicrystalline structure of truth and beauty. Instead, it is merely the latest description of a beach that will soon be changed, perhaps only a little by a high tide or wind, but perhaps more fundamentally by a storm or earthquake.

The first chapter also begins with a section that attempts to make students aware that science is not a unique area of human endeavour, carried out by passionless automata. Instead, the scientists whose work and ideas are presented here are, like the rest of humanity, subject to error, prejudice, and rivalry, and limited by the current assumptions and restrictions of science and society. In this book, we also try to evaluate conflicting theories and to give a reasoned judgment as to which is preferable. Whether a student ends up as a working scientist or within another area of our society, these are important lessons to learn and to apply when reading either the literature of science or its reporting in the newspapers and other media.

This new edition appears at what we believe is an historic turning point in the discipline of biogeography. Until recently, biologists were unable to document the dates at which new species appeared and diverged from one another. As a result, it was impossible to be sure of the relationship between these biological processes and events such as the separation of units of land by plate tectonics or climatic change. This in turn permitted the development of rival schools of thought, such as the panbiogeographers or vicariance biogeographers on the one hand, and the dispersalists on the other who, as explained in the first chapter, engaged in bitter controversy over their rival theories. The rise of molecular methods of investigation that provide reliable dates for the times of appearance and divergence of species has given us new confidence in the accuracy of our biogeographical analyses. This has led to the end of decades of sterile arguments on which interpretation of the evidence is the "correct" one. These have been replaced by deductions based on rigorous techniques of analysis of relationships in time and space, using increasingly complex and sophisticated computer algorithms.

At last it seems that biogeographical research is revealing, with increasing scope and detail, a single, consistent story of the history of the biogeography of the world today. This story shows how the results of the action and interaction of the two great engines of the world, plate tectonics and evolution, have led to the appearance of the patterns of life that we see on the planet in the past and today. As these two engines are, respectively, the paradigms of the earth sciences and the biological sciences, the demonstration that biogeography can show how the results of each are concordant not only strengthens each paradigm but also give biogeography a unique position in the natural sciences.

It is fortunate indeed that this transformation has taken place just when the increasing urgency for precise responses to the questions raised by climatic change has made the role of biogeography more important than ever before.

Acknowledgments

First, the authors wish to thank Dr. Richard Field, who refereed the original manuscript. His detailed, thorough, and constructive comments led to important improvements in the book, for which we are very grateful. Any remaining errors are, of course, our responsibility!

As noted earlier, biogeography involves the study of a very great range of data in the fields of both earth sciences and biological sciences, and it is nowadays impossible for any one person to cover all the literature in such a huge area. Our task in trying to identify the significant new references has been greatly aided by many people, but we would like to thank in particular the following:

Professor David Bellwood, School of Marine and Tropical Biology, James Cook University, Queensland, Australia

Professor Robert Hall, Earth Sciences Department, Royal Holloway College, University of London, UK

Professor Blair Hedges, Department of Biology, Pennsylvania State University, USA

Dr Isabel Sanmartín, Reál Jardín Botánico, Madrid, Spain

Entry numbers in **bold type** in the index indicate pages on which the concept involved is defined; these words or concepts are also shown in bold type on the page in question. These concepts also appear in the Glossary.

A History of Biogeography

History is the geography of time

This introductory chapter begins with an explanation of why the study of the history of a subject is important, and highlights some of the important lessons that students may gain from it. This is followed by a review of the way in which each of the areas of research in biogeography developed from its foundation to today.

Lessons from the Past

One of the best reasons for studying history is to learn from it; otherwise it becomes merely a catalogue of achievement. So, for example, it is often valuable to think about why a particular advance was made. Was it the result of personal courage in confronting the current orthodoxy of religion or science? Was it the result of the mere accumulation of data, or was it allowed by the development of new techniques in the field of research, or in a neighboring field? But the study of history also gives us the opportunity to learn other lessons—and the first of these is humility. We must be wary, when considering the ideas of earlier workers, not to fall into the trap of arrogantly dismissing them as in some way inferior to ourselves, simply because they did not perceive the "truths" that we now see so clearly. In studying their ideas and suggestions, one soon realizes that their intellect was no less penetrating than those that we can see at work today. However, compared to the scientists of today, they were handicapped by lack of knowledge and by living in a world in which, explicitly or implicitly, it was difficult or impossible to ask some questions.

First, less was known and understood. When Isaac Newton wrote that he had "stood on the shoulders of giants," he was acknowledging that, in his own work, he was building upon that of generations of earlier thinkers and was taking their ideas and perceptions as the foundations of his own. So, the further we go back in time, the more we see intellects that had to start afresh, with a page that was either blank or contained little in the way of earlier ideas or syntheses.

Second, we must be very aware that, for every generation of workers, the range of theories that one might suggest was (and is!) limited by what contemporary society or science views as acceptable or respectable. Attitudes to the idea of evolution (Ch. 6) or of continental drift (described below) are good examples of such inhibitions in the 19th and 20th centuries. The history of scientific debate is rarely, if ever, one of dispassionate, unemotional evaluation of new ideas, particularly if they conflict with one's own. Scientists, like all men and women, are the product of their upbringing and experience, affected by their political and religious beliefs (or disbeliefs), by their position in society, by their own previous judgments and publicly expressed opinions, and by their ambitions—just as "there's no business like show business," there's no interest like self-interest! A very good example of this is the use of the concept of evolution by the rising middle-class scientists of England as a weapon against the 19th-century establishment (see below, p. 10), while, at the individual level, the history of Leon Croizat and his ideas (see below, p. 17) provides an interesting study.

In our survey of the history of biogeography, we shall therefore see people who, like most of us, grew up accepting the intellectual and religious ideas current in their time, but who also had the curiosity to ask questions of the world of nature around them. Sometimes the only answers that they could find contradicted or challenged the current ideas, and it was only natural then to seek ways to circumvent the problem. Could these ideas be reinterpreted to avoid the problem, was there any way, any loophole, to avoid a complete and direct challenge and rejection of what everyone else seemed to accept?

So, to begin with, the reactions of any scientist confronted with results or ideas that conflict with current dogma is either to reject them ("Something must have gone wrong with his or my methods") or to view them as an exception ("Well, that's interesting, but it's not mainstream"). Sometimes, however, these difficulties and "exceptions" start to become too numerous, too varied or to arise from too many different parts of science. The scientist may then realize that the only way around it is to start again, to start from a completely different set of assumptions, and to see where that leads. Such a course is not easy, for it involves the tearing up of everything that one has previously assumed and completely reworking the data. And, of course, the older you get, the more difficult it is to do so, for you have spent a longer time using the older ideas and publishing research that explicitly or implicitly accepts them. That is why, all too often, older workers take the lead in rejecting new ideas, for they see them as attacking their own status as senior, respected figures. Sometimes these workers also refuse to accept and to use new approaches long after these have been thoroughly validated and widely used by their younger colleagues (see attitudes to plate tectonic theory, Ch. 5). Another problem is that the debate can become polarized, with the supporters of two contrasting ideas being concerned merely to try to prove that the opponents' ideas are false, badly constructed, and untrue (see dispersal vs. vicariance, pp. 15–20; punctuated vs. gradual evolution, p. 194). Neither side then stops to consider whether it is perhaps possible that both of the apparently conflicting ideas are true and that the debate should instead be about when, under what circumstances, and to what extent one idea is valid, and when the other is instead the more important. Also, only too often, scientists have rejected the suggestions of another worker, not because they were in themselves unacceptable, but because they rejected *other* opinions of that same author (e.g., Cuvier vs. Lamarck on evolution, see below, p. 9).

All of this is particularly true of biogeography, for it provides the additional difficulty of being placed at the meeting point of two quite different parts of science—biological sciences and earth sciences. This has had two interesting results. The first is that, from time to time, lack of progress in one area has held back the other. For example, the assumption of stable, unchanging geography made it impossible to understand past patterns of distribution. Nonetheless, it was a reasonable assumption until the acceptance of plate tectonics (continental drift) provided a vista of past geographies that had gradually changed through time. But it is also interesting to note that this major change in the basic approaches of earth sciences came in two stages.

To begin with, the problem was clearly posed and a possible solution was given. This was in 1912, when the German meteorologist Alfred Wegener (see below, p. 16) pointed out that many patterns in both geological and biological phenomena did not conform to modern geography, but that these difficulties disappeared if it was assumed that the continents had once lain adjacent to one another and had gradually separated by a process that he called **continental drift**. This explanation did not convince the majority of workers in either field of work, largely because of the lack of any known mechanism that could cause continents to move horizontally or to fragment. The fact that Wegener himself was not a geologist but an atmosphere physicist did not help him to persuade others of the plausibility of his views, for it was only too easy for geologists (who, of course, "knew best") to dismiss him as a meddling amateur. Most biologists, faced with the uncertainties of the fossil record, did not care to take on the assembled geologists.

The second stage came only in the 1960's, when geological data from the structure of the seafloor and from the magnetized particles found in rocks (see p. 153) not only provided unequivocal evidence for continental movements, but also suggested a mechanism for them. Only then did geologists accept this new view of world history (known as plate tectonics; see p. 155), and only then could biogeographers confidently use the resulting coherent and consistent series of palaeogeographical maps to explain the changing patterns of life on the moving continents. Such a theory, based on a great variety of independent lines of evidence, is known as a **paradigm**, and the theory of plate tectonics is the central paradigm of the earth sciences.

The moral of this story is, perhaps, that it is both understandable and reasonable for workers in one field (here, biologists) to wait until specialists in another field (here, geology) have been convinced by new ideas before they feel confident in using them to solve their own problems. This, in turn, leads to the second topic that results from the position of biogeography between biology and geology. That is the temptation for workers in one field, frustrated by lack of progress in some aspect of their own work, to accept uncritically and without proper understanding, new ideas in the other field that seem to provide a solution [1]. One must be particularly wary of new theories that are directed at explaining merely one difficulty in the currently accepted interpretations. This is because such suggestions sometimes simultaneously destroy the rest of the framework, without satisfactorily explaining the vast majority of the phenomena that were covered by that framework. For example, in the second half of the 20th century, some geologists suggested that the Earth had expanded, or that there had once been a separate "Pacifica" continent between Asia and North America. Some biological biogeographers welcomed these ideas as the solution to some detailed problems of the distribution of terrestrial vertebrates, even though they were not supported by geological data and had not been accepted by geologists.

All of this has important lessons for us today, for it would be naive to believe that the assumptions and methods used in biogeography today are in some way the final and "correct" ones that will never be rejected or modified. Similarly, every student should realize that those who teach science today have, of course, been trained to accept the current picture of the subject and may find it difficult to accept changes in its methodology. The price that we pay for gaining experience with age is an increasing conviction of the correctness of our own methods and assumptions! (On the other hand, it is interesting to note that, while in the physical sciences, major new discoveries are usually made by intuitive leaps early in the scientist's career, those in the biological sciences are more often made only later, after the accumulation of data and knowledge.) It is also worth noting that erroneous assumptions are far more dangerous than false reasoning because the assumptions are usually unstated, and therefore far more difficult to identify and correct. So, the past with its false assumptions and erroneous theories is merely a distant mirror of today, warning us in our turn not to be too sure of our current ideas. Sometimes the limitations and problems of a new technique only become apparent gradually, some time after it has been introduced.

But, of course, those of us who carry out research and publish our ideas in books such as this also have a responsibility to use their experience and judgment in trying to choose between conflicting ideas, showing which we prefer and why. For example, in this book the author who wrote the relevant chapter (Barry Cox) has criticized the methodology of a school of (mainly) New Zealand panbiogeographers (see p. 19). He could be wrong, however, and interested students should read around the subject and come to their own conclusions. After all, the purpose of learning a subject at this level is for students to develop their own critical faculties, not merely to acquire attitudes and opinions. Even over the past 35 years, we have seen attitudes to a new idea, the Theory of Island Biogeography, change quite considerably (see below and Chapter 8). How many of the explanations and assumptions in this book will still seem valid in 35 years' time? But that is also one of the pleasures of being part of science, of having to try continually to adapt to new ideas, rather than merely being part of some ancient monolith of long-accepted "truths."

Finally, this chapter is entitled "*A* History of Biogeography" rather than "*The* History of Biogeography" because any account is, inevitably, a wholly personal selection and interpretation. With a limited amount of space, the author must select for inclusion those workers and papers that he or she finds particularly interesting and important, and omit many that another author might have instead selected. Another biogeographer, with different interests and opinions, would have written a different "History," emphasizing different aspects of the subject.

Ecological vs. Historical Biogeography and Plants vs. Animals

The most fundamental split in biogeography is that between ecological and historical aspects of the subject. **Ecological biogeography** is concerned with the following types of questions. Why is a species confined to its present range in space? What enables it to live where it does, and what prevents it from expanding into other areas? What roles do soil, climate, latitude, topography, and interactions with other organisms play in limiting its distribution? How do we account for the replacement of species as one moves up a mountain or seashore, or from one environment to another? Why are there more species in the tropics than in cooler environments? Why are there more endemic species in environment X than in environment Y? What controls the diversity of organisms that is found in any particular region? Ecological biogeography is, therefore, concerned with short-term periods of time, at a smaller scale, with local, within-habitat or intracontinental questions, and primarily with species or subspecies of living animal or plant. (Subspecies, species, genus (plural: genera), family, order, and phylum (plural: phyla) are progressively larger units of biological classification. Each is known as a **taxon**: plural taxa.)

Historical biogeography, on the other hand, is concerned with different questions. How did the taxon come to be confined to its present range in space? When did that pattern of distribution come to have its present boundaries, and how have geological or climatic events shaped that distribution? What are the species' closest relatives, and where are they found? What is the history of the group, and where did earlier members of the group live? Why are the animals and plants of large, isolated regions, such as Australia or Madagascar, so distinctive? Why are some closely related species confined to the same region, while in other cases they are widely separated? Historical biogeography is, therefore, concerned with long-term, evolutionary periods of time, with larger, often sometimes global areas, and often with taxa above the level of the species and with taxa that may now be extinct.

Because of the different nature of plants and of animals, the way in which ecological and historical biogeography have been investigated and understood has differed in the two groups. Plants are static, and their form and growth are therefore much more closely conditioned by their environmental, ecological conditions than are those of animals. It is also far easier to collect and preserve plants than animals, and to note the conditions of soil and climate in which they live. But the fossil remains of plants are less common than those of animals, and they are also far more difficult to interpret, for several reasons. There are many more flowering plants than there are mammals—some 300 living families and 12,500 genera of plant, compared with 100 living families and 1,000 genera of mammal. Furthermore, although the leaves, wood, seeds, fruit, and pollen grains of flowering plants may be preserved, they are rarely found so closely associated that one can be sure which leaf belongs with which pollen etc. Finally, the taxonomy of flowering plants is based on the characteristics of their flowers, which are only rarely preserved. In contrast, the fossil bones of mammals are often associated as complete skeletons, which are easy to allocate to their correct family, and which provide a detailed record of the evolution and dispersal of these families within and between the continents through geological time.

For all these reasons, the biogeography of the more distant past was, until recently, largely the preserve of zoologists, while plant scientists were far more concerned with ecological biogeography—though studies of fossil pollen from the Ice Ages and postglacial times, which are easy to allocate to existing species, have been fundamental in interpreting the history as well as the ecology of this most recent past (see Chapter 12).

In following the history of biogeography, it would be easy merely to follow a path through time, recounting who discovered what and when. But it is more instructive instead to take each thread of the components of biogeography in turn, to follow the different contributions to its understanding, and to note on the way the lessons to be learned from how the scientists reacted to the problems and ideas of their time.

Biogeography and Creation

Biogeography, as a part of Western science, began in the mid-18th century. At that time, most people accepted the statements in the Bible as the literal truth, that the Earth and all living things that we see today had been created in a single series of events. It was also thought that these events had taken place only a few thousands of years before, and it was believed that God's actions had always been perfect. It followed that the animals and plants that had been created were perfect, had not changed (evolved) or become extinct, and that the world itself had always been as we see it today. The history of biogeography between then and the middle of the 20th century is the story of how that limited vision was gradually replaced by the realization that both the living world and the planet that it inhabits are continually changing, driven by two great engines—the biological engine of evolution and the geological engine of plate tectonics.

So, when the Swedish naturalist Linnaeus in 1735 started to name and describe the animals and plants of the world, he assumed that each belonged to an unchanging species, which had been created by God. But he soon found that there were species whose characteristics were not as constant and unchanging as he had expected. That might puzzle him, but he could only accept it. But there was a further problem, for, according to the Bible, the whole world had once been covered by the waters of the Great Flood. All the animals and plants that we see today must therefore have spread over the world from the point where Noah's Ark had landed, thought to be Mount Ararat in eastern Turkey. Linnaeus ingeniously suggested that the different environments to be found at different altitudes, from tundra to desert, had been colonized in turn by animals from the Ark as the floodwaters receded, progressively uncovering lower and lower levels of land. Linnaeus recorded in what type of environment each species was found, and so began what we now call ecological biogeography. He also recorded whereabouts in the world each species is found, but he did not synthesize these observations into any account of faunal or floral assemblages of the different continents or regions.

The first person to realize that similar environments, found in different regions of the world, contained different groupings of organisms, was the French naturalist Georges Buffon; this important insight has come to be known as **Buffon's Law**. In various editions of his multivolume *Histoire Naturelle* [2], published from 1761 onward, he identified a number of features of world biogeography and suggested possible explanations. He noted that many of the mammals of North America, such as bears, deer, squirrels, hedgehogs and moles, were found also in Eurasia, and he pointed out that they could only have traveled between the two continents, via Alaska, when climates were much warmer than today. He accepted that some animals, such as the mammoths, had become extinct. Buffon also realized that most of the mammals of South America are quite different from those of Africa, even though they live in similar tropical environments. Accepting that all were originally created in the Old World, he suggested that the two continents were at one time adjacent and that the different mammals then sought out whichever area they found most congenial. Only later did the ocean separate the two continents and the two now-different faunas, while some other differences might have been due to the action of the climate. Buffon also used the fossil record to reconstruct a history of life that clearly had extended over at least tens of thousands of years. Only the last part had witnessed the presence of human beings, and included earlier periods within which tropical life had covered areas that are now temperate or even subarctic.

Buffon strongly felt that one had to be guided by study of the facts, and this conviction drove him to accept that geography, climate, and even the nature of the species were not fixed, but changeable, and to suggest that continents might move laterally and seas encroach upon them. That was a truly courageous and visionary deduction to make in the late 18th century. So Buffon recognized, commented upon, and attempted to explain many phenomena that other, later, workers either ignored or merely recorded without comment. His observations on the differences between the mammals of the two regions were soon extended to land birds, reptiles, insects, and plants.

The Distribution of Life Today

As 18th-century explorers and naturalists revealed more and more of the world, so they also extended the horizons of biogeography itself, discovering a greater diversity of organisms. For example, in his second voyage around the world in 1772–1775, the British navigator Captain James Cook took the British botanist Joseph Banks and the German Johann Reinhold Forster, together with his son Georg Forster, who collected thousands of species of plants, many of them new to science. Forster found that Buffon's Law applied to plants as well as to animals, and also applied to any region of the world that was separated from others by barriers of geography or climate [3]. He also realized that there are what we now call gradients of diversity (see p. 124), there being more plant species closer to the equator, and progressively fewer as one moves toward the poles, and he made the first observations of island biogeography.

The concepts of ecological biogeography, botanical regions, and island biogeography had, then, all been recognized by the end of the 18th century. But it was still generally accepted that there could be little or no change in the nature of each species, or in the pattern of the geography of the world. The early naturalists therefore still struggled to explain how all these different floras had come into existence, widely scattered over the Earth's surface. Perhaps the most plausible explanation was that of the German botanist Karl Willdenow, who in 1792 suggested that, though there had been only one act of creation, it had taken place simultaneously in many places. In each area the local flora had been able to survive the Flood by retreating to the mountains, from which it was able later to spread downward to recolonize its own part of the world as the floodwaters receded. His book also included a chapter on the history of plants, and he noted that plants' growth habits were related to the conditions of their environment.

Despite the work of these two earlier botanists, the German Alexander von Humboldt is usually recognized as the founder of plant geography, perhaps because he was a far wealthier and more flamboyant figure. But Forster and Willdenow not only preceded Humboldt but also greatly influenced his life. It was Georg Forster who inspired Humboldt to become an explorer, while the slightly older Willdenow introduced him to botany and became his lifelong friend. Humboldt became famous for his 1799–1804 expedition to South America, during which he climbed to over 5,800 m (19,000 ft) on the volcano Chimborazo—a world height record that he held for 30 years. He noticed that the plant life on the mountain showed a zonation according to altitude, much like the latitudinal variation that Forster had described. Plants at lower levels are of tropical type, those of intermediate levels are of temperate type, and finally arctic types of plant are found at the highest levels. (Humboldt used the term *association* to describe the assemblages of plants that characterized each of these life zones; today, they are more commonly referred to as *formations* or *biomes*; see p. 103.) Humboldt believed that the world was divided into a number of natural regions, each with its own distinctive assemblage of animals and plants. He was also the first to insist that even biological observations had to include detailed, accurate, and precisely recorded data. He published a thorough account of his botanical observations in 1805, as part of a 30-volume series recording his findings in the New World [4].

Another early plant biogeographer was Augustin de Candolle of Geneva who in 1805, together with Lamarck, published a map showing France divided into five floristic regions with different ecological conditions. Candolle later went on to study the dispersal of plants by water, wind, or the actions of animals, pointing out that this would lead to the plants spreading until they encountered barriers of sea, desert, or mountains. He was also the first to realize that another limiting factor was the presence of other plants that competed with them. The result of these processes would be the appearance of regions that, though they might contain a variety of climatic zones and ecological environments, were distinct from one another because they contained plants that were restricted to that area, for which he coined the word "endemic" (see pp. 39, 57). The distinctions between these regions were thus partly dependent on their histories. Candolle went on to define 20 such regions, of which 18 were continents or parts of continents and two were island groups [5]. He also noted that some plants had apparently worldwide distributions, that species-pairs are to be found in Europe and North America, and that some taxa are found in both the north and the south temperate regions (what we now call bipolar distributions). Finally, he realized that other plants have strangely "disjunct" distributions (see p. 43) in locations that are widely separated from one another, such as the Proteas of southern Africa and Australia/Tasmania. Candolle also commented on Forster's contributions to island biogeography.

All in all, Candolle made a massive and varied intellectual contribution to the botany of the early 19th century. However, he did not provide any world maps to illustrate these concepts, and most of the maps that botanists published in the later 19th century, and even in the 20th century, continued to be primarily "vegetation maps"—maps of the relationships of the vegetation to temperature and climate. So, though the Danish botanist Joakim Schouw was the first to classify the world's flora and show the results on maps [6], these were mainly the distribution maps of particular groups of plants, rather than maps of regional floras. Grisebach's more detailed, colored map of 1866 was similarly a vegetation map. So all of these maps were primarily concerned with ecological biogeography, rather than with systematic studies of the distribution of organisms, which would have demanded an historical explanation. But then, it was only after biologists had become convinced of the reality of evolution that they could start to integrate into their thinking the impact of the fourth dimension—time.

Evolution—A Flawed and Dangerous Idea!

During the late 18th century, much of the leading work on biological and geological subjects had been carried out in areas of Europe that we now call Germany, but the French Revolution of 1789 led to a flowering of French science. To some extent, this was because the power of the Church, with its conservative influence on the generation and acceptance of new ideas, had been decisively broken. But the new government also carried out a complete reorganization of French science, centered on the new National Museum of Natural History, and liberally supported by the state, which became a powerhouse of ideas and debate in Europe. One of those employed in this new museum was Jean-Baptiste Lamarck. As an older worker, he had been brought up to believe that there was some underlying pattern and structure to every aspect of the physical and biological world—a mind-set common to many 18th-century inquirers into the phenomena of nature. It should therefore be possible to recognize a "scale of beings" in which different groups of organisms could be allocated to "lower" or "higher" places according to the level of "perfection" of their organization—with, of course, human beings at the apex of the resulting structure! In 1802, Lamarck suggested that the "lower" organisms might also be found earlier in time and that they might gradually change into the "higher" forms, due to an "inherent tendency of life to improve itself." So there was no need to suggest that fossil organisms were in reality extinct, for it was possible that they had evolved into different and perhaps still-living descendants.

All of this was strenuously opposed by one of the new, young appointees in the museum, the great Georges Cuvier, who founded the science of comparative anatomy. Cuvier used this new branch of science to prove that such great fossil mammals as the mammoths of Europe and North America and the giant ground sloth of South America, as well as many others, belonged to quite different species from those of today and were extinct. But, he believed, his detailed anatomical studies showed that even these creatures had been thoroughly and stably adapted to their environment. Their extinction must therefore have been due to a sudden catastrophic change in their environment. So, to Cuvier, Lamarck's theory of continual transformation was deeply unacceptable, for its suggestion that organisms were flexible and changeable challenged his own conviction that they were, on the contrary, irrevocably adapted to their existing environment. Cuvier was therefore opposed to Lamarck's views because they cast doubt on his belief in extinction (which was, perhaps, understandable). But this unfortunately also led him to reject the whole idea of evolution that Lamarck had championed—so throwing out the baby of evolution with the bathwater of extinction.

It is always very convenient if, in an argument, your opponent's views are championed by someone else of lesser ability. Lamarck's ideas were supported by another worker in the museum, Geoffroy St. Hilaire. Over the years 1818–1828, St. Hilaire suggested evolutionary homologies and links between such widely different animals as fish and cephalopods (octopus, squid, etc.), but his ideas were ridiculed by other zoologists. Similarly, his supposed evolutionary sequences of fossils placed them in an order that was contradicted by the sequence of the rocks in which they were found. So it was easy for Cuvier to make a devastating attack on St. Hilaire, and this had the effect of also discrediting Lamarck and the whole idea of evolution. In England, the case for evolution was further damaged in 1844, when the Scottish journalist Robert Chambers published a book, *Vestiges of the Natural History of Creation*, which contained astonishingly ignorant ideas. Chambers suggested, for example, that the bony armor of early fossil fishes was comparable to the external skeleton of arthropods (lobsters, crabs, insects, etc.) and that the fish might therefore have evolved from them. The progressively more detailed fossil record that was by then being revealed also gave no hint or indication that the major groups of organisms, traced back in time, converged toward a common, ancient ancestor. The fact that such people as St. Hilaire and Chambers supported the idea of evolution unfortunately gave the impression that it was associated with the lunatic fringe of science. And, by now, Lamarck's explanation of evolution as due to an "inherent tendency" seemed dreadfully old-fashioned.

When the geologist Robert Jameson translated Cuvier's ideas into English in 1813, he added notes suggesting that the most recent of Cuvier's continentwide catastrophes could be interpreted as the Biblical Flood. But Cuvier himself, and other scientists working in post-Revolutionary France, accepted that science and religion should not interfere in each other's affairs. Matters were very different in England. There, the Church of England had become closely integrated into the power structure of a still-hierarchical Establishment, and entry to the universities (and so to the professions) was barred to non-Protestants. So both the authorities of the state (monarchy, aristocracy, and wealthy landowners) and those of the Church (bishops and comfortable clergy) felt themselves threatened by the new-model social order of France, which they saw as encouraging a rising tide of atheism, republicanism, and revolution. For, in the first half of the 19th century, English society was undergoing fundamental changes, fueled by unemployment resulting from the end of the Napoleonic Wars and by the Industrial Revolution, which was driving people from the land and into overcrowded cities. In this conflict, the new ideas of evolution became a weapon that the rising middle class used in their attempts to gain entry to the universities and access to the professions and financial security. In response, to defend their own positions, the establishment portrayed evolution as atheistic, or even heretical.

Enter Darwin—and Wallace

So, in the early 19th century, evolution was seen as a slightly disreputable idea that also had links with a dangerously anarchic approach to the structure of society. It is therefore not surprising that the young Charles Darwin was cautious, secretive, and reluctant to publish his ideas when he began to suspect that the problems he found in trying to interpret the patterns of life could only be explained by invoking evolution. He was the son of a fairly wealthy country doctor, whose father had been an atheist who believed in evolution—so the family was not exactly mainstream. As a student at Cambridge, Darwin had become interested in geology and natural history, and in 1831 he was invited to join the crew of a Government ship, *H.M.S. Beagle*, to act as a companion to the captain and also as a naturalist for what became a six-year voyage to survey the coasts of South America [7]. Several experiences during this long voyage led him to wonder whether the idea of evolution might not, after all, contain some truth.

On the Galápagos Islands in the Pacific, isolated from South America by 960 km (600 mi) of sea, Darwin noticed that the mockingbirds on three islands were different from one another, suggesting that they had independently become different varieties on each island. Similarly, Darwin was shown that there were differences between the giant tortoises from the different islands: those that lived on well-watered islands and could feed on ground vegetation had low, rounded fronts to their shells, but those that lived on drier islands had notched fronts to the shells so that they could stretch their necks back to reach up to feed on the leaves of bushes. Darwin also noticed great flocks of finches, with a variety of sizes of beaks, but they all fed together, and he couldn't make up his mind whether there were any different varieties. (Only later, when Darwin's collections were studied back in England by the ornithologist John Gould was it realized that there were 13 different species of finch in the islands.) All of this suggested that species were not, perhaps, quite as unchanging as was then assumed. Equally disturbing were the fossils that Darwin had found in South America. The sloth, armadillo, and guanaco (the wild ancestor of the domesticated llama) were represented by fossils that were larger than the living forms, but were clearly very similar to them. Again, the idea that the living species were descended from the fossil species that had existed in the same part of the world was a straightforward explanation, but one that contradicted the view that each species was a fixed, unchanging product of creation and had no blood relationship with any other species.

As explained earlier, Darwin was not the first to suggest that organisms were related to one another by evolution; the British zoogeographer Alfred Russel Wallace was thinking along exactly the same lines. (In fact, Wallace was the first to realize and publish the significant fact that closely related species were often also found close to one another geographically, with the clear implication that the two were linked by an evolutionary process.) In the end, it was the receipt of a letter from Wallace, then working in the East Indies, that stimulated Darwin to finalize and publish his ideas after many years of agonizing over its possible hostile reception by the vociferously antievolutionary sections of British society. (It is interesting to note that, in the case of both workers, it was observation of the patterns of distribution of individual species of animal that led them to consider the possibility of evolution.) Their great discovery was to deduce the driving mechanism of evolution—natural selection.

Any pair of animals or plants produces far more offspring than would be needed simply to replace that pair. There must, therefore, be competition for survival among the offspring. Furthermore, these offspring are not identical to one another, but vary slightly in their characteristics. Inevitably, some of these variations will prove to be better suited to the mode of life of the organism than others. The offspring that have these favorable characteristics will then have a natural advantage in the competition of life and will tend to survive at the expense of their less fortunate relatives. By their survival, and eventual mating, this process of natural selection will lead to the persistence of these favorable characteristics into the next generation. (More detail on how this takes place is given in Chapter 6.)

The idea of natural selection was announced by short papers from both Darwin and Wallace, read at a meeting of the Linnean Society of London on June 30, 1858, and Darwin quickly went on to publish his great book the next year [8]. There can be no doubt that Darwin has to share with Wallace the credit for identifying natural selection as the mechanism of evolution and for the patterns of biogeography as evidence for evolution. However, the lion's share of the credit for the almost immediate acceptance of the reality of evolution has to be given to Darwin and his book *On the Origin of Species*. For Darwin had spent the 40 years after his return from the voyage of the *Beagle* in detailed research on many other areas of biology that provided evidence for evolution (see Box 6.2, p. 186), and published this research in 19 books and hundreds of scientific papers. The essentials of this work were given in his great book (which sold out immediately on publication and had to be reprinted twice in its first year) and were far more convincing in their variety and detail than the short papers read to the Linnean Society.

Darwin's theory of natural selection was extremely logical and persuasive. His studies on the ways in which animal breeders had been able to modify the anatomical and behavioral characteristics of dogs and pigeons provided a neat parallel to what he believed had happened in nature over long periods of time, and were even more convincing. But, said his critics, all these different breeds of dog or pigeon were still able to breed with one another, which did not support Darwin's suggestion that this was the way in which new species could appear. Nor could Darwin provide any explanation of precisely how the different characteristics were controlled and passed from generation to generation. In fact, the outlines of the ways in which all this took place had been discovered by the Austrian monk Gregor Mendel in 1866, but his work remained unnoticed until the beginning of the next century. So, our modern science of genetics was still a closed book. Also, Darwin did not understand the nature of species. It was generally assumed at that time that each species was innately stable and resisted innovation—which would have impeded the action of natural selection in trying to alter its characteristics. In fact, we now know that the continual appearance of changed characters or "mutations" (see Chapter 6) would quickly alter the nature of any species, and it is only the continual action of natural selection in weeding out most of these that gives the species the appearance of unchanging stability. So, like any scientist, Darwin was a child of his time, unaware of future discoveries that might have explained his difficulties.

Nevertheless, the idea of evolution and of natural selection as its mechanism was very quickly accepted and is now a part of the basic philosophy of biological science. Just as the theory of plate tectonics is the central paradigm of the earth sciences (see p. 3), so the theory of evolution by natural selection is the central paradigm of the biological sciences. Biogeography provides a striking example of the concordance of the implications of these two paradigms. For example, the dates that plate tectonics theory indicates for the different islands in the Hawaiian chain are similar to those that evolutionary studies indicate for their animals and plants. The way in which biogeography provides interlocking support for these two paradigms is overwhelming evidence for the correctness of each and gives it a unique position in the natural sciences.

World Maps—The Biogeographic Regions of Plants and Animals

Thanks to Darwin and Wallace, then, the engine that was responsible for the living world's reactions to changes in the physical world was at last understood and accepted. Its mechanism (genetics) was yet to be identified, and it would take another century before the mechanisms of the geological engine responsible for those changes was discovered. Nevertheless, it was now clear that some of the differences between the floras and faunas of the separate continents might have resulted from their having had separate evolutionary histories. The German botanist Adolf Engler (1879) was the first to make a world map (Fig. 1.1) showing the limits of distribution of distinct regional floras [9]—though his map also shows the different types of vegetation in each of his major areas. He identified four major floral regions, or realms, in the world, and attempted to trace the history of each of these back into what we now call the Miocene Epoch of the Tertiary Period, about 25 million years ago (see Chapter 5, Fig. 5.5). He also noted some of the plant families or genera that are characteristic or dominant in each realm. Apart from comparatively minor modifications [10, 11, 12], the system of plant regions accepted today (Fig. 1.2a) is very similar to that of Engler—though no one has yet provided any systematic comparison and contrast of the composition of the floras of these different realms [13]. He was also surprisingly perceptive in realizing that, scattered over the islands

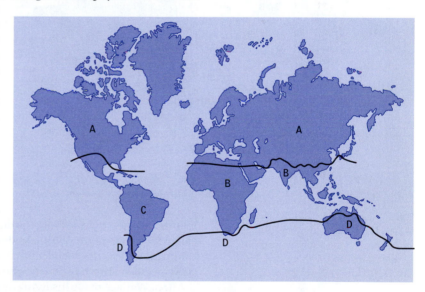

Fig. 1.1 The world's botanic realms according to Engler [9]. A, the Northern Extratropical Realm; B, the Palaeotropical Realm, stretching from Africa to the East Indies; C, the South American Realm; D, the Old Ocean Realm, stretching from coastal Chile via southernmost Africa, the islands of the South Atlantic and Indian Ocean, to Australia and part of New Zealand.

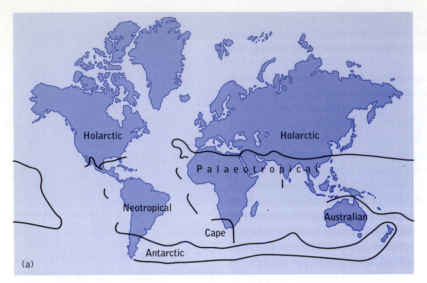

Fig. 1.2 (a) Floral kingdoms, according to Good [11] and Takhtajan [12]. (b) Zoogeographic regions, according to Sclater [16] and Wallace [18]. From Cox [13].

and lands of the southernmost part of the world, lay the remains of a single flora, which he called the Ancient Ocean flora. (It was over 80 years before acceptance of the movement and splitting of continents at last explained this very surprising pattern of distribution.)

Zoogeography, too, had been developing from the early 19th century onward, but with a different emphasis. Because such dominant groups as the birds and mammals are warm-blooded, they are largely insulated from the surrounding environmental conditions and are often found in a great variety of environments. So, unlike the plants, they do not show a close correlation to local ecology. Even such early zoogeographers as Prichard in 1826 [14] and Swainson in 1835 [15] were therefore free to concern themselves with distribution at the world level, and recognized six regions that corresponded with the continents. This was first formalized by the British ornithologist Philip Sclater in 1858 [16], who based his system on the distribution of the most successful group of birds, the passerines or "songbirds." He believed that all species had been created within the area in which they are found today, so that comparison of the different local bird faunas might identify where the centers of creation might have been. (He even thought that these might reveal where the different races of human being had been created.) As was normal in those days, he gave classical names to the six continental areas that he identified but, though he listed or described the areas included in each region, he gave no maps to illustrate his views.

Alfred Wallace made his living by collecting bird skins, butterflies, and beetles in the East Indies, and selling them to naturalists. (He had already made extensive collections in the Amazon rainforest.) These travels and collections had led him to become, like Darwin, interested in their patterns of distribution. He immediately accepted Sclater's scheme, including his names for the regions, and expanded it to include the distribution of mammals and other vertebrates (Fig. 1.2b). Because of the pattern of barriers of ocean, desert, and mountain between the zoogeographic regions, the only area where there is a significant overlap between the faunas of adjacent regions is precisely where Wallace was working: in the East

Indies chain of islands between Asia and Australia. Wallace became fascinated by the unexpectedly abrupt north-south demarcation line that separated the more western islands, which had an overwhelmingly Oriental fauna, from those to the east that were, equally overwhelmingly, Australian. His map, and the "Line" that has been named after him, has been largely accepted by zoogeographers ever since (cf. Fig. 11.9, p. 348).

Although he should always be remembered as the joint discoverer of natural selection, in many ways Wallace's greater claim to fame is as a profound thinker and contributor to the fundamentals of zoogeography. His books *The Malay Archipelago*, *The Geographical Distribution of Animals*, and *Island Life* [17, 18, 19] were read by many people and were very influential. Wallace identified or commented on many aspects of biogeography that still occupy us today. These include the effects of climate (especially the most recent changes), extinctions, dispersal, competition, predation and adaptive radiation; the need to be knowledgeable about past faunas, fossils, and stratigraphy, as well as about those of today; many aspects of island biogeography (see below, see p. 29); and the possibility that the distributions of organisms might indicate past migrations over still-existing or even now-vanished land connections. He and Buffon were truly the giants in the development of zoogeography.

Getting Around the World

The final acceptance of evolution gave a new importance to biogeography and posed new problems, which persisted over the century that elapsed before the mechanics of its geological counterpart, continental drift or plate tectonics, were revealed. If Darwin (and Wallace) were correct, new species arose in a particular place and dispersed from there over the pattern of geography that we see today, except where this had become modified by comparatively minor changes in climate or sea level. This concept of **dispersalism** therefore assumed that, where a taxon, or two related taxa, are found on either side of a barrier to their spread, this is because they had been able to cross that barrier after it formed.

But this was inadequate to explain many of the facts of world biogeography, especially some that were revealed by the rapidly expanding knowledge of patterns of distribution in the past. One might be able to invoke floating islands of vegetation, mud on the feet of birds, or violent winds, to explain dispersal between islands or otherwise isolated locations today. But even Darwin's old friend, the botanist Joseph Hooker, who had traveled and collected widely in the Southern Hemisphere continents and islands, found these explanations quite unconvincing. Hooker became one of a group who instead believed that the many similarities between the plants and animals of the separate southern continents, and of India, could only be explained by their having once been connected; this could have been by narrow land bridges or by wider tracts of dry land, across the present Atlantic and Indian Oceans. But even by the end of the 19th century, this had become dismissed as a fanciful explanation for which there was no geological evidence.

The past, too, was providing more and more examples of puzzling patterns of distribution. For example, 300 million years ago, the plant *Glossopteris* existed in Africa, Australia, Antarctica, southern South America and, most

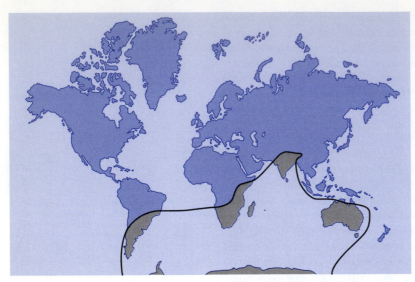

Fig. 1.3 Distribution of the *Glossopteris* flora (shaded area).

surprisingly of all, India (Fig. 1.3). A linkage between all these areas at that time was also suggested by the fact that all of them contained deposits of coal and traces of a major glaciation, contemporary in all those continents. Such facts, together with similarities in the outline of the Atlantic coasts of the Americas, Europe, and Africa, and similarities in the nature and dates of the rocks, and detailed stratigraphy of the rocks along these coastlines, were what led the German meteorologist Alfred Wegener to present his theory of continental drift in 1912 [20]. Wegener suggested that all of today's continents had originally been part of a single supercontinent, **Pangaea** (Fig. 1.4). But, as noted at the beginning of this chapter, in the absence of any known mechanism that might split and move whole continents, his suggestions were not accepted by either geologists or biologists. Biogeographers were instead driven back on progressively more desperate defenses of dispersal as the only possible explanation of the patterns of distribution.

This was particularly true of what the botanist Leon Croizat called the New York School of Zoogeographers, a group of vertebrate zoologists founded by Walter Matthew. In his 1924 paper on "Climate and Evolution" [21], Matthew suggested that all the patterns of mammal distribution could be explained if the different groups had originated in the challenging environments of the Northern Hemisphere. From there, they had dispersed across the intermittently open Bering land bridge between Asia and North America, and southward to the various continents of the Southern Hemisphere. Probably the most influential later member of this School was George Simpson, who not only wrote many papers on mammalian palaeontology and biogeography [22], but also several important books on evolutionary theory. He had no doubt that the patterns of distribution of mammals could be explained perfectly well without invoking continental drift. (This was largely true, for the radiations of the families of living mammals took place well after the fragmentation of Pangaea; only the presence of marsupials, but not placentals, in Australia provided an obvious problem.) Together with such other workers as the herpetologist Karl Schmidt, George Myers (who worked on freshwater fishes), and the zoogeographer Philip Darlington (who in 1957 wrote a major and influential

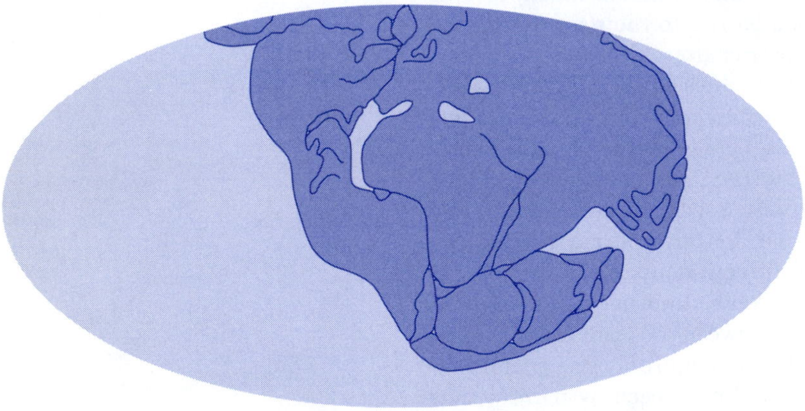

Fig. 1.4 How today's landmasses were originally linked together to form a single supercontinent, Pangaea, according to Wegener. (Compare this with Fig. 10.1 to see the modern, plate-tectonic reconstruction of Pangaea.)

textbook on zoogeography [23]), they provided a powerful and united body of opinion that was wholly opposed to the idea of continental drift and equally fervently supportive of the idea of dispersal.

Some idea of the lengths to which these workers were driven in trying to explain the facts of distribution is shown by Darlington's statement, in discussing the distribution of *Glossopteris*: "The plants may have been dispersed partly by wind, and, since they were frequently associated with glaciation, they may have been carried by floating ice, too. I do not pretend to know how they really did disperse, but their distribution is not good evidence of continuity of land" [24, p. 193]. Surely, one might think, this distribution, scattered across continents separated by thousands of miles of ocean (Fig. 1.3) *was* evidence of continuity of land, but Darlington gave no reason why he thought that it was not *good* evidence.

It is not surprising that such attitudes provoked opposition, and this surfaced most strongly in the person of Leon Croizat. Born in Italy in 1894, Croizat was overwhelmed by the effects of fascism, the 1914–1918 war, and the Great Depression. After spending periods as an artist in New York and Paris, he became a botanist, at first in New York and eventually in Venezuela, where he lived from 1947 until his death in 1982. Croizat rightly felt that the dispersalists were going to extremes in their refusal to countenance any other explanation for the patterns of distribution that one could see in the world today, such as the widely disjunct distributions of many taxa, especially in the Pacific and Indian oceans. He amassed a vast array of distributional data, representing each biogeographical pattern as a line, or **track**, connecting its known areas of distribution. He found that the tracks of many taxa, belonging to a wide variety of organisms, could be combined to form a **generalized track** that connected different regions of the world. These generalized tracks (Fig. 1.5) did not conform to what might have been

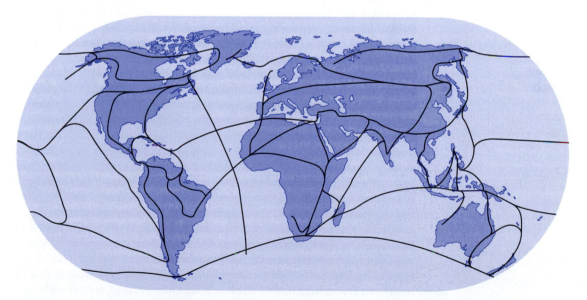

Fig. 1.5 Croizat studied the distribution patterns of many unrelated taxa, and for each he drew lines or "tracks" on the map linking the areas in which they are found. In many cases, these lines were similar enough in position to be combined as "generalized tracks," shown here.

expected if these organisms had evolved in a limited area and had dispersed from there over the modern pattern of geography, as other biologists then believed. Croizat felt that it would be surprising if any single taxon had managed by chance to cross the intervening gaps, and incredible that a considerable variety, with different ecologies and methods of distribution, should have been able to do so. His method, which he called **panbiogeography**, argued that all of the areas connected by one of these tracks had originally formed a single, continuous area that was inhabited by the groups concerned. This theory therefore rejected both the concept of origin in a limited area and that of dispersal between subsidiary locations within that area. However, having rejected the facile use of dispersal as an explanation of each and every example of a transbarrier pattern of distribution, Croizat then went to the other extreme and completely rejected dispersal in any shape or form—though, confusingly, he used the word "dispersal" in a different sense, as describing the pattern of distribution of a taxon.

Instead, Croizat believed that the organisms had *always* occupied the areas where we now see them, together with the intervening areas, and that they had colonized all these areas by slow spread over continuous land. So, the flora of such isolated island chains as the Hawaiian Islands, or the scattered patterns of distribution of plants along the Pacific margins of North and South America, had arisen because land had once linked all these areas, or because islands containing the plants had moved to fuse with the mainland. Croizat believed that any barriers, such as mountains or oceans, that exist today within the pattern of distribution of the taxa, had appeared *after* that pattern had come into existence, so that these taxa had never needed to cross them—a concept that came to be known as **vicariance**. To this extent, Croizat's theorizing anticipated the way in which plate tectonics would provide a geological contribution to the spread of organisms.

Croizat published his ideas in the 1950s and 1960s, his major presentation being his 1958 book *Panbiogeography* [25]—but little attention was paid to his work. This was partly because of the dominance of the New York School, with its acceptance of dispersalism, but also because of several weaknesses in Croizat's own work. He concentrated on the patterns of distribution of living organisms, was scornful of the significance of the fossil record, and paid little attention to the effects of changes in geography or climate. In addition, because the idea of the stability of modern geography seemed to have successfully weathered Wegener's heresies, Croizat's theories of the movement of islands or of massive extensions of land into the Pacific and Atlantic cast him into that same mold, of passionate amateurs. And, even after the theory of plate tectonics had become well documented and widely accepted, Croizat refused to accept it and never integrated it into his methodology. He also became increasingly embittered by the way in which his work was largely ignored.

Ironically, the recognition of some of Croizat's perceptions and methods began in New York, where there arose a new generation of biogeographers who had not been brought up under the influence of the old New York School. For Croizat was correct, and ahead of his time, in believing that in many cases speciation had taken place *after* a barrier had emerged within an existing area of distribution of a taxon. But, unfortunately, the pendulum now swung to the opposite extreme—instead of "dispersal explains everything," their attitude was "vicariance explains everything," and dispersal is merely random noise in the system. Even more unfortunately, Croizat's supporters also inherited his confrontational approach, and the argument between the supporters of dispersal and the supporters of vicariance became increasingly bitter. (One problem that underlay this whole argument may have been that the available evidence was, in the majority of cases, inadequate for anyone to be able to prove whether dispersal or vicariance had been the cause—and biogeographers were only too aware of this. Quite often, those who shout the loudest are those who are least secure of their case, and are trying to silence their own doubts as well as those of their opponents!)

Perhaps the most enthusiastic of Croizat's supporters has been a group of biogeographers, most of whom work in New Zealand, where the origins of the fauna and flora provide particularly difficult problems. These panbiogeographers accepted his generalized tracks running across the ocean basins, referring to them as **ocean baselines** (Fig. 1.6), and viewed them as more useful and important than the conventional system of continental zoogeographical and plant geographical regions. Their methodology also considers the area where a taxon is most diverse in numbers, genotypes, or morphology as the center from which the track for that particular taxon radiated—a dangerous assumption. The writer (Barry Cox) has reviewed the history and development of the New Zealand school of panbiogeographers [26], one of whom (John Grehan) has responded to these criticisms [27]. The American biogeographer Jack Briggs has also written a useful review of these issues [28].

The long and bitter argument about dispersal vs. vicariance has only been ended in the last few years, with the appearance of new molecular techniques of establishing the patterns of relationship of organisms and the time that has elapsed since the origin of each lineage. This now allows us to compare the timing of biological events and of the geological or climatic

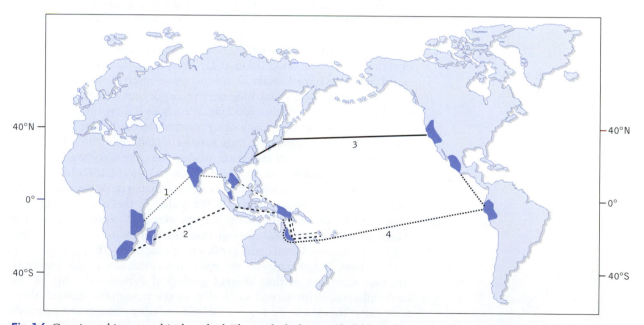

Fig. 1.6 Craw's panbiogeographical method. The tracks link areas (dark blue) where related taxa are found. Tracks 1 and 2 are examples of an Indian Ocean baseline, while tracks 3 and 4 similarly are examples of a Pacific Ocean baseline. After Craw [55].

events that might have been associated with them. The result has been, rather ironically, to show the prevalence of dispersal to an extent far greater than the most optimistic dreams of the dispersalists! Even more ironically, the area most affected by this reversal of opinion has been the homeland of the modern panbiogeographers, New Zealand. Instead of being an ancient Noah's Ark, carrying a cargo of survivors of the biota of the continent from which it split in the distant past, it now seems increasingly likely that all of its animals and plants arrived there by comparatively recent overseas dispersal, mainly from Australia (p. 351). As expressed by a recent reviewer [29], rather than being a "Time Capsule of the South Seas," New Zealand is now seen more as the "Fly-Paper of the Pacific"! Interestingly, some of the previous proponents of vicariance are still deeply antipathetic to plate tectonic, molecular, and phylogenetic approaches, and some have turned to other frameworks that release them from these restraints, such as the concept of an expanding Earth.

The Origins of Modern Historical Biogeography

A century after Darwin had published his theory, acceptance of his ideas had revolutionized approaches to nearly every aspect of the biological sciences. These ideas had implicitly suggested that the contents of each biogeographic unit might have changed and diversified through time, and discoveries of the fossil record had in many cases documented these changes. But as long as the Earth's geography was assumed to have been stable, problems remained in the explanation of at least some of the patterns of disjunct distribution. Some of these could be explained by patterns of extinction. For example, the presence of fossil camelids and tapirs in North America and Asia showed that the disjunct distribution of these groups today, in South America and Southeast Asia, did not have to be explained by some theory of the rafting of early members of these groups across the Pacific. On the other hand, the patterns of distribution shown by the ancient *Glossopteris* flora, or of the Antarctic Floral Kingdom today, still provided a major puzzle. How could organisms have dispersed across oceans to reach these scattered locations? As already mentioned, Wegener's theory of continental drift had provided an explanation to this conundrum early in the century, but he had been unable to suggest any convincing mechanism that might have caused the movement and splitting of huge masses of land. As a result, his theory had been rejected by most geologists, while most biogeographers had reluctantly felt obliged to follow their lead. It was only in the 1960s that strong new evidence of the mechanism for Wegener's theory, now renamed "plate tectonics," led to the acceptance of the reality of this phenomenon (see Chapter 5). It was only now that geologists were able to provide a series of palaeogeographic maps that showed, from the Silurian onward, the changing patterns of association of the various tectonic plates [30].

Until now, biogeographers had tried to analyze the biogeography of the past according to the different geological Periods—the life of the Carboniferous, Permian and so on. But as the new maps showed, there were major changes in the patterns of land and ocean within these periods of geological time. There would therefore have been corresponding changes in the likely biogeographic patterns, dooming to failure any attempt to detect a single pattern of biogeography for the time in question. However, the new maps also made it possible to identify stretches of time (<u>not</u> corresponding to the geological Periods) within which the geographical patterns had remained constant. As the British biogeographer Barry Cox realized [31], these maps therefore provided the potential basis for appropriate biogeographical analysis, if to them one added the patterns of the shallow "epicontinental" seas that lay on the edges of the continental plates—for these, too, are biological barriers. All that a palaeobiogeographer then had to do was to summate the faunas and floras from every locality within each of the resulting palaeocontinents. For the first time, the results made perfect sense, with elements of these faunas and floras showing clear evidence of endemicity (cf. Chapter 10, p. 312) (Fig. 1.7).

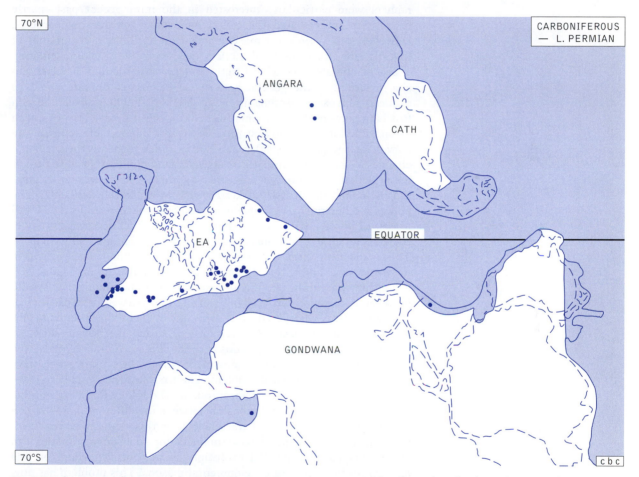

Fig. 1.7 Palaeogeographic map of the Carboniferous—Lower Permian period of time, as reconstructed in 1973. Seas and oceans stippled.

The small circles show the positions of all the localities containing early terrestrial vertebrates. The one indicated in northern South America was later shown to belong to a later period of time, while those in northern India and in Siberia are doubtful fragments. The map thus strongly suggests that the earliest land vertebrates evolved in Euramerica.

The four different floras recognized by palaeobotanists (the Angaran, Cathaysian CATH, Euramerican EA, and Gondwana floras) are also found to have lived on different palaeocontinents. This explains the previously puzzling fact that the *Glossopteris* flora, found in Gondwana, is found scattered over five of today's continents. From Cox [31].

The theory of plate tectonics was soon accepted by nearly all biogeographers but, perhaps unsurprisingly, some of the older ones held out against it. For example, Philip Darlington [24] rejected the idea of a general union of southern continents into a single supercontinent. He felt that such a geography would not have provided enough adjacent water for the development of the ice sheets that appeared to have covered it about 300 million years ago.

At long last, biogeographers could build up a coherent, increasingly detailed set of pictures of the geography of the world over many millions of years. Now they could start to analyze the changing patterns of distribution of living organisms over that period of time and discover the historical roots of the patterns of biogeography that are seen in the world today. The results of this are reviewed in Chapters 10 and 11. But some biogeographers were particularly interested in the more recent past—partly because it encapsulated the origin and spread of our own species. However, this period of time demanded quite different techniques of investigation, for it included the Ice Ages, with their major, oscillating effects on climate and sea level. In the end, it was found that the most reliable evidence of the general patterns of change in climate could be deduced from oxygen isotope studies of the fossil skeletons of plant microfossils in cores of the sediments in the deep ocean floors. The American Cesare Emiliani was the first, in 1958, to provide reliable temperature curves for the past 700,000 years. But in order to relate these general changes to more local climatic changes on land, biologists had to turn to fossil pollen, a technique pioneered by the Swedish worker Gunnar Erdtmann and the British worker Harry Godwin in the 1930s [32]. The pollen of different species of plant is often easily recognizable and is well preserved in sediments found in peats and lake deposits, so that study of these shows clearly how the vegetation of the area has gradually changed. The results of all of these studies, and their implications for the origins of our own species and of civilization, are dealt with in Chapters 11 and 12.

Biogeographers now had the tools with which, they thought, it should be possible to construct satisfying correlations between the patterns of geography and climate, on the one hand, and of life on the other hand. But at first this was still disappointingly difficult to achieve, because different taxonomists had different opinions as to the taxonomy (and therefore, the pattern of evolution) of the organisms involved. Two innovations have transformed this biological problem. The first, known as cladistics (see Chapter 6, p. 197), provided a rigorous methodology for analyzing the patterns of evolutionary relationship between the different members of a group. Still, as long as the characters used in this evaluation were morphological ones, the problems remained, for such characteristics can show convergent or parallel evolution, or may be dependent on one another for functional or developmental reasons. This problem has now been reduced by the development of molecular systematics, which uses more abstract and fundamental characteristics of the organisms, which lie in the detailed molecular makeup of their DNA and proteins (see Chapter 6, p. 198). This not only provides more confidence in the accuracy of our reconstructions of the patterns of evolutionary divergence of the group under study, but also indicates the times at which the different branching events took place. This in turn allows us to make an informed decision as to whether a particular event was due to vicariance or dispersal (in those cases where the two explanations involve different periods of time).

These two advances have permitted major improvements in our methods of establishing biological relationships and is revolutionizing our understanding of biogeography at all levels (see Chapter 7).

The Development of Ecological Biogeography

As we have seen, ecological biogeography began with the simple observations of men such as Linnaeus, who recorded in what type of environment each plant was found, while Forster recognized latitudinal gradients of diversity, later matched by Humboldt's altitudinal gradients, and Candolle pointed out the importance of competition in limiting the distribution of plants. But the full development of this field of inquiry came much later, mainly in the 20th century, for it depended on the rise of modern science with its techniques of experimental, physiological studies. Unlike that of historical biogeography, its history was not complicated by the need to counter the attitudes of antagonistic philosophies or religion, nor by having to wait until data from another field, such as the earth sciences, could be understood. The development of ecological biogeography was, however, strongly dependent on the increasing application of chemical and physical concepts and techniques to the understanding of plant and animal function, and hence distribution. The birth of the science of genetics in the 20th century, leading ultimately to the development of molecular genetics, also expanded the horizons of ecological biogeography.

It was, of course, obvious to the earliest botanists that the distribution of plants was closely linked to climate. In trying to structure the results of this relationship, they could focus either on the demands of the environment on the physiology of the plants, or on the type of vegetation that resulted. Candolle, in 1855, was the first to contribute to this field of inquiry, recognizing three physiologically different types of plant that resulted from their adaptations to different levels of heat and moisture. He called these **megatherms**, **mesotherms** and **microtherms**, which, respectively, required high, moderate, and low levels of heat and moisture, and **hekistotherms**, which live in the polar regions. Later, he added **xerophytes**, which can tolerate low levels of moisture.

Botanists soon also started to analyze the effects of the geology of the area in which the plants lived, and the interacting role of climate and of the plants themselves in breaking down the native rocks and converting them into soils of different characteristics. The American botanist E. W. Hilgard showed, in 1860, how climate and plant life combined to gradually break down the native rock into smaller fragments and provide an increasing component of soil as a product of the biological activity, while the Russian V. V. Dokuchaev analyzed the mineralogical and physical attributes of the soils that resulted from the breakdown of different types of rock.

The alternative focus, on the type of vegetation that resulted from the action of the climate, had begun with Engler's map (see p. 13), which had shown the limits of various types of vegetation, but he had used a confusing system of classification. The first clear and simple system of categorizing the different types of vegetation was produced by the German botanists Hermann Wagner and Emil von Sydow in 1888 [33]. Their system was amazingly precocious, for it recognized 9 of the 10 categories that are still commonly used, such as tundra, desert, grassland, conifer forest, and rainforest; only the Mediterranean type of scrubland was not identified. The many maps and systems produced by various workers since then have contributed different details and variations in emphasis, but have added little to Wagner and Sydow's basic system, though various terms have been coined to describe its elements. The earliest of these, introduced by Clements and Shelford in 1916, were **plant formation** or, with the addition of its animals, a **biome**. Tansley in 1935 [34] added the climatic and soil aspects of the complex, calling it an **ecosystem**, which became the basic unit of ecology. "Biome" has remained the usual term for classification at the macroscale level, but is used in a variety of ways. If the main emphasis is on vegetation structure, ecophysiology, and climate, then biomes can be seen to be the reactions of the living world to these conditions, and the "same" biome can be found in different continents. If, instead, the emphasis is on the taxonomic or phylogenetic aspect of its plant components, then the biomes become regional, as in Takhtajan's "floristic regions" [12]. On the whole, the word "biome" is best used in the former, nontaxonomic, sense.

Living Together

The rise of ecology as a scientific discipline during the early part of the 20th century led to new approaches to biogeographic studies. Ecophysiology, the study of the ecological implications of plant and animal physiology, played an important role in these developments. The German botanist and plant physiologist Julius von Sachs had injected a strong physiological approach into the debates concerning adaptation to the environment that were prominent at the end of the 19th century [35]. Environmental stresses were seen as limiting factors in plant distribution patterns, and the morphology, anatomy, and physiology of plants often reflected their capacity to cope with these stresses. Known as **plant form**, these features were recognised as a more effective way of defining the formations and biomes than any taxonomic or evolutionary system of classification. It was from this line of thinking that the Danish botanist Christen Raunkiaer developed his proposal of **life forms** of plants, based on their means of survival from one growing season to the next (see p. 103). A plant's growing points, he argued, are the most sensitive to environmental stress during an unfavorable period (be it cold or dry) and the position in which those growing points are held therefore provide an indication of the degree of stress to which it is exposed. He classified plants according to the height of their growing points above ground (or below ground). Plants growing in the unstressed conditions of the wet tropics could develop forms with their buds high above ground, while those in the polar regions or in deserts survived only if their buds were close to the ground or, in the case of the dry lands, below the surface. Annuals were a special case, as they survived an unfavorable period as dormant seeds.

The concept of life form has been highly influential in plant geographical studies and generally fits well with observed facts. Plant formations, or biomes, are indeed characterized by the proportions of the different life forms of plants present—what Raunkiaer described in 1934 as the "biological spectrum" of the vegetation. There are other important adaptations, however, apart from those associated with growing points of plants, or their means of surviving from one year to the next. Evergreen and **deciduous** leaf characters (deciduous plants shed their leaves during a cold or dry season of the year), rooting characteristics, drought and flooding physiology, and symbiotic nitrogen fixation are all important aspects of coping with environmental stresses that are unrelated to bud positions.

In the latter part of the 20th century, the concept of **plant functional types** emerged, incorporating and moving beyond that of life forms. It is an approach that can actually be traced back over 2,000 years to the work of the Greek botanist Theophrastus around 300 BC, but its use in recent times has drawn strongly upon the idea of **guilds**, a concept borrowed from animal ecology [36]. A guild is a group of animals, not necessarily related taxonomically, which all make use of the same resource or overlap significantly in their environmental requirements. It is a concept that has been used in a rather varied manner, sometimes being applied to organisms that respond in the same way when disturbed or have a particular management system applied. In one respect, all green plants are part of a single guild in that they all obtain energy directly from the sun, but with respect to other resources, such as water, nutrient elements, pollinators, seed dispersal vectors, and so on, plants have different ways of coping with their environments. They can thus be classified as different functional types. The concept owes much to the work of Philip Grime, who developed the idea that plants have a range of survival strategies available to them [37]. It is an approach that is proving useful in studies such as the nature of stability and resilience in communities, and is also being used in predicting the outcome of global change on vegetation.

The use of the word **community** has itself generated much debate in ecological biogeography. (Communities and ecosystems are examined in more detail in Chapter 4.) We observe organisms mixed together in groups, or assemblages, whose relative stability suggests that the different species are in an equilibrium, tolerating or perhaps even encouraged by the presence of others—perhaps because the different species may coevolve and adapt to the presence of those others. The American plant ecologist Frederic Clements was the first to suggest, in the early 20th century, that such integrated communities resemble individual organisms in their degree of internal organization, and may similarly behave as units in their patterns of distribution. The community concept was very convenient for biogeographers, as it facilitated the precise classification of vegetation, which is needed for it to be effectively mapped. But the voices of many ecologists were raised against it. Henry Gleason formally set out the alternative approach in his Individualistic Hypothesis, stating that each species was distributed according to its own ecological requirements, and what we regard as a community is really little more than a chance assemblage of species with compatible ecological tolerances.

Emphasis on the concept of community led to the development of a distinct branch of plant geography, **phytosociology**, in which the plant communities are classified and may be arranged in a hierarchy—which is undoubtedly convenient, but may be unrealistic. Highly detailed systems of plant community classification have been established by using the techniques of phytosociology, pioneered by the botanist J. Braun-Blanquet [38]. The classification of vegetation, rather like the classification of organisms, is based on the idea that relatively sharp lines can be drawn around each defined unit. Field ecologists, however, soon recognized that in the case of vegetation there is usually gradual change from one type to another, leading to gradients along a continuum. Only where there are abrupt changes in the environment does one find sharp boundaries in vegetation. Recent developments in classification have therefore been based on the idea of defined reference points between which there may be a whole range of intermediates. Classification is necessary for the purpose of mapping, but in a situation of continuous variation any system has to be considered relatively fluid.

Vegetation varies not only in space, but also in time, adding to the complexity involved in classification. Increasing amounts of data from fossil pollen grains in lake and peat sediments have shown quite clearly that the distribution patterns of different plant species change quite independently of one another during periods of climatic change. What we currently regard as a community will change in its composition as the environment changes, and assemblages of the past will never be fully repeated. The community, therefore, is a convenient, but artificial, concept. Changes in assemblages of plants and animals are constantly taking place, and these sometimes follow a predictable pattern. The American botanist Henry Cowles, working in the Chicago region, showed that vegetation develops over the course of time, passing through several different assemblages of plants to finally reach what came to be known as the **climax vegetation** of the region, governed mainly by climate. This climax he regarded as both predictable and stable [39].

The linked concepts of **succession** and **climax**, first developed by Henry Cowles, have also been questioned over the last 100 years. Ecosystems certainly develop over time, and we can make some generalizations about this (see Chapter 4). But it is difficult to show that this somehow involves a predictable process, ending in a predetermined climax that is governed by climatic factors. The climax itself is never static, but is in a constant state of change, so the idea of equilibrium has to be more dynamic than Cowles' original concept. An alternative approach is that of chaos theory, a concept which assumes that the outcome of a process is highly dependent on the initial conditions. If that is true, the development and outcome of successions may be determined by relatively minor differences in such original conditions as the availability of organisms and soil and weather conditions. So, although climate may in very general terms determine the end point (i.e., the biome) of succession, its detailed composition and nature will be affected by many other factors, including chance.

The ecosystem concept was one of the most influential ideas to emerge from ecological studies in the 20th century, and it has proved extremely useful in biogeographical studies. One of its most valuable features is that it can be applied at any scale, from a rock pool on the shore to the entire Earth. The concept owes much to the work of Raymond Lindemann, who in 1942 put forward a formal account of energy flow in nature. The idea was expanded by the work of American ecologists Howard and Eugene Odum, and named by the British botanist Arthur Tansley. It allows any selected portion of nature to be viewed as an entity, within which energy flows and elements cycle. The concept has recently proved especially valuable when applied on a large scale, where the global circulation of elements can be studied, and the relationships between human and natural processes can be identified. In the early 1960s the first landscape-scale ecosystem was subjected to monitoring and manipulative management at Hubbard Brook, a forested mountainside in New Hampshire [40]. The budgets of chemical elements were examined in the undisturbed ecosystem, and again following deforestation, thus establishing an experimental approach to the study of large-scale ecosystems.

Ecophysiology, which examines how plants and animals vary in their physiological processes in response to the environment, also developed in new directions in the 20th century. Subtle differences between plants in their photosynthetic systems may provide some species with the capacity to survive in stressful environments. Similarly, animals vary in their capacities to cope with abiotic stresses, such as cold or high altitude, and in their tolerance to human-produced toxins. Thus the explanation for the presence of a particular species in a given locality (one of the main questions underlying biogeography) may relate closely to the physiological capacity of the species to cope with local environmental stress. This area of research is now entering a new phase as it seeks to understand physiological processes at a molecular level. Molecular biology holds the clue to many biogeographical problems and will undoubtedly be used increasingly to advance biogeographical science. Its value in determining taxonomic relationships is casting a new light on many controversial areas of historical biogeography, and its applications in physiological ecology will allow us to increase our understanding of the current distribution patterns of species and of their environmental limitations.

Advances in physiological research, together with ecological and behavioral studies, will help biogeographers to understand more fully the environmental requirements and the niches of organisms within ecosystems. The concept of the niche is complex, broadly being the role played by an organism in its particular setting. A very large number of variables contributes to the niche, including physical factors, chemical factors, food requirements, predation and parasitism, and competition from similar organisms. The concept of the niche was first devised by G. E. Hutchinson in the 1950s and has established itself as a valuable contribution to ecology and biogeography. Perhaps it is best viewed as a kind of conceptual envelope that has many dimensions relating to each requirement of an organism. An organism cannot survive outside these limits, so a full knowledge of those limits could be used to predict its theoretical geographical range [41]. Such knowledge, however, demands the accumulation of very large databases and very complex analyses, and both are becoming increasingly available to researchers as a result of the development of fast and capacious computers.

The application of niche theory in ecological biogeography lays emphasis on the environmental factors that control the survival of a species in an area, but does not really take into account the availability of a species and its dispersal capacity. An alternative approach has been developed called neutral theory, which is based on the idea that the assemblage of species in a site is entirely a matter of chance [42]. The neutral theory claims that the arrival of a species is a stochastic process and that the best predictive models are based on this concept of chance dispersal. Certainly, the part played by chance needs to be taken into account when trying to explain the composition of communities.

Computers were first applied to problems in ecology and biogeography in the 1960s, and their use has expanded to the point where almost all such studies make use of them. Complex statistics, such as multivariate analyses, as used in niche research and community analysis, are vital analytical tools and can be performed rapidly and routinely on computers small enough to be carried in the field. Global positioning systems, using satellites to establish the precise location of an observer on the ground, have also revolutionized the mapping of distribution patterns in remote areas. Technological advances in the last half century must, therefore, be regarded as major steps forward in the history of ecological biogeography.

All of the above avenues of inquiry, using increasingly sophisticated methods of experimentation and analysis, are now used in modern research on ecological biogeography, as explained in Chapters 2, 3, and 4. They are also now used in trying to cope with the problems and questions that arise from humanity's use, and abuse, of an increasingly crowded planet, as explained in Chapter 14.

Marine Biogeography

As explained at the beginning of Chapter 9, p. 265, the biogeography of the oceans is similar to that of the continents because it is concerned with the biota of vast areas of the surface of the globe. But it is also very different because of the nature of the environment and of the organisms that it contains. We ourselves are terrestrial and air-breathing, so the oceans are a far more challenging environment for us to study and census, and they also contain little in the way of obvious demarcations between biogeographical regions or zones. As a result, marine biogeography has been relatively slow to develop, and we still have a great deal to learn about it.

Though earlier naturalists had published limited studies on the faunas of particular regions, the first worldwide survey, based on the distribution of corals and crustaceans, was that of the American scientist James Dana, who later became an eminent geologist. His brief paper, published in 1853, divided the surface waters of the globe into several different zones based on their mean minimum temperature. Three years later, the British zoologist Edward Forbes [43] published the first comprehensive work, recognizing 5 depth zones and 25 faunal provinces along the coasts of the continents. He was the first to recognize the enormous Indo-Pacific faunal region, and stated that the coastal faunas varied according to the nature of the coast, seabed, local currents, and depth, and placed the 25 faunal provinces in nine latitudinal belts. Forbes also later published a little volume on the natural history of European seas which made important contributions to marine zoogeography and ecology.

In 1880, the British zoologist Albert Günther published a book on fishes in which he recognized 10 different regions in the distribution of shore fishes, and the German Arnold Ortmann published a similar work based on the distribution of crustaceans such as crabs and lobsters. However, the great landmark in early studies of marine zoogeography was the 1911 *Atlas of Zoogeography* [44] assembled by three British zoologists (John Bartholomew, William Clark, and Pery Grimshaw). Their 30 maps of the distributions of fishes were based on the patterns of distribution of 27 families. An influential review and synthesis of all the relevant literature was carried out by the Swedish worker Sven Ekman; it was published initially in German in 1935, followed by an English translation in 1953 [45]. This divided the faunas of the shallow seafloors into seven (mainly climatic) areas, and included the perception of the unity of the faunas of the Indian and West Pacific oceans, as well as the unity of the faunas of the East Pacific and Atlantic oceans. Ekman suggested that the Panama barrier must formerly have been absent, that the island-free East Pacific acted as a barrier to the dispersal of organisms, and commented on the phenomenon of "bipolarity", where a species is found on either side of the equatorial regions, but not within them (see p. 293).

Ekman's work was extended by the American marine zoologist Jack Briggs in 1974. In his book, *Marine Zoogeography* [46], Briggs used the patterns of endemicity of coastal faunas to identify locations where there appears to be a zone of unusually rapid faunal change, and then used this to distinguish 23 zoogeographic regions. Out in the oceans themselves, our knowledge of the distribution of plankton was greatly increased thanks to the work of the Dutch oceanographer Siebrecht van der Spoel and his co-workers. Their *Comparative Atlas of Zooplankton* [47] included over 130 maps of examples of different types of distribution, categorizing these and the different types of seawaters, the physical properties of the waters, and diagrams of the relationships between the faunas of the different oceans.

The greatest of the more recent advances in our knowledge of marine biogeography have come partly from our increasing ability to explore the depths of the sea but also, surprisingly, from our ability to establish sensing and recording satellites in space. Our now-possible journeys into the deepest part of the oceans led to the discovery in 1977 of what is probably the last of the ecosystems of the world to be found, as well as perhaps the weirdest—the strange hydrothermal vent faunas (see p. 284). But, far more importantly, space satellites such as NIMBUS have enabled scientists to monitor and record the changing patterns of planktonic life in the oceans continually and comprehensively. This has allowed the British marine biologist Alan Longhurst to propose a system of biomes and provinces within the oceans [48]. These provide for the first time a framework for their regional ecology that integrates their physical features with our increasing knowledge of the annual periodicity in the life, movements, and reproduction of the plankton. We shall have great need of such studies in our efforts to comprehend and manage the life of the oceans, which we are increasingly affecting, while we also increasingly need it to feed the rapidly growing population of our planet.

Island Biogeography

As mentioned earlier (p. 7), Georg Forster was the first biologist to remark on some of the particular features of island biogeography; he noted that island floras contain fewer species than the mainland, but that the number of species varies according to the size and ecological diversity of the island.

Another early contributor was Candolle, who pointed out that the age, climate, and degree of isolation of an island, and whether or not it was volcanic, would also affect the diversity of its flora. Nevertheless, the sheer variety and volume of the works on island biogeography published by Alfred Wallace marks him as the real founder of studies on this subject. His travels around the islands of the East Indies led him to make many profound observations on the reasons for their differing faunas and floras. He realized that the origins of the islands would affect the nature of their biota (that is to say, their faunas and floras). Those of islands that had once been a part of a neighboring continent were likely to contain most of the elements of the fauna and flora that they had inherited from the mainland. In contrast, islands that had arisen independently, as volcanic or coral-atoll islands, would only have organisms that had been able to cross the intervening stretch of sea. Wallace also pointed out that their distance from the mainland, or from one another, would affect the diversity of their biota. Finally, he realized that the diversity of islands made them good natural experiments, in each of which the processes of colonization, extinction, and evolution had taken place independently, and so provided abundant material for comparative studies. These fundamental perceptions, as well as the sheer number of his books and research papers, leave no doubt that Alfred Wallace was the father of island biogeography.

But in Wallace's day, and for nearly a century afterward, island biogeography remained the preserve of the naturalist. There were so many islands whose biota needed to be described, for they are fertile breeding grounds for evolutionary innovation. Hundreds of papers were published on the plants of this group of islands, on the animals of that group of islands, or on the distribution of animals or plants over the islands of this or that part of the world. But each group of organisms or plants was treated as unique, with its own special history. Relatively few studies contained any attempts to be analytical and to identify underlying phenomena or processes that might explain some of this myriad diversity. An exception was Philip Darlington's observation in 1943 that larger islands contain a greater number of individuals, and a greater diversity of species, than smaller islands, the species diversity increasing by a factor of 10 for every doubling of island area.

Although science always tries to provide a unifying theory that can integrate a mass of data, it can only produce such an analysis once it has developed the tools to do so. It may be significant that such an integrated, synthetic approach to island biogeography only appeared after sophisticated mathematical techniques had been used to analyze biological phenomena in the new field of population genetics. The groundbreaking little paperback book, *The Theory of Island Biogeography* [49], published in 1967, was written by two American biologists: the mathematical ecologist Robert MacArthur and the taxonomist–biogeographer Edward Wilson. Other workers, such as the Swedish worker Olof Arrhenius in 1921, and the Americans Eugene Munroe in 1948 and Frank Preston in 1962, had noted the relationship between the area of an island and the number of species that it contains. But MacArthur and Wilson's book was on a quite different level, for it was a sustained (181 pages of text) exploration not only of the basic concepts but also of the ecological evidence and implications of the theory. The book put forward two main suggestions: that the changing, and interrelated, rates of colonization and immigration would eventually lead to an equilibrium between these two processes, and that there is a strong non-linear correlation between the area of the island and the number of species it contains. The arguments for these ideas were mathematical, with detailed equations and graphs, and the results were very persuasive. Here, at last, it seemed as though biologists would be able to move beyond the raw data to understand the relationships between the simple biological processes. Even more importantly, in a world increasingly worried about the effects of human activity, the concept of an equilibrium of numbers promised to allow predictions as to what would happen under given circumstances, and so to optimize designs for conservation areas.

Over the years that followed the publication of the *Theory of Island Biogeography*, many papers were written that interpreted individual biota in terms of the theory. These papers were in turn taken as providing such a wide measure of support for the theory that it became almost uncritically accepted as a basic truth. In turn, therefore, results that did not conform to expectations based on the theory were reexamined in search of procedural or logical faults, or for unusual phenomena that might explain the "anomalous" result. Sometimes they were simply ignored, rather than being seen to cast doubts on the applicability or universality of the theory. Unfortunately, this is far from unique as an example of the way in which new theories can come to so dominate the scientific field that critical evaluation, and even the concept that it may hold some of the truth, but not necessarily all of it, become forgotten. This can happen especially when the field in question has either been seen as extremely difficult to interpret, as in the case of this theory, or when the field has previously been dominated by another, equally dominant and intolerant concept, as in the case of the confrontation between dispersalist and vicariance schools of biogeography.

The story of the rise of the theory and of the later mounting wave of criticism has been told in the fascinating book *The Song of the Dodo— Island Biogeography in an Age of Extinction*, by the American science writer David Quammen (see 'Further Reading'). It now seems clear that the theory cannot predict equilibrium levels for the biota of any island and that it is valid only in relating island area to biotic diversity. But MacArthur and Wilson nevertheless revolutionized the study of island biogeography, for they led the way in introducing mathematical techniques, and in providing a standard format for analysis and comparison. As we shall see later, the ecology of island faunas and floras is far more fragile than that of the continents. We therefore greatly need to understand them, for, as a result of their number, diversity, and role as natural laboratories for evolutionary change, they contain a high proportion of the biotic diversity that we now desperately need to conserve. For example, though New Guinea contributes only 3% of the world's land area, it contains some 10% of its species of terrestrial organism.

Biogeography Today

As explained earlier in this chapter, the first aspect of biogeography to be recognized by scientists, during the 18th century, was its ecological component. Inevitably, its historical component could only become recognized as a field of research after the scientific community accepted the reality of evolution itself in the middle of the 19th century. Until quite recently, these two approaches to biogeography remained largely independent of one another. Ecologists began with the study of living species or subspecies, and with the factors that control, or alter, their patterns of distribution today. But if they attempted to extend their conclusions into the past, they soon ran into difficulties. This was because they were working at a scale of detail, both in geographical terms and in taxonomic terms, that could not be perceived in the historical record. Only in the study of the comparatively recent past, such as the Ice Ages, could the biogeographer be confident of the ecological preferences of the organisms under study, because they were closely related to those alive today. Only for that period of time was the fossil record sufficiently detailed for the palaeontologist to be confident of the nature and taxonomic level of the changes that were taking place. And only for that period of time were the records of changes in the environment sufficiently detailed, in both time and space, for it to be possible to make plausible correlations between the environmental changes and any biogeographical changes. For the more distant past, it was not possible to establish precisely when any evolutionary changes had taken place, and therefore impossible to correlate these to any ecological changes that might have occurred at that time.

The lack of integration between historical biogeography and ecological biogeography continued until the 1990s, when it was rapidly transformed by developments in two areas of study. The development of techniques of analysis of the details of the molecular structure of their genes provided an enormous quantity of data on the molecular characteristics of the organisms (see Chapter 6, p. 178), showing precisely how they differed from one another. At the same time, as it became easier and cheaper to obtain this data, the number of organisms whose molecular characteristics had been analyzed rapidly increased. So great was the quantity of data that it would have been impossible to make any sense of it, had it not been for the parallel development of techniques of computer analysis. This, together with the use of cladistics, made it possible to work out the patterns of relationship between the different members of a group. But even more importantly for biogeographers, these techniques made it possible to show precisely when two different lineages had diverged from one another. Now, for the first time, biogeographers could start to correlate the patterns of evolutionary divergence of the organisms and the patterns of change in the environment, over the timescales with which historical biogeographers worked. These advances have also made it possible to discover when related groups that live in different biomes diverged from one another. This in turn allows us to start to work out the history of the assemblage of the different components of the biomes—again, permitting an important linkage between historical and ecological biogeography (see p. 212). It seems possible, and even likely, that the combination of cladistics and molecular analysis will allow us to solve many of the current problems in biogeography. So, today, the old distinction between the two approaches has largely disappeared. And at last it seems that biogeographical research is revealing, with increasing scope and detail, a single, consistent story of the history of the biogeography of the world today.

Ecological biogeography has also raised its level of research from the mainly local to larger scales of analysis, and is developing rapidly both in its establishment of a firm theoretical base and in its practical application to current global problems. In 1995 James H. Brown of the University of New Mexico proposed a new type of research program, which he termed **macroecology** [50], dealing with ecological questions that demanded large-scale analysis. Range changes in response to climate change, patterns of diversity, and analysis of ecological complexity, all lend themselves to statistical and mathematical analysis on a larger scale than normally used by experimental ecologists. This is not a new discipline, but a fresh approach to old problems, and one that is increasingly appropriate in days of rapid global change.

During the latter part of the last century, it was progressively recognized that the human impact on the landscape was virtually ubiquitous. Throughout the world, landscapes have been so modified that they can effectively be considered cultural landscapes, a term that became increasingly used from the 1940s onward [51]. An entirely new discipline of **landscape ecology** appeared, pioneered by Richard Forman of Harvard University [52]. One of the main emphases of the study on the ecology of cultural landscapes was the predominance of fragmentation, as reflected in the title of Forman's classic book, *Land Mosaics*. Landscape ecology needed to examine the ecological consequences of habitat fragmentation on animal and plant populations (Fig. 1.8), and so this discipline began to develop in a new direction, leading to the concept of **metapopulations**. A metapopulation consists of a series of separated subpopulations between which genetic exchange may be limited. Clearly, this is an important area of research in the study of gene flow in populations, and hence in the process of evolution. Not just populations, but whole communities are fragmented as a result of human agricultural and industrial activities, so one can conceive of metacommunities of organisms that can be highly complex in their spatial dynamics

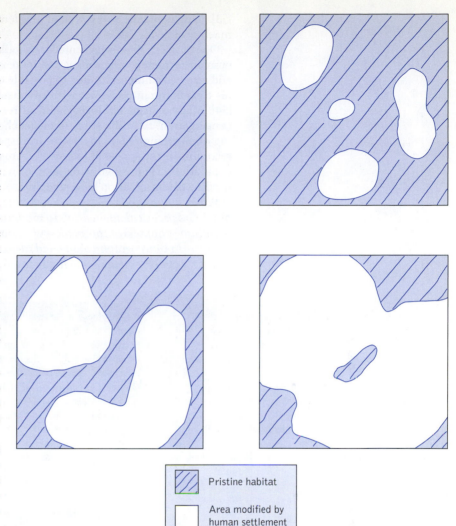

Pristine habitat

Area modified by human settlement

Fig. 1.8 The impact of human settlement and disturbance on a natural habitat is progressive, leading to an increasing degree of fragmentation of the original habitat into isolated units. For some species, especially those animals that are of limited mobility and those plants that have limited seed dispersal, this can result in reduced gene flow. Genetic impoverishment can lead to an increased risk of local extinction, and the loss of such a species is not always compensated for by reinvasion.

[53]. This is precisely the type of problem one can assign to the area of macroecology.

One of the problems presented by habitat fragmentation and the development of metapopulations is the increased danger of genetic isolation and impoverishment, leading to possible extinction. Biogeographic studies thus come into contact with the discipline of wildlife conservation [54]. Many aspects of biogeographical research have a direct bearing on conservation, from the study of biogeochemical cycles and the monitoring of changing ranges of species in response to climate change, to the recording of the spread of invasive organisms and their impact on native populations. Thus, a growing body of work can be classified under the heading of conservation biogeography.

Biogeography today is thus developing both in its theoretical aspects and in its practical application to modern environmental problems. The remaining chapters of this book review our knowledge and techniques of analysis in biogeography today and identify those areas in which important new developments seem likely to take place.

Summary

1 Examining the history of biogeography helps us to understand the nature of the subject today and how biogeographers carry out their work within the current framework of the theories and assumptions of science and society.

2 The early biogeographers were inevitably preoccupied with the immense task of documenting the distributions of animals and plants on the surface of the planet, and trying to establish how these vary according to latitude, altitude, and climate.

3 Increasing knowledge of the fossil record showed how the world's faunas and floras had undergone great changes, which could only have taken place over long periods of time. It was difficult to reconcile this with the doctrines of the Church that life on Earth was a comparatively recent creation and that species were unchanging. By providing a plausible explanation of how and why these changes might have taken place, Darwin's idea of evolution by natural selection was a major step in getting the general public to accept this very different view of the world's history.

4 However, as long as it was assumed that landmasses had always been stable in their positions, it was still very difficult to understand the patterns of life in the past, and biologists were driven to sometimes-bizarre theories to explain these. It was only in the 1960s that the discovery of plate tectonics provided the key to understanding how the Earth's geography, as well as its living cargo, had varied through time.

5 Finally, two advances have transformed the whole field of research into the history of organisms and of their patterns of distribution. The first was the conception and acceptance of cladistic taxonomy. This gave biologists a rigorous system for establishing patterns of relationship that could then be used as a framework onto which patterns of distribution could be applied. Second, the use of molecular methods has provided biologists, for the first time, with reliable procedures for the analysis of relationships and the dating of divergences between lineages.

6 Meanwhile, ecological biogeographers were establishing a framework for the description of the varied types of vegetation, were progressively coming to understand how climate affects the form of plants, and how, together with the local geology and soil, climate affects the development and succession of plant communities.

7 Because of the alien nature of its environment, the study of marine biogeography is far more difficult than that of the land. The general outlines of the distribution of shallow-sea marine faunas were documented in the 18th and early 19th centuries, along with the recognition of faunal zones controlled by latitude and depth. But the huge extent of the open oceans made it difficult to understand the dynamics of the annual changes in their faunas and floras until the recent introduction of satellite-based mapping and modern techniques of marine exploration. Even today, we have much to learn about the organisms of the oceans and about the processes that underlie their biogeography.

8 Islands, too, posed problems for the biogeographer because each is a unique natural "experiment" in the evolution of floras and faunas. The radical concepts of the *Theory of Island Biogeography,* published by MacArthur and Wilson in 1963, introduced a major attempt to provide a framework for understanding this bewildering mass of data. The subsequent history of attitudes to the theory,

from initial almost uncritical acceptance through subsequent criticism and evaluation, provides a fascinating study of science at work today.

9 The introduction of molecular methods of analysis of the genetic basis of the taxonomy of living organisms, and its application to a large and increasing number of species, together with the development of powerful methods of computer analysis of the resulting mass of data, has allowed us to extend our application and understanding of ecological biogeography into the past, blurring the old distinction between ecological and historical biogeography.

Further Reading

Lomolino MV, Sax DF, Brown JH (eds.). *Foundations of Biogeography. Classic Papers with Commentaries.* Sunderland, MA: Sinauer Associates, 2005. (This gives detailed references to, and translations of many of the 18th- and 19th-century works referred to in this chapter, as well as reprints and commentaries on later works.)

Quammen D. *The Song of the Dodo—Island Biogeography in an Age of Extinctio*n. London: Pimlico/Random House, paperback, 1996.

Website

http://www.wku.edu./~smithch/biogeog/ Smith CH. Early Classics in Biogeography, Distribution and Diversity Studies to 1950. This is a free-access website, with a sister site listing classic works published 1951–1975, with links to many other relevant bibliographies including later publications and biographies of listed scientists. Approximately one-third of the entries are linked either to free online copies or to the archive service JSTOR. Extremely useful.

References

1 Cox CB. New geological theories and old biogeographical problems. *J Biogeogr* 1990; 17: 117–130.

2 Buffon G. *Histoire Naturelle Générale et Particulière.* Paris: 1867.

3 Forster JR. *Observations Made during a Voyage Round the World, on Physical Geography, Natural History, and Ethnic Philosophy.* London: G. Robinson, 1778.

4 Humboldt A. de, Bonpland A. *Voyage de Humboldt et Bonpland aux régions équinoxiales du Nouveau Continent.* 30 vols, 1805–1834.

5 Candolle A. de. Essai elementaire de géographie botanique. *Dictionnaire des Sciences Naturelles*, 18. Paris: 1820.

6 Schouw JF. *Grundzüge der einer allgemeinen Pflanzengeographie.* Berlin: 1823.

7 Darwin C. *Journal of the Researches into the Geology and Natural History of Various Countries Visited by* H.M.S. Beagle, *under the Command of Captain Fitzroy, R.N. from 1832 to 1836.* London: Henry Colburn, 1839.

8 Darwin C. *On the Origin of Species by Natural Selection.* London: John Murray, 1859.

9 Engler A. *Versuch einer Entwicklungsgeschichte der Pflanzenwelt.* I, II, Leipzig: Engelmann, 1879, 1882.

10 Diels L. *Pflanzengeographie.* Leipzig: 1908.

11 Good R. *The Geography of the Flowering Plants.* London: Longman, 1947.

12 Takhtajan A. *Floristic Regions of the World.* Berkeley: University of California Press, 1986. (An English translation of the original 1978 book in Russian.)

13 Cox CB. The biogeographic regions reconsidered. *J Biogeogr* 2001; 28: 511–523.

14 Prichard JC. *Researches into the Physical History of Mankind.* London: Sherwood, Gilbert & Piper, 1826.

15 Swainson W. Geographical considerations in relation to the distribution of man and animals. In *An Encyclopaedia of Geography*, Ed. H. Murray, pp. 247–268. London: Longman, 1836.

16 Sclater PL. On the general geographical distribution of the members of the Class Aves. *J. Proc. Linn. Soc., Zool.* 1858; 2: 130–145.

17 Wallace AR. *The Malay Archipelago.* New York: Harper, 1869.

18 Wallace AR. *The Geographical Distribution of Animals.* London: Macmillan, 1876.

19 Wallace AR. *Island Life.* London: Macmillan, 1880.

20 Wegener A. *Die Entstehung der Kontinente und Ozeane.* Braunschweig: Vieweg, 1915.

21 Matthew WD. Climate and evolution. *Ann. N.Y. Acad. Sci.* 1915; 24: 171–318.

22 Simpson GG. *Evolution and Geography*. Eugene: Oregon State System of Higher Education, 1953.

23 Darlington PJ. *Zoogeography*. New York: Wiley, 1957.

24 Darlington PJ. *Biogeography of the Southern End of the World*. Cambridge, *MA*: Harvard University Press, 1965.

25 Croizat L. *Panbiogeography*. Caracas: published by the author, 1958.

26 Cox CB. From generalized tracks to ocean basins—how useful is panbiogeography? *J Biogeogr* 1998; 25: 813–828.

27 Grehan JH. Panbiogeography from tracks to ocean basins: evolving perspectives. *J Biogeogr* 2001; 28: 413–429.

28 Briggs JC. Darwin's biogeography. *J Biogeogr* 2009; 36: 1011–1017.

29 McGlone MS. Goodbye Gondwana. *J Biogeogr* 2005; 32: 739–774.

30 Smith AG, Briden JC, Drewry GE. Phanerozoic world maps. In: Hughes NF. *Organisms and Continents through Time. Special Paper in Palaeontology* 1973; 12: 1–47.

31 Cox CB. Vertebrate palaeodistributional patterns and continental drift. *J Biogeogr* 1974; 1: 75–94.

32 Godwin H. *Fenland: Its Ancient Past and Uncertain Future*. Cambridge: Cambridge University Press, 1978.

33 Wagner H, Sydow E. von. *Sydow-Wagners Methodischer Schul Atlas*. Gotha, 1988.

34 Tansley AG. The use and abuse of vegetational concepts and terms. *Ecology* 1935; 16: 284–307.

35 Sachs J von. *Lectures on the Physiology of Plants*. Oxford: Oxford University Press, 1887.

36 Smith TM, Shugart HH, Woodward FI (eds.). *Plant Functional Types*. Cambridge: Cambridge University Press, 1997.

37 Grime JP. *Plant Strategies, Vegetation Processes, and Ecosystem Properties*. Chichester: Wiley, 2001.

38 Braun-Blanquet J. *Plant Sociology: The Study of Plant Communities*. New York: McGraw-Hill, 1932.

39 Moore PD. A never-ending story. *Nature* 2001; 409: 565.

40 Likens GE, Bormann FH, Pierce RS, Eaton JS, Noye MJ. *Biogeochemistry of a Forested Ecosystem*. New York: Springer-Verlag, 1977.

41 Hirzel AH, Le Lay G. Habitat suitability modelling and niche theory. *Journal of Applied Ecology* 2008; 45, 1372–1381.

42 Hubbell SP. *The Unified Neutral Theory of Biodiversity and Biogeography*. Princeton, NJ: Princeton University Press, 2001.

43 Forbes E. Map of the distribution of marine life. In: Johnston W, Johnston AK, eds. *The Physical Atlas of Natural Phenomena*, plate 31. Edinburgh: W Johnston & AK Johnston, 1856

44 Bartholomew JG, Clark WE, Grimshaw PH. *Bartholomew's Atlas of Zoogeography*, vol. 5. Edinburgh: Bartholomew, 1911.

45 Ekman S. *Zoogeography of the Sea*. London: Sidgwick & Jackson, 1935.

46 Briggs, JC. *Marine Zoogeography*. New York: McGraw-Hill, 1974.

47 van der Spoel S, Heyman RP. *A Comparative Atlas of Zooplankton*. Berlin: Springer, 1983.

48 Longhurst A. *Ecological Geography of the Sea*. New York: Academic Press, 1998.

49 MacArthur RH, Wilson EO. *The Theory of Island Biogeography*. Princeton, NJ: Princeton University Press, 1967.

50 Brown JH. *Macroecology*. Chicago: University of Chicago Press, 1995.

51 Birks HH, Birks HJB, Kaland PE, Moe D. (eds.). *The Cultural Landscape, Past Present and Future*. Cambridge: Cambridge University Press, 1988.

52 Forman RTT. *Land Mosaics: The Ecology of Landscapes and Regions*. Cambridge: Cambridge University Press, 1995.

53 Holyoak M, Leibold MA, Holt RD. *Metacommunities: Spatial Dynamics and Ecological Communities*. Chicago: University of Chicago Press, 2005.

54 McCullough DR. *Metapopulations and Wildlife Conservation*. 2005. Washington, DC: Island Press, 1996.

55 Craw R. Panbiogeography: method and synthesis in biogeography. In: Myers AA, Giller PS, eds. *Analytical Biogeography*, pp. 405–435. London: Chapman & Hall, 1988.

Patterns of Distribution

*M*ost organisms can be assigned to a particular species, and each species has a certain geographical range within which it is found. One of the questions biogeography tries to answer is what determines the range of each species. In this chapter we examine many of the main factors that determine the ranges of species, whether climatic, geological, or historical. No species lives in isolation from other species, so sometimes range limitation may be due to biological factors, such as competition for food or space, predation, or parasitism. The factors that influence the limits of a species can thus interact in a complex pattern. Understanding how a species reacts to these factors, however, will prove increasingly important in predicting the biogeographical outcome of global environmental changes in the future.

The basic units with which biologists have to operate are individual organisms, whether animals, plants, or microbes. In the majority of cases these individuals can be sorted into groups that have most features in common, which we call species. Species are thus reasonably distinguishable groups of organisms within which interbreeding can occur and to which it is normally confined. Having said this, however, we must bear in mind that there may be considerable variation within any given species, both in visual features, such as form, size and color, and in physiological and biochemical features that may affect the preferred environment, food, or climatic tolerance. Such variations may become so distinctive that they justify the description of subspecies or races as convenient systems for the classification of individuals. Studies of molecular genetics, especially the detailed structure of genetic codes as encrypted in DNA (see Chapter 6), have greatly aided the process of measuring the degree of variation within species and also the differences between species [1].

Species are classified into higher units, such as genera and families, again according to the degree of similarity encountered, but such divisions are not always clear and they may not always reflect adequately the degrees of difference or similarity that actually occur between organisms. Now that it is possible to analyze the genetic constitution of plants and animals, some old arguments are being settled, but other new ones are arising. One group of birds that has caused dispute among ornithologists for many years is the herring gull complex. The herring gull (*Larus argentatus*) has until recently been regarded as a **polytypic species,** that is, one in which there are many races or subspecies. But analysis of the molecular genetics of these supposed races has revealed that some should be regarded as separate species [2]. The American populations, for example, are sufficiently distinct in their genetics to be given species status as *Larus smithsonianus*, even though the adults cannot be separated from European herring gulls on the basis of their plumage (juveniles do have some distinctive plumage features). Asian herring gulls show closer similarities to the American species than to the European, and there may be as many as five or six separate species among this group. The Mediterranean "race," which differs from American and north European birds in having yellow rather than pink legs, is also now regarded as a separate species, *L. michahellis*. Such controversy serves to illustrate that the classification of living organisms is by no means cut and dried, and is essentially a convenient system that **taxonomists** (those who make a close study of classification) have constructed and are constantly modifying as new information becomes available. It also illustrates how the process of evolution is a continuing process and we cannot expect species boundaries to be static.

Biogeographers, therefore, are dependent on taxonomists to define the units that they study. When they examine the geographical distributions of species, they find that there are further complications in that no two species are identical in their geographical ranges. Some correspond fairly closely, but others differ totally. When we use terms such as *distribution* and *range*, we must also be careful about the spatial scale we are considering. Two species may be widespread within a given geographical area, such as the British Isles, or the state of North Carolina, and yet occupy different types of habitats (such as woodland or grassland). Even within a habitat, species may occupy different **microhabitats**, such as forest canopies or forest floors. In a New Zealand forest, for example, one may find both the brown kiwi (*Apteryx australis*) and the fantail (*Rhipidura fuliginosa*), a kind of flycatcher. But they occupy different microhabitats, for the kiwi is confined to the forest floor whereas the fantail nests in canopy branches. Therefore, scale, in both horizontal and vertical dimensions, is an important consideration when studying distribution patterns [3].

Limits of Distribution

Whether the species' distribution is considered on a geographical, habitat, or microhabitat scale, it is surrounded by areas where the species cannot maintain a population because different physical conditions or lack of other factors, such as food resources, will not permit survival. These areas can be viewed as barriers that must be crossed by the species if it is to disperse to other favorable, but as yet uncolonized, places. Any climatic or topographic factor, or combination of factors, may provide a barrier to the distribution of an organism. For example, the problems of locomotion or of obtaining oxygen and food, are quite different in water and air. As a result, organisms that are adapted for life on land are unable to cross oceans, unless they can sustain flight for the appropriate distance. Their eventual death will be due, in varying proportions, to drowning, starvation, exhaustion and lack of freshwater to drink. Similarly, land is a barrier to organisms that are adapted to life in sea or freshwater because they require supplies of oxygen dissolved in water rather than as an atmospheric gas, and because they desiccate rapidly in air. Cetaceans (whales and dolphins) use atmospheric oxygen, but have no powers of locomotion on land, and so are confined to aquatic environments. Mountain ranges, too, form effective barriers to dispersal because they present extremes of cold too great for many organisms. Once again, flying organisms such as birds may be able to overcome such physical barriers. The bar-headed goose (*Anser indicus*), for example, is able to migrate over the Himalayas, flying at heights up to 10,175 m (33,382 ft). Their blood contains hemoglobin that is particularly efficient at oxygen absorption. Various other factors, such as the amount of rainfall, the rate of evaporation of water from the soil surface, and light intensity, are all critical factors limiting the distribution of many plants. But in all these cases, the ultimate barriers are not the hostile factors of the environment but the species' own physiology, which has become adapted to a limited range of environmental conditions. In its distribution, a species is therefore the prisoner of its own evolutionary history.

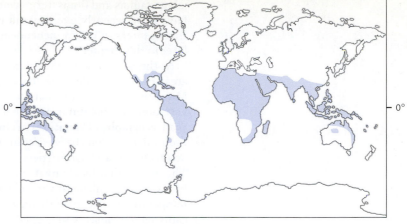

Fig. 2.1 World distribution map of the palm family (Palmae), a pantropical family of plants.

Take the palms (family Arecaceae), for example. Figure 2.1 shows the global distribution of this plant family, and it can be seen that members are found in all areas of the tropics and in many subtropical regions too. When one comes to the temperate areas, however, such as Europe, very few species of palm can be regarded as native. Indeed, there are only two truly native palms in Europe. One of these, *Chamaerops humilis*, is a very small species that grows in sandy soils in southern Spain and Portugal, eastward to Malta (Fig. 2.2). The second species, *Phoenix theophrasti*, is found on certain Mediterranean islands, mainly on Crete. So a family that is extremely successful and widespread in the tropics has failed to achieve similar success in the temperate regions. The real problem with the palms is the way they grow: they have only a single growing point at the apex of their upright stems, and if this is damaged by frost then the whole stem perishes. This weakness has even limited the use of palms as domesticated crop plants, for species such as the date palm (*Phoenix dactylifera*) cannot be grown in areas with frequent frosts. Even in the deserts of northern Iran where the summers are hot and dry, the date palm is a rare sight because of the intense cold in the high-altitude deserts of Iran during winter (Fig. 2.3). Perhaps the most successful palms in the temperate regions are the *Serenoa* species, which reach 30°N in the United States, and *Trachycarpus* species, particularly *T. martianus*, which grows to an altitude of 2,400 m (8,000 ft) in Nepal [4]. But the family as a whole is limited geographically by its sensitivity to frost.

Some plants and animals are confined in their distribution, sometimes, though not always, to the areas in which they evolved; these plants and animals are said to be **endemic** to that region. Their confinement may be due to physical barriers to dispersal, as in the case of many

Fig. 2.2 The dwarf palm *Chamaerops humilis*, one of the two native palms found in Europe.

Fig. 2.3 The date palm, *Phoenix dactylifera*, at its most northerly site in the Great Kavir Desert of Iran.

island faunas and floras (termed **palaeoendemics**), or to the fact that they have only recently evolved and have not yet had time to spread from their centers of origin (**neoendemics**). These will be discussed in detail later in the chapter.

At the habitat level, the microhabitats of organisms are surrounded by areas of small-scale variation of physical conditions, or **microclimate.** Microclimate is a term that covers the temperature, humidity, and light variations in a habitat in a manner similar to, but on a much smaller scale than, geographical variations in climate. Animals may also be restricted in their microhabitats because of limitations in food availability. These various factors may form barriers restricting species to their microhabitats. The insects that live in rotting logs, for instance, are adapted by their evolution to a microhabitat with a high water content and relatively constant temperatures. The logs provide the soft woody materials and the microorganisms that insects may need for food, and also afford good protection from predators. Around the logs are areas with fewer or none of these desirable qualities, and, for many animals, attempts to leave their microhabitat would result in death by desiccation, starvation, or predation.

The Niche

The demands that an organism places on its environment in terms of physical and chemical conditions, space, and food supply help to define what ecologists call its **niche**. But the concept of the niche goes beyond the basic physics and chemistry of its habitat and covers all aspects of how the organism makes a living. It includes the food an animal requires, but also encompasses the way in which it acquires that food. The kestrel is a bird that hunts small mammals by day, while an owl performs a similar activity by night. They overlap in their food requirements but obtain their food under quite different conditions. In the case of plants, they may have similar requirements for water and chemical elements from the soil, but may root at different depths and thus tap slightly different resources. In this way they differ in their niche.

One can illustrate this differentiation in niche by returning to the palms. On Lord Howe Island off the east coast of Australia there are two closely related endemic species of palm [5]. One of these, *Howea forsteriana*, flowers approximately seven weeks before the other species, *H. belmoreana*. *Howea belmoreana* also prefers more acidic soils than *H. forsteriana*. The two species thus differ in their niches, and these differences enable them to coexist on the island. Their niches can be seen to be multidimensional, in the sense that there are several requirements in which the two species vary, both in terms of the chemical environment and in the timing of their life cycles. One can think of these as separate axes of variation. Species may coincide in their requirements on one or more axes, but are unlikely to coincide on all axes. No two niches will be identical.

Ecologists have developed two ways of looking at the niche. There is the theoretical, or ideal type of niche, usually called the fundamental niche, which is the sum of all the niche requirements under ideal conditions when the species is given unimpeded access to resources. In the real world such conditions are unlikely, usually because other species compete for those resources (that is, have overlapping niches) and may perform better in their acquisition. The result is that the observed distribution of the organism is confined by species interactions, and the outcome, the realized niche, is that the species is found over a smaller range than would have been predicted. These concepts are important in biogeography, especially when attempts are made to model potential niches as an aid to predicting distribution patterns. This will become increasingly clear as the theme of this chapter develops.

One additional complication is the role of sheer chance in the distribution of organisms. The arrival of a wind-borne insect or a seed at a particular point in space cannot be predicted with certainty, and the first arrival may well be at an advantage over those arriving later. Chance events are said to be stochastic, and these random elements within ecology and biogeography may be of great significance [6, 7]. A consideration of the role of random factors in biogeography has led to the development of **neutral theory**, which will be considered when we examine the ways in which species become assembled together in communities (see Chapter 4).

Overcoming the Barriers

There are, therefore, many dimensions to the niche, which restrict the habitats within which a species is found. Habitats, such as the rotting log mentioned earlier, are often scattered or spatially fragmented, leading to individual organisms becoming confined in their distributions. If an organism is to spread, it needs to overcome spatial and physical barriers to gain access to new locations where its niche requirements can be satisfied. This may prove difficult, but a few inhabitants of rotting logs do occasionally make the dangerous journey from one log to another. Few environmental factors are absolute barriers to the dispersal of organisms, and these factors vary greatly in their effectiveness. Most habitats and microhabitats have only limited resources, and the organisms living in them must have mechanisms enabling them to find new habitats and new resources when the old ones become exhausted. These mechanisms often take the form of seeds, resistant stages, or (as in the case of the insects of the rotting-log microhabitat) flying adults with a fairly high resistance to desiccation. There is plenty of evidence that many geographical barriers are not completely effective. Organisms may extend their distribution by taking advantage of temporary, seasonal, or permanent changes of climate or distribution of habitats that allow them to cross barriers normally closed to them. The British Isles, for instance, lie within the geographical range of about 220 species of birds, but a further 50 or 60 species visit the region as casual vagrants. These birds do not breed in Britain, but one or two individuals are seen by ornithologists every few years. They come for a variety of reasons: some are blown off course by winds during migration; others are forced in certain years to leave their normal ranges when numbers are especially high and food is scarce. Many of these "accidentals" have their true home in North America, such as the ring-necked duck (*Aythya collaris*), a few of which are seen every year. Some, however, come from eastern Asia, such as the olive-backed pipit (*Anthus hodgsoni*) or even from the South Atlantic, such as the black-browed albatross (*Diomedea melanophris*).

It is possible, though not very likely, that a few of these chance travelers may in time establish themselves permanently in Europe, as did the collared dove (*Streptopelia decaocto*), which since about 1930 has spread from Asia Minor and southern Asia across central Europe and into the British Isles and Scandinavia, marking perhaps the most dramatic natural change in distribution recorded for any vertebrate in recent times. This species is now common around the edges of towns and settlements in western Europe, and seems to depend for food largely on the seeds of weed species common in farms and gardens, together with the bread that humans often put out for garden birds. Several factors may have interacted to permit this extension of range of the collared dove. Increased human activity during the last century, involving extensive changes in the environment, has produced new habitats and food resources, and it is possible, too, that small changes in climate may have significantly favored this species. It is, however, considered unlikely that the collared dove would have been able to take advantage of these changes without a change in its own genetic makeup, perhaps a physiological one permitting the species to tolerate a wider range of climatic conditions or to utilize a larger variety of food substances. Its behavior patterns have also changed, from nesting largely on buildings to nesting in trees, which may have favored it in temperate Europe [8]. Since its introduction to the Bahamas in 1974, the collared dove has also spread rapidly through North America [9], as will be discussed later, so this is one organism that has proved remarkably successful once dispersal barriers have been overcome.

Biogeographers commonly recognize three different types of pathway by which organisms may spread between one area and another. The first and easiest pathway is called a **corridor**; such a pathway may include a wide variety of interconnecting habitats, so that the majority of organisms found at either end of the corridor would find little difficulty in traversing it. The two ends would therefore come to be almost identical in their **biota** (i.e., the fauna plus the flora). For example, the great continent of Eurasia that links western Europe to China has acted as a corridor for the dispersal of animals and plants, at least until the recent climatic changes of the Ice Ages. In the second type of dispersal pathway, the interconnecting region may contain a more limited variety of habitats, so that only those organisms that can exist in these habitats will be able to disperse through it. Such a dispersal route is known as a **filter**; the exclusively tropical lowlands of Central America provide a good example. Finally, some areas are completely surrounded by totally different environments, so that it is extremely difficult for any organism to reach them. The most obvious example is the isolation of islands by wide stretches of sea, but the specially adapted biota of a high mountain peak, of a cave, or of a large, deep lake is also extremely isolated from the nearest similar habitat from which colonists might originate. The chances of such a dispersal are therefore extremely low, and largely due to chance combinations of favorable circumstances, such as high winds or floating rafts of vegetation. Such a dispersal route is therefore known as a **sweepstakes** route. It differs from a filter in kind, not merely in degree, for the organisms that traverse a sweepstakes route are not normally able to spend their whole life histories en route. Such organisms are alike only in their adaptations to traversing the route, such as those aerial adaptations of spores, light seeds, or flight in the case of insects and birds that enable them to disperse from island to island. Such a biota is therefore not a representative sample of the ecologically integrated, balanced biotas of a normal mainland area, and is said to be **disharmonic.**

A discussion of some patterns of distribution shown by particular species of animals and plants will reveal how varied and complex these may be and will help to emphasize the various scales or levels on which such patterns may be considered. In fact, the number of examples that we can choose is quite limited because the distribution of only a very small number of species has been investigated in sufficient detail. Even among well-known species, chance finds in unusual places are constantly modifying known distribution patterns, sometimes demanding changes in the explanations that biogeographers give to explain these patterns.

Some existing patterns are continuous, the area occupied by the group consisting of a single region or of a number of regions that are closely adjacent to one another. These patterns can usually be explained by the distribution of present-day climatic and biological factors; the detailed distributions of several species of dragonfly provide good examples. Other existing patterns are discontinuous or **disjunct,** the areas occupied being widely separated and scattered over a particular continent, or over the whole world. The organisms that show such a pattern may, like the magnolias, be evolutionary relics, the scattered survivors of a once-dominant and widespread group now unable to compete with newer forms. Others, the climatic relics or habitat relics, appear to have been greatly affected by past changes in climate or sea level. Finally, as will be shown in Chapters 10 and 11, the disjunct patterns of some living (**extant**) groups, and of many extinct groups, have resulted from the physical splitting of a once-continuous area of distribution by the process of continental drift (see Chapter 5).

A Successful Family: The Daisies (Asteraceae)

The daisy family provides an example of the way in which we need to invoke different explanations for distribution patterns at different geographical and taxonomic scales. The daisy family is extremely large (having over 22,000 species) and extremely successful, if you measure biogeographical success by the areal extent of distribution. It is a **cosmopolitan** family, which means that it is found throughout the world. In fact, the term *cosmopolitan*, when used of the flowering plants, is usually a slight exaggeration, since very few species of flowering plants have managed to establish themselves in Antarctica; even the Asteraceae have not achieved that, but they are present on all other continents. With the exception of the isolated cold regions of Antarctica, there has clearly been no insuperable barrier to the geographical spread of the family during its evolutionary history.

When we look at those areas of the world where members of the Asteraceae are most abundant and diverse, we find that the mountainous

Fig. 2.4 Giant tree-groundsels from East African mountains. Family Asteraceae, genus *Senecio*, subgenus *Dendrosenecio*. (a) Branched form; (b) unbranched form.

regions of the tropics and subtropics, together with some of the semiarid regions of the world and those with Mediterranean climates (hot, dry summers and mild, wet winters), are the richest in members of this family. The equatorial rainforests are actually rather poor in daisy family species. Often, biogeographers use such information in trying to reconstruct the evolutionary origins of a group. A great deal of generalization is involved, but it does seem that this family has been most successful away from the competition of tall trees, in the more drought-prone habitats where their general adaptability and very diverse fruit-dispersal systems have given them many advantages.

Taking just one genus from within the family, the groundsel genus *Senecio*, we find that it reflects the whole family in many ways, being large (about 1,250 species) and widely dispersed (cosmopolitan apart from Antarctica). Many members are efficient weeds, being short-lived, having efficiently dispersed air-borne fruits, and possessing wide ecological tolerances of climate and soils. Some taxonomists prefer to split this very large genus up into subgenera, and one of these, the subgenus *Dendrosenecio*, is remarkable both for its form (Fig. 2.4) and its restricted distribution pattern. This subgenus consists of just 11 species, often referred to as giant tree-groundsels, which are stocky, woody plants up to 6 m (20 ft) in height, often with branched upper sections bearing terminal clusters of tough, leathery leaves. Botanists refer to thick-stemmed plants of this type as **pachycaul**. In distribution, this subgenus is restricted to East Africa and, examining the distribution pattern on a more detailed scale, only on the high mountains of East Africa (Fig. 2.5) above the forest limits of bamboos and tree-heathers [10]. If we focus in from the taxonomic level of subgenus to species, we find that each of the major mountains of East Africa has its own group of endemic species of *Dendrosenecio*, with never more than three species on any particular mountain (see Fig 2.5).

Detailed analysis of the genetic material, the DNA, of the tree-groundsel species by Eric Knox at Kew in London, and Jeffrey Palmer at Indiana University [11], has shown that each species is more closely related to its neighboring species on its own mountain than to the species of other mountains, despite the fact that in form, such as branching pattern (as shown in Fig. 2.4), it may more closely resemble the giant groundsels from other mountains. It seems that chance has led to the colonization of each mountain peak (with the exception of one, Mount Meru, which is devoid of giant groundsels), and that the chance invader has in the course of time evolved into two or three separate species. (The time involved, incidentally, cannot be very long since Mount Kilimanjaro is only a million years old, which is quite young by geological standards.)

If we take the spatial scale of analysis one step lower and look at separate species on just one of the mountains, then additional factors come into play in the interpretation of distribution patterns. On Mount Elgon (4,300 m), situated on the border between Uganda and Kenya, north of Lake Victoria (see Fig. 2.5), two species of the tree-groundsels are found, *Senecio elgonensis* and *S. barbatipes*. In the open, alpine zone where these trees are found, *S. elgonensis* predominates below 3,900 m (13,000 ft) and *S. barbatipes* above this level, so there is an altitudinal differentiation in their ranges on the mountain. Precisely what

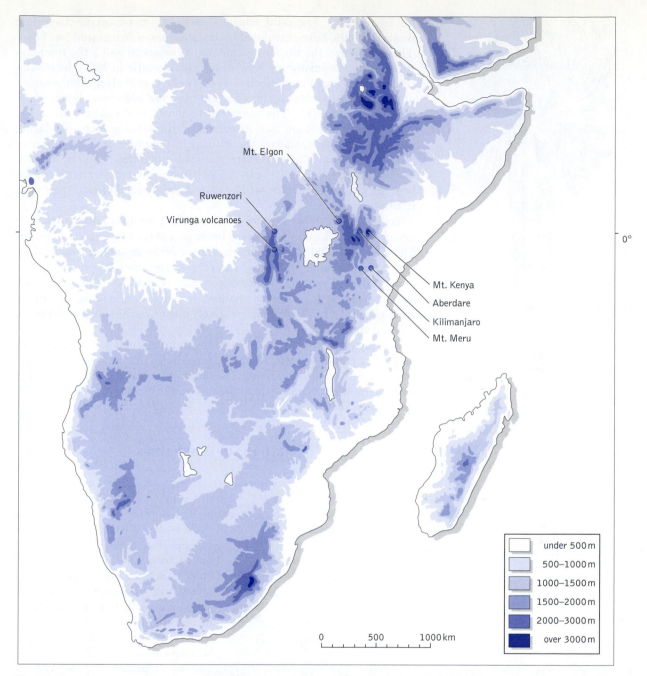

Mt. Elgon

Ruwenzori

Virunga volcanoes

Mt. Kenya

Aberdare

Kilimanjaro

Mt. Meru

0°

	under 500 m
	500–1000 m
	1000–1500 m
	1500–2000 m
	2000–3000 m
	over 3000 m

0   500   1000 km

Fig. 2.5 Map of eastern and southern Africa showing the high mountain peaks on which tree-groundsels are found. Mount Meru is an exception, having no tree-groundsels.

features of the morphology or the physiology of the two species lead to these climatic preferences is not known. There are no detailed meteo-rological measurements for the mountain, but temperature differences with altitude and, in particular, the frequency of frost during the night, are likely to be the most important factors affecting the distribution of

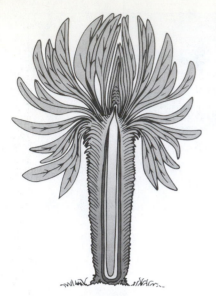

Fig. 2.6 Cross section of a tree-groundsel showing its thick central pith, surrounded by wood and cortex, together with the outer layer of dead leaves and leaf bases, forming an insulating sheath that protects the living tissues from frost.

the two species. The giant senecios are more frost-tolerant than most tropical plants, being insulated by thick layers of leaves and leaf bases (Fig. 2.6). When the night air temperature drops to −4°C, the temperature within the insulating layer of leaves only falls to 2°C. The insulation is made even more effective because the leaves alter their position during the night, closing together and trapping additional layers of air around the stem [12]. The temperature-sensitive dividing cells of the main trunk system are thus protected from frost. Different insulating efficiencies, or temperature sensitivities of the individual species, may affect altitudinal limits, perhaps via seed production or germination. The two species may also be in competition with one another for space or some other resource, as will be illustrated by further examples later in this chapter.

Taking a final and even more detailed look at the distribution of *S. elgonensis* in a small valley within the lower part of the alpine zone of Mount Elgon, we find that the population is most dense around the valley floor (Fig. 2.7) where a damp area fed by water seepage exists. The species is evidently affected at this habitat scale by the availability of deeper, moist soils, preferring these to the shallow, free-draining soils of the alpine slopes and ridges, where drought is likely during the hot conditions of the tropical alpine day.

This analysis of distribution patterns at increasingly detailed levels of scale and taxonomy within the Asteraceae demonstrates the way in which we must invoke different factors to account for the distribution patterns of organisms depending on both the taxonomic and the geographical scale we are using.

Patterns of Dragonflies

One group of species whose distributions are quite well known, at least in western Europe, are the Odonata, dragonflies and damselflies [13]. The common blue damselfly, *Enallagma cyathigerum*, is possibly one of the most abundant and widely distributed dragonfly species (Fig. 2.8a). The adults are on the wing in midsummer, around bodies of freshwater. The female lays eggs in vegetation below the surface of the water, and the larvae hatch in a week or two. The larvae live on the bottom of the pond, stream, or lake and feed on small crustaceans and insect larvae until they reach a size of 17-18 mm, which may take from 2 to 4 years, depending on the quality and quantity of food available. In the May or June after reaching full size, the larvae climb up the stems of emergent vegetation (plants rooted in the mud with stems and leaves emerging from the water surface), cast their larval skins, and emerge as winged adults. *E. cyathigerum* is found in a wide

Fig. 2.7 Diagram of a cross section of a small valley on Mount Elgon, Uganda, showing the higher density of the tree-groundsels in the valley bottom.

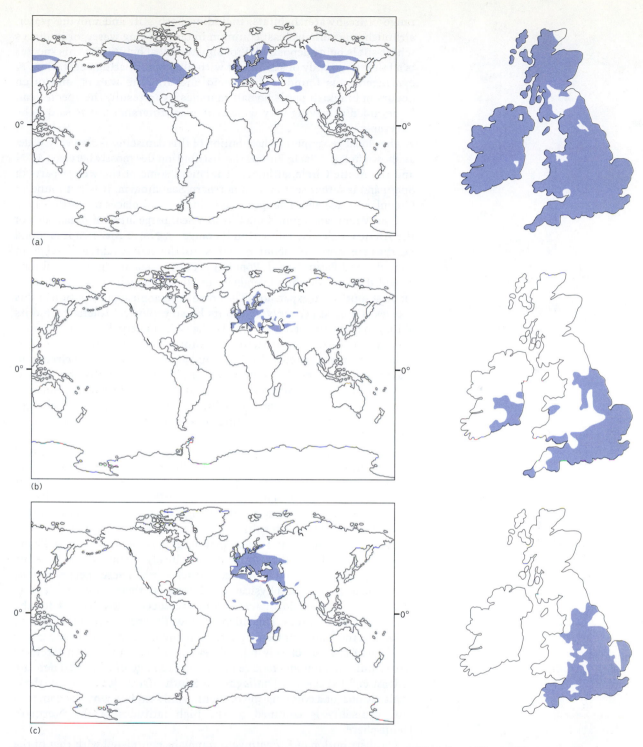

Fig. 2.8 The distribution on a world scale of three species of dragonfly. (a) Common blue damselfly, *Enallagma cyathigerum*; (b) ruddy sympetrum, *Sympetrum sanguineum*; and (c) emperor dragonfly, *Anax imperator*.

range of freshwater habitats, including both still and moving water, although it is perhaps least common in fast-moving water, or in places where silt is being deposited. Probably the ideal habitat for this species is a fairly large body of still water with plenty of floating vegetation. *E. cyathigerum* is found in both acid and alkaline waters, and often occurs in brackish pools on salt marshes. As a result, the species can be regarded as having very wide ecological tolerances; it is said to be **eurytopic**.

The global geographical distribution of this damselfly is also very wide, as shown in Fig. 2.8a. In Europe the distribution lies mostly between 45°N and the Arctic Circle, although it includes some of the wetter parts of Spain and is rather scattered in northern Scandinavia; it is not found in Greenland or Iceland. The species is found in a few places in North Africa, in Asia Minor, and around the Caspian Sea. Large areas of Asia south of the Arctic Circle also fall within its range. In North America it is found everywhere north of about 35–40°N to the Arctic Circle, except in Labrador and Baffin Island. Populations also occur in the ideal habitats provided by the swampy Everglades of Florida, which mark the species' farthest southward expansion. The broad geographical distribution of this species is almost certainly due to its broad ecological tolerance and its ability to make use of a wide range of habitats in very different climates. This type of distribution pattern, a belt around the Northern Hemisphere, is shown by many species of animals and plants, and is termed **circumbo-real**, meaning around the northern regions. The frequency of this pattern in very different organisms suggests that the two northern landmasses may once have been joined, enabling certain species to spread right around the hemisphere. This line of argument will be elaborated in Chapter 5.

As *E. cyathigerum* is so successful, one might ask why it has not spread farther southward. One reason may be the relative scarcity of watery habitats in the subtropical regions immediately to the south of its present range, the arid areas of Central America, North Africa, and central Asia. The species is perhaps not robust enough for the long migrations that would be needed to reach suitable habitats in the Southern Hemisphere (there are very few wind belts that might assist such a migration). Another possibility is that other species already occupy all the habitats that *E. cyathigerum* could colonize further south. These species may be better adapted to the physical conditions of their habitats than *E. cyathigerum* and could therefore compete successfully with it for the available food resources. This might exclude the species from these areas. In fact, in the Southern Hemisphere there are many species of the genus *Enallagma* and the closely related genus *Ischnura* that seem likely to have similar habitat and food requirements to *E. cyathigerum*. There are at least eight species of *Enallagma* in South Africa alone. Any of these explanations, or a combination of them, would explain why the common blue damselfly is confined to the high latitudes of the Northern Hemisphere.

The distribution of *E. cyathigerum* may be contrasted with that of the beautiful dragonfly *Sympetrum sanguineum*, sometimes called the ruddy sympetrum (Fig. 2.8b). As its name "sanguineum" suggests, it has a blood red body. This species has a limited distribution in western Europe, parts of Spain, a few places in Asia Minor and North Africa, and around the Caspian Sea; it is not found in eastern Asia or North America. The reason for the limited distribution of this dragonfly is almost certainly that the larva has very precise habitat requirements. It is found in ditches and ponds with still waters, but only where certain emergent plants, namely, the bulrush or cattail (*Typha latifolia*) and horsetails (*Equisetum* species), are growing. Why the larva should have these very specific requirements is not clear because it is certainly not a herbivore in its diet, feeding on insect larvae and crustaceans, but so far the larvae have never been found away from the roots of these plants. *S. sanguineum* could therefore be described as a **stenotopic** species—one with very limited ecological tolerance. The fact that it can colonize only a very few habitats must certainly limit its distribution, but other, unknown, factors are also at work, for the species is often absent even from waters in which cattails or horsetails are present.

The northern distribution of these two species may be contrasted with that of the emperor dragonfly, *Anax imperator* (Fig. 2.8c). The adults of this species are 8–10 cm long, and the larva is found typically in large ponds and lakes and in slow-moving canals and streams. It is a voracious predator and can eat animals as large as fish larvae. The distribution covers a band of Europe between about 50°N and 40°N but, unlike the other two species, it is well distributed on the North African coast and the Nile Valley and stretches across Asia Minor to northwest India. It even spreads across the Sahara Desert down into Central Africa where there are suitable habitats, such as Lake Chad and the lakes of East Africa, and it is found in most parts of South Africa except the Kalahari Desert. (It is possible that the South African population may belong to a separate subspecies from the European forms.)

The distribution of *A. imperator* is therefore confined to the Old World and does not extend far into Asia. It appears to be basically a Mediterranean and subtropical species whose good powers of flight and fairly broad ecological tolerance have enabled it to cross the unfavorable dry areas of North Africa to new habitats in southern Africa. No doubt favorable habitats for *A. imperator* do occur in the other landmasses (although there may be potential competitors there, of course). The dragonfly cannot now reach them, however, because land connections are not available except in the north of its range, where the species is not very successful.

In most tropical dragonfly species, the larvae emerge from the water and metamorphose to the adult at night; they are very vulnerable to predators at the time of emergence, and darkness probably affords some protection from birds. But the process of metamorphosis is inhibited by cold temperatures, and in northern Europe many species are compelled by low night temperatures to undergo at least part of their emergence in daylight, when birds eat large numbers of them. This probably imposes a northern limit to the distribution of many species of dragonfly, including *A. imperator*, which would explain why this species has not been able to invade the Americas or eastern Asia.

Magnolias: Evolutionary Relicts

The magnolias (family Magnoliaceae, genus *Magnolia*) have a very interesting modern distribution, as shown in Fig. 2.9. Of the 80 or so species of the genus *Magnolia*, the majority are found in Southeast Asia and the remainder, about 26 species, in the Americas, ranging from Ontario in the north, through Mexico, down into the northern regions of South America [14]. Their distribution is clearly disjunct, being separated into two main centers in this case. Unlike the palms, we cannot explain their distribution pattern simply in terms of the climatic sensitivities of the plants concerned, for the magnolias are reasonably hardy; they can be cultivated well into the north of the temperate area. Nor would climatic constraints explain why they are not found in intermediate tropical and subtropical regions, as are the palms.

To understand the distribution of the magnolias, we need to look at their evolutionary history. Fossils of magnolia-like leaves, flowers, and pollen grains are known from Mesozoic times, the age of the dinosaurs. Indeed, botanists regard the magnolia family as one of the most primitive families of flowering plant groups. Its showy flowers were attractive to the rapidly evolving insects and together they coevolved into a most successful team in which the insect visited the flowers for food and, in doing so, ensured the passage of pollen from one plant to another, thus taking the chance and the waste out of the highly risky wind-pollination process. The magnolias spread and must have formed a fairly continuous belt around the tropical, subtropical and temperate parts of the world, for their fossil remains have been found through Europe and even in Greenland. For perhaps as long as 70 million years, the magnolias remained widespread, right up to the last 2 million years, during which they have been lost from areas such as Europe that would link their current isolatedcentres of distribution.

Being small, slow-growing shrubs and trees, they were not strong competitors for the more robust and fast-growing tree species. When the climatic fluctuations of the last 2 million years began to disturb their stable woodland environment, they succumbed to the competitive pressures imposed by more robust and faster growing trees, and thus became extinct across much of their former range. Only in two parts of the world have they managed to escape and survive, as evolutionary **relicts**. (The word "relict" was originally applied to a widow and implies being left behind, which is precisely what has happened to the magnolias.)

It is interesting that another genus of the magnolia family, the tulip trees, genus *Liriodendron*, have a very similar distribution to that of the *Magnolia* genus and in all probability the tulip trees share a similar fossil

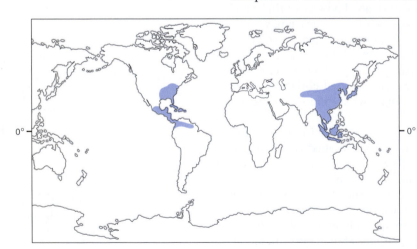

Fig. 2.9 World distribution map of the magnolias, illustrating a disjunct distribution.

history. But in the case of the tulip trees, only two species have survived, *L. tulipifera* being a successful component of the deciduous, temperate forests of eastern North America, and *L. chinense* surviving only in a restricted area of Southeast Asia (Fig. 2.10).

The Strange Case of the Testate Amoeba

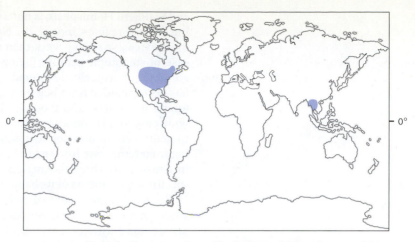

Fig. 2.10 World distribution of the tulip trees (*Liriodendron* species). Only two species now survive, in widely separated localities, although it was once a widespread genus.

Very small organisms, especially microbes, tend to have very wide global geographical distributions; many are cosmopolitan [15]. The reason for this is their effective dispersal, suspended in air currents. The rust and smut fungi, for example, can travel many thousands of kilometers carried along in the atmosphere, and their abundance ensures that some will land in locations where conditions will be suitable for their survival and population growth. In the case of the rusts and smuts, that generally is on the leaves of the plant species that they parasitize. But the tendency to develop a cosmopolitan distribution does not apply to all microbes, as has been demonstrated for the testate amoeba *Nebela vas*, the mainly Southern Hemisphere distribution of which is shown in Fig. 2.11.

Testate amoebae are tiny protozoans that live in wet habitats, often in the spongy mosses of bogs and marshes. They differ from other amoebae in having a permanent tough shell, which enables them to survive periods of drought. Other amoebae are capable of producing a cyst when subjected to adverse conditions, but the testate species, rather like snails, carry a cyst around with them just in case. Due to the small size, the problems of finding them, and the difficulties involved in identifying testate amoebae, information on their distribution patterns is naturally less abundant than for dragonflies or magnolias. But studies by Humphrey Smith and David Wilkinson [16] collating records from around the world have revealed that some species have surprising geographical distributions, including *Nebela vas*, as shown in the map.

Sometimes the distribution pattern of species that are not easily recognized simply reflects the geographical locations of experts in the field and the intensity of field survey. But the restriction of this species to the tropics

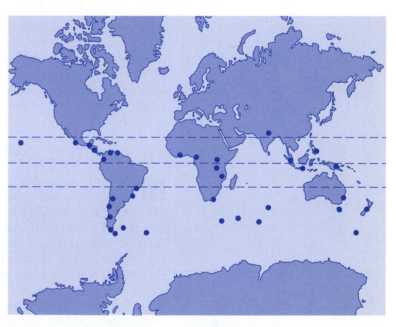

Fig. 2.11 Locations around the world where the testate amoeba *Nebela vas* has been recorded. Note its predominantly Southern Hemisphere and tropical distribution. This may be explained by former linkage of the landmasses in the supercontinent of Gondwana. From Smith and Wilkinson [16].

and Southern Hemisphere is not a consequence of more thorough searching in those regions. Indeed, the Northern Hemisphere has probably seen more extensive survey work than regions farther south [17]. Nor is this a case of stenotopic ecology (being ecologically fastidious as opposed to eurytopic or ecologically tolerant) on the part of the protozoan because it has been recorded from bog mosses, forest floors, and even high-altitude forests of bamboo and rhododendron. *Nebela vas* is even found over a wide range of pH conditions, from 3.8 to 6.5, thus ranging from very acid to neutral environments. Its climatic requirements are also broad, ranging from temperate lowland to high-altitude tropical sites and extending into regions that are sub-Antarctic. So this is not a species that has highly specific requirements of its habitat.

The longitudinal range of *Nebela vas* is wide, covering the whole of southern America from Costa Rica to Tierra del Fuego, Africa south of the Sahara, and Australasia. Its island records range from tropical Hawaii and Java south to the sub-Antarctic island of South Georgia. But this tiny amoeba has not been recorded north of the Tropic of Cancer apart from one location in the Himalayas of Nepal. What can possibly account for such an odd distribution pattern?

The one feature that links all the regions occupied by *N. vas* is the fact that they were once part of a huge supercontinent called Gondwana. It was not until the late Cretaceous, around 90–100 million years ago, that this great southern continent finally broke up into the pattern with which we are familiar today (see Chapter 7). The strange distribution pattern of this protozoan may well be due to its evolution and spread within Gondwana prior to its fragmentation.

This example illustrates the possibility that some organisms owe their distributions to events from the deep geological past. Others can be explained by more recent historical changes.

Climatic Relicts

Many species of animals and plants which in the past were widely distributed have been affected by relatively recent climatic changes and survive now only in a few "islands" of favorable climate. Such species are called **climatic relicts**. They are not necessarily species with long evolutionary histories, since many major climatic changes have occurred quite recently. The Northern Hemisphere has an interesting group of **glacial relict** species whose distributions have been modified by the northward retreat of the great ice sheets that extended as far south as the Great Lakes in North America, and to Germany in Europe, during the Pleistocene Ice Ages (the last glaciers retreated from these temperate areas about 10,000 years ago). Many species that were adapted to cold conditions at that time had distributions to the south of the ice sheets almost as far as the Mediterranean in Europe. Now that these areas are much warmer, such species survive there only in the coldest places, usually at high altitudes in mountain ranges, and the greater part of their distribution lies far to the north in Scandinavia, Scotland, or Iceland. In some cases, species even appear to have become extinct in northern regions and are represented now only by relict populations at high altitude in the south, such as in the Alpine ranges. The places where relicts have managed to survive through a time of stress are called **refugia**.

An example of a climatic relict is the springtail *Tetracanthella arctica* (Insecta, Collembola). This dark-blue insect, only about 1.5 mm long, lives in the surface layers of the soil and in clumps of moss and lichens, where it feeds on dead plant tissues and fungi. It is quite common in the soils of Iceland and Spitzbergen, and has also been found further west in Greenland and in a few places in Arctic Canada. Outside these truly Arctic regions it is known to occur in only two regions; in the Pyrenean Mountains between France and Spain, and in the Tatra Mountains on the borders of Poland and Czechoslovakia (with isolated finds in the nearby Carpathian Mountains) (Fig. 2.12). In these mountain ranges the species is found at altitudes of around 2,000 m in arctic and subarctic conditions. It is hard to imagine that the species can have colonized these two areas from its main center further north because it has very poor powers of distribution (it is quickly killed by low humidity or high temperatures) over land and is not likely to have been transported there accidentally by humans. Springtails are capable of survival in the surface layers of ocean waters and could be transported around the Arctic in this way, but that would be of no help in reaching the land-locked mountains of Europe. The likely explanation for the existence of the two southern populations is that they are remnants of a much wider distribution in Europe in the Ice Ages. But it is surprising that *T. arctica* has not been found at high altitudes in the Alps, despite careful searching by entomologists. Perhaps it has simply not yet been noticed, or perhaps it used to occur there but has since died out. One interesting feature of this species is that whereas representatives from the Arctic and the Tatras have eight small eyelets (ocelli) on either side of the head, specimens from the Pyrenees have only six. This suggests that the Pyrenean forms have undergone some evolutionary changes since the end of the Ice Ages while they have been isolated from the rest of the species, and perhaps they should be classified as a separate subspecies.

A plant example of a glacial relict (Fig. 2.13) is the Norwegian mugwort (*Artemisia norvegica*), a small alpine plant now restricted to Norway, the Ural Mountains, and two isolated localities in Scotland. During the last glaciation and immediately following it, the plant was widespread, but it

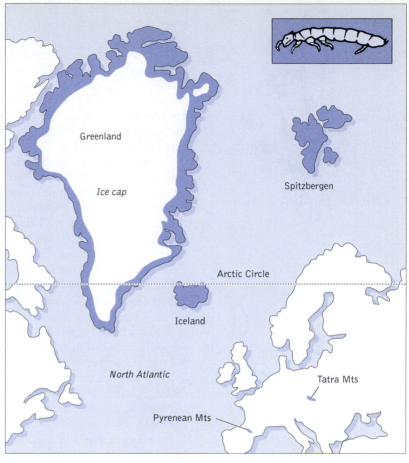

Fig. 2.12 The springtail *Tetracanthella arctica*, and a map of its distribution. It is found mostly in northern regions, but populations exist in the Pyrenees and in mountains in central Europe. These populations were isolated at these cold, high altitudes when the ice sheet retreated northward at the end of the Ice Age.

(a)

(b)

Fig. 2.13 The Norwegian mugwort, *Artemisia norvegica*: (a) the plant; (b) distribution map showing its restricted range in only two mountainous areas of Europe.

became restricted in distribution as forest spread. Relict distributions, however, are rarely as extreme as that of the dung beetle described in Box 2.1.

There are probably several hundred species of both animals and plants in Eurasia that are glacial relicts of this sort, and they include many species that, in contrast to the springtail, have quite good powers of dispersal. One such species is the mountain or varying hare, *Lepus timidus*, a seasonally variable species (its fur is white in the winter and bluish for the rest of the year), which is closely related to the more common brown hare, *L. capensis*. The varying hare has a circumboreal distribution, including Scandinavia, Siberia, northern Japan, and Alaska, being replaced in Canada by the snowshoe hare (*L. americanus*), a closely related species with similar seasonal variations in its pelt. The southernmost part of the main European distribution is in Ireland and the southern Pennine Mountains of England, but there is a glacial relict population living in the Alps that does not differ in important features from those in the more northerly regions. There is, however, an interesting complication. *L. timidus* is found throughout Ireland, thriving in a climate that is no colder than that of many parts of continental western Europe. The brown hare is absent from Ireland, so the varying hare has no competition in that isolated region. There seems to be no climatic reason why this hare should not have a wider distribution in many parts of the world, but it is probably excluded from many areas by its inability to compete for food resources and breeding sites with its close relatives, the brown hare (*L. capensis*) in Europe and various other species of hare (*Lepus* spp.) and rabbit (*Sylvilagus* spp.) in North America. Relict populations of the varying hare survive in the Alps because, of the two species, it is the better adapted to cold and snowy conditions [18].

This is an example of a species that has distinct fundamental and realized niches as a consequence of its competitive interaction with closely related and ecologically similar species.

Climatic relicts are not confined to the temperate regions. The last 2 million years of geological history, when the currently temperate zone was being subjected to glaciation, have seen considerable changes in the vegetation of the tropics. Many areas now occupied by rainforests were

---

Displaced Dung Beetle

Concept

Box
2.1

One very remarkable example of a glacial relict is the dung beetle species *Aphodius holdereri* (Fig. 2.14). This beetle is now restricted to the high Tibetan plateau (3,000–5,000 m) (10,000–16,400 ft), having its southern limit at the northern slopes of the Himalayas. In 1973 G. Russell Coope, of London University, found the fossil remains of at least 150 individuals of this species in a peaty deposit from a gravel pit at Dorchester-on-Thames in southern England [19]. The deposit dated from the middle of the last glaciation, and subsequently 14 sites have yielded remains of this species in Britain, all dated between 25,000 and 40,000 years ago. Evidently, *A. holdereri* was then a geographically widespread species, possibly ranging right through Europe and Asia, but climatic changes have severely restricted the availability of suitable habitats for its survival. Only the remote Tibetan mountains now provide *Aphodius holdereri* with the extreme climatic conditions within which it is able to survive, free from the competition of more temperate species of dung beetle.

replaced by drier vegetation of a tropical woodland or grassland type, the savannah. Fragments of rainforest undoubtedly remained in the most favorable of locations, and this may account for the disjunct distribution of certain rainforest species at the present time. Termites (Isoptera) are important arthropods in tropical forests that make a living by attacking dead wood, leading to its decomposition. Many of the termite species found in Southeast Asia are stenotopic and are particularly sensitive to environmental disturbance, such as the removal of the forest canopy. Recovery from disturbance is slow because they are poor dispersers and find it difficult to reinvade regions where populations have been eliminated. The distribution pattern of this group of termites could therefore provide an indication of long-term stability of the forests of those areas, climatic refugia for both forests and termites. Using this approach, Freddy Gathorne-Hardy and co-workers have identified regions of Sumatra, Brunei, northern Sarawak, and eastern Kalimantan that served as rainforest refugia during the main glacial advances in the higher latitudes [20]. Evidence from other sources, including geological and botanical data, helps to confirm these conclusions, so analysis of termite assemblages in the area provides a clue to the existence of a climatic refugium.

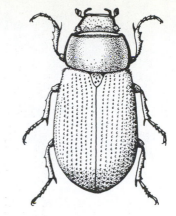

Fig. 2.14 *Aphodius holdereri*, a dung beetle now found only in the high plateau of Tibet.

The strawberry tree (*Arbutus unedo*) in Europe is a good example of what may be termed a **postglacial relict** (Fig. 2.15), for its current locations reflect climatic changes that have taken place since the glaciation ended. *Arbutus unedo* is disjunct, having its main center of distribution in the Mediterranean region but with outliers in western France and western Ireland. The Irish population is particularly surprising because it lies far north of the tree's limits on the mainland of Europe. The Ice Age closed with a sudden warming of the climate, and the glaciers retreated northward; behind them came the plant and animal species that had been driven south during glacial times. Warmth-loving animals, particularly insects, were able to move northward rapidly, but plants were slower in their response because their rate of spread is slower. Seeds were carried northward, germinated, grew, and the mature plants finally flowered and sent out more seeds to populate the bare northlands. As this spread of vegetation continued, melting glaciers produced vast quantities of water that poured into the seas, and the ocean levels rose. Some of the early colonizers reached areas by land connections that were later severed by rising sea levels.

The maritime fringe of western Europe must have provided a particularly favorable migration route for southern species during the period following the retreat of the glaciers. Many warmth-loving plants and animals from the Mediterranean region, such as the strawberry tree, moved northward along this coast and penetrated at least as far as the southwest of Ireland, before the English Channel and the Irish Sea had risen to form physical barriers to such movement. The nearness of the sea, together with the influence of the warm Gulf Stream, gives western Ireland a climate that is wet, mild, and frost-free, and this has allowed the survival of certain Mediterranean plants that are scarce or absent in the rest of the British Isles. Perhaps this explanation also accounts for the presence of the cold-tolerant varying hare in Ireland but the absence of the warmth-demanding brown hare, which arrived after any land bridges were severed by the rising sea.

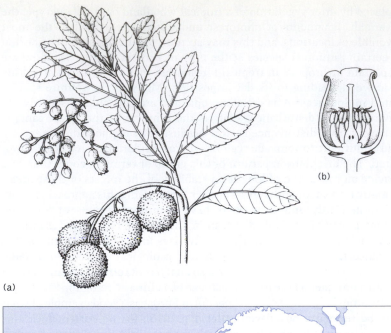

(a)

(b)

(c)

Fig. 2.15 The strawberry tree, *Arbutus unedo*: (a) plant showing leathery leaves, and swollen fruit, which are red in color; (b) cross section of a flower; (c) map of European distribution, showing relict population in Ireland.

Like many Mediterranean trees and shrubs, the strawberry tree is **sclerophyllous**, which means it has hard, leathery leaves (Fig. 2.15). This is a plant adaptation often associated with arid climates and seems out of place in the west of Ireland. Flowering in many plant species is triggered by a response to a particular day length, a process called **photoperiodism**. *Arbutus unedo* flowers in late autumn, as the length of night is increasing; this adaptation is again associated with Mediterranean conditions, since at this season the summer drought gives way to a warm, damp period. The flowers, which are cream-colored, conspicuous, and bell-shaped, have nectaries that attract insects, and in Mediterranean areas they are pollinated by long-tongued insects such as bees, which are plentiful in late autumn. In Ireland, however, insects become increasingly scarce in the autumn, and pollination is therefore much less certain. Thus, the strawberry tree reached Ireland soon after the retreat of the glaciers and has since been isolated there as a result of rising oceans. Although the climate has steadily grown colder since its first colonization, *A. unedo* has so far managed to hold its own and survive in this outpost of its range, despite having features in its structure and life history that seem ill adapted to western Ireland.

A further example of a disjunction that has taken place in relatively recent times is the gorilla. The western gorilla (*Gorilla gorilla*) is found in an area of lowland tropical rainforest in the extreme west of tropical Africa. It is considered to be represented by two subspecies, *G. g. gorilla* in the far west of its range, and *G. g. diehli* to the eastern side of the range. The eastern gorilla (*G. beringei*), as its name implies, inhabits regions farther to the east of Africa, but is not limited to lowland forest, being also found in mountains (Fig. 2.16). There are thus two populations of the eastern gorilla, which are regarded as separate subspecies, *G. b. beringei* in the mountains and *G. g. grauen* in the eastern lowland forest. The two gorilla species and their constituent populations have diverged as a result of the changing patterns of vegetation in Central Africa over the last 2 million years or so [21], during which forest has alternated with more open savannah vegetation. The pattern of disjunction exhibited by the gorillas is reflected in the distributions of many African plants and animals [22].

Fig. 2.16 Distribution map of the gorilla (*Gorilla* species), a mammalian genus with a disjunct distribution. The two populations are now regarded as distinct species, *G. gorilla*, the western lowland gorilla, and *G. beringei*, the eastern gorilla, which consists of two subspecies, the mountain gorilla (*G. b. beringei*) and the eastern lowland gorilla (*G. b. grauen*).

Endemic Organisms

Because each new species of organism evolves in one particular, restricted area, its distribution may be limited by the barriers that surround its area of origin. Other species may evolve in one region, spread to other locations, and then become extinct in all but a restricted area where it survives. Species restricted in this way are said to be **endemic** to that area. As time goes by, increasing numbers of organisms may evolve within an area, or become confined there. The percentage of its biota that is endemic is therefore a good guide to the length of time for which an area has been isolated.

As these organisms continue to evolve, they will also become progressively different from their relatives in other areas. Taxonomists often recognize this by giving higher taxonomic rank to the organisms concerned. So, for example, after 2 million years the biota of an isolated area might

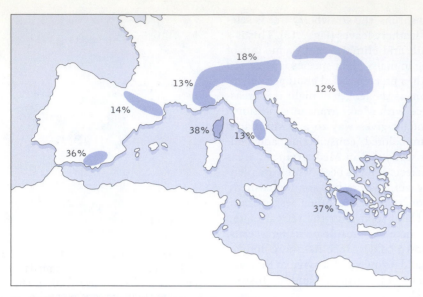

Fig. 2.17 The percentage of endemic plants in the floras of the mountain ranges of southern Europe. From Favarger [15].

contain only a few endemic species. After 10 million years, the descendants of these species might be so unlike their nearest relatives in other areas that they might be placed in one or more endemic genera. After 35 million years, these genera might appear to be sufficiently different from their nearest relatives as to be placed in a different family, and so on. (The absolute times involved would, of course, vary depending on the rate of evolution of the group in question.) Therefore, the longer an area has been isolated, the higher the taxonomic rank of its endemic organisms is likely to be, and vice versa.

Figure 2.17 shows the proportion of the mountain flora in various European mountain ranges that are endemic to their particular area. It is evident that the more northerly of the mountain ranges shown have a lower proportion of their flora that is endemic, whereas the southern, Mediterranean mountains have higher proportions [23]. The mountains of southern Spain and Greece have more endemics than the Pyrenees and the Alps. This could be interpreted to mean that the southern mountains have been isolated for longer periods. But montane plants, such as the glacial relicts described earlier, are now limited in range because of the increasing warmth of the last 10,000 years. The northern mountains may be poorer in endemics simply because local glaciation there was more severe and some of the species that still survive further south consequently became extinct. On the other hand, the richness of the southern mountains could be explained by the fact that the geographical barriers between the northern montane blocks are less severe (less distance, no sea barriers); hence migration and sharing of mountain floras are more likely than in the south, where barriers are considerable. The interpretation of patterns of endemicity must be undertaken with care.

In general, two major factors influence the degree of endemism in an area: isolation and stability. Thus, isolated islands and mountains are often rich in endemics. The island of Australia, for example, has long been isolated from outside influences, until the arrival of European people with their associated invasive organisms. Although not particularly stable in its climate, it is extensive, so on a simple area basis it should be expected to have a wide range of endemics. But Australia also contains few physical barriers to movement during times of change, so extinction by local isolation has not been an important factor. As a result, Australia is rich in endemics, many of which have a long geological history. This kind of fossil endemism is called **palaeoendemism**, in contrast to **neoendemism** resulting from recent surges in the evolutionary process and the generation of new species that have not yet had an opportunity to spread beyond their current limits.

California, for example, is rich in neoendemics, including such plant genera as *Aquilegia* and *Clarkia*, which are undergoing rapid evolution. California is also isolated from much of the North American continent by the high mountains of the Sierra Nevada, and by the Mojave and Sonoran deserts, so the evolution taking place there has not been able to disperse at all easily. The richness of the flora of California, however, is typical of many regions of the Earth with a Mediterranean-type climate, including the Mediterranean Basin itself, Chile, the southern tip of South Africa, and the southwestern extremity of Australia. Much debate has surrounded the high floral richness of these regions, and it may well be that the long history of recurrent fires has created conditions under which small, isolated populations of plants have diversified, leading to a high density of species, many with restricted distributions [24].

Physical Limitations

The geographical range of a species is not always determined by the presence of topographic barriers preventing its further spread. Often a species' distribution is limited by a particular factor in the environment that influences its ability to survive or reproduce adequately. These factors in the environment include physical factors such as temperature, light, wetness, and dryness, as well as biotic factors such as competition, predation, or the presence or absence of suitable food. All of these factors contribute to the niche of the organism, as described early in this chapter.

Of these factors, there is often one that is particularly important and that may be overriding in determining survival and hence distribution. This is called the **limiting factor**. Anything that tends to make it more difficult for a species to live, grow, or reproduce in its environment may prove to be a limiting factor for the species in that environment. To be limiting, such a factor need not necessarily be lethal for a species; it may simply make the working of its physiology or behavior less efficient, so that it is less able to reproduce or to compete with other species for food or living space. For instance, we suggested earlier that a northern limit may be set to the distribution of certain dragonflies by low night-time temperatures. In the more southerly parts of those northern regions, temperatures are not so low that they kill dragonflies directly, but they are low enough at night to force the insects to metamorphose during the day, when they are more vulnerable to predatory birds. In this case, then, the limiting factor of temperature does not operate directly but is connected with a biotic environmental factor, that of predation. Many other limiting factors act in a similar way.

Environmental Gradients

Many physical and biotic factors affect any species of organism, but most can be considered as forming a gradient. For example, the physical factor of temperature affects species over a range from low temperatures at one extreme to high temperatures at the other, and this constitutes a temperature gradient. These gradients exist in all environments and affect

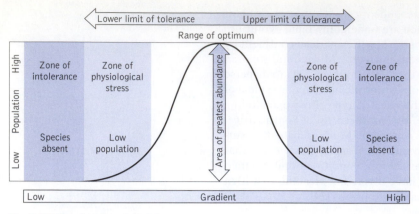

Fig. 2.18 Graphical model of the population abundance maintained by a species of animal or plant along a gradient of a physical factor in its environment.

all the species in each environment. As seen earlier, different species vary in their tolerance of environmental factors, being either eurytopic (ecologically tolerant) or stenotopic (ecologically intolerant), or intermediate between the two, but most species can function efficiently over only a relatively limited part of any given environmental gradient. Within this range of optimum the species can survive and maintain a large population; beyond it, toward both the low and the high ends of the gradient, the species suffers increasing physiological stress: it may stay alive, but because it cannot function efficiently, it can maintain only low populations. These areas of the gradient are bordered by the upper and lower limits of tolerance of the species to the environmental factor. Beyond these limits the species cannot survive because conditions are too extreme. Individuals may live there for short periods but will either die or, if they are mobile, pass quickly through to a more favorable area (see Fig. 2.18). A species may not achieve its full potential distribution in the field because of competitive interactions with other organisms (the realized as opposed to the fundamental niche). When under conditions of physiological stress, a species easily succumbs to such competition.

The grey hair-grass (*Corynephorus canescens*) is widespread in central and southern Europe and reaches its northern limit in the British Isles and southern Scandinavia (Fig. 2.19). Examination of the factors that may be responsible for maintaining its northern limit shows that both flowering and germination are affected by low temperature [25]. This grass has a short life span (about 2–6 years), so it relies on seed production to maintain its population. Any factor interfering with flowering or with germination could therefore limit its success in competitive situations. At its northern limit, low summer temperature delays its flowering with the result that the season is already well advanced when the seeds are shed. Seed germination is slowed down at temperatures below 15°C, and seeds sown experimentally after October have a very poor survival rate. This may explain why its northern limit in western Europe so closely matches the 15°C July mean isotherm. Other factors, however, must be in operation to prevent its spread in southern and central Britain and southern Ireland. Its eastern limit may also be determined by a separate

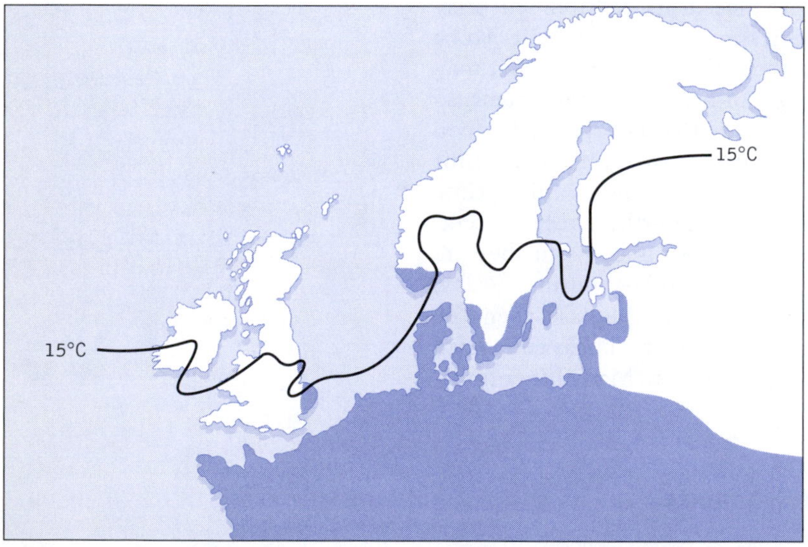

Fig. 2.19 Distribution of the grey hair-grass (*Corynephorus canescens*) in northern Europe (shaded) and its relationship to the 15°C July mean isotherm.

factor, possibly the duration and the severity of winter conditions in northeastern Europe. Range limitation by physical factors can also be found in migratory organisms, such as some birds, as shown by the example in the Box 2.2.

Many plants have their seeds adapted to a specific temperature for germination, and this often relates to conditions prevailing when germination is most appropriate for the species. P. A. Thompson of Kew Gardens, UK, has devised a piece of apparatus for examining the effect of temperature on germination [27]. It consists of a metal bar, one end of which is maintained at 40°C and the other at –3°C; between is a gradient of temperatures. Groups of seeds of the species to be examined are placed along the bar and kept moist, and a record is kept of the number of days required for 50% of the seeds within each group to germinate. The results are expressed on graphs, and the lowest point on the U-shaped curve shows the optimum temperature for germination.

In Fig. 2.21 the germination responses of three members of the catchfly family are shown, together with their geographical ranges. The catchfly (*Silene secundiflora*) is a Mediterranean species, so the optimum time for germination is the autumn, when the hot, dry summer is over and the cool, moist winter is about to begin. Its optimum germination occurs at about 17°C. The ragged robin (*Lychnis flos-cuculi*) occurs throughout temperate Europe, and for this species the cold winter is the least favorable period for growth. Hence there are advantages to be gained by germinating only with the onset of warmer conditions in the spring. Optimum germination occurs at about 27°C. The third species, the sticky catchfly (*Silene viscosa*), is an eastern European steppe species. The invasion of open grassland is an opportunistic business; each chance that offers itself must be taken, so any temperature limitation is likely to be an unacceptable restriction on a plant in its struggle for space. Wide tolerance of temperature is thus an advantage, and *S. viscosa* seeds germinate well over the wide temperature range 11–31°C.

Fig. 2.20 Northern boundary (solid line) of the distribution of the eastern phoebe (*Sayornis phoebe*) in North America in December/January, compared with the –4°C January minimum isotherm (dashed line). From Root [26].

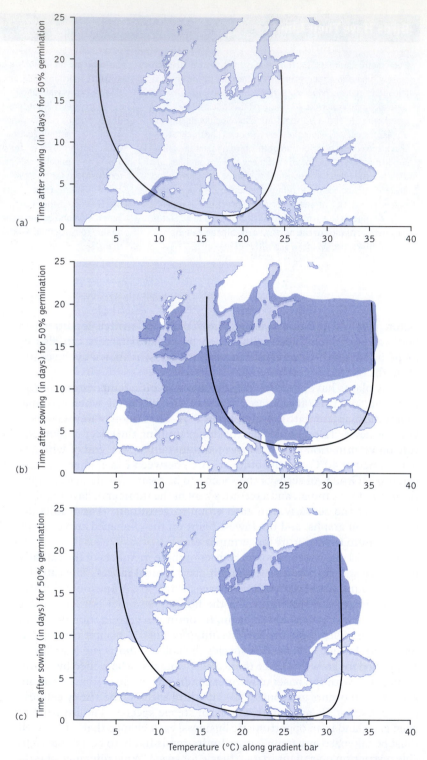

Fig. 2.21 Distribution maps of three members of the plant family Caryophyllaceae, together with their germination responses to temperature: (a) *Silene secundiflora*, (b) *Lychnis flos-cuculi*, and (c) *Silene viscosa*. From Thompson [27].

Germination is not the only process affected by temperature, however. Most metabolic activities in plants and animals are assisted by the activity of enzymes, proteins that act as catalysts in biochemical interactions. All enzymes become deactivated at very high or very low temperatures, but different enzymes vary in their optimum temperature for operation. Photosynthesis, in which atmospheric carbon dioxide is reduced and fixed into organic materials, is critical to the function of green plants and, like all processes mediated by enzymes, it is sensitive to temperature. In most plants, the first product of photosynthesis is a sugar containing three carbon atoms, and the fixation process is catalyzed by the enzyme ribulose bisphosphate carboxylase/oxidase (rubisco for short). These plants are known as **C3 plants**. In some green plant species, however, there is a supplementary mechanism at work in which carbon dioxide is temporarily fixed into a four-carbon compound, which is later fed into the conventional C3 fixation process in specialized cells around the bundles of conducting tissue in the leaf. These are called **C4 plants**, and this process uses an additional enzyme, phosphoenolpyruvate (PEP) carboxylase [28].

For a variety of biochemical reasons, the C4 mechanism is most advantageous under conditions of high light intensity and high temperature, whereas it may be disadvantageous at low light and low temperature. J. R. Ehleringer has calculated that, at a latitude of about 45° in an area like the Great Plains of North America, the relative advantages and disadvantages of each system are roughly in balance [29]. South of that latitude the C4 system should prove superior, and north of this point C3 plants should be at an advantage during the hot, dry summer (Fig. 2.22). If one examines the proportion of C3 and C4 species among the grasses (a plant family in which both photosynthetic systems occur) in different locations through North America (Fig. 2.23), then it is apparent that the C4 system does indeed occur in more than 50% of the grasses south of 40°N and in fewer than 50% of grasses at sites north of the line. Competitive interaction between species of grass has therefore led to the selection of the photosynthetic mechanism most appropriate to the needs of any given locality. Those C4 species found north of the critical line are often associated with particular circumstances that favor them. For example, they may have their maximum growth rate in late summer when temperatures are highest, while the C3 species grow best in the cooler conditions of spring and early summer [31]. As the climate changes and becomes warmer it is possible that there will be a shift in the balance of C3 and C4 species, but the situation is complicated by the effects that changing atmospheric carbon dioxide concentrations will have on the efficiency of the enzymes, so the situation is a complex one. This will be discussed further in Chapter 14.

The C4 photosynthetic system in plants is a mechanism for coping with high temperature and high illumination. One of its great advantages is that it is efficient at foraging for carbon dioxide molecules at low concentrations. This means that it does not require the stomata (leaf pores) to be open for gaseous exchange over extended periods, and thus water loss is limited. There is, however, another type of ecosystem in which this efficiency at carbon foraging can be highly advantageous, namely, shallow pools of water, where the diffusion of dissolved carbon dioxide is much slower than in the atmosphere. On a bright sunny day, the carbon

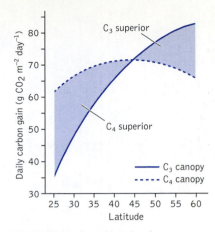

Fig. 2.22 Predicted levels of photosynthesis for C3 and C4 species over a range of latitudes in the Great Plains during July. The C4 advantage is lost in latitudes higher than 45°N. From Ehleringer [29].

Fig. 2.23 Proportion of C4 species in the grass flora of various parts of North America. From Teeri and Stowe [30].

available for photosynthesis can become depleted in aquatic environments, and the enzyme system (PEP carboxylase) that drives the C4 method of carbon acquisition is a very useful asset under such circumstances. Studying seasonal pools in California, Jon Keeley [32] has discovered that a number of the plants that inhabit such locations adopt the C4 photosynthetic system, or a closely related system called **Crassulacean Acid Metabolism (CAM)** that lacks the anatomical complexity of C4. In addition, some plants are able to switch between C3 and CAM according to whether the site is submerged in water, or is seasonally converted into a terrestrial habitat. The biochemistry of the switch is relatively simple, but the advantages gained in providing such plants with the adaptability to compete under changing environmental conditions are great. Broad tolerance to environmental variables can be very valuable to an organism, especially when those variables are subject to sudden alterations.

Although temperature is one of the most important environmental factors because of its effect on enzymes and hence the metabolic rate of organisms, many other physical factors in the environment can become limiting. A whole family of factors is related to the amount of water present in the environment. Aquatic organisms obviously require water as the basic medium of their existence, but most terrestrial animals and plants, too, are limited by the wetness or dryness of the habitat. They are often also limited by the humidity of the atmosphere, which in turn affects its "drying power" (or, more precisely, the rate of evaporation of water from the ground and from animals and plants). Light is of fundamental importance because it provides the energy that green plants need to fix carbohydrates during photosynthesis, thus obtaining energy for themselves (and ultimately for all other organisms).

Light in its daily and seasonal fluctuation also regulates the activities of many animals. The concentrations of oxygen and carbon dioxide in the water or air surrounding organisms are also important. Oxygen is essential to most animals and plants for the release of energy from food by respiration, and carbon dioxide is vital because it is used as the raw material in the photosynthesis of carbohydrates by plants. Many other chemical factors of the environment are of importance, particularly soil chemistry where plants are concerned. Pressure is important to aquatic organisms; deep-sea animals are specially adapted to live at high pressures, but the tissues of species living in more shallow waters would be easily damaged by such pressures.

In marine environments, variation in the salinity of the water affects many organisms because many marine organisms have body fluids with much the same salt concentration as seawater (about 35 parts per thousand), in which their body tissues are adapted to function efficiently. If they become immersed in a less saline medium (in estuaries, for instance), water moves into their tissues due to the physical process called **osmosis**, by which water passes across a membrane from a dilute solution of a salt to a concentrated one. If the organisms cannot control the passage of water into their bodies, the body fluids are flooded and their tissues can no longer function. This problem of salinity is an important factor in preventing marine organisms from invading rivers, or freshwater ones from invading the sea and spreading across oceans to other continents.

In a coastal rock pool, the salinity can change very quickly. Once isolated from the main body of the sea, such a pool can become increasingly saline as a result of evaporation. But if there is rain, then the salinity can be rapidly lowered, placing any organisms present under great osmotic stress. An estuary is rather more predictable because the salinity varies regularly both in space and in time. The distance from the sea influences salinity as the input of seawater becomes less, but salinity at any given location will vary with time because of the impact of tidal flows. The crustacean genus *Gammarus* is found in estuaries, but is represented by different species according to the nature of the salinity conditions (Fig. 2.24). Each species has its optimum set of conditions for salinity, but also has its distribution limits, which result from a combination of its reduced tolerance and also competition from other species that may perform more efficiently under the new conditions [33]. This regular change in physical or chemical conditions through space thus creates a sequence of replacement of one species by another, both among animals and plants. This is known as **zonation** and is common where habitats gradually merge from one type to another.

The natural world, as we have seen, does not consist of a set pattern of environmental factors, but is composed of a complex pattern of varying factors. All organisms have a tolerance range within which they can cope. Some species have broad tolerance ranges for particular factors, while others have narrow ranges of tolerance. We can ask the question, which is the more favorable of the two options and will lead to the greater success of an organism? But the answer is not a simple one, because it depends on the circumstances and the species. A highly specialized species that is able to exploit certain resources efficiently within narrow environmental limits may be successful in the sense that it is able to compete effectively against other species and thus survive. A species with broad tolerance, on the other hand, will have the capacity to cope in a greater variety of habitats. It may not be technically as efficient as the narrow-tolerance species, but it can make up for this by exploiting the wider range of options open to it. We might expect in general, therefore, that eurytopic species will have wider geographical distributions than related stenotopic species.

Kevin Gaston and John Spicer [34] have examined this relationship in their studies of various species of *Gammarus*. They considered comparable pairs of species. For example, *G. zaddachi*, as shown in Fig. 2.24, is an estuarine species with the capacity to tolerate a limited range of salinities. A similar species, *G. duebeni* (not shown in the diagram), is even more tolerant of salinity variation, occurring even in rock pools, where evaporation may increase the salinity. When we look at their global

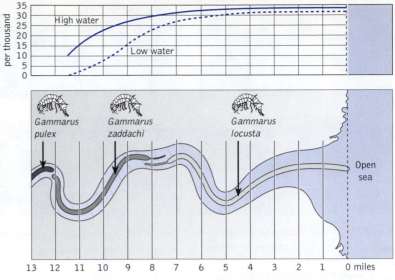

Fig. 2.24 Distribution along a river of three closely related species of amphipod (Crustacea), relative to the concentration of salt in the water. *Gammarus locusta* is an estuarine species and is found in regions where the salt concentration does not fall below about 25 parts per thousand (ppt). *G. zaddachi* is a species with a moderate tolerance of saltwater and is found along a stretch of water between 11 and 19 km (8–12 mi) from the river mouth, where salt concentrations average 10–20 ppt. *G. pulex* is a true freshwater species and does not occur at all in parts of the river showing any influence of the tide or saltwater [33].

Fig. 2.25 Global distribution maps of four species of the amphipod genus *Gammarus*: (a) *G. duebeni* (eurytopic) has a wide distribution pattern, (b) *G. zaddachi* (less tolerant) is more confined, (c) *G. oceanicus* (eurytopic) is widespread, and (d) *G. locusta* (stenotopic) is more confined. From Gaston and Spicer [34].

distribution patterns, it is the more tolerant species, *G. duebeni*, that has the greatest geographical range, being found on both sides of the Atlantic, whereas *G. zaddachi* is confined to northern Europe and Iceland. Similarly, when we look at *Gammarus locusta*, a strictly marine species, and compare it with another marine species, *G. oceanicus*, we find that the more tolerant *G. oceanicus* also has the widest geographical range, as shown in Fig. 2.25. So, if we measure the success of an organism by its geographical distribution range, then the broad-tolerant species seem to have the advantage, at least as far as *Gammarus* species are concerned.

Interaction of Factors

The environment of any species consists of an extremely complicated series of interacting gradients of all the factors, biotic as well as physical, and these influence its distribution and abundance. Populations of the species can live only in those areas where favorable parts of the environmental gradients that affect it overlap. Factors that fall outside this favorable region are limiting ones for the species in that environment. The species must also, of course, be available to invade the area when given the opportunity.

Some of the interactions between the various factors in an organism's environment may be very complex and difficult for the ecologist to interpret or for the experimentalist to investigate. This is because a series of interacting factors may have more extreme effects on the behavior and physiology of a species than any factor alone. To take a simple example, temperature and water interact strongly on organisms, because both high and low temperatures reduce the amount of water available to an organism in an environment. High temperatures cause evaporation and low temperatures cause freezing, but it may be very hard to discover whether an organism is being affected by the direct effects of heat or cold, or by lack of water. Similarly, light energy in the form of sunlight exerts a great influence on organisms because of its importance in photosynthesis and in vision, but it also has a heating effect on the atmosphere and on surfaces, and therefore raises temperatures. A shade-preferring organism may be seeking low light intensities, or may simply be avoiding high temperatures or the low humidity associated with high temperatures. In natural situations it is often almost impossible to tell which of many possible limiting factors is mainly responsible for the distribution of a particular species.

An example of the complexity of interaction between various environmental factors was studied by the American ecologist M. R. Warburg, in his work on two species of woodlice (sowbugs or slaters; Crustacea,

Isopoda) living in rather dry habitats in southern Arizona [34]. One species, *Armadillidium vulgare*, is found mostly in grasslands and scrubby woodland and is also widely distributed in similar habitats elsewhere in North America and in Europe. The other, *Venezilla arizonicus*, is a rather rare species, confined to the southwestern United States and found in very arid country, with stony soil and a sparse vegetation of cactus and acacia. Warburg investigated the reactions of these two species to three environmental factors: temperature, atmospheric humidity, and light. His experimental techniques involved the use of a simple apparatus, the choice chamber or **preferendum** apparatus, in which animals may be placed in a controlled gradient of an environmental factor. The behavior of the animals, particularly the direction in which they move and their speed, can then be used to suggest which part of the gradient they find most satisfactory. This is termed the preferendum of these particular animals in this particular gradient. Warburg's method for testing the interactions of light, temperature, and humidity on the woodlice was the classic scientific approach of isolating the effects of each factor separately and then testing them two or three at a time in all possible combinations. For instance, he might set up a gradient of temperature between hot and cold and test the reactions of the animals to this, either with the whole gradient at a low humidity (dry) or the whole gradient at a high humidity (wet), or with the hot end of the gradient dry and the cold wet, or with the cold end dry and the hot end wet. He might then test the effect of light on these four situations by exposing each in turn to constant illumination, constant darkness, or one end of the gradient in darkness and the other in light. Such work is extremely time-consuming and requires great patience. It is similar to Thompson's work on seed germination discussed earlier, but permits the examination of several factors in combination.

Warburg found that, in general, *A. vulgare* prefers low temperatures (around 10–15°C) and high humidities (above 70% relative humidity, i.e., the air is 70% saturated with water vapor) and is rather weakly attracted to light. This accords well with what is known of the species' habitat and habits—it lives in fairly humid, cool places and is active during the day. *V. arizonicus*, in contrast, prefers lower humidities (around 45%, higher temperatures (20–25°C) and will generally move away from the light. Again, this accords well with the species' habits, since it lives in rather dry, warm places and is active at night. The reactions of the species change, however, and become harder to interpret when they are exposed to more extreme conditions. For instance, at high temperatures (35–40°C), *A. vulgare* tends to choose lower humidities, regardless of whether these are in light or dark. One of several possible explanations for this behavior is that at these high temperatures the species' physiological processes can be maintained only if body temperatures are lowered by permitting loss of water vapor from the body surface, which is more rapid at lower humidities. The normal reaction of *V. arizonicus*, on the other hand, changes if the species is exposed to very high humidities; it then tends to move to drier conditions even if these are in the light. Warburg concludes that, for these two species, light is not really an important physiological factor and acts mostly as a 'token stimulus,' a clue to where optimum conditions of humidity and temperature may be found. For *V. arizonicus*, which lives in a dry or xeric habitat, darkness indicates the likely presence of the high temperatures and low humidities it prefers. For *A. vulgare*, in its cooler, more humid, or **mesic** habitat (a habitat lacking extreme conditions), there is little risk of desiccation except in the most exposed situations, and the species can afford to be relatively indifferent to light.

Warburg's study indicates the great complexity in the reactions of even relatively simple invertebrate animals, such as Crustacea, to the physical factors of their environment. If we also analyze the biotic factors of the animal's environment, their food supply, and the impact of predators, the picture becomes even more complex. Other studies of *A. vulgare* in California, for example, indicate that the species shows quite strong preferences for different types of food (mostly various types of dead vegetation), and these also influence its distribution.

Knowing the physical requirements of a species can be of great economic significance, especially when an animal or plant is to be taken from one part of the world to another. The acacia trees of Australia, for example, are useful because many of them can grow in hot, dry conditions and they can provide a valuable resource of fuel wood for human populations in the dry regions of the world. But there are many acacia species, and each has its own peculiar climatic requirements. Trevor Booth and his colleagues at Canberra, Australia, have been documenting as many as possible of these requirements, based on the analysis of species distribution in Australia [36]. The climate of the region in which it is intended to introduce the acacia tree is then analyzed, and a computer match can be generated in which the ideal acacia for a particular region is selected. This has saved much wasted money and effort in avoiding the old trial-and-error style of forestry. Figure 2.26 shows those parts of Australia with a climate most similar to that of a proposed site for acacia introduction in Zimbabwe. The best acacia match for such requirements was found to be *Acacia holosericea*, the distribution map of which clearly corresponds very closely to this climatic requirement.

Defining the parameters of a species' niche in this way can also lead to important applications in conservation. The leopard (*Panthera pardus*), for example, is a scarce and threatened species in western Asia, and it is valuable to be able to map the areas where the species would be able to survive. As in the case of the acacia trees, one cannot conduct simple preferendum experiments like those possible with woodlice, so data on the environmental preferences of the organism have to be obtained by field observation. Ecologists from Russia and Georgia [37] have located populations of wild leopards in western and central Asia and have documented various features of the habitats where they are found. They noted various aspects of climate, terrain features, such as vegetation and tree cover, and the proximity to human activity. Pooling their observations, they were able to construct a model that described the conditions tolerated by the leopard. The cats were found to avoid deserts, urban developments, and regions with prolonged snow cover. The ecologists could then produce maps showing where the requirements of the leopard were all met, and thus they could highlight sites appropriate for survey work to enumerate leopard populations and locate sites suitable for leopard conservation. From this point they could then examine the likelihood of leopard movements between these regions to maintain genetic flow.

This study did not examine in detail certain other factors that could influence the presence and survival of leopards, such as the availability of prey or the intensity of hunting and poisoning by people. Potential distributions of species cannot be fully understood without reference to the influence of other organisms and their distributional requirements.

Species Interaction

Physical factors evidently play an important part in determining the distribution limits of many plants and animals, but organisms also interact with one another, and this can place constraints on geographical ranges. One species may depend strictly on another for food, as in the case of some butterflies, which may be limited to a single food plant, or a parasite may be limited to a specific host. Some species may be unable to colonize an area because of the existence of certain efficient predators or parasites in that area, or because some other species that is already established there can compete more efficiently for a particular resource that is in demand. These are biotic factors, and they are often responsible for limiting the geographical extent of a species within its potential physical range.

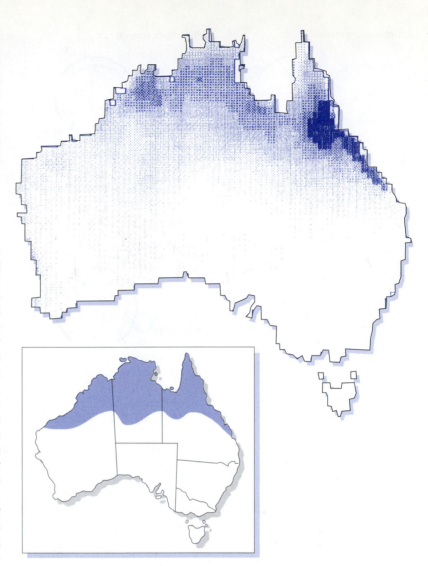

Fig. 2.26 Computer analysis of the climate of Australia to identify the region most similar to that of Kadoma, Zimbabwe, where it was planned to introduce a species of acacia tree for wood fuel production. The inset shows the distribution pattern in Australia of *Acacia holosericea*, a species that seems to match the climatic requirements and would therefore be suitable for introduction. From Booth [36].

When a species is prevented from occupying an area by the presence of another species, this is termed **competitive exclusion**. It is not always easy to observe this ousting of one species by another in nature, but an example of its occurrence is in the barnacle species that occupy the rocky seashores of western Europe and northeastern North America. Adult barnacles are firmly attached to rocks and feed by filtering plankton from the water when the tide is in. Two common species are *Chthamalus stellatus*, which is found within an upper zone of the shore just below the high-tide mark, and *Balanus balanoides*, which occupies a much wider zone below that of *C. stellatus*, down to a low water mark. The distribution of the two species does not overlap by more than a few centimeters.

This situation was analyzed by the ecologist J. H. Connell [38], who found that when the larvae of *C. stellatus* ended their free-swimming

Spartina patens removal

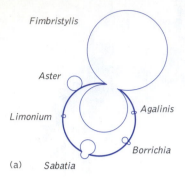

Spartina patens removal

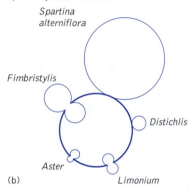

Fimbristylis removal

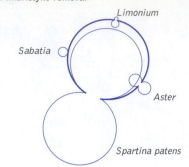

Spartina alterniflora removal

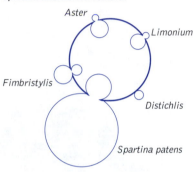

Fig. 2.27 Graphic illustration of the effect of removing a single plant species from a salt marsh community. The sizes of the circles represent the abundance of the plant species concerned, and the heavy circle refers to the species that has been removed. Circles intruding into the heavy circle denote the responses of different species to the perturbation of removal. (a) High-marsh site, where the removal of *Spartina patens* results in mainly the expansion of *Fimbristylis*, and the removal of *Fimbristylis* results in the expansion of *Spartina*. The two species seem to be in competition, and the effects of removal are roughly reciprocated. (b) Low-marsh site where two species of *Spartina* predominate. The removal of *S. patens* results in no response by the other *Spartina* species, whereas the removal of *S. alterniflora* does permit some expansion of *S. patens*. Competition here is thus not reciprocal. From Silander and Antonovics [39].

existence in the sea and settled down for life, they did so over the upper part of the shore above mean tide level. The larvae of *B. balanoides* settled over the whole zone between high and low water, including the area occupied by the adults of *C. stellatus*. Despite overlapping patterns of distribution of the larvae, different distributions of the adults of the two species result from two separate processes. One process acts on the zone at the top of the shore. The young *B. balanoides* are eliminated from this region because they cannot survive the long period of desiccation and the extremes of temperature to which they are exposed at low tide. *C. stellatus* are more resistant to desiccation and survive. Lower down the rocks, the *B. balanoides* persist because they are not exposed for so long, and here the larvae of *C. stellatus* are eliminated by direct competition from the young *B. balanoides*. These grow much faster and simply smother the *Chthamalus* larvae or even prise them off the rocks. Connell also performed experiments on these species and found that, if adult *B. balanoides* were removed from a strip of rock and young ones prevented from settling, the *C. stellatus* were able to colonize the full length of the strip right down to low-tide level. This showed that the competition with *B. balanoides* was the main factor limiting the distribution of *C. stellatus* to the upper part of the shore. This example illustrates the difference between the fundamental niche and the realized niche of an organism.

The example of the barnacles is a relatively simple one because only two species are involved. In most communities of animals and plants many species interact, and this makes it extremely difficult to sort out the full picture of the relationships between species. One approach to the problem is to remove one species from the community and to observe the reaction on the part of the others. This method has been tried in salt-marsh communities in North Carolina by J. A. Silander and J. Antonovics [39], who removed selected plant species and recorded which of the other plants present in the community expanded into the spaces left behind (Fig. 2.27). They found a great range of responses. The removal of one grass species, *Muhlenbergia capillaris*, resulted in an equal expansion on the part of five other plants, suggesting that this grass was in competition with many other species. In the case of the sedge *Fimbristylis spadiceae*, however, removal led to the expansion of only one other plant, the chord grass species *Spartina patens*. The reciprocal experiment in which *Spartina* was removed similarly led to *Fimbristylis* taking full advantage of the new opportunity. In this case we seem to have only two species that are competing for this particular niche.

Selective removal of species in this way is somewhat artificial, however, and can result in disturbance to the physical environment that alters the very nature of the habitat, so it can only provide a preliminary guide to the relationships of species in the community.

Invasion

The capacity to spread is important to all organisms. The success of a species can, in part, be measured by its geographical distribution, and the ability to move into new areas is one of the attributes required in order to achieve this. Habitat may alter, or be lost, so a species needs to be capable of moving to more appropriate sites. Climate may change, and species will then need to alter their ranges to cope with the new conditions. Often this involves overcoming physical barriers that may seem insuperable. Mosses and ferns, for example, may seem to have little hope of long-distance dispersal, because most are relatively small and very static. Both demonstrate an alternation of generations in which spores are produced, however, and these spores are of dust-like dimensions, usually less than 30 microns in diameter, and therefore have the capacity to be carried thousands of miles in the atmosphere. Some ferns, and many mosses, also produce gemmae—organs of vegetative propagation consisting of just a few cells that can be dispersed in the same way as spores. Species with this capacity have proved particularly effective in long-distance movements, such as between islands in the South Pacific [40]. Among animals, the small flightless springtails, just 1 or 2 mm in length, may also seem unlikely candidates for long-distance movements (see p. 53). But these small organisms have hydrophic, unwettable surfaces and, despite being terrestrial soil inhabitants, they can walk on the surface of water. Indeed, experiments have shown that they can survive 16 days on the surface of agitated seawater. Given this capacity, they are able to travel many hundreds of miles in the oceans. They can also cope with freezing and can live for as long as four years at a temperature of −22°C, so they could be incorporated into sea ice and be carried over considerable distances in the polar regions [41]. The dispersal capacity of organisms, therefore, can be much greater than might be expected.

All organisms need to disperse to ensure survival. Environmental conditions, including climate, are constantly changing, so species need to be able to move to new areas if local conditions deteriorate for them. But range expansion for organisms with little or no mobility of their own presents problems, and these have been overcome in a variety of ways. Plants often use wind or water. A survey of the plants of the Southeast Asian island of Rakata showed that 49% had arrived by aerial dispersal, 17% by flotation on the sea, and the remaining 34% by animal transport, either attached to the external surface or carried in their guts [42]. This will be discussed in greater detail in Chapter 8.

Overcoming barriers to dispersal, such as mountain ranges and oceans, has been greatly assisted by humans in the last few centuries [43]. *Ecological imperialism* is a term sometimes used to characterize the

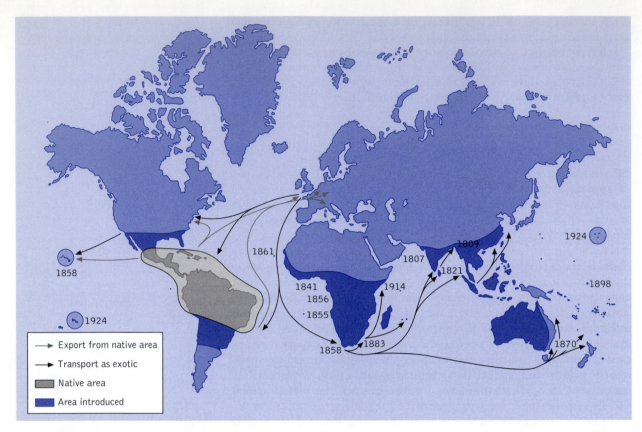

Fig. 2.28 The spread around the world of the weed species *Lantana camara* as a result of human introductions. Its transport from its native range in South and Central America was due to its horticultural attractiveness. But it has subsequently become a serious invasive pest in many tropical and subtropical parts of the world. From Cronk and Fuller [43].

wave of biological invasions that has occurred in the wake of human invasions, either by accidental transport or deliberate introduction. A typical example is the white sage, or tickberry (*Lantana camara*). This is a shrub native to Central and South America that has very attractive red and yellow flowers, initially endearing it to gardeners from Europe. It did not prove a pest in northern Europe, but when it was carried to warmer temperate regions and to other tropical and subtropical countries, it became an invasive plant especially of disturbed soils. The map in Fig. 2.28 shows the history of its spread with human aid during the 19th century, when imperialism was at its height. The consequences of this dispersal are now being felt acutely in areas such as South Africa, India, and the Galápagos Islands, where native species of plant are threatened by the vigorous and competitive growth of *Lantana*. Its remarkable success in so many parts of the world results from many attributes, all of which contribute to its invasive capacity. It is spread by birds (especially the mynah birds of India) that disperse its fruit; it flowers profusely (hence its appeal to gardeners); it grows rapidly; it fragments easily and its vegetative parts take root and grow; it is toxic to many grazing animals, including mammals and insects; its chemical content can even poison other plants; it has broad ecological tolerance to environmental factors (see p. 48). These characters are typical of many invasive plant species.

An aggressive interloper may prove a threat to native organisms, and there are many examples of the displacement of native species by an invader. The European starling (*Sturnus vulgaris*), for instance, was introduced into Central Park, New York, in 1891. Since then, it has spread widely and is now present throughout the United States (Fig. 2.29). It is mostly found in urban areas, and in the east has partly displaced the bluebird (*Sialia sialis*) and the yellow-shafted flicker (*Colaptes auratus*). These species nest in tree holes or in man-made holes, and starlings can occupy and hold most of the limited supply of these nest sites. In the towns, then, the starling successfully competes with the native species for living space. But when flocks of starlings invade the countryside, they compete for food, insects, and seeds with the meadow larks (*Sturnella* spp.), birds that also have declined in some areas.

Even more rapid in its colonization of North America has been the Eurasian collared dove (*Streptopelia decaocto*). This bird was deliberately introduced into the Bahamas in 1974 and had spread to the mainland of North America by 1986. Since then its progress has been remarkably rapid, as shown by the map (Fig. 2.30), moving north into the Carolinas and up the Mississippi River system into Montana and beyond [44]. The secret of the collared dove's success is its adaptability and, like the starling, its willingness to avail itself of human settlements and gardens. Its arrival in North America is a consequence of human introduction, but the species had already shown itself a capable invader even without such aid. In 1900 it was restricted to Asia, especially the subtropics, but then began its westward spread, moving from the Middle East into Egypt and Turkey [45]. It continued westward through central Europe, reaching Britain in the 1950s and becoming an abundant inhabitant of suburban habitats. It is difficult to establish what caused its sudden expansion, but it appears to have changed its behavior pattern, becoming more closely associated with people and also tolerant to more temperate climates. There are concerns that the expansion of this dove may have a detrimental effect on similar native species, such as the turtle dove (*Streptopelia turtur*) in Europe and the mourning dove (*Zenaida macroura*) in North America. In Europe the turtle dove has certainly declined recently, but this could be due to other factors, such as drought and land-use changes in its wintering grounds in Africa and a high intensity of hunting in the

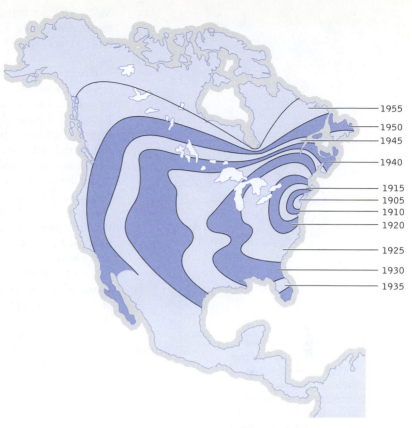

1955
1950
1945
1940
1915
1905
1910
1920
1925
1930
1935

Fig. 2.29 Map of North America showing the range extension of the European starling (*Sturnus vulgaris*) following its introduction to the continent late in the 19th century.

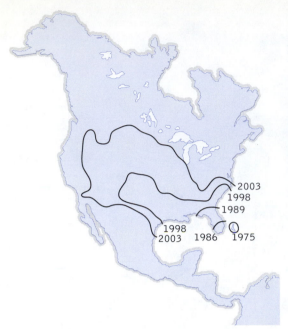

Fig. 2.30 Map of North America showing the range extension of the Eurasian collared dove (*Streptopelia decaocto*) since its introduction to the Bahamas in the 1970s. Its spread in North America follows a similarly rapid extension of range in Europe over the last century.

Mediterranean region. The mourning dove may well prove more susceptible because its niche in North America is more closely linked with human settlements than the turtle dove.

The invasion of North America by European species has not been a one-way process. Some North American species have been successful as invaders in new regions, such as Europe and Australia. An example is the American gray squirrel (*Sciurus carolinensis*), which was introduced into the British Isles in the 19th century. Between 1920 and 1925 the native red squirrel (*S. vulgaris*) suffered a dramatic decline in numbers in Britain, largely due to disease. The spread of the gray squirrel has been accompanied by the disappearance of the red squirrel from many areas, particularly those in which the red squirrel's numbers were reduced by disease and those into which the gray squirrel first spread and established itself. Where the gray squirrel has replaced the native red, it probably has done so by virtue of its superior adaptability to the niche of herbivore at canopy level in deciduous woodland. In the few locations where the gray squirrel has not succeeded in invading, such as the Isle of Wight, an island off the south coast of Britain, the red squirrel still thrives, so disease is not the sole cause of red squirrel decline. The presence of the new invader is evidently also involved, probably because it competes more robustly for food or habitat.

Dispersal to a new area does not ensure that an invader will persist there. When an organism arrives, it must be able to establish itself, possibly by outcompeting and displacing native species at its point of arrival. The resident community may present a degree of **biotic resistance** to any invasive species. It must also be able to survive the pressures of predation and parasitism in its new environment. In the case of plants, one of the most important factors is its ability to cope with the many pathogens that exist in the soil and that often threaten its survival soon after arrival [46]. Some invasive organisms may prove successful simply because the parasites and predators that have evolved with them in their native regions are not found in the new location. Tamarisk shrubs (*Tamarix* spp) are native to Europe and western Asia but have become widespread in the western United States, especially along riverbanks, since their introduction to that country [47]. As they spread, they displace native willows and cottonwoods, but have the advantage of an absence of predators. It is hoped that the introduction of Chinese leafbeetles (*Diorhabda elongata*) will solve the problem, but the use of such **biological control** methods carries dangers that the introduced predator will find new sources of food among native species. Invasive species can thus become "pest" or "weed" species, but these terms can be applied only in relation to human attitudes, causing problems with agriculture, horticulture, transport, industry, or nature conservation. Invaders are likely to be successful, and pest problems are thus likely to arise only if the biotic resistance of the invaded region is inadequate to prevent the establishment of the alien.

It is not easy to predict whether an introduction is likely to prove invasive or whether it may become a pest species. We can make some predictions on the basis of the physical environment, including climatic conditions, and this may help in determining what areas of an invaded land may be at risk. Take the garlic mustard (*Alliaria petiolata*), for example [48]. This is a European herb that has reputed medicinal properties and was probably brought into North America by the early settlers for this reason. In Europe it grows in dense shade beneath a forest canopy, and hence it found the woods of New England very much to its liking, so it soon spread through the region. It now has a range that extends from Ontario to Tennessee and continues to expand into the Midwest. In order to determine which areas are at risk of invasion, ecologists have studied its distribution pattern in its native Europe and have established what climatic factors limit its range. These have been used in the construction of a computer model that can predict the likely outcome of its expansion in North America, based on climatic matching (Fig. 2.31). This type of work can provide advance warning of future problems; in this case there are evidently areas of the western United States that are still liable to be invaded by this aggressive plant. This example illustrates the value of computer-based models of species' niches in order to predict potential ranges, as in the case of the acacia trees and the leopard mentioned earlier.

An invasive species is thus faced with the problem of coping with the physical environment of its new home, and the resistance presented by the native species as they compete for resources. This biotic resistance is not accounted for in the climate modeling described for the potential weed, garlic mustard (see Box 2.3). It is possible that the native flora in some of the areas that the invader could occupy will not succumb to its competition and will resist its invasion. In other words, it may not fulfill the fundamental niche, of which the climate pattern is a partial representation. Its realized niche may prove smaller. An analysis of data concerning the introduced birds of the world, however, has revealed that the success of an invader is more often determined by the physical, abiotic aspects of the region invaded than by resistance from the native species [49].

There is much discussion among biogeographers and ecologists regarding what makes a community resistant to invasion. In his classic book on the ecology of invasions first published in 1958 [50], Charles Elton proposed that more complex communities—that is, those with a greater diversity of species—would be more resistant to invaders than simple communities with few species. But this assumes that the community is in a state of equilibrium and is saturated with species, which is probably rarely the case. Perhaps an invader is successful because it fills a vacant niche in the community, but if this were the case, then no other species would be displaced by its arrival.

When an invasive species takes hold in an area, the consequences may extend beyond its immediate competitors. In North America various alien species of honeysuckle (*Lonicera* spp.) and buckthorn (*Rhamnus* spp.) have become established and are now widespread, especially in the east. *Lonicera maackii*, for example, is now well established in 25 states east of the Rocky Mountains. These shrubs are highly favored by certain songbirds, especially the American robin (*Turdus migratorius*), as nesting locations because they produce a suitable branch structure and have a flush of leaves providing cover early in the spring. But compared with the native shrubs, such as the hawthorns (*Crataegus* spp.), they lack spines and they

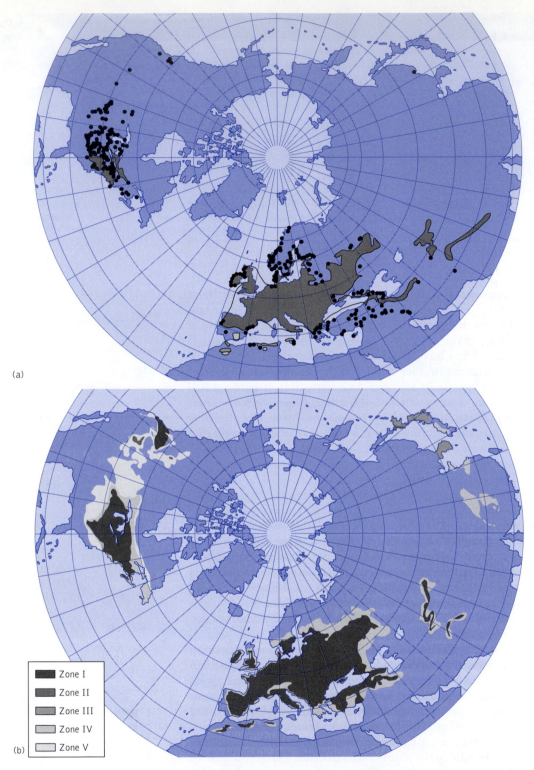

(a)

(b)

Zone I
Zone II
Zone III
Zone IV
Zone V

Fig. 2.31 (a) Distribution of the garlic mustard (*Alliaria petiolata*) in Europe and Asia, where it is native, and in North America, where it has been introduced and is proving invasive. (b) The modeled potential range of the species based on its climatic requirements. There are clearly extensive areas of North America that are at risk of further invasion. The zone grades refer to the proportion of the year that is supportive of the species, where zone I = 100%, II = >96%, III = >92%, IV = >88%, and V = <88%. From Welk et al. [48].

Region	Native Species	Alien Species	Percentage of Aliens
Hawaii	1,143	891	44
British Isles	1,255	945	43
New Zealand	2,449	1,623	40
Australia	15,638	1,952	11
United States	17,300	2,100	11
Continental Europe	11,820	721	6
Southern Africa	20,573	824	4

Table 2.1 The proportion of introduced plant species in the floras of various regions of the world. Note the high proportion of alien species on islands. Data from Lovei [52].

encourage nesting close to the ground. The result is that robin nests in the invading shrubs have a much higher rate of predation than is the case with those in the native vegetation [51]. The impact of an invader can thus extend far into the web of ecological interactions within the ecosystem.

As we have seen, invasion and displacement of species in recent times are often consequences of human transport of organisms from one part of the world to another. Human colonists of new lands have often carried with them the familiar plants and animals from the old country. Where the climate of the new lands have proved appropriate for their survival, these species have often gained a permanent foothold and thus extended their geographical range. Table 2.1 shows the numbers of alien plant species that have established themselves in different parts of the world, compared with the number of species that were already present. It can be seen that the localities with the highest proportion of newcomers are islands; these seem to be particularly sensitive to invasion by aliens [52].

New sets of environmental conditions can lead to shifts in the balance of communities and give rise to waves of invasion. One concern about the current rate of atmospheric change and its climatic consequences is that some species will inevitably suffer and others benefit in any given area. In the Mojave Desert of California, for example, elevated levels of atmospheric carbon dioxide are likely to favor nonnative annual grasses, such as cheatgrass (*Bromus tectorum*) at the expense of native species [53].

Despite these dramatic examples of invasion and competitive displacement, it is most likely that, in natural situations, species that compete for food or other resources have evolved means of reducing the pressures of competition and of dividing up the resources between them. This is mutually advantageous since it reduces the risk of either species being eliminated and made extinct by competition with the others. This is an advantage not only to the species directly involved but to the whole community of species in the habitat, since it results in more species being able to cohabit an area, depending on as many different sources of food as possible. Over evolutionary timescales, therefore, one might expect the species richness of an area to increase as this process of division of the resources gradually takes place. This assumes, however, that the environment is relatively stable, allowing organisms time to evolve and equilibrate. In such communities, competition would occur between many different species, each with its own specialized adaptations, so that no single species could become so numerous as to totally displace others. This equilibration could result in a greater degree of stability for the community, and stable communities are strongly resistant to the invasion of any new species that might disrupt the highly evolved pattern of competition within them. There is certainly evidence that disturbed, unstable habitats are more susceptible to invasion by nonnative species than stable, undisturbed habitats [54], but this idealized scheme, together with the functional consequences of high species diversity, are hotly debated. Evidence suggests that only a smaller proportion of nonnative species arriving at a site ever manage to establish themselves; an even smaller proportion manages to persist over the course of time; and only a fraction of these can be regarded as a threat to native species. In the light of these observations, some ecologists consider the dangers associated with invasive species to be overrated [55].

Reducing Competition

An organism may find considerable advantage in avoiding competition, whether with other species or other members of its own species. Many different ways of reducing competition between organisms have evolved. Sometimes species with similar food or space requirements exploit the same resources at different seasons of the year, or even at different times of day. A common system among predatory mammals and birds is for one species (or a group of them) to have evolved specialized night-time activity, while another species or group of species are daytime predators in the same habitats. Many species of owl hunt at night, judging the location of their prey mostly by ear, while the hawks and falcons are daytime hunters with extremely keen eyesight, especially adapted for judging distances accurately. Thus, both groups of predators can coexist in the same stretch of country and prey on the same limited range of small mammals. Many bats are night-active insectivores, avoiding competition for prey with insectivorous birds during the day and also avoiding the predatory attention of day-active hawks and falcons. Cases of this sort are described as **temporal separation** of species, and this is an effective method of tapping food resources among several species. Among plants this process can be seen operating in deciduous forest habitats, where many woodland floor herbs flower and complete the bulk of their annual growth before the leaf canopy emerges on the trees. In this way the light resources of the environment are used most efficiently.

A different type of temporal separation is shown within the complex grazing community of the East African savanna [56]. During the wet season all of the five most numerous grazing ungulates (buffalo, zebra, wildebeest, topi, and Thomson's gazelle) are able to feed together on the rich forage provided by the short grasses on the higher ground. At the beginning of the dry season, plant growth ceases there. The herbivores then descend to the lower, wetter ground in a highly organized sequence. First are the buffalo, which feed on the leaves of very large riverine grasses, which are little used by the other species. The zebra, which are highly efficient at digesting the low-protein grass stems, move down next. By trampling the plants and eating the grass stems, they make the herb layer suitable for the next arrivals, topi and wildebeest. These two are found in slightly different areas. The jaws and teeth of the topi are adapted for cropping the short, mat-forming grasses common in the northwestern part of the Serengeti. Those of wildebeest are instead adapted to eating the leaves of the upright grasses more common in the southeastern Serengeti. These two species reduce the amount of grass, facilitating the grazing of the last species, Thomson's gazelle, which prefers the broader-leaved dicotyledonous plants to the narrow-leaved monocotyledonous grasses. The whole community therefore interacts in a complex manner, utilizing the pasture in a highly organized and efficient fashion.

Probably much more common than temporal allocation of resources, however, are cases where the resources of a habitat are divided up between species by the restriction of each of them to only part of the available area, to specialized microhabitats. This is called **spatial separation** of species; it means that each species must be adapted to live within the fixed set of physical conditions of its particular microhabitat. It also means that such a species is not adapted to live in other microhabitats and may find it difficult to invade them even if they were for some reason vacant and their food resources untapped.

An example of spatial separation has been described in the extensive marshlands of the Camargue in southern France, where various wading bird species have different preferences for the available feeding areas. The greater flamingo (*Phoenicopterus ruber*) has very long legs and is thus able to wade into deep water where it can sieve planktonic organisms with its highly specialized bill. In shallower water the avocet (*Recurvirostra avosetta*) and the shelduck (*Tadorna tadorna*) feed in a similar way, by sweeping, side-to-side actions of their necks. On the water's edge it is the Kentish plover (*Charadrius alexandrinus*) that feeds predominantly, being restricted to these regions by its shorter legs.

The spatial patterns of distribution and feeding of predatory birds, such as these waders, sometimes reflect the patterns of their preferred food species. For example, the oystercatcher (*Haematopus ostralegus*) has a strong predilection for the bivalve mollusc *Cardium edule*, the cockle, and this is found mainly on sandy and muddy shores just below the mean high-water mark of neap tides [57]. Therefore, this is the favorite feeding zone of the oystercatcher. Similarly, the mud-dwelling crustacean *Corophium volutator* is a favored food species for the redshank (*Tringa totanus*). Since it thrives best in the upper regions of mudflats, usually above the mean high-water mark of neap tides, this is often where large numbers of feeding redshanks can be found.

One can find separation of feeding behavior even within a single species. In the oystercatcher, for example, even though the cockle is the main prey animal, individual birds within a flock may use different techniques for extracting the food from the protective cover of its shell [58]. Some adopt a hammering technique, beating their bills against one of the valves until it breaks, while others use a more subtle approach, forcing their bills between the valves and prising them apart. Even among the hammerers there is further specialization, some attacking only the upper valve and others only the lower valve. These behavioral techniques are reflected in the bill structures of the individuals demonstrating such specializations. Careful observation of tagged birds has demonstrated that the hammerers have blunter bills than the stabbers (Fig. 2.32), but this is

Fig. 2.32 Bill shape variation within the oystercatcher (*Haematopus ostralegus*), a shorebird. The upper bird is a "stabber" and extracts its shellfish prey by pushing the bill between the valves of the shell and prising it open. The lower bird is a "hammerer," which breaks the shell by violent blows of the bill on one of the prey's valves.

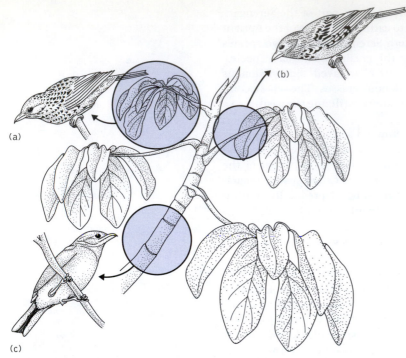

Fig. 2.33 Three species of tanager that coexist in the same forest on the island of Trinidad in the West Indies. All feed on insects, but they exploit different microhabitats within the canopy and thus avoid direct competition. The speckled tanager (a) takes insects from the underside of leaves; the turquoise tanager (b) obtains its insects from fine twigs and leaf petioles; and the bay-headed tanager (c) preys upon insects on the main branches.

quite possibly an effect of bill wear and tear in the case of the hammerers. Thus, there is not necessarily a genetic basis for the beak-shape separation of specialists. But the advantage of this behavior is that different oystercatchers are seeking prey with different weaknesses (either thin shells, or weak muscles holding the valves together), so at least some of the competition for food has been removed.

This type of feeding specialization within a species can under some circumstances lead to the development of distinctive races. In the case of oystercatchers, for example, it is possible that certain differences in shape of bill could have a genetic basis, which would equip some individuals for a slightly different diet or hunting technique. Some ornithologists recognize different subspecies of oystercatcher based on their bill structure [59]. In such a situation, evolution may lead to the splitting of a population, especially if the structural differences were accompanied by behavioral differences in, for example, mate choice. A species could divide in two even in the absence of any geographical barrier to breeding. The subject will be discussed at greater length in Chapters 6 and 7.

Speciation in the absence of any geographical barrier has led to the high level of specialization found within some groups of species, such as the tanagers of Central America (Fig. 2.33). Three closely related species of tanager—the speckled tanager (*Tangara guttata*), the bay-headed tanager (*T. gyrola*), and the turquoise tanager (*T. mexicana*)—may be found coexisting and feeding alongside one another without any apparent competitive interaction. The reason for this harmony is that each feeds in a slightly different location in the forest canopy. The speckled takes insects from the underside of leaves, the turquoise from fine twigs, and the bay-headed from main branches. Each occupies its own niche, and there is little overlap between them.

Even where overlap of niches does occur, animals may often coexist because they have their own distinctive location or way of life that they share with no other. This can be seen in Fig. 2.34, which displays diagrammatically the niches of various primate species in the tropical forest of Ghana [60]. Each species has its own preferred position in the forest canopy, some preferring undisturbed forest and others coping with exploited and cleared areas. There is considerable overlap in their tolerances, but each has its own individual specialist location where it can hold its own in the face of competition for resources from other species.

Migration

Environmental conditions alter with seasons, especially in the higher latitudes, and some animals alter their distribution patterns in concert with the seasons. This is called **migration**. It should not be confused with range expansion, or **spread**, of a species because it consists of the temporary occupation (usually seasonal) of a region while conditions are suitable and then mass movement to an alternative region when the seasonal conditions demand it. Only motile organisms can partake in migration, but even microscopic plankton are capable of changing the depth at which they live, depending on conditions, and this can be regarded as a form of vertical migration.

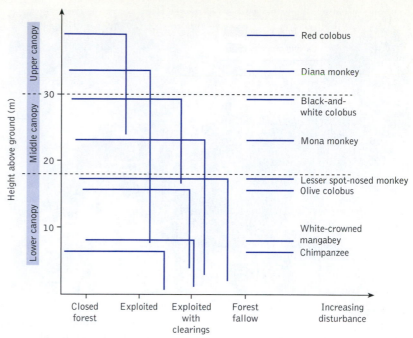

Fig. 2.34 Diagram to illustrate the limits to different niche requirements of a range of primate species in the tropical forest of West Africa in relation to canopy height and degree of human disturbance. Each species indicated occupies the space below and to the left of the lines. Although the demands of the various species overlap, each has a particular height in the canopy or a type of site where it is most efficient as a competitor, and therefore more successful. From Martin [60].

Migration often takes the form of latitudinal movements in order to take advantage of long summer days and high productivity in the high latitudes, and then to retreat to lower latitudes to avoid the stresses of the winter season. The movements of caribou (*Rangifer tarandus*) in North America illustrate such migratory patterns [61]. The females give birth to calves in the early summer, and the various distinct herds of the North American population migrate to the north as the snows melt, to find the most appropriate location where calving takes place. Each herd has a traditional location (Fig. 2.35), usually reflecting the high quality of vegetation production that will best ensure survival of the young calves. Consequently, the spring sees a northward migration of caribou herds heading for the breeding grounds. In the fall, as the tundra becomes colder, productivity falls, and the snow accumulation begins, the herds move south again. Migration is energetically expensive and also exposes animals to the risks of predation, in the case of caribou by wolves that follow the herds. But the benefits of migration in terms of food availability and quality must outweigh these costs.

Birds are among the most mobile of animals, and many species resort to migration in order to maximize food supply, especially during the breeding season. The white-fronted goose (*Anser albifrons*) has a circumpolar distribution pattern, breeding in the long summer days of the Arctic and subarctic in North America, west Greenland, and Siberia. The geese spend their winter in southern parts of North America and Central America, in Europe and the Persian Gulf, and in Japan and eastern China, depending on their breeding locations (Fig. 2.36). Perhaps the most extraordinary of all migrant birds, however, is the Arctic tern (*Sterna paradisaea*) which, as its name suggests, nests in the Arctic and yet travels to the Antarctic during the Northern Hemisphere winter (Fig. 2.37). This bird must enjoy more daylight in the course of its life than any other organism!

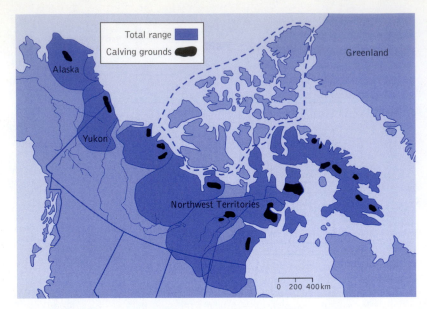

Fig. 2.35 Ranges of caribou herds in North America, also showing their calving grounds to which they migrate each spring. Caribou are also located on the islands enclosed within the dashed line. From Sage [61].

Even small songbirds, such as thrushes and warblers, undertake seasonal migrations. Swainson's thrush (*Catharus ustulatus*), for example, migrates from Canada and the Pacific Northwest to Central and South America each fall, returning in the spring. This journey costs a lot of energy, and very recently it has proved possible to trap and mark birds along the route to determine just how much energy is expended [62]. It takes a Swainson's thrush about 42 days to fly from Panama to Canada, but during this time the actual travel involved consists of 18 nights of flying. The rest of the time is taken up with resting at stopover locations along the route. Over the 4,800-km (3,000-mile) journey, 4,450 kJ of energy is expended, so the cost is just a little less than 1 kJ for each kilometer (1.6 kJ per mile). What is surprising is that only 29% of the energy lost is expended on the actual flight; the remainder is lost during the stopover rests, mainly because so much time is spent recuperating at these locations. The fact that resting periods are so energetically costly underlines the importance of choosing the right weather conditions for the migration. If conditions during the stopovers are cold and energetically costly, it could result in the failure of the bird to survive the

Fig. 2.36 Breeding grounds, migration routes, and wintering grounds of the white-fronted goose (*Anser albifrons*). From Mead [73].

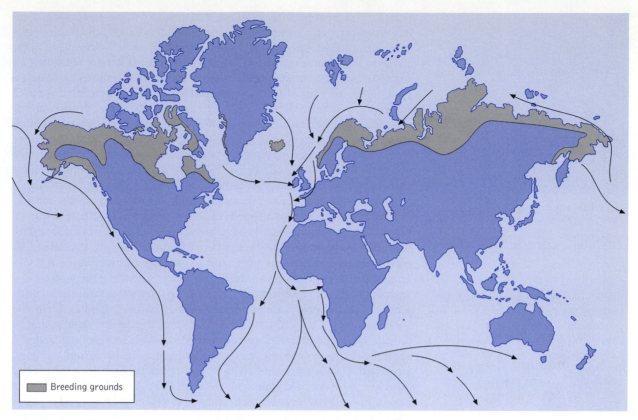

Fig. 2.37 Breeding grounds, migration routes, and wintering area of the arctic tern (*Sterna paradisaea*). From Mead [73].

migration. But, once established in the northern breeding locations, the long days provide ample time for food acquisition, so the bird is better able to feed its young.

Migration, then, provides the species with a means of changing its distribution pattern with the season. It also means that the organism belongs to different communities and ecosystems at different times of its life. A salmon, for example, spends much of its life as part of the oceanic biome, but then moves upriver and into tributary streams for its breeding. There it spends the last days of its life as a temporary member of a fresh-water community. When it dies, the nutrients that it contains then become part of this inland ecosystem and can contribute significant quantities of certain nutrients, such as nitrogen, to the other organisms that share the ecosystem [63]. The implications of migratory behavior, therefore, extend beyond mere distribution patterns.

Predators and Prey, Parasites and Hosts

Predators may be another biological factor influencing the distribution of species, as may the presence and abundance of parasites in a habitat, but their effects have been much less studied than those of competition. The simplest influences that predators might have is to eliminate species by eating them or, alternatively, to prevent the entry of new ones into a habitat. There is very little evidence that either of these processes is common in nature. One or two experimental studies have shown that predators sometimes eat all the representatives of a species in their environment, particularly when the species is already rare. However, all such studies have been made in rather artificial situations in which a predator is introduced into a community of species that have reached some sort of balance with their environment in the absence of any predator; such communities are not at all like natural communities, which already include predators. In general, it is not in the interests of predatory species to eliminate a prey species because if they do they will destroy a potential source of food. Probably most natural communities have evolved so that there is a great number of potential prey species available to each predatory species. Thus, no species is preyed upon too heavily, and the predators can always turn to alternative food species if the numbers of their usual prey should be reduced by climatic or other influences.

Prey switching of this type has been described on the island of Newfoundland, where the gray wolf (*Canis lupus*) and the lynx (*Lynx lynx*) were major predators in the last century, but where the wolf is now extinct as a result of human persecution. The lynx was a rare animal until a new potential prey animal was introduced to the island in 1864, the snowshoe hare (*Lepus americanus*). The hares multiplied rapidly, and so did the lynxes in response to the newly available food source. But the snowshoe hare population crashed to low levels in 1915, and the lynx, faced with starvation, switched its attention to caribou calves, which had once been a major food source for the wolf. The snowshoe hare has now developed a 10-year cycle of high- and low-population levels, and the lynx has continued to switch between hare and caribou depending on whether the hare is in a peak or a trough [64]. This pattern of prey switching allows the lynx to maintain a fairly stable population level and, as a consequence, it also permits the recovery of snowshoe hare populations.

This type of behavior on the part of a predator may thus serve to prevent the extinction of its prey. Sometimes the relationship is even more complex; the predator may prevent the invasion of more efficient or voracious predators, which could reduce the prey population yet further. A good example is the Australian bellminer (*Marlorina melarlophrys*) [65]. This is a communal and highly territorial bird that occupies the canopy of eucalypt forests and that feeds largely on the nymphs, secretions, and scaly covers of psyllid plant bugs. Yet the psyllid bugs survive well under this predation and, what is even more remarkable, they seem to require the attentions of the bell miners, for when these birds are removed, the populations of bugs crash and the eucalypt trees become healthier. It seems that the aggressive behavior of the miners toward other birds prevents their entering bellminers' territories and eating all the psyllids. While the bellminers stay the psyllids are safe, but the trees suffer!

Parasites generally reduce the rate of growth of populations because of their negative influence on general fitness, survival, and fecundity. Like predators, however, they rarely cause the complete extinction of the host species; this would hardly be in their interests. Detailed studies concerning the influence of parasites on the range and geographical spread of species are often based on introduced species. It has been found that introduced species often have lower parasite loads than native species. One study showed that there were 40% fewer parasites associated with introduced species of plants and animals than were found infesting the native species [66]. In addition, the parasites that the invaders carry are often the least virulent because these tend to be more prevalent among hosts simply because they cause less harm. If the invader is also less susceptible to the local parasites than the native species, then its success potential is further enhanced. So an invasive organism may be at an advantage with respect to parasite limitation, and this may be a clue to the success of many invasive organisms.

As mentioned earlier, competition may prevent two species from living together in a habitat and may modify the distribution of species because the resources of the habitat are inadequate to support both of them. Probably the most important effect of predators and of parasites and disease (which are effectively "internal" predators) on the distribution of species is that, by feeding on the individuals of more than one species, they reduce the pressures of competition between them. Thus, by reducing pressures on the resources of the habitat, predators may allow more species to survive than would be the case if the predators were not there. This possibility was first demonstrated experimentally by classical studies on flour beetles and their susceptibility to a sporozoan parasite [67]. When two flour beetles, *Tribolium castaneum* and *T. confusum,* were kept together in a single container, *T. castaneum* inevitably drove *T. confusum* to extinction because it proved a more effective competitor for the food resources. On the introduction of the parasite *Adelina tribolii* to the system, however, the competitive *T. castaneum* proved more susceptible to the parasite and hence its performance was impeded, with the result that the two species were able to coexist. Thus the presence of a parasite in this situation enabled the weaker competitor to survive and thereby increased the species diversity of the system. The experiment was repeated more recently using a different parasite, *Hymenolepis diminuta,* and the result was very different [68]. The weaker competitor, *T. confusum,* proved more susceptible to this parasite than its dominant companion, so the presence of the parasite drove it to extinction even more quickly. So, one cannot generalize about the influence of parasitism on coexistence as it depends on the differential impact of the parasite on the host species involved. A parasite may increase diversity, but it may also reduce it.

A general conclusion, then, is that the presence of predators in a well-balanced community is likely to increase rather than reduce the numbers of species present, so that, overall, predators broaden the distribution of species. Only a few experiments similar to Paine's (described in Box 2.4) have been performed, and so one must be cautious about applying this conclusion to all communities. As in the case of parasites, much depends on the susceptibility to the predator of the various component species in the community. There is some independent evidence, however, that herbivores, which can loosely be regarded as predators of plants, may similarly increase the number of plant species that can live in a habitat. In the nineteenth century, Charles Darwin noticed that in southern England, meadowland grazed by sheep often contained as many as 20 species of plants, while neglected, ungrazed land contained only about 11 species. He suggested that fast-growing, tall grasses were controlled by sheep grazing in the meadow, but that in ungrazed land these species grew tall so that they

Many studies of natural communities have confirmed the hypothesis that predators may increase the number of different species that can live in a habitat. The American ecologist Robert T. Paine made an especially fine study on the animal community of a rocky shore on the Pacific coast of North America [69]. The community included 15 species, comprising acorn barnacles, limpets, chitons, mussels, dog whelks, and one major predator, the starfish *Pisaster ochraceus*, a generalist that fed on all the other species. Paine carried out an experiment on a small area of the shore in which he removed all the starfish and prevented any others from entering. Within a few months 60–80% of the available space in the experimental area was occupied by newly settled barnacles, which began to grow over other species and to eliminate them. After a year or so, however, the barnacles themselves began to be crowded out by large numbers of small, but rapidly growing, mussels. When the study ended these completely dominated the community, which now consisted of only eight species. The removal of predators thus resulted in the halving of the number of species, and there was additional evidence that the number of plant species of the community (mainly rock-encrusting algae) was also reduced because of competition from the barnacles and mussels for the available space.

shaded the small slow-growing plants from the sun and eliminated them. A large-scale experiment occurred in the chalk grassland areas of Britain, when the disease myxomatosis caused the death of large numbers of rabbits in the 1950s. The resulting reduction in grazing allowed considerable invasion by coarse grasses and scrub. As a result, many of these areas are much less rich in species than they were under heavy "predation."

On the Washington coast, Robert Paine, who had investigated the effects of predation on animal communities (see Box 2.4), performed another series of experiments in which he removed the herbivorous sea urchin *Strongylocentrotus purpuratus*, which grazes on algae [70]. Initially, there was an increase in the number of species of algae present; the six or so new species were probably ones that were normally grazed too heavily by the sea urchin to survive in the habitat. But over two or three years the picture changed as the community of algae gradually became dominated by two species, *Hedophyllum sessile* on exposed parts of the shore and *Laminaria groenlandica* in the more sheltered regions below low-water mark. These two species were tall and probably shaded out the smaller species, as did the tall grasses studied by Darwin. The total number of species present was in the end greatly reduced after the removal of the herbivores.

In the floristically rich grasslands of the European Alps, it has been found that the presence of a semiparasitic plant, the yellow rattle (*Rhinanthus minor*), is associated with increased plant diversity. Conservationists have therefore taken to including this species in seed mixtures when rehabilitating areas of damaged grassland as a means of maintaining plant biodiversity [71]. The parasite taps the roots of the more competitive grasses for part of its nutrition, thus reducing their productivity and giving smaller and less competitive plant species an opportunity to survive.

The activities of carnivorous predators in a community can also have an effect on the plants since, by limiting to some extent the number of their herbivorous prey, they prevent overgrazing and thus reduce the risk of rare species of plants being eliminated. Such interactions may be quite complex, however, as in the case of the Hawaiian damselfish, which is a predator of herbivorous fish in coral reef habitats. In an experimental study of the influence of this fish [72], plates were constructed which were suitable for algal colonization, and these were placed in three types of location: (1) within cages which excluded all herbivorous fishes; (2) uncaged, but within the territories of the carnivorous damselfish; and (3) uncaged and placed outside damselfish territories. The diversity of the colonizing algae was highest on the uncaged plates inside damselfish territories and least in the uncaged samples outside the territories. In other words, where there was no grazing at all the algal diversity was higher than when there was intense grazing, but aggressive algae became dominant and excluded some smaller plant species. The highest diversity was found in sites where grazing was controlled to an extent by the predation of the damselfish upon the grazers. The presence of some light grazing suppressed the more robust algae and thereby allowed colonization of more delicate species.

The complicated sets of interactions between predator, grazer, and plant can lead to the development of a finely balanced and diverse community, as shown by this coral reef example. In all of these experiments, based on the manipulation of communities of organisms, it has been found that any one species exerts an influence over many other components of the community, not just its prey organism. The removal of this one species could create effects far in excess of what may originally have been expected. Influential species of this kind are known as **keystone species**. Identifying the keystone species in an ecosystem is clearly a very important task, especially if biodiversity is to be maintained. The loss of a keystone species can cause an avalanche of local extinctions.

Although one can study the requirements of individual species of plants and animals in isolation from one another, this chapter has demonstrated that the interaction between different species is often critical in determining the extent to which the potential range of an organism is actually achieved. Species have both negative and positive effects on one another, so it is important for biogeographers to consider whole assemblages of plants and animals as well as looking at the individual components. This will be the subject of the next chapter.

Summary

1 Patterns of plant and animal distributions over the surface of the planet are very varied and can be accounted for by a large number of different causes, sometimes in complicated combinations.
2 The causes of patterns vary according to the taxonomic level that we are dealing with.
3 The causes of patterns also vary with the spatial scale at which we are considering the organism, whether global, regional, or local—the habitat scale.
4 Factors that need to be considered in the explanation of patterns include geological history, climate and microclimate, availability of food, chemistry of the environment, competition, predation, and parasitism.
5 Human beings have created some global experiments in biogeographical patterns as a result of species introductions.
6 It is possible to study the physical requirements of potentially invasive organisms and thus create predictive models of their likely future success and their threat to human interests as pest species. This approach has only been recently attempted, but is proving very promising.
7 The spatial and temporal separation of organisms, as well as the specialization of groups within populations, can lead to the formation of new species and an increase in biodiversity.
8 Migration constitutes a special kind of dynamic pattern that is developed in response to diurnal or, more frequently, seasonal changes in food supply.
9 Interactions between species, such as competition and predation, create delicate balances in species assemblages that affect both individual species ranges and the biodiversity of communities.

Further Reading

Beeby A, Brennan A-M. *First Ecology: Ecological Principles and Environmental Issues*. 3rd ed. Oxford: Oxford University Press, 2008.

Begon M, Harper JL, Townsend CR. *Ecology: Individuals, Populations and Communities*, 3rd ed. Oxford: Blackwell Science, 1996.

Lomolino, MV, Riddle BR, Brown JH. *Biogeography*, 3rd ed. Sunderland, MA: Sinauer Associates, 2006.

Bullock, JM, Kenward, RE, Hails, RS (eds.). *Dispersal Ecology*, Oxford: Blackwell Publishing, 2002.

References

1 Beebee T, Rowe G. *An Introduction to Molecular Ecology*. Oxford: Oxford University Press, 2004.

2 Collinson JM, Parkin DT, Knox AG, Sangster G, Svensson L. Species boundaries in the Herring and Lesser Black-backed Gull complex. *British Birds* 2008; 101: 340–363.

3 Rosen BR. Biogeographic patterns: a perceptual overview. In: Myers AA, Giller PS, eds. *Analytical Biogeography*, pp. 23–55. London: Chapman & Hall, 1988.

4 Gibbons M. *A Pocket Guide to Palms*. London: Salamander, 2003.

5 Savolainen V, Anstett M-C, Lexer C, Hutton I, Clarkson JJ, Norup MV, Powell MP, Springate D, Salamin N, Baker WJ. Sympatric speciation in palms on an oceanic island. *Nature* 2006; 441, 210–213.

6 Hubbell SP. *The Unified Neutral Theory of Biodiversity and Biogeography*. Princeton, NJ: Princeton University Press, 2001.

7 Leibold MA. Return of the niche. *Nature* 2008; 454, 39–40.

8 Cramp S, ed. *Handbook of the Birds of Europe, the Middle East and North Africa IV*. Oxford: Oxford University Press, 1985.

9 Baughman M, ed. *Reference Atlas to the Birds of North America*. Washington, DC: National Geographic, 2003.

10 Hedberg O. Features of Afroalpine plant ecology. *Acta Phytogeogr Suecica* 1995; 49, 1–144.

11 Knox EB, Palmer JD. Chloroplast DNA variation and the recent radiation of the giant senecios (Asteraceae) on the tall mountains of eastern Africa. *Proc Natl Acad Sci USA* 1995; 92, 10349–10353.

12 Crawford RMM. *Plants at the Margin: Ecological Limits and Climate Change*. Cambridge: Cambridge University Press, 2008.

13 Dijkstra K-DB. *Field Guide to the Dragonflies of Britain and Europe*. London: Collins, 2007.

14 Dandy JE. Magnolias. In: Horai B, ed. *The Oxford Encyclopedia of Trees of the World*, pp. 112–114. Oxford: Oxford University Press, 1981.

15 de Wit R, Bouvier T. "*Everything is everywhere*, but, *the environment selects*"; what did Baas Becking and Beijerinck really say? *Environmental Microbiology* 2006; 8, 755–758.

16 Smith HG, Wilkinson DM. Not all free-living microorganisms have cosmopolitan distributions—the case of *Nebela* (*Apodera*) *vas* Certes (Protozoa: Amoebozoa: Arcellinida). *Journal of Biogeography* 2007; 34, 1822–1831.

17 Charman D. *Peatlands and Environmental Change*. Chichester: John Wiley & Sons, 2002.

18 Harris S, Yalden DW. *Mammals of the British Isles*. 4th ed. London: The Mammal Society, 2008.

19 Coope GR. Tibetan species of dung beetle from Late Pleistocene deposits in England. *Nature* 1973; 245, 335–336.

20 Gathorne-Hardy FJ, Syaukani, Davies RG, Eggleton P, Jones DT. Quaternary rainforest refugia in south-east Asia: using termites (Isoptera) as indicators. *Biol J Linn Soc Lond* 2002; 75, 453–466.

21 Jolly D, Taylor D, Marchant R, Hamilton A, Bonnefille R, Buchet G, Riollet G. Vegetation dynamics in central Africa since 18,000 yr BP. Pollen records from the interlacustrine highlands of Burundi, Rwanda and western Uganda. *J Biogeogr* 1997; 24, 495–512.

22 Hamilton AC. *Environmental History of East Africa*. London: Academic Press, 1982.

23 Favarger C. Endemism in the montane floras of Europe. In: Valentine DH, ed. *Taxonomy, Phytogeography and Evolution*, pp. 191–204. London: Academic Press, 1972.

24 Cowling RM, Rundel PW, Lamont BB, Arroyo MK, Arianoutsou M. Plant diversity in Mediterranean-climate regions. *Trends Ecol Evol* 1996; 11, 362–366.

25 Marshall JK. Factors limiting the survival of *Corynephorus canescens* (L.) Beauv. in Great Britain at the northern edge of its distribution. *Oikos* 1978; 19, 206–216.

26 Root T. Energy constraints on avian distributions. *Ecology* 1988; 69, 330–339.

27 Thompson PA. Germination of species of Caryophyllaceae in relation to their geographical distribution in Europe. *Ann Bot* 1978; 34, 427–449.

28 Raven PH, Evert RF, Eichorn SE. *Biology of Plants*. 6th ed. New York: W. H. Freeman, 1999.

29 Ehleringer JR. Implications of quantum yield differences on the distribution of C3 and C4 grasses. *Oecologia* 1978; 31, 255–267.

30 Teeri JA, Stowe LG. Climatic patterns and the distribution of C4 grasses in North America. *Oecologia* 1976; 23, 1–12.

31 Williams GJ, Markley JL. The photosynthetic pathway type of North American shortgrass prairie species and

some ecological implications. *Photosynthetica* 1973; 7, 262–270.

32 Keeley JE. Photosynthetic pathway diversity in a seasonal pool community. *Functional Ecology* 1999, 106–118.

33 Spooner GM. The distribution of *Gammarus* species in estuaries. *J Mar Biol Assoc* 1974; 27, 1–52.

34 Gaston KJ, Spicer JI. The relationship between range size and niche breadth: a test using five species of *Gammarus* (Amphipoda). *Global Ecology and Biogeography* 2001; 10, 179–188.

35 Warburg MR. Behavioural adaptations of terrestrial isopods. *Am Zool* 1968; 8, 545–599.

36 Booth T. Which wattle where? Selecting Australian acacias for fuelwood plantations. *Plants Today* 1988; 1, 86–90.

37 Gavashelishvili A, Lukarevskiy V. Modelling the habitat requirements of leopard *Panthera pardus* in west and central Asia. *J. Applied Ecology* 2008; 45, 579–588.

38 Connell J. The influence of interspecific competition end other factors on the distribution of the barnacle *Chthamalus stellatus*. *Ecology* 1961; 42, 710–723.

39 Silander JA, Antonovics J. Analysis of interspecific interactions in a coastal plant community—a perturbation approach. *Nature* 1982; 298, 557–560.

40 Dassler CL, Farrar DR. Significance of gametophyte form in long-distance colonization by tropical epiphytic ferns. *Brittonia* 2001; 53, 352–369.

41 Coulson SJ, Hodkinson ID, Webb NR, Harrison JA. Survival of terrestrial soil-dwelling arthropods on and in seawater: implications for trans-oceanic dispersal. *Functional Ecology* 2002; 16, 353–356.

42 Thornton I. *Island Colonization: the Origin and Development of Island Communities*. Cambridge: Cambridge University Press, 2007.

43 Cronk QCB, Fuller JL. *Plant Invaders: The Threat to Natural Ecosystems*. London: Earthscan, 2001.

44 Baughman M., ed. *Reference Atlas to the Birds of North America*. Washington, DC: National Geographic Society, 2003.

45 Hengeveld R. *Dynamics of Biological Invasions*. London: Chapman & Hall, 1989.

46 Klironomos JN. Feedback with soil biota contributes to plant rarity and invasiveness in communities. *Nature* 2002; 417, 67–70.

47 Knight J. Alien versus predator. *Nature* 2001; 412, 115–116.

48 Welk E, Schubert K, Hoffmann MH. Present and potential distribution of invasive garlic mustard (*Alliaria petiolata*) in North America. *Diversity and Distributions* 2002; 8, 219–133.

49 Blackburn TM, Duncan RP. Determinants of establishment success in introduced birds. *Nature* 2001; 414, 195–197.

50 Elton C. *The Ecology of Invasions by Animals and Plants*. Chicago: University of Chicago Press, 2000.

51 Schmidt KA, Whelan CJ. Effects of exotic *Lonicera* and *Rhamnus* on songbird nest predation. *Conservation Biology* 1999; 13, 1502–1506.

52 Lovei GL. Global change through invasion. *Nature* 1997; 388, 627–628.

53 Smith SD, Huxman TE, Zitzer SF, Charlet TN, Housman DC, Coleman JS, Fenstermaker LK, Seemann JR, Nowak RS. Elevated CO2 increases productivity and invasive species success in an arid ecosystem. *Nature* 2000; 408, 79–82.

54 Chytry M, Maskell LC, Pino J, Pysek P, Vila M, Font X, Smart SM. Habitat invasions by alien plants: a quantitative comparison among Mediterranean, sub continental and oceanic regions of Europe. *Journal of Applied Ecology* 2008; 45, 448–458.

55 Davis MA. *Invasion Biology*. Oxford: Oxford University Press, 2009.

56 Bell RHV. The use of the herb layer by grazing ungulates in the Serengeti. In: Watson A. ed. *Animal Populations in Relation to Their Food Resources*, pp. 111–127. Oxford: Blackwell Scientific Publications, 1970.

57 Hale WG. *Waders*. London: Collins, 1980.

58 Sutherland WJ. Why do animals specialize? *Nature* 1987; 325, 483–484.

59 Cramp S., ed. *Handbook of the Birds of Europe the Middle East and North Africa. Vol. 3*. Oxford: Oxford University Press, 1983.

60 Martin C. *The Rainforests of West Africa*. Basel: Birkhäuser Verlag, 1991.

61 Sage B. *The Arctic and Its Wildlife*. London: Croom Helm, 1986.

62 Wikelski M, Tarlow EM, Raim A, Diehl RH, Larkin RP, Visser GH. Costs of migration in free-flying songbirds. *Nature* 2003; 423, 704.

63 Ben-David M, Hanley TA, Schell DM. Fertilization of terrestrial vegetation by spawning Pacific salmon: the role of flooding and predator activity. *Oikos* 1998; 83, 47–55.

64 Bergerud AT. Prey switching in a simple ecosystem. *Sci Amer* 1983; 249 (6), 116–124.

65 Lyon RH, Runnalls RG, Forward GY, Tyers J. Territorial bell miners and other birds affecting populations of insect prey. *Science* 1983; 221, 1411–1413.

66 Thomas F, Renaud F, Guegan J-F. *Parasitism and Ecosystems*. Oxford, Oxford University Press, 2005.

67 Park T. Experimental studies of interspecies competition. I. Competition between populations of the flour beetles, *Tribolium confusum* Duval and *Tribolium castaneum* Herbst. *Ecological Monographs* 1948; 18, 265–308.

68 Yan G, Stevens L, Goodnight CJ, Schall JJ. Effects of a tapeworm parasite on the competition of *Tribolium* beetles. *Ecology* 1998; 79, 1093–1103.

69 Paine RT. Food web complexity and species diversity. *American Naturalist* 1966; 100, 65–75.

70 Paine RT, Vadas RL. The effect of grazing in the sea urchin *Strongylocentrotus* on benthic algal populations. *Limnol Oceanogr* 1969; 14, 710–719.

71 Moore PD. Parasite rattles diversity's cage. *Nature* 2005; 433, 119.

72 Hixon MA, Brostoff WN. Damselfish as keystone species in reverse: intermediate disturbance and diversity of reef algae. *Science* 1983; 220, 511–513.

73 Mead, C. *Bird Migration*. Feltham, Middx., UK: Country Life, 1983.

Communities and Ecosystems

We saw in the last chapter that all species have certain limits to their distribution patterns, sometimes determined by climate or by other factors in the environment. Physical barriers, such as oceans and mountain ranges, can also confine species to certain regions. The evolutionary history of a species is partially responsible for determining what part of the world it is able to occupy, but no organism lives in total isolation from all others. Different organisms interact with one another in both the long and the short term, competing for resources and sometimes excluding one another from certain areas. Over evolutionary time, this can lead to populations specializing in certain ways, perhaps in the way they obtain food, or the type of food they eat, or the type of microclimate in which they perform best. One animal may feed exclusively on a specific source of food, so that the consumer is associated with its food species in its distribution. Species can thus become dependent on one another. Alternatively, species may simply have similar environmental requirements and histories, and hence tend to be found together. The outcome is an assemblage of organisms that appears bound together in a community. This chapter will examine the concepts underlying the community and will also consider the interactions between the living, biotic community and the nonliving environment—a combination that has come to be called the ecosystem.

The Community

A **community** is an assemblage of different species. It is a concept that can be applied at different spatial scales, and it may be limited to specific groups of organisms that are distinguished by their shared taxonomy or their role in the total assemblage. Thus we can speak of the bird communities of cliffs, or the microbial community in a soil, or the plant community in a meadow. So, although the term *community* can refer to the total assemblage of living species found in a site, interacting in a whole range of different ways and forming a complex grouping of plant, animal and microbial components [1], it can also be used in a more limited sense. When it is used of animals that play a particular role in the community, such as plant-sucking insects, then the term **guild** is applied rather than community.

Some animal species, such as those herbivores with very specific food requirements, may form close, dependent unions with certain plants, but others may have requirements for certain spatial architectural conditions that are best supplied by particular assemblages of plants. Again, the outcome of such associations is the existence of communities of plants and animals in nature which, within a particular area, may be repeated in similar topographic and environmentally comparable sites, and which may be very predictable in their species composition.

The study of plant communities has developed as an independent area of ecology (sometimes termed **phytosociology**, or vegetation science) mainly because the plant components of biological communities are often the most evident of their features. In terrestrial sites, the plants generally contribute the most biomass, are static and so can easily be sampled and counted, and are relatively easy to identify when compared with such groups as the invertebrate animals. The green plants are also all involved in the same basic exercise—that is, trapping solar energy and fixing atmospheric carbon dioxide—and so can be treated as a distinctive set of organisms in the general community.

The idea of discrete plant communities that can be described, and even named and classified, is very attractive to the inherent neatness of the human mind and is certainly valuable in the process of mapping areas and assessing their value for conservation and in determining plans for management. Whether plant communities have an objective reality as discrete entities was energetically debated throughout the 20th century, often turning upon the views of the two American ecologists, Frederic Clements and Henry Gleason, who began the discussion in the 1910s and 1920s. Essentially, Clements regarded the plant community as an organic entity in which the positive interactions and interdependencies between plant species led to their being found in distinct associations that were frequently repeated in nature. He felt that the community assembled itself in a manner comparable to the embryology of an organism and could thus be conceived as an integrated entity. The view proved both attractive and pragmatically useful, forming the basis for early attempts at describing and classifying vegetation by such ecologists as Braun Blanquet in France and Arthur Tansley in Britain.

Gleason's argument, on the other hand, emphasized the individual ecological requirements of plant species, pointing out that no two species have quite the same needs. Very rarely do the distributional or ecological ranges of any two species coincide precisely, and the degree of association between ground flora and canopy is often weaker than one might assume from casual observation. The application of statistical techniques to the problem soon demonstrated that, although species often overlap in frequently occurring assemblages, the composition of these groups varies geographically as the physical limits of species are encountered. Studies of the past history of plant and animal species over the last 10,000 years or so have also demonstrated that species come into contact at certain times in their history, but have also been periodically separated as the climate changed. Often the associations we now observe are of relatively recent origin and should be regarded as transitory, a moment in history when certain species happen to have coincided in their distribution patterns. The concept of the community, according to this school of thought, must

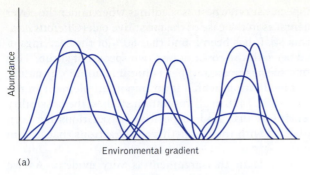

(a)

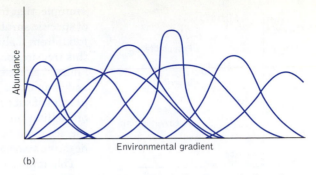

(b)

Fig. 3.1 Diagrammatic representation of two models of vegetation. (a) The model of Clements in which species' requirements coincide, leading to the separation of distinct "communities."

(b) The "individualistic" model of Gleason in which each species is distributed independently and no clear "communities" are apparent.

be looked upon as useful but somewhat artificial; vegetation is actually a continuum both in space and in time.

These two approaches to the description of plant communities can be expressed graphically in the form shown in Fig. 3.1, where the distributions of individual species are depicted along an environmental gradient. This may represent any environmental factor such as soil moisture, acidity, altitude, and so on. All species have their optimum for a given factor (see Fig. 2.18) and have their ecological limits within which survival and growth are possible; some may have a narrow range of tolerance, and others may have a wide range of tolerance and hence, potentially, a wider distribution. If the Clements model is correct, then we would expect that a number of species might coincide in their optima and limits, as shown in Fig. 3.1a. But if the individualistic concept of Gleason is more realistic, the expected pattern should be as shown in Fig. 3.1b, where there is little coincidence between species. Most field studies have demonstrated that the Gleason model is closer to the truth, but the two concepts are not totally mutually exclusive. It is possible that species may show a degree of clustering without attaining a complete separation of discrete units. Much depends on whether certain sets of environmental variables are encountered more frequently than others. Also, the existence of sharp boundaries between habitats in nature (as at the edge of a steep-sided lake) may lead to "communities" being apparently more discrete. Many modern systems of vegetation classification, such as that devised by John Rodwell for the description of the vegetation types of Britain [2], are based on the idea of frequently repeated combinations being selected and described, which can then be used as reference points. This type of scheme allows the possibility of a wide range of intermediate types, which is what one would expect if Gleason's ideas are valid.

The existence of discrete communities depends on positive interactions between species, and this is not always easy to account for, especially in the case of plants that are basically all seeking the same resources of sun, space, water and soil nutrients. It is much easier to visualize competitive, negative interactions between species than mutually beneficial ones. The Darwinian approach to ecology has also led us to be suspicious of any hint of altruism in a hypothesis involving the interaction of species. But positive relationships can occur [3]. Often we find, for

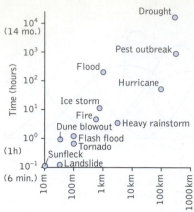

Fig. 3.2 Disruption of habitats by disturbance creates a mosaic of patches in different stages of recovery. This leads to an uneven landscape where separate "communities" of organisms may be perceived. The spatial scale of patches varies with the type of disruption as shown here. Note the log/log scale on the axes. From Forman [5].

example, that tree species survive best as seedlings when under the cover of specific shrub plants, as in the case of the coast live oak (*Quercus agrifolia*) from California [4]. It has been found that 80% of the seedlings of this tree are located beneath the cover of just two species of shrub. This "nurse" effect of one species upon another is not uncommon in nature and is an example of **facilitation**, which is an important element in the process of succession (see p. 142). In all vegetation studies, the possibility of facilitation interactions that lead to positive associations between species must be balanced with the competitive interactions that lead to negative associations.

The importance of scale in this argument is very evident. As we increase the size of the area under study, increasing numbers of species are recorded together and thus appear positively associated, while reducing the sampling area must inevitably lead to more negative associations. In addition, vegetation itself can often be visualized as a mosaic [5]. Patches of vegetation may be in different stages of recovery from disturbance, or regional catastrophe, or from the death of an old tree, or human clearance, or the vegetation may simply reflect underlying environmental patterns of geology, soils, or hydrology. All of these can affect the pattern of species distributions and associations, both in space and in time, and each type of disturbance leads to a different spatial scale of vegetation pattern (Fig. 3.2). If the boundaries between the different elements in the mosaic are sharp and well defined, then the distinction between communities will also be more clear-cut. Perhaps this is why the community concept, especially in vegetation studies, has proved more useful in countries like France and Switzerland, which have very old cultural landscapes that have evolved over millennia of intensive human fragmentation and cultivation. In countries where such intensive agricultural activity is a relatively recent event, such as in America and Australia, landscapes still persist that are occupied by a continuum of vegetation rather than a clear patchwork mosaic.

The Ecosystem

The concept of community encompasses only living organisms and ignores the physical and chemical environment in which those organisms live. If we include these features, including the underlying rock and soil, the water moving through the habitat, and the atmosphere permeating the soil and surrounding the vegetation, then we have an even more complex, interactive system that is called the **ecosystem**. Whereas the idea of community concentrates on the different species found in association with one another, the concept of the ecosystem is largely concerned with the processes that link different organisms to one another.

Two fundamental ideas underlie the ecosystem concept; these are **energy flow** and **nutrient cycling**. Energy, initially fixed from solar radiation into a chemical form by green plants, moves into herbivores as a result of their feeding on plants, and then moves on into carnivores as the herbivores are themselves consumed. Since herbivores rarely consume all the available plant material, and since carnivores do not eat every individual of their prey organisms, some living tissues are allowed to die naturally,

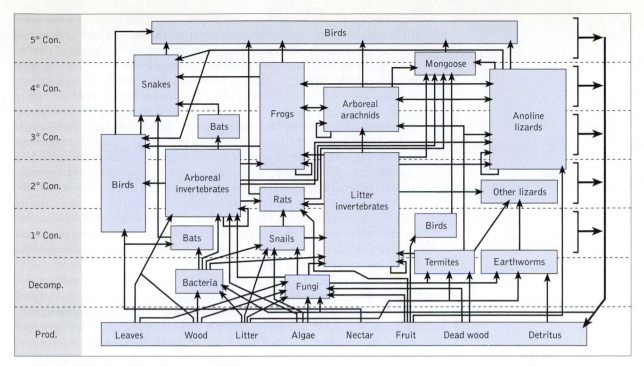

Fig. 3.3 Food web of a tropical rainforest derived from observations at El Verde, Puerto Rico [7]. Organisms are grouped first taxonomically (into boxes) and then arranged in a series of layers (trophic levels) according to their feeding positions in the system. These are labelled as primary, secondary, tertiary con (consumer), up to 5 trophic levels. Several taxonomic groups are represented in more than one trophic layer (e.g., birds) because of the varied feeding habits of their component species. Detritus from dead organisms in this diagram is brought back into the base layer with the primary producers (green plants).

which itself may involve parasites and pathogens [6], before being eaten. These dead tissues constitute an energy resource that can be exploited by scavengers, detritus feeders, and decomposers. Thus, we can classify all the organisms in any community in terms of their feeding relationships. In practice, a series of complex feeding webs are usually formed, relating each species to many others, whether as feeder or food.

Foodwebs can be very complicated, especially in ecosystems with a high species diversity. Figure 3.3 shows the food web of a tropical rainforest at El Verde, Puerto Rico [7], and the complexity of interactions even within this simplified diagram are immediately apparent. Here the major groups of organisms are not split into their individual species because to do so would make the system too complicated to be able to visualize. But the real ecosystem has an abundance of species within each of the boxes shown, each one with its own particular way of life, or niche, and each occupying one or more specific locations in the interactive net of feeding relationships. The arrows shown in a diagram of this sort indicate the linkages between consumers and those they consume: grazer and plant, parasite and host, predator and prey. But they can also be regarded as routes along which energy flows. The base level in the diagram consists of the plants and their products that have captured the energy of sunlight in the process of photosynthesis, and have stored this energy in organic materials by taking carbon dioxide gas from the atmosphere and with it building large molecules. These compounds include the sugars of nectar, the lignins of wood, and the cellulose from leaves, and they are available to grazing animals or to detritus feeders and decomposers. The carbon derived from the atmosphere is also combined with elements obtained from the soil; nitrogen is combined to form amino acids and hence proteins and phosphorus is used in phospholipids that form an essential component of living cell membranes. All of these form a food resource for consumers and decomposers. From the primary consumers, energy is passed through the feeding web, sometimes being stored for a short while in the bodies of the living and dead organisms, often being lost in the process of respiration as the animals conduct themselves in an energetic manner, and eventually all being dissipated in the respiratory heat output of the ecosystem. In Fig. 3.3 the dead residues of animals and plants are shown returning to the detritus that is available to decomposers and that leads back into the food web.

It is convenient to conceive of the ecosystem as a series of layers, as represented here, since organisms obtain their energy in a kind of stratified sequence as it passes from one animal to another. But it is an oversimplification to place whole groups of animals within particular layers; hence birds are represented three times on the diagram according to whether they eat fruit (primary consumers), herbivorous and carnivorous invertebrates (secondary and tertiary consumers), or whether they feed as predators at the top of the food web. For these top predators, the energy they eventually receive has passed through many organisms and many feeding levels, termed **trophic levels**. Since energy is lost at each transfer from one trophic level to the next, the energy reaching the top carnivores is more limited than that available at the base of the food web. This is one reason there will be fewer top carnivores than animals lower down the system.

Ecosystems have inputs and outputs of energy, but generally these are relatively simple. Solar energy is the principal source of energy for most ecosystems, and respiration and consequent energy dissipation as heat is the main loss. But some ecosystems are exceptional. Among these are the deep-sea vents where bacteria capture the energy released from chemical reactions involving inorganic materials such as iron and sulphur compounds and from the methane that belches out of the vents. Some ecosystems, such as streams running through forests, and mudflats in estuaries, may receive most of their energy second-hand from other ecosystems in the form of plant detritus. Therefore ecosystems can exchange energy with one another.

Whereas energy is eventually dissipated as low-grade heat, nutrient elements, such as calcium, nitrogen, potassium and phosphorus, are not irretrievably lost in this way. They are cycled within ecosystems along the same paths as the energy, but are eventually returned to the soil from which they can be reused by plants. Movements of elements between ecosystems occurs, and in terrestrial ecosystems the **hydrological cycle** (the movement of water from oceans through atmospheric water vapor, through precipitation and back to the oceans via streams and rivers) plays a major role in both delivering and removing elements to and from the ecosystem. Apart from the nutrients arriving in rainfall, the other major source for most ecosystems is the gradual degradation of underlying rocks (weathering), which replenishes elements removed by plants and by water percolating through the soil. The water movement through the soil also leaches away unbound nutrients and takes them from one ecosystem to another, often from a terrestrial ecosystem to an aquatic one. Knowledge of the quantities and flow rates of ecosystems helps us to manage them efficiently [8]. If we wish to crop the ecosystem at any given trophic level (e.g., hay at the first trophic level or sheep at the second trophic level), then we must ensure that the rate of removal of nutrients can be compensated for by natural inputs from rainfall and weathering. If this is not so, then we must either reduce the level of exploitation or add those elements in deficit as fertilizers. An element in short supply may limit the rate of ecosystem productivity.

The ecosystem concept therefore involves all the complexity of species interactions within the system, but views these in relation to the processes in which the ecosystem is engaged, including productivity, energy flow, and nutrient cycling. It is a concept that can be applied at a variety of scales, whether to a pool of water in a rotting log, to the whole forest, or even to the entire Earth. In the case of the entire planet, it is essentially a closed system for nutrients but still has energy inputs and outputs. The ecosystem concept has proved a very useful one, not only in the assistance it provides in understanding the relationships between organisms and the interactions with the physical environment, but also because it gives us a basis for their rational use as a resource for the support of human populations.

Ecosystems and Species Diversity

An important question for biogeographers, ecologists, and conservationists is, how many species are really necessary to keep an ecosystem functioning? Is there a basic minimum number of species that is needed for any given ecosystem to operate, above which other species are simply excess to requirements? In other words, are some species redundant? If this were the case, we could remove species from an ecosystem without any effect on ecosystem function until all the redundant species had been extracted. Following this the ecosystem would begin to suffer as a result of further losses (Fig. 3.4a). A possible alternative model is based on the supposition that all species are equally important to ecosystem function, so the loss of each species renders it a little less efficient, as shown diagrammatically in Fig. 3.4b. This is a linear relationship model. The removal of any species from this ecosystem would render it less efficient in functional terms. A third option (Fig. 3.4c) is that certain species play key roles in the ecosystem, and when they are lost there is a sudden drop in the capacity of the ecosystem to function. This is sometimes referred to as the rivet hypothesis. Some species could be lost from such an ecosystem, and it would result in very little, if any, change in the ecosystem function. But if one of the key species ("rivets") is removed, then the entire structure is weakened, just as would happen if certain rivets were removed from the paneling of a ship's hull. There is also the possibility, of course, that the entire structure could collapse if a final and vital rivet were removed. This

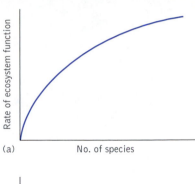
(a)

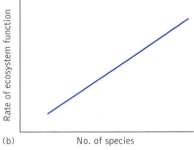
(b)

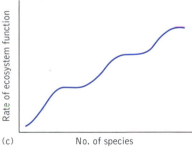
(c)

Fig. 3.4 The possible relationships between the number of species in an ecosystem and its rate of function (e.g., primary productivity, decomposition, nutrient turnover, etc.). (a) Redundant species hypothesis, where at high species density some species can be lost without affecting ecosystem function. (b) Linear model in which every species is equal in functional importance and the loss of any species reduces the efficiency of ecosystem function accordingly. (c) Rivet hypothesis, in which certain species are critical in supporting ecosystem function ("rivets"). Loss of these species has a disproportionate effect on ecosystem function.

concept links with that of keystone species (see Chapter 2), which are species that play a particularly significant role in the function of an ecosystem and on whose presence and activities many other species depend. Such a species may be large, like a beaver in a pond, or very small, such as the nitrogen-fixing lichens on an alpine scree. Keystone species would act as vital rivets in the model of ecosystem stability. There is a final option, sometimes called the idiosyncratic model, in which the stability and functioning of the ecosystem are entirely independent of how many species are present. The consequences of any species loss in that case would be entirely random. Figure 3.4 shows the expected outcome of reducing or increasing the number of species in an ecosystem if any of the main non-random hypotheses were true [9].

These models have been tested both in the laboratory and in the field, but as yet there does not seem to be a simple answer regarding which is closer to the natural situation. John Lawton and his research team at Imperial College, London [10], used miniature ecosystems in controlled-environment chambers ("ecotrons") to test the hypotheses. Keeping the soil conditions and microclimates constant, they have varied the number of animal and plant species between totals of 9 and 31. They found that the species-rich ecosystems were more effective at primary productivity than the species-poor ones (possibly because they developed more complex vegetation architecture) and also in the accumulation of inorganic nutrients into plant tissue. This suggests that diversity does improve ecosystem function in a manner similar to that presented in the linear hypothesis model (Fig. 3.4b). But it is possible that the continued addition of species could eventually bring the system to a saturation point (as in Fig. 3.4a), so the experiment does not settle the question. It does, however, lead us to reject the possibility that ecosystem function is unrelated, or randomly related, to diversity.

Lawton's experiments using ecotrons and Tilman's work in the field (see Box 3.1) could be criticized on the basis of the inevitably small scale on which they operate. A field study of a natural ecosystem in the savanna grasslands of the Serengeti National Park in East Africa by Sam McNaughton and his colleagues from Syracuse University uses a larger scale approach and has provided an example of how more animal species could result in better ecosystem function [13]. Comparing sites that have been grazed by the herds of large mammalian herbivores of the tropical grasslands with other sites where these animals had been excluded by the erection of fences, they demonstrated that several nutrients, such as nitrogen and sodium, were cycled more efficiently in the grazed ecosystems. Therefore, grazing animals actually enrich the nutrient availability in the ecosystems they occupy. Again, it is probable that some species are more effective than others in this role, and their loss would be particularly serious for ecosystem function.

Species are constantly evolving within ecosystems, and the resulting diversification is likely to have an impact on ecosystem function. This is difficult to measure, of course, but an experimental study with three-spine sticklebacks (*Gasterosteus aculeatus*) in British Columbia [14] has shown that several aspects of lake ecosystem function change in the course of diversification, including the community structure of prey

Are Some Species More Equal Than Others?

David Tilman and his co-workers at the University of Minnesota [11] have approached the question of diversity and ecosystem function by experimenting with communities they have assembled in the field rather than in a laboratory ecotron, and they have broadly come to similar conclusions to the Lawton group; more species means better ecosystem function, such as productivity and nutrient retention. They go one step further, however, by pointing out that not all species are equally important in an ecosystem (where "important" means that its removal has a greater impact than might be predicted by random selection). Also, how important a species is depends in part on the other species present. For example, plants from the pea family (Fabaceae) often play a significant role in the nutrient cycling of ecosystems because they have symbiotic relationships with nitrogen-fixing bacteria in their roots that ultimately make extra nitrates available to other species as a result of death, decomposition, and element recycling. The loss of members of the Fabaceae to an ecosystem therefore can be disproportionately harmful to nutrient cycling and, consequently, productivity. But the importance of a pea-family species depends on whether other members of the family are present; if not, its loss may have very high impact. Effectively, a species of this type can be regarded as a keystone species (see Chapter 2). Therefore, not all species are equal in terms of the impact resulting from their loss, and if a species is the only representative of a particular **functional group**, such as producer, decomposer, or nitrogen-fixer, then clearly its importance is even greater. It is possible that the range of functional groups available in an ecosystem is more important to its efficient functioning than is its biodiversity in taxonomic terms [12].

organisms, the primary production of the system, and even the degree of light penetration through the water. As species evolve and diversify, so the ecosystem itself evolves and changes.

A related question, but perhaps even more difficult to answer, is whether diversity has an influence on the stability of an ecosystem. It was the British ecologist Charles Elton who first proposed (in the 1950s) that a more complex and rich ecosystem should also be more stable, meaning that it was less prone to violent fluctuations, such as those caused by epidemic disease or pest outbreaks [15]. It seemed a reasonable proposal, since experience showed that such instability, as defined in these terms, was often a feature of simple ecosystems, as in the case of agricultural monocultures. It was also argued that a complex food web could provide a buffer against any perturbation in which certain species might become scarce. With a complex web there are more opportunities for prey-switching, as with the lynx and hare relationship (see p. 84).

But the development of mathematical modeling as an approach to understanding populations did not provide the expected results. Such models have generally shown that a species in a diverse ecosystem is no less subject to fluctuations caused by unfortunate events, such as drought or disease, than is a species in a simple ecosystem. It remains possible, however, that although individual species may still fluctuate, the function of the entire ecosystem could be less vulnerable to such chance events if the system is diverse and complex than if it is species-poor and simple. Thus, individual species may rise and fall in abundance, but the ecosystem survives because many other contenders are available to replace the unfortunate sufferer in the event of catastrophe. The American ecologist David Tilman [16] has illustrated this possibility in his field experiments with natural vegetation plots. Some of these plots suffered an unplanned natural disturbance in the form of drought, and the plots with higher species numbers suffered lower declines in biomass than species-poor plots. Therefore, although richness does not guarantee the success, or even survival, of individual species, it does provide an ecosystem with greater capacity to cope with disaster. There are clear lessons here for both conservationists and agriculturalists. On the conservation side, the loss of global biodiversity that we are currently experiencing may well be affecting the functioning of the entire biosphere, as will be discussed in the next chapter. On the agricultural side, the use of multicropping systems rather than single-species stands in agriculture should provide advantages both in terms of productivity and in stability of the system, which is a particular concern in marginal areas.

Some confusion can arise because of the different ways in which the term *diversity* is used. Very often it is simply used as an alternative to the number of species present within an ecosystem, the **species richness**, but an ecosystem could be regarded as more diverse if the representation of the various species is reasonably equal, rather than dominated by one or two species. **Evenness** can thus contribute to diversity. But most studies regarding the relationship between diversity and stability have concentrated on richness rather than evenness. Experiments with microbial microcosms, in which the richness and evenness of the original community were varied, showed that the functioning of the system (in this case the process of denitrification) was sustained more effectively when placed under stress (increasing salinity) if the original composition was even [17]. This could be an important consideration when managing ecosystems. For example, by selectively harvesting one particular species from an ecosystem, the evenness and therefore the stability of that system might be put at risk.

One final point in this debate needs to be clarified, and that is what precisely do we mean by **stability**? Is a stable ecosystem one that is difficult to deflect from its current composition or function? This approach defines stability in terms of **inertia**, or resistance to change. Alternatively, a stable ecosystem could be defined as one that rapidly returns to its original state following disturbance. This uses the concept of **resilience** as a basis for defining stability. Both ideas, of course, are inherent in the concept that most people have of stability. Perhaps the most effective way of combining the two ideas is by employing predictability as a measure of stability [18]. A stable ecosystem should behave in a predictable manner no matter what fate may cast in its path, and biodiversity does appear to render an ecosystem predictable by providing a kind of "biological insurance" against the failure of certain sensitive species when exposed to particular stresses.

At present there is no general consensus on the question of how many species are needed to maintain ecosystem functions, or how diversity affects stability. Continuing research is revealing the sheer complexity of the interactions within an ecosystem, sometimes described as an ecological network, but how complexity relates to stability seems to depend on what species are lost. The species with the most complex interactions with other species may be the ones whose removal results in the greatest threat to the ecosystem [19]. The question of ecosystem function is equally vexed. The more ecosystem "functions" that are considered, the more species are required to sustain them [20]. Any generalization is thus likely to prove an oversimplification.

Various experiments are combining the concepts of community and ecosystem and are asking questions regarding the ways in which community composition affects ecosystem operation. These are very important questions in biogeography because we need to be able to appreciate how the presence or absence of particular species in an area will affect the behavior and survival of the remaining species. As we begin to realize the extent to which the world is changing, largely as a consequence of the activities of our own species, we need to be able to predict the outcome of biogeographical alterations in species distributions and assemblages.

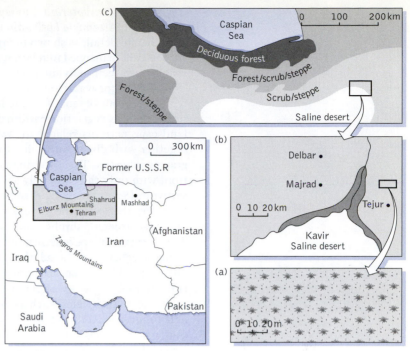

Fig. 3.5 The effect of sampling scale on the appropriateness of different types of vegetation description, illustrated by the vegetation of northern Iran. (a) At a scale of tens of meters, individual shrubs are apparent, and their specific identity can be determined. Bushes of the desert shrub *Zygophyllum eurypterum* tend to be evenly spaced, and smaller woody plants of *Artemisia herba-alba* (shown as dots) occur between them. (b) At a scale of kilometers, different groupings of species form distinct assemblages ("communities") having their own constituent species that can cope with various environmental stresses (drought, salinity, grazing, etc.). The saline desert is almost devoid of vegetation except for a halophyte, *Salsola* sp. The moister alluvial margins and riverside bears a community of the salt bush *Haloxylon persicum* and shrubs of tamarisk (*Tamarix* sp.). Most of the land surface is covered with *Zygophyllum* scrub, except where villages have caused vegetation loss by overgrazing and firewood gathering. (c) At a scale of hundreds of kilometers, vegetation is perceived in terms of physiognomic form (deciduous forest, forest-scrub-steppe, scrub-steppe, saline desert, etc.) rather than as species assemblages.

Biotic Assemblages on a Global Scale

Assemblages of species of plant, animal, and microbe can be viewed in different ways and at different scales. We may adopt a taxonomic approach, identifying all the organisms that are found in an area, analyzing their associations and defining clusters of species that can be classified and named as communities. As has been pointed out, such definitions are likely to be somewhat artificial because they usually lack sharp boundaries between them, but they may be very useful, enabling biogeographers to map the biota within regions. It is also possible to view communities in a different way, based on how their living and nonliving components interact. Using this ecosystem view, the taxonomic identity of species is of lesser interest than their respective roles within the system. Thus, each species falls within a particular functional type, initially determined by its feeding systems, but also by other aspects of its mode of life, such as its nitrogen-fixing capacities.

The application of the community concept on a global scale is complicated by the restriction of individual species to certain areas. The use of lists of species and the study of species interrelationships are useful only within a limited region. Broader-based studies require a different approach. The vegetation of northern Iran in central Asia can serve as an example (Fig. 3.5). Much of the semidesert region of Iran is occupied by sparse scrub, as shown in the photograph in Plate 3.1a. The vegetation over large areas of flat, sandy expanses is relatively uniform, as seen in the photograph, and is composed largely of the deciduous woody shrub,

Zygophyllum eurypterum, interspersed with a smaller woody species of wormwood, *Artemisia herba-alba*. If a sample plot of about 50 m × 50 m is studied in detail, as shown in Fig. 3.5a, the individual plants of the two species can be mapped in relation to one another, and it is found that the spacing between individuals is extremely even, the outcome of intense competition for water among the plants [21]. Upon observing the same region at a scale of tens of square kilometers rather than square meters, a more complex vegetation pattern emerges. Now it is possible to make out landscape patterns related to water movements in the alluvial plain, together with disturbances due to human settlements and variations in grazing intensity. The low-lying salt desert is practically devoid of vegetation, with only the extremely salt-tolerant plants surviving, such as the saltworts (*Salsola* spp.). The fringe of moister regions surrounding the salt flats, together with the valleys of temporary streams, are occupied by other salt-tolerant shrubs, *Haloxylon persicum* and species of *Tamarix*. This vegetation type is shown in Plate 3.1b. The surroundings of the rural villages Delbar and Majrad, though lying within the scrub belt, are actually stripped bare of shrubs because of intensive sheep grazing and collection of fuel wood by the villagers. A few grazing-resistant species, including toxic species, such as *Peganum harmala* and *Ephedra* spp., are able to survive. At this scale of observation, it is therefore possible to pick out "communities" of species that share an ability to exist under certain types of environmental stress (salinity, drought, grazing, etc.). These communities are characterized by particular species, most of which are confined to this region of the world. Similar desert scrub vegetation, however, is found in the Mojave Desert of California, which closely resembles the Iranian scrub in general appearance, but has different species, such as creosote bush (*Larrea tridentata*), which belongs to the same plant family as *Zygophyllum*, together with white bur-sage (*Ambrosia dumosa*), which is in the same family as *Artemisia*. The Californian scrub is thus ecologically equivalent but consists of different species of plants.

When considering vegetation on a global scale, therefore, the use of species for vegetation classification is inappropriate. More useful is a classification by the **physiognomy** (general form and lifestyle) of the vegetation. This permits a comparison of the vegetation of similar climatic locations in different parts of the world. The need for a physiognomic approach to the classification and mapping of vegetation is evident if we move up a level of scale on the Iranian map, shown in Fig. 3.5c, and observe the northern part of Iran, where the units fall into groups such as temperate deciduous forest, forest-scrub, scrub-steppe, and saline desert. This level of classification is defined on the general nature of the vegetation, expressed in terms of its gross structure and appearance, and this is controlled by major factors such as climate and type of soil. Similar patterns to that of Iran occur in different parts of the world, so that a very similar type of vegetation may evolve in each of these areas, as in the case of the desert scrub of California. This comparability of vegetation form may lead to the evolution of similar types of animal in each area. Both regions have tortoises, toads, small burrowing rodents, such as kangaroo rats, jack rabbits and hares, quails and sandgrouse, buzzards and hawks, eagles and vultures, coyotes and foxes. The species are all different, but they represent **ecological equivalents** in different parts of the world.

A general classification of terrestrial, large-scale ecosystems has gradually evolved, which includes the following major types: deserts; the cold tundra of high latitudes and high altitudes; northern coniferous forest or taiga; temperate forest, including temperate rainforest; tropical rainforest; tropical seasonal forest; tropical grassland and open woodland, usually termed savanna; temperate grassland (known as prairies in North America, steppes in Eurasia, pampas in South America, and veld in South Africa); and finally the chaparral of areas with a Mediterranean climate. This classification was first based simply on vegetation, and the resulting units were called **formations**, but in modern usage animal life is included in their descriptions and definitions, and they are called **biomes**.

For reasons already outlined, any system of classifying the world's biota into units must avoid the simple use of particular groups of species for their definition. Although the rainforests of Brazil are comparable in many respects to those of West Africa or Southeast Asia, the actual assemblage of species is quite different. Their similarity is due mainly to the fact that they are structurally comparable, being dominated by tall trees arranged in a series of layers, most of which are evergreen and broadleaved. There are other similarities, such as the presence of canopy-dwelling primates, and the fact that pollination is often brought about by birds, bats, and so on. Essentially, we are comparing the communities in terms of the functional types of plants and animals present rather than their taxonomic affinities. Since vegetation forms the basic template for life within the biomes, it is natural that early ideas on global community classification came from botanists.

The concept of a **life form** among plants, for example, was first put forward by the Danish botanist Christen Raunkiaer in 1934. He observed that the most common or dominant types of plants in a climatic region had a form well suited to survive in the prevailing conditions. Thus, in arctic conditions the most common plants are dwarf shrubs and cushion-forming species that have their buds close to ground level. In this way they survive the winter conditions when wind-borne ice particles have an abrasive effect on any elevated shoots. In warmer climates, buds are usually carried well above the ground and the tree is an efficient life form, but periodic cold or drought may necessitate the seasonal loss of foliage and the development of a dormant phase. This has resulted in the evolution of the deciduous habit. More prolonged drought results in a different type of vegetation, with shrubs that have smaller above-ground structures. Some plants of areas with seasonal drought survive the unfavorable period as underground organs (e.g., bulbs or corms), or as dormant seeds. Animals also show distinct life forms adapted to different climates with cold-resistant, seasonal, or hibernating forms in cold regions, and forms with drought-resistant skins or cuticles in deserts. Nevertheless, animal life forms are usually far less easy to recognize than are those of plants. Consequently, biomes are distinguished and defined primarily by the plant life forms they contain.

The plant life forms recognized by Raunkiaer consisted of the following.
1 Phanerophytes; woody plants with buds more than 25 cm above ground level. These he subdivided into different size classes.
2 Chamaephytes; plants with buds above the soil surface but held less than 25 cm high, including both woody and herbaceous types, often cushion-shaped in form.

3 Hemicryptophytes; plants with their perennating buds at ground level, either with leafy stems in the growing season or with a basal rosette of leaves.

4 Geophytes; plants with their buds below the soil surface, having stem tubers, corms, buds, rhizomes, root tubers, and so on.

5 Helophytes; marsh plants.

6 Hydrophytes; aquatic plants.

7 Therophytes; plants that survive an unfavourable season as seeds.

Although Raunkiaer stressed the importance of perennating organs in his classification of plant life forms, he also recognized that leaf size and form were important. Large, undivided leaves were particularly associated with the hot, wet conditions of the tropics, and small, thick leaves with tough surface layers (**sclerophylls**) were more characteristic of drier regions. He analyzed the flora of different parts of the world into component functional types and found that each region had its own distinctive **biological spectrum**. Tundra regions were characterized by an abundance of chamaephytes and a lack of therophytes; temperate grasslands were typified by hemicryptophytes; tropical rainforests contained a preponderance of phanerophytes; deserts were rich in therophytes and geophytes; and so on.

Using this physiognomic approach to vegetation, it proved possible to define and characterize units, initially referred to as plant formations since they were based on purely botanical criteria. Global maps were constructed by plant geographers who used these plant formations as their basic units, and these were found to have a broad correspondence with climatic zones. Indeed, in the early days the vegetation of understudied areas of the world was often taken as a basis for predicting climate, so it was inevitable that vegetation maps and climate maps had a general similarity to one another.

The life-form approach to vegetation classification is limited simply because it depends solely on one aspect of the plant's response to environmental conditions, namely, the protection of its perennating organs during unfavorable seasons. Many other aspects of a plant's form, structure, and life history could be taken into account, and the development of the idea of functional types has endeavored to take such features into account. Evergreen or deciduous foliage, the association of nitrogen-fixing microbes, mechanisms of pollination, and fruit dispersal, all contribute to a plant's capacity to cope with the stresses and strains of the environment and to establish a niche within a competitive community. There are also characteristics of plant life histories that can be regarded as **strategies** for survival, although the word does suggest a degree of volition on the part of the plant, which is, of course, not intended. Particularly important among such strategies are the reproductive processes (fast versus slow breeding), the allocation of resources to robust growth, and individual longevity. Shade tolerance, fire tolerance, and nutrient requirements are additional factors that contribute to the survival and competitive strategy of a plant. The concept of strategies was first developed by the British ecologist Philip Grime [22] when he was examining the ways in which communities are assembled, but has subsequently been expanded and applied to communities on the biome scale [23].

There is no real agreement among biogeographers about the number of biomes in the world. This is because it is often difficult to tell whether a particular type of vegetation is really a distinct form or is merely an early stage of development of another, and also because many types of vegetation have been much modified by the activities of human beings. This is very apparent from the global vegetation map in Plate 3.2a [24]. Satellite imagery has increased the accuracy with which vegetation can be mapped on a global scale, but the very considerable impact of humans, especially in the temperate zone and around the edges of the arid regions, means that the expected close relationship between vegetation patterns and climate patterns is no longer clear.

There is, however, an overall latitudinal zonation of biomes, which is apparent as we move from the rainforests of the equatorial regions to the dwarf-shrub tundra of the Arctic. A similar trend is reflected in altitudinal terms if we examine the changes in biomes with altitude in the Himalayan Mountains (Fig. 3.6). Ascending from the northern Indian savanna and thorn scrub, we move up through subtropical monsoon forest, largely occupied by drought-deciduous tree species. A zone of scrub follows, which owes its existence largely to the impact of human activity and that of domesticated animal grazing. Above this is temperate deciduous forest with oak and rhododendron, its general form and appearance being very similar to that found in western Europe or the eastern part of the United States. Coniferous forest lies above the oak zone, dominated mainly by the deodar cedar, and above this is the alpine birch and juniper scrub that leads on up to the tundra and the permanent snows of the high mountains. The altitudinal sequence thus broadly mirrors the latitudinal zones but in a much shorter distance.

The link between biomes and climate becomes even more apparent if biomes are mapped against climatic variables, such as precipitation and temperature [25], as has been done in Fig. 3.7. Again, the divisions between biomes are not as sharp as those indicated here, but it is plain that each biome occupies a region where specific climatic requirements are met. Using satellite data, it has proved possible to elucidate the precise climatic requirements of particular biomes as defined by plant life forms and functional types. Each biome is said to fit within a certain **climatic envelope**, which is the sum of all the climatic variables that limit that biome. This approach has led to a refinement of conventional biome definitions, particularly in the case of forests and grassland. Ian Woodward of the University of Sheffield in England has proposed the following classification of these vegetation types [26].

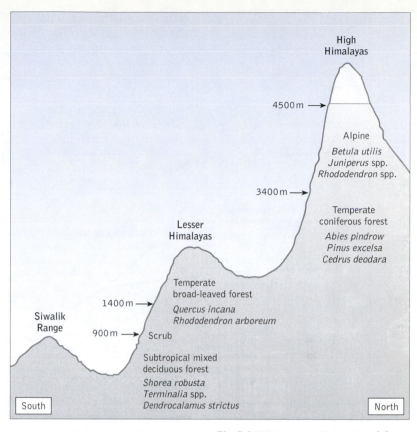

Fig. 3.6 Diagrammatic section of the western Himalayas in northern India, showing the approximate altitudinal limits of the major vegetation types. With increasing altitude, one passes through vegetation belts similar to those found on passing from lower into higher latitudes. The scrub zone (900–1400 m) is strongly modified by human deforestation and subsequent grazing.

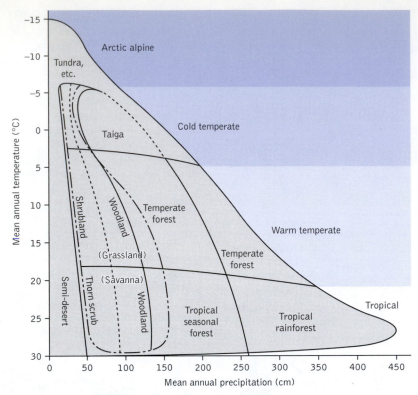

Fig. 3.7 The distribution of the major terrestrial biomes with respect to mean annual precipitation and mean annual temperature. Within regions delimited by the dashed line, a number of factors, including geographical location, seasonality of drought, and human land use may affect the biome type that develops. From Whittaker [25].

1 Evergreen needle-leaf forests—tall (over 2 m high), dense (over 60% cover) forests of evergreen trees with narrow leaves (e.g., boreal coniferous forests).

2 Evergreen broadleaf forests—tall, dense forests of evergreen trees with broad leaves (e.g., tropical rainforests).

3 Deciduous needle-leaf forests—tall, dense forests of narrow-leaved trees that seasonally lose their leaves (e.g., larch).

4 Deciduous broadleaf forests—tall, dense forests of broad-leaved trees that seasonally lose their leaves (e.g., beech, maple, some oaks).

5 Mixed forests—tall, dense forests with an intermixture or a mosaic of deciduous and evergreen trees.

6 Woody savannas—trees exceed 2 m high, but cover only 30 to 60% of the land surface, intermixed with herbaceous vegetation.

7 Savanna—trees exceed 2 m high, but are widely scattered, covering only 10 to 30% of the surface, the remainder being dominated by herbaceous vegetation.

8 Grasslands—land with herbaceous cover and with less than 10% tree or shrub cover.

9 Closed shrublands—lands with woody vegetation less than 2 m tall and with a shrub cover (evergreen or deciduous) of more than 60%.

10 Open shrublands—lands with woody vegetation (evergreen or deciduous) less than 2 m tall and with a shrub cover of between 10 and 60%.

This classification system has the objective advantage that it can be easily recognized from satellite imagery. Each of these types can then be assigned a particular climatic envelope. To understand the global pattern of biome distribution, therefore, it is necessary to appreciate underlying patterns of climates and their global causes.

Patterns of Climate

The **climate** of an area is the whole range of weather conditions, including temperature, rainfall, evaporation, sunlight, and wind, that it experiences through all the seasons of the year. Many factors are involved in determining the climate of an area, particularly latitude, altitude, and location in relation to seas and landmasses. The climate in turn is a very important factor in determining the species of plants and animals, and even the life forms or functional types that can live in an area.

Climate varies with latitude for two reasons. The first reason is that the spherical form of the Earth results in an uneven distribution of solar energy with respect to latitude. As the angle of incidence of the Sun's rays approaches 90º, the area over which the energy is spread is reduced, so that there is an increased heating effect. In the high latitudes, energy is spread over a wide area; thus, polar climates are cold (Fig. 3.8a). The precise latitude that receives sunlight at 90º at noon varies during the year; it

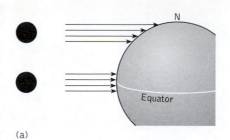

(a)

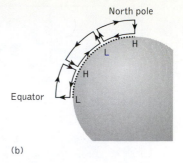

(b)

Fig. 3.8 Patterns of climate.
(a) Due to the spherical shape of the Earth, polar regions receive less solar energy per unit area than the equatorial regions.
(b) The major patterns of circulating air masses (cells) in the Northern Hemisphere: H, high pressure; L, low pressure.

is at the equator during March and September, at the Tropic of Cancer (23°28′N) during June, and at the Tropic of Capricorn (23°28′S) during December. These two tropics mark the limits beyond which the Sun is never overhead. The effect of this seasonal fluctuation is more profound in some regions than in others.

Variations in climate also result from the pattern of movement of air masses (Fig. 3.8b). Air is heated most strongly over the equator and therefore rises (causing a low-pressure area) and moves toward the pole. As it moves toward the pole, it gradually cools and increases in density until it descends, where it forms a subtropical region of high pressure, known as the Horse Latitudes. Air from this high-pressure area either moves toward the equator, or else moves poleward. The poleward-heading air mass eventually meets cold air currents moving south from the polar regions where air has been cooled and descends (causing a high-pressure area). Where these two air masses meet, a region of unstable low pressure results, in which the weather is very variable.

This idealized picture is complicated by the Coriolis effect (named in honor of the French mathematician Gaspard Coriolis, who analyzed it), which results from the west-east rotation of the Earth. This force tends to deflect a moving object to the right of its course in the Northern Hemisphere and to the left in the Southern Hemisphere. As a result, the winds moving toward the equator come to blow from a more easterly direction. These trade winds, coming from both the Northern and the Southern Hemispheres, therefore meet at the equator, and this region is known as the **intertropical convergence zone (ITCZ)**. Where these easterly winds have passed over oceans they become moist, and this moisture is deposited as rain, generally over the easterly portions of the equatorial latitudes of the continents. Similarly, the winds that move poleward from the high-pressure Horse Latitudes come to blow from a more westerly direction and provide rain along the westerly regions of the higher latitudes of continents. The Horse Latitudes themselves are regions in which dry air is descending, and arid belts form along these latitudes of the continents.

The distribution of oceans and landmasses modifies this picture yet further. Because heat is gained or released more slowly by water than by landmasses, heat exchange is slower in maritime regions, while at the same time humidity is higher. In summer, therefore, continental areas tend to develop low-pressure systems as a result of the heating of landmasses and the conduction of this heat to the overlying air masses. Conversely, in winter the reverse situation occurs, continental areas becoming cold faster than the oceans and high-pressure systems developing over them (Fig. 3.9). One effect of this process is that the continental low-pressure systems draw in moist

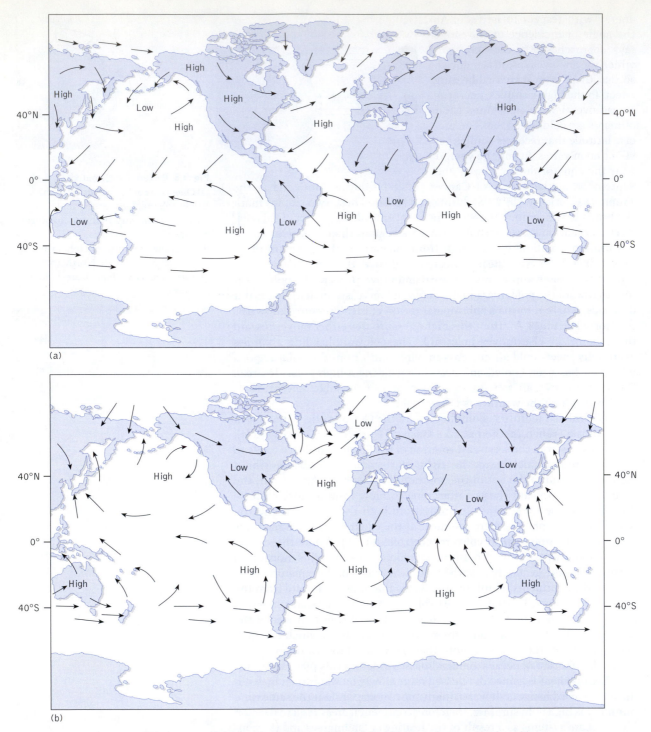

Fig. 3.9 General pattern of air movement across the Earth's surface and distribution of areas of high and low pressure (a) in January and (b) in July. Note the reversal of winds in the Indian Ocean in July, taking monsoon rains into northern India and East Africa.

air from neighboring seas, for example, from the Indian Ocean to East Africa and India, causing summer monsoon rains. The winter of these areas, on the other hand, is usually dry.

The global circulation patterns of the oceans is also of great importance in determining world climate patterns (Fig. 3.10). Warm, surface waters from the equatorial regions of the Atlantic Ocean are carried north-eastward toward Iceland and Norway, warming this section of the Arctic Ocean, keeping it ice-free through the summer and bringing the whole of western Europe mild conditions. As the water body cools, it becomes denser and sinks, reversing its flow pattern as it heads back to the southern Atlantic. From here, the cold, dense water may flow up into the Indian Ocean where it receives new warmth, or it may continue around the southern latitudes and eventually move northward into the Pacific Ocean and pick up heat there. Once heated, the less dense waters move along the ocean surface and back into the Atlantic. This **thermohaline oceanic circulation**, sometimes referred to as an oceanic conveyor belt, is responsible for dispersing much of the tropical warmth to higher latitudes, particularly in the Atlantic seaboard. Without it, the high latitudes would be much colder. Indeed, one of the main features of the glacial episodes in the Earth's recent history has been the shutting down of the global oceanic heat conveyor. Its effect on biome distribution is apparent in that boreal coniferous forest, for example, extends to much higher latitudes in Scandinavia than it does in Alaska.

In addition to the heating and cooling effects of landmasses and oceanic currents, climate is also affected by altitude. On average, the air temperature falls by 0.6°C for every 100 m (330 ft) rise in height, but this varies considerably according to prevailing conditions, especially the aspect and steepness of the slope and the wind exposure. Because of this tendency for temperature to fall with increasing altitude, the organisms inhabiting high tropical and subtropical mountains, such as the Himalayas in northern India (see Fig. 3.6), may be more like the flora and fauna of colder regions than that of the surrounding lowlands. However, although temperature in general falls as one ascends such mountains, other environmental conditions do not precisely mirror those found at higher latitudes. For example, the seasonal variations in daylength typical of high-latitude tundra areas are not found in the "alpine" regions of tropical mountains. Also, the high degree of insolation resulting from the high angle of the Sun produces considerable diurnal fluctuations in temperature in tropical alpine locations that are not found in high-latitude tundra regions. It is not surprising, therefore, that the altitudinal zonation of plants and animals should not precisely reflect the global, latitudinal zonation. Also, where a species is found in both types of tundra, the arctic and alpine races of the species often differ in their physiological make-up as a consequence of these climatic differences.

Fig. 3.10 The oceanic conveyor belt carrying warm, low-salinity surface waters northward into the North Atlantic and deep, higher salinity, cold waters from west to east into the Pacific.

As we have seen, individual species of plants and animals as well as life forms are affected by a whole range of physical factors in their environment, many of these being directly related to climate. Biogeographers, therefore, have long sought a means of portraying climates in simple, condensed form that would give at a glance an indication of the main features that might be of critical importance to the survival of organisms in the area. Mean values of temperature and rainfall may be of some use, but one also needs to know something of seasonal variation and of extreme values if the full implications of a particular climatic regime are to be appreciated. It is with this aim in view that Heinrich Walter, of the University of Hohenheim in Germany, devised a form of climate diagram, which is now widely used by biogeographers [27]. An explanation of the construction of these diagrams is given in Fig. 3.11, and a selection of climate diagrams displaying the climates of some of the major biomes of the Earth is given in Fig. 3.12.

Simple summaries of the climate of a region can be achieved using climate diagrams, as described in Box 3.2 and illustrated in Figs. 3.11 and 3.12.

Modeling Biomes and Climate

The general link between plant formations or biomes, defined in terms of life forms or functional types, and climate is evident, but modern biogeographers require more robust methods in the study of connections between vegetation and climate, especially if they are to be in a position to predict the future pattern of biomes in the event of climate change. As we have seen, much effort is now being expended in defining biological units, or biomes, more precisely, and fitting them to specific climatic envelopes. We now have much more detailed information about the physiology of different plant types than was available to Raunkiaer, including their tolerance of cold or heat and their ability to cope with drought or flooding. When all of this information is put together, it is possible to define much more precisely the range of plant functional types that can be useful in classifying, mapping, and understanding vegetation and its relation to climate. It is also possible to escape the notion that biomes are fixed in their makeup. Instead, we can admit the possibility of regional variations and transitions between biomes. In other words, we can adopt a more Gleasonian approach to global vegetation. Studies of vegetation history, which are discussed in Chapter 12, have further shown that the composition of biological assemblages is constantly changing. We therefore need to introduce a time variable into our concept of the biome, which gives it a dynamic rather than static status.

One of the most elaborate attempts to describe the complex relationships between vegetation and climate was that of E. O. Box [28], who compiled a list of 90 plant functional types. He was able to assess the climatic tolerances and requirements of his functional types, involving temperature, precipitation, and the seasonal variation in these. Precipitation alone, however, can be misleading because other conditions, such as high temperature and intense solar radiation, can lead to the rapid evaporation of water. Therefore a moisture index expressing the ratio of precipitation to potential **evapotranspiration** (the combination of evaporation of water from surfaces and the upward movement and loss of water from plants) is

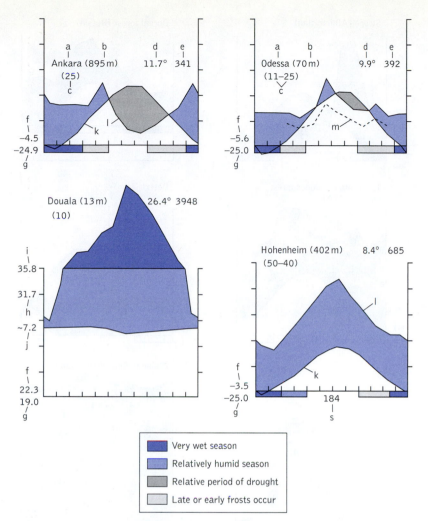

Fig. 3.11 Climate diagrams convey many aspects of seasonal variation in climate at a site in a manner that is easily assimilated visually, and that concentrates on those climatic features of greatest relevance to the determination of vegetation in the area. Key to the climate diagrams: Abscissa: months (Northern Hemisphere January–December, Southern Hemisphere July–June); Ordinate: one division = 10°C or 20 mm rainfall. a, station; b, height above sea level; c, duration of observations in years (if there are two figures, the first indicates temperature and the second precipitation); d, mean annual temperature in °C; e, mean annual precipitation in millimeters; f, mean daily minimum temperature of the coldest month; g, lowest temperature recorded; h, mean daily maximum temperature of the warmest month; i, highest temperature recorded; j, mean daily temperature variations; k, curve of mean monthly temperature; l, curve of mean monthly precipitation; m, reduced supplementary precipitation curve (10°C = 30 mm precipitation); s, mean duration of frost-free period in days. Some values are missing from some diagrams indicating that no data are available for the stations concerned. After Walter [27].

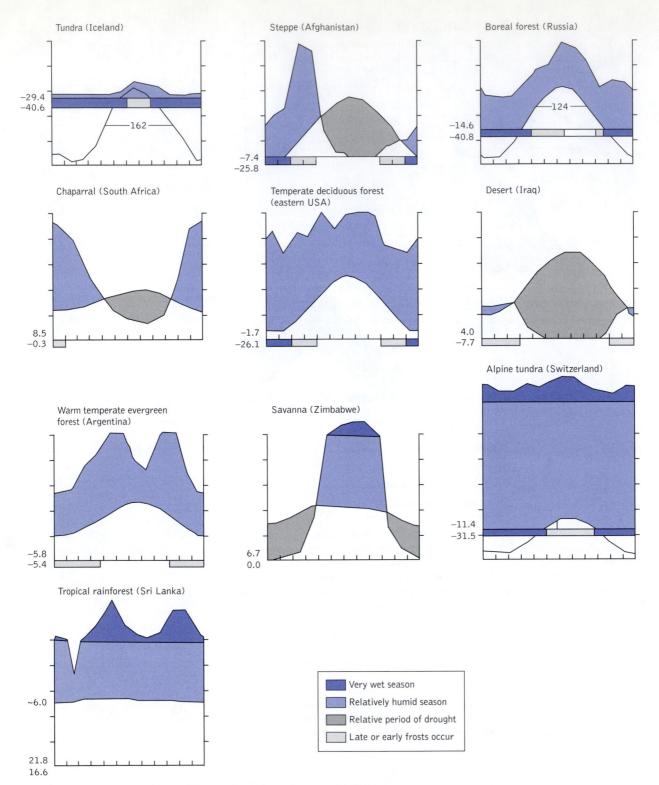

Fig. 3.12 Representative climate diagrams from the major terrestrial biomes.

required. Once these data are available, it is possible to inspect a set of climatic conditions and predict the plant functional types that would be found, together with their relative abundance and importance. The predictions from such a model [29] can then be checked against observations in the field to see how closely the model matches reality. It must be borne in mind, of course, that the predictions on this model will show potential vegetation on the assumption that climate is the determining factor; it makes no allowance for human modification of the natural habitats. It is essentially the study of ideal climatic envelopes.

One of the most effective and widely used models relating global vegetation to climate is that constructed by Colin Prentice, working at that time from the University of Lund, Sweden, and his colleagues [30]. They have sought to maintain a simple approach in their efforts to build a computer model of world vegetation. Thus, they use just 13 functional types based not simply on morphology but also on physiology, especially the temperature tolerances of the plants. They regard minimum temperature as particularly important, for this seems to determine woody plant distribution, as Raunkiaer correctly asserted. Most broad-leaved evergreen tropical trees, for example, are killed if exposed to frost. Temperate deciduous trees, such as many of the oaks, are damaged by temperatures below −40°C, while many of the boreal evergreen, needle-leaved conifers, such as spruce and fir, can cope with temperatures between −45° and −60°C. Some of the needle-leaved deciduous conifers, such as larch, are able to tolerate even lower temperatures.

On the other hand, many temperate trees need to be chilled during the winter if they are to bud effectively and produce flowers the following spring, so this requirement needs to be built into the model. It means that some of the needle-leaved evergreens will demand winters in which the coldest month has an average temperature of below −2°C. Warmth in the growing season, however, is also needed, and even cold-dwelling trees need about 5°C for adequate photosynthetic activity. Lower-growing plants, on the other hand, such as dwarf shrubs and cushion chamaephytes, may survive effectively where the average temperature of the warmest month is only 0°C. To allow for the wetness of an area, Prentice and his colleagues used a moisture index of the type developed by Box, building in an allowance for seasonal variation in wetness.

The Prentice model works by taking into account all of these, and several other climatic variables, and determining which of the functional types could survive, and then assessing which is likely to be dominant. In this way the potential biome for the area is deduced. The outcome of the model is shown in Plate 3.2b, which depicts the potential biome distribution in both the Old and the New World. The model is essentially Gleasonian in that it begins from the requirements of the plant functional types and assembles groupings on the basis of which types could tolerate given climatic conditions in different parts of the world. There is no initial assumption about how many biomes exist or how they should be composed. In fact, 17 terrestrial biomes emerge, as can be seen from the maps. The model predictions of the distribution of these biomes is in very reasonable agreement with the actual information available (shown in Plate 3.2a). We have to bear in mind, however, that the information is incomplete (hence the gaps in the real-world maps), and often reflects the consequences of human cultural activities rather than potential vegetation under the existing climatic conditions.

The importance of this approach to the global classification and mapping of biomes is that it provides us with a robust model on which we can project new circumstances and observe their possible outcome. Climatic scenarios of different types can be imposed, and the consequences for biome distribution can be estimated. Since we know the relationships of certain types of agriculture to these biomes, we can also examine the impact of climate change on agricultural potential of different parts of the world.

Biomes in a Changing World

It is now generally accepted that the climate is changing, largely as a consequence of human activities in the biosphere. Thus, the locations of the boundaries around the biomes (Plate 3.2a) can be expected to change in response to rising global temperature and other consequent climatic changes. Theoretical studies based on predictive models of rising global carbon dioxide and the development of a greenhouse effect have been extensively used to estimate the future trends in global temperature, but the models are by their very nature complex and many variables are still uncertain. It is now firmly established, however, that global temperatures have been rising for the last 150 years, having increased by about 1.5°C in that time [31]. Biogeographers are faced with the problem of predicting the consequences of climate change in terms of changing distribution patterns of individual species of organisms and hence the restructuring and movement of entire biomes. This subject is discussed in greater detail in Chapter 14, but the process of modeling the relationships between climate and vegetation, and the establishment of climate envelopes for entire biomes is fundamental to such predictive processes.

As we have seen, it is possible to model present-day vegetation in relation to climate and to produce quite a close fit. The next stage is to manipulate the climate in the model to mimic conditions that are believed to have existed in the relatively recent past (say, 6,000 to 21,000 years ago (e.g. [32])), when very different sets of climatic conditions were in operation (see Chapter 12) and an ice age was in the process of departing. If the model can produce vegetation maps that correspond with what we know of former vegetation from the fossil record, then it should prove robust enough to project into the future.

But the problems facing modelers should not be underestimated. One of the most significant difficulties is determining the extent of feedback mechanisms [33]. If, for example, the global vegetation became more productive as a result of elevated atmospheric carbon dioxide levels, how would this affect the greenhouse system? If vegetation becomes more sparse in areas subject to drought, will this result in more energy being reflected back into space? Will certain biomes become more prone to fire? If so, how will they change? If the rise in temperature of the Earth influences the flow of the global oceanic conveyor system, how would this affect future climates? How will global patterns of precipitation change?

One oceanic complication has already become apparent, namely, the phenomenon known as the **El Niño Southern Oscillation (ENSO)** in the

Pacific Ocean (see Chapter 13). The consequences of El Niño are felt in many parts of the world, involving drought in Amazonia and South Africa, a lower level of flooding in the Nile, and a weak summer monsoon in India. Recently, El Niño events have been occurring more frequently and have been stronger in their effects, which may well be a consequence of rising global temperature and its impact on the atmospheric/oceanic circulation patterns of the Pacific Ocean. We may well expect alterations in biome patterns as a consequence of such changes.

Perhaps our whole attitude to the world's biomes is too static for the present situation of rapid change. The picture of the relationship between biomes and climate depicted in Fig. 3.7 is comfortable and reassuring, but it is also based on the idea of equilibrium. It assumes that there is time for vegetation to settle down in its climatic mold. This is unlikely to be the case in these days of rapidly changing climate, and we need to develop more dynamic attitudes and models to cope with the understanding of the consequences of our activities on the world's biotic assemblages [34]. Life on Earth is both diverse and dynamic. Conservationists must first locate those parts of the Earth where life is at its most varied and then predict the outcome of future changes in order to determine the most appropriate approach to diversity conservation. This is the subject of the next chapter.

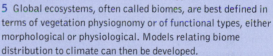

1 The idea of communities of organisms that occur in discrete units, which are predictable in terms of species composition, is attractive and useful to biogeographers, but nature often exhibits a gradual and continuous change in species assemblages depending on the individual requirements of species.

2 In a landscape consisting of a fragmented mosaic of different habitats, communities are more likely to have distinct boundaries and therefore to be recognizable in nature.

3 The ecosystem is a useful way of considering biotic (animal, plant, and microbial) assemblages in relation to the nonliving world. It is a concept based on the ideas of energy flow through a series of feeding (trophic) levels and the circulation of elements between living organisms and the nonliving world.

4 The use of the ecosystem concept and the notion of functional types of organisms (producers, decomposers, nitrogen-fixers, etc.) within the community provide a way of investigating the implications of biodiversity for natural systems. It allows us to ask the question, are all species really necessary for the survival of an ecosystem or are some redundant? Current research suggests that some species can be lost without necessarily destabilizing an ecosystem. More critical is the maintenance of a balance of functional types.

5 Global ecosystems, often called biomes, are best defined in terms of vegetation physiognomy or of functional types, either morphological or physiological. Models relating biome distribution to climate can then be developed.

6 Climate–biome models provide a means of predicting the outcome of climate change on the Earth's biogeography and will have implications both in conservation and agriculture. But predictions of biome shifts are only as good as the climatic predictions that underlie them.

Further Reading

Archibold OW. *Ecology of World Vegetation*. London: Chapman & Hall, 1995.

Crawley MJ. *Plant Ecology*, 2nd ed. Oxford: Blackwell Science, 1997.

Grime JP. *Plant Strategies, Vegetation Processes, and Ecosystem Properties*. 2nd ed. Chichester: Wiley, 2001.

Loreau M, Naeem S, Inchausti P. *Biodiversity and Ecosystem Functioning*. Oxford: Oxford University Press, 2002.

Ohgushi T, Craig TP, Price PW. *Ecological Communities*. Cambridge: Cambridge University Press, 2007.

Smith TM, Shugart HH, Woodward, FI. *Plant Functional Types: Their Relevance to Ecosystem Properties and Global Change*. Cambridge: Cambridge University Press, 1997.

Woodward FI, Lomas MR. Vegetation dynamics: simulating responses to climatic change. *Biol Rev Camb Phil Soc* 2004; 79, 643–670.

References

1 Ohgushi T, Craig TP, Price PW, eds. *Ecological Communities* Cambridge: Cambridge University Press, 2007.

2 Rodwell JS. *British Plant Communities, Vol. I. Woodlands and Scrub*. Cambridge: Cambridge University Press, 1991.

3 Callaway RM. Positive interactions among plants. *Bot Rev* 1995; 61, 306–349.

4 Callaway RM, D'Antonio CM. Shrub facilitation of coast live oak establishment in central California. *Madroño* 1991; 38, 158–169.

5 Forman RTT. *Land Mosaics: The Ecology of Landscapes and Regions*. Cambridge: Cambridge University Press, 1995.

6 Thomas F, Renaud F, Guegan J-F. *Parasitism and Ecosystems*. Oxford: Oxford University Press, 2005.

7 Reagan DP, Waide RB. *The Food Web of a Tropical Rain Forest*. Chicago: University of Chicago Press, 1996.

8 Vitousek P. *Nutrient Cycling and Limitation*. Princeton, NJ: Princeton University Press, 2004.

9 Naeem S, Loreau M, Inchausti, P. Biodiversity and ecosystem functioning: the emergence of a synthetic ecological framework. In: Loreau M, Naeem S, Inchausti P, eds. *Biodiversity and Ecosystem Functioning*, pp. 3–17. Oxford: Oxford University Press, 2002.

10 Naeem S, Thompson LJ, Lawler SP, Lawton JH, Woodfin RM. Empirical evidence that declining species diversity may alter the performance of terrestrial ecosystems. *Phil Trans R Soc London* B 1995; 347, 249–262.

11 Symstad AJ, Tilman D, Wilson J, Knops MH. Species loss and ecosystem functioning: effects of species identity and community composition. *Oikos* 1998; 81, 389–397.

12 Smith TM, Shugart HH, Woodward FI. *Plant Functional Types*. Cambridge: Cambridge University Press, 1997.

13 McNaughton SJ, Banyikwa FF, McNaughton MM. Promotion of the cycling of diet-enhancing nutrients by African grazers. *Science* 1997; 278, 1798–1800.

14 Seehausen O. Speciation affects ecosystems. *Nature* 2009; 458, 1122–1123.

15 Elton CS. *The Ecology of Invasion by Animals and Plants*. Chicago: University of Chicago Press, 2000.

16 Chapin FS, Walker BH, Hobbs RJ, Hooper DU, Lawton JH, Sala OE, Tilman D. Biotic control over the functioning of ecosystems. *Science* 1997; 277, 500–504.

17 Wittebolle L, Marzorati M, Clement L, Balloi A, Daffonochio D, Heylen K, De Vos P, Vertraete W, Boon N. Initial community evenness favours functionality under selective stress. *Nature* 2009; 458, 623–626.

18 Naeem S, Li S. Biodiversity enhances ecosystem reliability. *Nature* 1997; 390, 507–509.

19 Montoya JM, Pimm SL, Sole RV. Ecological networks and their fragility. *Nature* 2006; 442, 259–264.

20 Hector A, Bagchi R. Biodiversity and ecosystem multifunctionality. *Nature* 2007; 448, 188–190.

21 Moore PD, Bhadresa R. Population structure, biomass and pattern in a semi-desert shrub, *Zygophyllum eurypterum* Bois. and Buhse, in the Turan Biosphere Reserve of northeastern Iran. *J Appl Ecol* 1978; 15, 837–845.

22 Grime JP. *Plant Strategies and Vegetation Processes*. New York: John Wiley, 1979.

23 Shugart HH. Plant and ecosystem functional types. In: Smith TM, Shugart HH, Woodward FI, eds. *Plant Functional Types*, pp. 20–43. Cambridge: Cambridge University Press. 1997.

24 Prentice IC, Cramer W, Harrison SP, Leemans R, Monserud RA, Solomon AM. A global biome model based on plant physiology and dominance, soil properties and climate. *J Biogeogr* 1992; 19, 117–134.

25 Whittaker RH. *Communities and Ecosystems*, 2nd ed. New York: Macmillan, 1975.

26 Woodward FI, Lomas MR, Kelly CK. Global climate and the distribution of plant biomes. *Phil Trans R Soc Lond B* 2004; 359, 1465–1476.

27 Walter H. *Vegetation of the Earth*, 2nd ed. Heidelberg: Springer-Verlag, 1979.

28 Box EO. *Macroclimate and Plant Forms: An Introduction to Predictive Modeling in Phytogeography*. The Hague: Dr W. Junk, 1981.

29 Cramer W, Leemans R. Assessing impacts of climate change on vegetation using climate classification systems. In: Solomon AM, Shugart HH, eds. *Vegetation Dynamics and Global Change*, pp. 190–217. London: Chapman & Hall, 1992.

30 Haxeltine A, Prentice IC. BIOME 3: An equilibrium terrestrial biosphere model based on ecophysiological constraints, resource availability, and competition among plant functional types. *Global Biogeochem Cycl* 1996; 10, 693–709.

31 Cowie J. *Climate Change: Biological and Human Aspects*. Cambridge: Cambridge University Press, 2007.

32 Claussen M, Brovkin V, Ganopolski A, Kubatzki C, Petoukhov V. Modelling global terrestrial vegetation-climate interaction. *Phil Trans R Soc London B* 1998; 353, 53–63.

33 Woodward FI, Lomas MR, Betts RA. Vegetation-climate feedbacks in a greenhouse world. *Phil Trans R Soc London B* 1998; 353, 29–39.

34 Woodward FI, Beerling DJ. The dynamics of vegetation change: health warnings for equilibrium "dodo" models. *Global Ecol Biogeogr Lett* 1997; 6, 413–418.

Patterns of Biodiversity

Biodiversity is a term that encompasses all of the living things that currently exist on Earth. It includes all animals and plants that zoologists and botanists have discovered and described, and also all those that remain undiscovered and still await a scientific description. In addition, biodiversity includes all fungi, bacteria, protozoa, and viruses that, on the whole, are even less well known than the animals and plants. But biodiversity goes even beyond this and includes the genetic variation found within each species, and some would even extend the use of the term to cover the great range of habitats that exist on Earth and that support all living things. In this chapter we examine what is known of the Earth's biodiversity and whether one can discern patterns in biodiversity distribution.

As far as we know, the Earth is unique in the universe in that it supports living creatures. As astronomical research continues, the discovery of evidence for life existing on some other distant planet in another solar system becomes statistically greater, but one thing is sure: life is an exceedingly rare commodity in the universe and should therefore be greatly valued. **Biodiversity** is an expression of the great variety of living things on our planet, but it is far more than a simple count of species.

When we take a species and analyze its composition, we find that it consists of a series of populations, sometimes adjacent to one another and sometimes fragmented and isolated, and that within these populations there is often great variation between individuals. Biodiversity includes the whole range of populations, together with all the genetic variations found within each species.

Just as we can regard a species as a collection of component populations, so we can interpret communities of organisms as assemblages of many populations of a variety of different species, all interacting (see Chapter 3). When these communities are placed in the setting of their nonliving environment, they make up ecosystems. Our concept of biodiversity should therefore include the rich variety of ecosystems that occupy the Earth, many of which have an important human component. If we wish to conserve biodiversity, and most ecologists believe that this is basically a sensible aim, indeed a responsibility, then the conservation of whole ecosystems and their habitats is the most appropriate starting point.

Why should the conservation of biodiversity be regarded as important? Why do we need, or why do we want to maintain the biotic richness of the

Earth? Needing and wanting are very different experiences. We need something when it is useful to us, and one could argue that the other living organisms of the Earth are indeed useful. Some provide food, or materials for building homes and making fabrics; others are a source of pharmaceuticals. Many organisms are part of our general support system on Earth, being involved in the maintenance of the gaseous balance of the atmosphere, the healthy structure of the soil, and even the modification of our climate. So there is a strong utilitarian argument for maintaining the Earth's biodiversity. Even those species that are not currently regarded as useful may one day be found to be so.

A further argument, however, may be associated with "wanting" rather than "needing." We may want to have birds and flowers in the world simply because we enjoy having them there; this is a kind of aesthetic argument. We may regret that we shall never have the opportunity to see a passenger pigeon, a dodo, or a great auk. Or we may take a view that is even less human-centered and claim that the organisms have rights of their own to exist, perhaps even as great a right as ours. This is an ethical argument that lies beyond the remit of science, but this does not negate its validity. It is our responsibility, therefore, as a species with a high impact on the Earth's ecosystems, to ensure that extinction is minimized. Conservation itself then becomes in part a matter of ethics.

If we accept any or all of these arguments, then it leads us to a position where we need to know whether the Earth is actually losing species and, if so, how fast? We also need to be concerned about whether human beings are contributing substantially to the rate of loss and whether we can do anything about it. But to understand extinction rates and their causes, we have to go back further still and ask just how many species there are on Earth, so that we can calculate how rapidly we may be losing them.

How Many Species Are There?

No one likes to lose things, but the outcome of any loss can best be measured on the basis of what is left. The loss of a dollar will be felt more severely by a pauper than by a millionaire. The importance to humanity of the loss of species from the Earth can only be judged, therefore, if we can view it in proportional terms and from the perspective of what remains. To evaluate the importance of current extinction rates we need to appreciate how many species occupy the Earth.

One of the most surprising things about science is how little we know. You might assume, for example, that biologists would have a reasonably good idea of how many species of living organisms exist on Earth, but this is not so. The question is still hotly debated, and the estimates that biologists have made range between 3 and 500 million species! What they are all agreed upon is that only a very small proportion of the total is currently known to science and has been adequately described. Some parts of the world have been little studied, especially the tropical regions where the diversity of species is particularly high and the depths of the ocean, which are difficult to survey. Many very abundant species are extremely small and so may have gone unnoticed in the past. Many species are difficult to identify, so very careful study is needed to distinguish them. The numbers of expert scientists in the field of **taxonomy** (the study of plant and animal classification) are relatively few, so the task of counting species is much more difficult to accomplish than might at first appear.

The confusion about the number of species present on Earth may seem surprising to nonbiologists, but the sheer wealth of species makes it difficult to be sure that all of those that have been described are valid and are not duplicates, or that those described as a single species do not, in fact, consist of a number of species that we have ignorantly lumped together. The herring gull, formerly considered a single species, *Larus argentatus*, for example, is found in coastal areas right around the world in the temperate zone of the Northern Hemisphere (termed a **circumboreal** distribution), but it occurs in a variety of different forms in different places, and close study of the molecular genetics of this gull has revealed that the supposed species actually consists of several species closely resembling one another. The northern European herring gull remains as *L. argentatus*, but the American herring gull is now *L. smithsonianus*. The southern European, yellow-legged gull is *L. michahellis*, the Armenian gull is *L. armenicus*, and the Caspian gull of Asia Minor is *L. cachinnans*. Thus one species has overnight become at least five species! Some doubt still surrounds the relationships of the similar white-headed gulls of Siberia. One of the problems here is actually defining a species. The convenient idea that two species cannot interbreed is not actually workable because interbreeding between several animal and plant species that are morphologically distinct is often possible: If populations of an organism are sufficiently distinct in their outward form, or their physiology, or their behavior patterns, and if they are distinct in their molecular genetic patterns, then they are regarded as distinct species (see Chapter 6). But the definition is quite plastic, and the lines drawn around species will vary from time to time as more information becomes available. Thus the task of determining how many species there are on Earth becomes even more of a problem.

On the other hand, of the approximately 1.8 million species of organisms that have been described so far, many may have been named twice, or even more times! Among beetles, for example, 40% of the species described have only ever been recorded at the site of their first description. This is very unlikely to be a true reflection of their distributions, and it is quite likely that many of the species are in fact duplicates, having been given different names in different localities. This so-called alias problem will inflate the numbers of species described [1], but the likelihood is that there are far more species remaining to be found, so that the true number of species still living on the Earth must greatly exceed the number currently described. Some of the problems involved in counting species are explained in Box 4.1.

The diversity of microbes, the bacteria, fungi, viruses, and the like, is particularly difficult to estimate because these groups are not nearly as well known as, say, mammals or flowering plants. Of the bacteria, for example, only about 4,000 species have so far been described (see Table 4.1), and this may well represent only about one-tenth of 1% of the total, so much remains to be done in this area. The study of biodiversity among microbes is complicated by the range of genetic and biochemical variation found among wild populations [6,7]. It is also made more difficult by the fact that bacteria can survive deep in geological deposits, far beyond those surface layers of the Earth that were once supposed to represent the

Counting Species

A conservative estimate for the possible number of species on Earth is 12.5 million [2], but the tropical ecologist Terry L. Erwin [3] has proposed that the total is far greater, perhaps as high as 30 million for tropical insects alone. He came to this conclusion as a result of the study of beetles on a single tree species, *Luehea seemannii*, in Panama, which he sampled by "fogging." This is an efficient technique for stunning the insects in a canopy by smoking them with an insecticide. The dazed insects fall from the tree and are collected in trays placed beneath the canopy. Erwin examined just 19 individual trees of *L. seemannii* in the Panamanian forests and managed to obtain 1,200 species of beetles alone from this analysis. This large number is not entirely surprising, since beetles are extraordinarily successful insects and may comprise as much as 25% of the total number of species of living organisms. But this study does illustrate the remarkable richness of beetles in the tropical forest.

From these data Erwin made a number of assumptions about the numbers of beetle found specifically on particular tree species, the numbers of tree species found, and the proportions of different organisms in the forest in relation to one another. He extrapolated from the information gathered and came to the conclusion that, if this number of beetles is truly representative of the forest richness, then one might predict a total of 30 million species of insect on Earth. The uncertainty of many of his assumptions, however, should make us very cautious in accepting this figure uncritically. Other entomologists, such as Nigel Stork and Kevin Gaston from the University of Sheffield, England [4], have checked Erwin's estimates using data from studies in the tropical forests of Borneo. Stork has generated estimates ranging from 10 million to 80 million for the arthropods (a group of invertebrate animals including the insects). Another independent estimate [5] supports the lower end of this scale, placing tropical arthropods at 6 to 9 million. The range of error in all estimates is still so wide that there is bound to be a great deal of discrepancy in the figures arrived at, but the world's wealth of species is likely to exceed 10 million.

limits of the biosphere [8]. Their capacity to live in extreme environments and their great potential in the service of humankind makes them particularly interesting and important to humanity, so it is in our interests to improve our knowledge of microbial diversity. But even the concept of the species has to be reconsidered when dealing with microbes [9].

An example of the immensity of the problem facing those studying microbial diversity is afforded by the Rio Tinto in southern Spain [10]. This river was known to the ancient Phoenicians as the River of Fire because of its deep red color, caused by the high concentration of iron and other metals dissolved in its highly acidic (pH 2) waters. These conditions are now often associated with highly polluted waters, usually as a result of human mining activities, and the Rio Tinto has indeed been affected in this way for over 5,000 years. But even before the times of human pollution,

Table 4.1 The number of described species in selected groups of organisms, together with the likely total numbers on Earth, and the percentage of the group that is currently known. Data from Groombridge [12].

Group	No. of Described Species	Likely Total	%
Insects	950,000	8,000,000	12
Fungi	70,000	1,000,000	7
Arachnids	75,000	750,000	10
Viruses	5,000	500,000	5
Nematodes	15,000	500,000	3
Bacteria	4,000	400,000	1
Vascular plants	250,000	300,000	83
Protozoans	40,000	200,000	20
Algae	40,000	200,000	20
Molluscs	70,000	200,000	35
Crustaceans	40,000	150,000	27
Vertebrates	45,000	50,000	90

this river was rich in metals because of the high metal content of the rocks over which it flows. Recent research into the biodiversity of the river has employed molecular techniques to examine the range of microorganisms present and has found a great wealth of previously undetected species. Over 60% of the biomass of the river consists of highly tolerant strains of microscopic algae that have evolved and adapted over the millennia of exposure to these extreme conditions.

Among the fungi, some 70,000 species have been described so far, but David Hawksworth, formerly head of the International Mycological Institute, England, believes that the true total could be around 1.6 million species [11]. In areas of the Earth where the fungi have been thoroughly investigated, each higher plant species supports about five or six fungal species. Therefore, if the total flowering plants (once all have been described) is assumed to be about 300,000, the total number of fungi must be in the region of 1.5 million or more.

From this example it can be seen that the estimation of possible numbers of organisms is derived by a process of extrapolation. If we have certain facts and make further assumptions about proportional representation of different groups, then we can begin to project from what we know into the misty realms of uncertainty. The outcome is not satisfactory, but it is the best we can do so far. Table 4.1 provides some idea of what we presently know about the richness of the Earth's diversity of species for a few groups of organisms, and also approximately what proportion of each group is currently thought to have been described [12]. The vertebrates (animals with backbones) are reasonably well known, and relatively few new species may be expected in this group; the same is true for flowering plants. In recent surveys, 4,327 species of mammal and 9,672 species of bird were listed, and it is unlikely that these totals will grow very substantially even with further survey work and research into their classification, although new species do continue to be described at a rate of about 100 species per decade in the case of mammals [13]. In fact, over 40 new species of primates have been described since 1990, including two new monkey species from the Brazilian Amazon region in 2002. But the spiders and mites (arachnids), the algae and the nematode worms, among others, are still very poorly understood and many more species can be expected in these groups. There is the additional frustration, from the biologist's point of view, that smaller organisms tend to be both more abundant and more diverse than are larger ones, thus adding to the workload at the most difficult end of the spectrum.

All these difficulties in estimating the biodiversity of the Earth apply equally to individual sites and habitats. Even listing all the species present at a site, apart from the genetic and habitat variation that contribute to biodiversity, would be a long, costly, time-consuming, and finally inaccurate process. An alternative approach, illustrated earlier by its use with fungi, is to assess the richness of certain groups of well-known organisms that are easily observed and identified (such as higher plants, mammals, birds, or butterflies) and to assume they have a consistent proportional relationship to the less easily observed and identified groups. This method should work well where species are closely dependent on one another (as host and parasite or as food and feeder), as in the case of gall-forming organisms associated with the plant species in the scrub grasslands of South Africa (Fig. 4.1) [14]. Here, there is a strong linear relationship between the

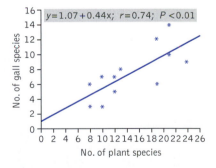

Fig. 4.1 Graph showing the relationship between the richness of gall-forming species and the number of woody shrub species in the Aynsberg Nature Reserve, South Africa. Data from Wright & Samways [14].

two groups. This is not always the case, however. The work of John Lawton and colleagues in the Mbalmayo Forest Reserve in Cameroon [15], covering birds, butterflies, beetles, ants, termites, and nematodes, has shown little overall relationship between the abundance of one group and another. But other workers have analyzed data from numerous biological surveys of different sites in the United States and Canada and have found that it is possible to ignore a proportion of the total species without significant loss of data [16]. Eliminating from a survey the 10% of species that are the most difficult and time-consuming to identify, for example, resulted in no significant change to the assessed pattern of biodiversity.

Another attempt to find a shortcut to the estimation of the total number of living things has been the use of body size. It is quite obvious that there are fewer large organisms than small ones, but is there a simple relationship between species number and body size? Since we know the numbers of large-bodied species reasonably well, we could then extrapolate from what we know of them to estimate how many small creatures exist on Earth. If we consider animals of 5–10 m in length (e.g., elephants, whales), there are fewer than 10 species that come into that size range. In the range 1–5 m (e.g., horses, deer), there are perhaps a few hundred. In the 0.5- to 1-m range (e.g., fox), there may be up to 1,000 species, while rat-sized organisms (0.1–0.5 m) could run to 10,000. This body-size/abundance relationship continues to hold as we move down the size range, so that at 0.005–0.01 m (ant-sized), there could be around a million species. But for smaller creatures (below 0.005 m, such as fleas, mites, and smaller), the number of known species declines from the straight-line logarithmic relationship that holds at higher levels of size. Is this lack of apparent diversity in the smaller organisms a consequence of our own ignorance of their taxonomy and their diversity, or are we wrong to assume that they should follow the same linear pattern as that of larger species? If we assume that the linear graph represents a real pattern of size/diversity relationship, then we can calculate an expected number of small organisms and estimate that total biodiversity should reach around 10 million.

Researchers at the University of Minnesota have tested the assumption that small creatures must be more diverse than they appear by examining single ecosystems in great detail and working out the relationship between size and diversity for these specific sites. Figure 4.2 shows the results of analyzing in detail the invertebrate species of a North American grassland ecosystem in relation to their body size (measured here as volume rather than length) [17]. It can be seen that in a wide range of insect groups there are indeed fewer species at the small end of the spectrum than the linear model predicts. In this case we can be assured that it is not because small species have been overlooked. We must therefore conclude that the linear model itself is at fault and that this is not an appropriate way to calculate total biodiversity. Perhaps studies of this kind will result in the establishment of a more reliable model than the linear one that can in future be used for extrapolation and biodiversity estimation. There is also a need to investigate the size/diversity relationship among plants. At present there is evidence that there are considerably more species of small plants rather than large ones, possibly because they have narrower niches or higher fecundity [18], but no precise model of the relationship is currently available.

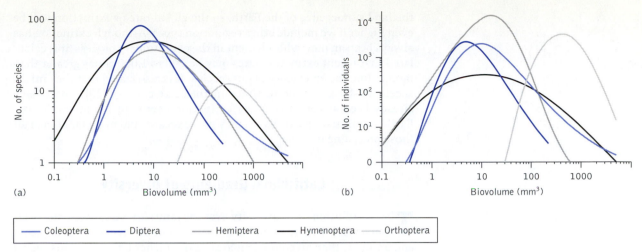

Fig. 4.2 The numbers of species (a) and individuals (b) of various insect groups in grassland in Minnesota in relation to their body size. The groups shown are Orthoptera (grasshoppers), Hymenoptera (wasps and bees), Hemiptera (bugs), Coleoptera (beetles), and Diptera (flies). After Siemann et al. [17].

It is clear from this account of the problems in assessing how many species are present on Earth that rates of extinction can only be vaguely estimated. Although many known extinctions in the recent past have resulted from human hunting or persecution, the organisms involved have been large, conspicuous, and easily targeted. Nothing is known of the smaller creatures, especially the microbes. It is likely, however, that extinction of small organisms is mainly due to a loss of habitat, and some habitats are richer than others. Therefore, calculating possible extinction rates depends on what kind of habitats are being destroyed. Assessing rates of extinction is also made difficult because we can rarely be sure that a species is actually lost, that no isolated members remain. There is still a possibility that the ivory-billed woodpecker (*Campephilus principalis*) survives somewhere in the southern United States, even though it has been regarded as extinct. In the case of plants, extinction is even more difficult to record. Some plants are able to survive for decades or even centuries as buried seeds, so they can unexpectedly reappear after long episodes of presumed extinction. Many of the plant species of the Atlantic forests of Brazil, now occupying only about 10% of their original cover, were last recorded back in the 1850s and have not been seen since. But it is difficult to be sure that they have finally gone and that not even dormant seeds survive. There are occasions when plants are discovered that have formerly been known only in a fossil state. Even as recently as 1994, a new gymnosperm tree species was discovered in a deep gorge near Sydney, Australia [19]. This species has been given the name *Wollemia nobilis* and, like *Ginkgo*, closely resembles a fossil plant of the Cretaceous that was presumed to be long extinct.

Despite the problems associated with recording the process accurately, extinction is undoubtedly occurring all around us. The biologist Edward O. Wilson [2] has calculated that the loss of species from the tropical forest area alone could currently be as high as 6,000 species per year. This amounts to 17 species each day, and the tropical forests cover only 6% of the land surface area of the Earth, so the global rate of extinction will be even higher if we include other vegetation types. Although extinction has always been an inevitable element in the evolutionary process, it is calculated that recent extinction rates may be 100 to 1,000 times greater than rates before the emergence of our species. It is also feared that they might accelerate by a further 10 times in the next century [20]. There have already been at least five major extinction events in the history of the Earth, but this sixth extinction may prove greater and more rapid than all those preceding it.

Latitudinal Gradients of Diversity

The distribution of biodiversity over the land surface of the planet is far from even [21]. The tropics contain many more species, both of plants and animals, than an equivalent area of the higher latitudes. This seems to be true for many different groups of animals and plants, as can be seen from Fig. 4.3, which illustrates the number of breeding birds and mammals found in various Central and North American countries and states.

The tropical country Panama, only 800 km (500 mi) north of the equator and a close neighbor of Costa Rica, has 667 species of breeding birds, three times the number found in Alaska, despite the much greater area of Alaska.

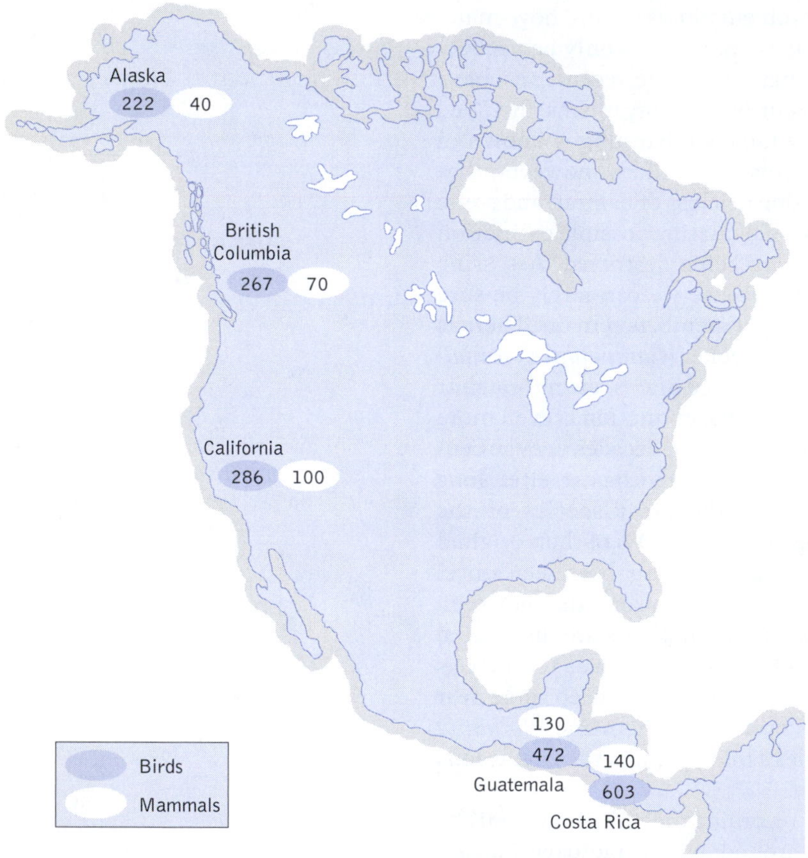

A similar pattern is seen in the number of mammal species at different latitudes in North America (Fig. 4.3). Considering just the forest areas from southern Alaska (65°N) in the north through Michigan (42°N) into the tropical forest of Panama (9°N), these have 13, 35, and 70 mammalian species, respectively. Breaking the mammals down into their component groups by diet and by taxonomy, we find that the bats account for a large part of the difference between the three locations. Moving from the north, one-third of the increase in species from Alaska to Michigan is due to the larger number of bats, and so also is two-thirds of the increase between Michigan and Panama. Yet more information becomes available if we consider diet among the mammals. Much of the tropical diversity among mammals is due to the greater predominance of a fruit-eating way of life and to the greater number of insectivores, many of which eat

Fig. 4.3 Numbers of breeding bird and mammal species in different parts of Central and North America.

Alaska 222 40

British Columbia 267 70

California 286 100

Guatemala 130 472 140

Costa Rica 603

Birds
Mammals

insects that in turn feed on the fruits of the forest. This may also account for the diversity of frogs in the lower latitudes (Fig. 4.4). Therefore diet is evidently an important aspect underlying the diversity gradient found among animals.

As has been discussed, insect diversity is difficult to measure because the insect groups are generally not fully described. Butterflies are among the best recorded of the insects, and Fig. 4.5 shows the latitudinal gradients of richness in just one group of butterflies, the swallowtails. The high number of species found in the tropics is again apparent, and here it can be seen that this applies to all tropical areas of the world. One easily explained anomaly in the gradients can be seen in the African/European section, where there is a dip in species richness that coincides with the North African desert region.

Diet may be an important factor in determining animal diversity, but what of plants? They also show a general trend toward increasing diversity in the tropics (Fig. 4.6), but they do not vary in their diet because they all use solar energy for their photosynthesis.

The relationship between plant species richness and latitude, however, is not at all a simple one. David Currie and Viviane Paquin of the University of Ottawa constructed a map of the richness of tree species across North America [24], as is shown in Fig. 4.7. From this map it can be seen that the contours of richness do not simply follow the lines of latitude, especially in the areas south of Canada. Patches of low diversity occur in the Midwest, and an exceptionally high diversity of trees is seen in the Southeast. When these workers examined the possible environmental factors that may be associated with this pattern, the one that correlated most closely was the sum of evaporation (directly from the ground) and transpiration (from the surface of vegetation) combined to give a value for the loss of water from the land surface (**evapotranspiration**). Evidently, those regions with the highest evapotranspiration are able to support the highest diversity of tree species. But evapotranspiration itself also correlates closely with the potential **productivity** of a region (the amount of plant material that accumulates by photosynthesis in a given area in a given time), so perhaps plant diversity is essentially determined by how much photosynthesis can be carried out in a given

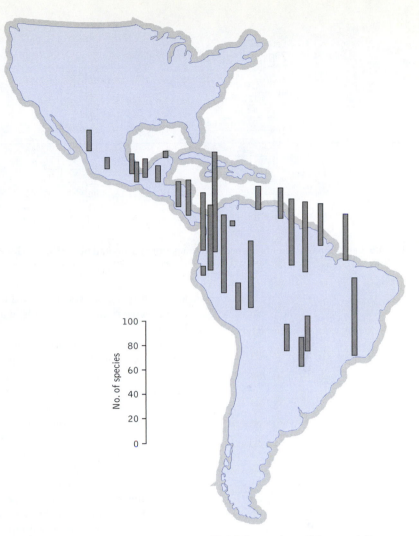

Fig. 4.4 Numbers of frogs in different parts of Central and South America. Data from Groombridge [12].

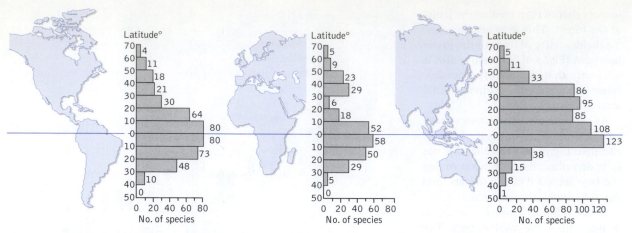

Fig. 4.5 Latitudinal gradients of species richness for swallowtail butterflies in three different parts of the world. Data from Collins & Morris [23].

site. Figure 4.8 shows this relationship between primary production and tree species richness, and indeed a good correlation between the two can be seen.

In general, the equatorial regions are the areas in which highest productivity is possible because of the prevailing climate, which is hot, wet, and relatively free from seasonal variation. Figure 4.9 illustrates this by displaying the world distribution of mean annual primary productivity. From this map it can be seen that very high productivity is concentrated in the equatorial belt and that this drops off as one moves toward higher latitudes. The picture is complicated by the arid belt in northern Africa and central Asia, of which more will be said in due course, but the general trend is decreasing productivity at higher latitudes. An examination of the North American part of this map shows a good correlation with the tree-richness map (see Fig. 4.7), especially with regard to the high diversity of tree species and high productivity in the southeastern United States. This approach to explaining the latitudinal gradient of species diversity suggests that the critical factor is how much energy is captured by the vegetation. This has come to be known as the **energy hypothesis** and is particularly well supported by plant-based data [25].

When developing the hypothesis to account for animal diversity gradients, several factors have to be considered. Higher plant productivity in general means greater energy stores available for consumers. Higher productivity often results in higher biomass and more complex vegetation architecture, which can result in greater opportunities for animals. And warm, moist climates in which high productivity occurs can also lead to greater metabolic rates in organisms. This latter approach has led to the development of what has been termed a **metabolic theory** to account for the latitudinal gradients in biodiversity [26]. Researchers from the University of Ottawa, Canada, have tested the metabolic theory by examining the relationship between temperature and species richness for a wide range of animal groups in North America, including amphibians, reptiles, tiger beetles, butterflies, and blister beetles, as well as trees [27]. They found that the metabolic theory failed to fully account for the observed patterns and suggest that predictive models need to take water availability into account in addition to temperature. Using a water-energy model for predicting the richness of mammals, birds, amphibians, reptiles, and plants in Europe, Robert Whittaker and colleagues from Oxford University [28] have shown good correlations. Thus the general climatic theory of biodiversity determination seems to hold up well in this case.

Productivity, therefore, seems to be involved, but perhaps its influence is indirect. Where conditions are most suitable for plant growth— that is, where temperatures are relatively high and uniform and where there is an ample supply of water— one usually finds large masses of vegetation, a high biomass. This leads to a complex structure in the layers of plant material. In a tropical rainforest, for example, a very large quantity of plant material builds up above the surface of the ground. There is also a large mass of material developed below ground as root tissues, but this is less apparent and, in the case of the tropical rainforest, confined to the upper layers of soil. Careful analysis of the above-ground material reveals that it is arranged in a series of layers, the precise number of layers varying with the age and nature of the forest. The arrangement of the biomass of the vegetation into layered forms is termed its **structure** (as opposed to its **composition**, which refers to the species of organisms forming the community). Structure is essentially the architecture of vegetation and, as in the case of some tropical forests, can be extremely complicated. Figure 4.10 shows a profile of a mature floodplain tropical forest in Amazonia [29] expressed in terms of the percentage cover of leaves at different heights above the ground. There are three clear peaks in leaf cover at heights of approximately 3, 6, and 30 m above the ground, and the very highest layer, at 50 m, corresponds to the very tall, emergent trees that stand clear of the main canopy and form an open layer of their own. So this site contains essentially four layers of canopy.

Forests in temperate lands are simpler, often with just two canopy layers, so they have much less complex architecture. Structure, however, has a strong influence on the animal and plant life inhabiting a site. It forms the spatial environment within which an animal feeds, moves

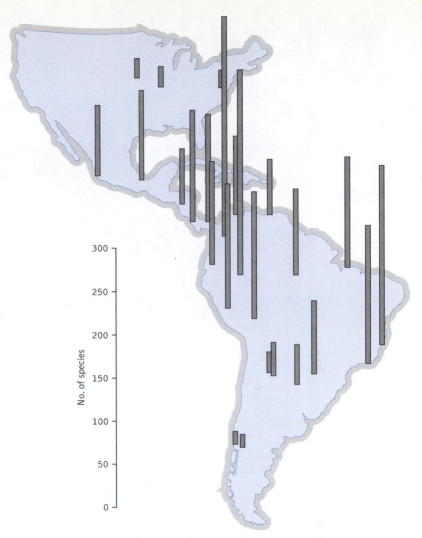

Fig. 4.6 Latitudinal gradients of tree species richness in the Americas. Data from Duellman [22].

Fig. 4.7 Number of tree species (i.e., any woody plant over 3 m in height) found in different parts of North America. The contours indicate areas where particular numbers of tree species were recorded within large-scale quadrats (mean area of 70,000 km²). Data from Currie & Paquin [24].

around, shelters, lives, and breeds. It even affects the climate on a very local level (the **microclimate**) by influencing light intensity, humidity, and both the range and extremes of temperature. Figure 4.11 shows a profile through an area of grassland vegetation that has a very simple structure, and it can be seen that the ground level has a very different microclimate from that experienced in the upper canopy of the grasses. Wind speeds are lower, temperatures are lower during the day (but warmer at night), and the relative humidity is much greater near the ground. The complexity of microclimate is closely related to the complexity of structure in vegetation, and, generally speaking, the more complex the structure of vegetation, the more species of animal are able to make a living there per unit area of land surface. This is illustrated in Fig. 4.12, which relates the number of bird species found in woodland habitats to the number of leaf canopy layers that can be detected [30]. The high plant biomass of the tropics leads to a greater spatial complexity in the environment, and this will lead to a higher potential for diversity in the living things that can occupy the region. The climates of the higher latitudes are generally less favorable for the accumulation of large quantities of biomass. Hence the structure of vegetation is simpler, and the animal diversity is lower.

One important extension to this line of argument is worth pursuing. Complexity, or the conception of complexity, depends on the size of the observer. It was stated earlier that grassland has a relatively simple structure, but this is the case only if one views it from a human perspective. From an ant's point of view, on the other hand, a grassland environment may be highly complex. For this reason, an area of grassland offers a home to far more ants than it does to humans, cows, or bison. As discussed earlier, the Earth as a whole can support far more small creatures than it can large ones. In part this can be explained by the greater number of opportunities offered to very small organisms even within habitats of relatively simple structure. In habitats of more complex structure, of course, small organisms find even more microhabitats and different ways of making a living. The latitudinal diversity of small organisms, therefore, tends to follow the same patterns as larger ones, being richest in the tropics.

There is one group of very small animals, however, that goes against the rules. The aphids (such as plant-feeding greenflies), of which about

4,000 species are known, are less diverse in the tropics than they are in the temperate regions [31]. Most aphid species feed on only one type of plant and are rather poor at locating that plant from a distance, relying on the sheer chance of air-flow patterns to carry them from one suitable host to another. They do best, therefore, where populations of particular plant species are dense, as is the case in agricultural crops. Unfortunately for the aphids, tropical plant assemblages consist of many species, each of which is present only at low density, so aphids are not well suited to such conditions. Aphid diversity is therefore inversely related to plant diversity, so they transgress the general pattern of latitudinal gradients.

The richness of the tropics as far as animal life is concerned may thus be a consequence not simply of the high productivity of these latitudes, but also of their great structural complexity resulting from their high biomass, which can support many species of small animal.

In a sense, we may be asking the wrong question when we try to analyze the factors that contribute to the high species diversity of the tropics, for this is really the perspective of biologists viewing the problem from the temperate regions, which happens to be where most ecologists live and work. From a tropical angle, the question should be why the higher latitudes have lower diversities than the tropics [32]. Perhaps it is simply a matter of land area? The tropics contain a larger surface area of land than higher latitudes (a fact that is not always evident when we examine commonly used projections of the Earth's curved surface since this tends to exaggerate the areas of land in the higher latitudes). Some biogeographers regard the latitudinal gradients of diversity as a reflection of this effect [33]. But an analysis of the data by Klaus Rohde [34] does not support this explanation. Although area may contribute to biodiversity, it is certainly not the major story; otherwise large landmasses would always be richer.

Overall, the proposal that the latitudinal gradient in biodiversity, demonstrated by so many plant and animal groups, is largely determined by energy availability and hence climate, operating through the productivity of vegetation, is supported by many observations. There are, however, alternative or additional possible causes that may have an influence on global patterns.

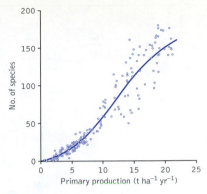

Fig. 4.8 Number of tree species in North American sites (see Fig. 4.7) plotted against the primary productivity of those sites. A distinct positive relationship can be observed. Data from Currie & Paquin [24].

Is Evolution Faster in the Tropics?

Evolution, the process by which new species are generated, is discussed in detail in Chapter 6. It essentially involves genetic variation and subsequent selection of the most suitable genetic combinations within a particular environment. The important question, as far as the explanation of latitudinal gradients of diversity is concerned, is whether evolution (either the generation of genetic variation or its selection) proceeds more rapidly under tropical conditions. If it does, then the richness of the tropics could simply be due to the continual evolution of new forms there [35].

One can approach the question historically and ask whether the tropics have acted as a center of evolution for groups of organisms in the past. Plants and their fossils provide a useful group in which to investigate this possibility. If we examine the modern distribution patterns of the various families of flowering plants, we find that they tend to center on the tropics.

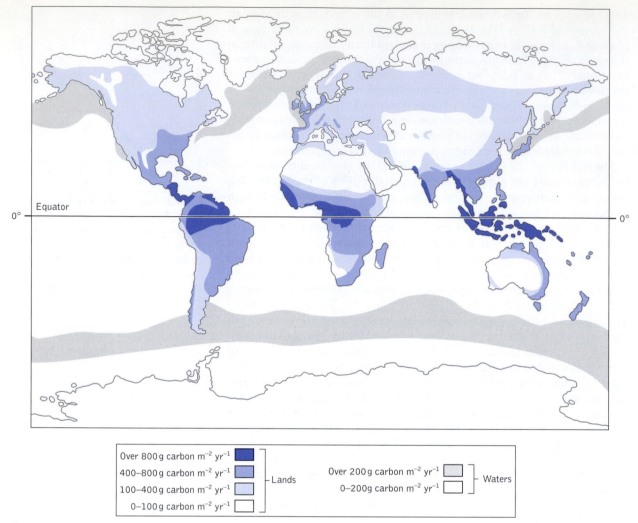

Equator

0°   0°

| Over 800 g carbon m^{-2} yr^{-1} | ⬛ |
| 400–800 g carbon m^{-2} yr^{-1} | 🟦 | — Lands
| 100–400 g carbon m^{-2} yr^{-1} | 🟦 |
| 0–100 g carbon m^{-2} yr^{-1} | ⬜ |

| Over 200 g carbon m^{-2} yr^{-1} | ⬜ |
| 0–200 g carbon m^{-2} yr^{-1} | ⬜ | — Waters

Fig. 4.9 World distribution of plant productivity. The data displayed here are simply estimates of the amount of organic dry matter that accumulates during a single growing season. Full adjustments for the losses due to animal consumption and the gains due to root production have not been made. Map compiled by H. Leith.

Very roughly, about 30% of flowering plant families are widespread in distribution, about 20% mainly temperate and about 50% mainly tropical. These figures have led to the suggestion that the tropics have been a center for the evolution of many of the flowering plant (angiosperm) groups. This proposal can be examined by looking at the fossil record to see whether the tropics have always been richer than the temperate latitudes as far as flowering plants are concerned. This approach has been attempted by Peter Crane and Scott Lidgard of the Field Museum of Natural History in Chicago [36], and some of their results are represented in Fig. 4.13. This figure shows an analysis of fossil plant material covering the period from 145 million years ago to 65 million years ago, and depicts the relative abundance of the angiosperms in different latitudes during this period of

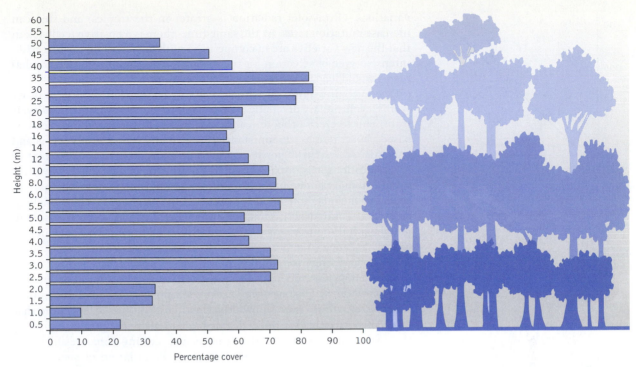

Fig. 4.10 Profile of a tropical rainforest with the percentage leaf canopy cover recorded at different heights above the ground. Note the stratification of the leaf cover into distinct layers. See Fig. 4.19 for further details of canopy structure development in rainforest. From Terborgh & Petren [29].

time. From this data it is clear that the flowering plants first rose to some prominence in the tropics and that their predominance in the tropics was maintained throughout this period of time as they gained importance in the plant kingdom. The latitudinal gradient in diversity for flowering plants goes back a very long way, right back, in fact, to the evolutionary origins of the group. Tropical biodiversity is clearly an ancient phenomenon and may be related, in this case, to tropical origins and enhanced tropical rates of evolution and diversification.

This line of argument leads back to the metabolic theory. In an environment where energy is abundant and temperature is consistently high, such as the tropics, metabolic rates in organisms tend to be faster. Fecundity is greater, and generation times may be shorter. Consequently, genetic modification by mutation (see Chapter 6) is faster, so that species are constantly generating new

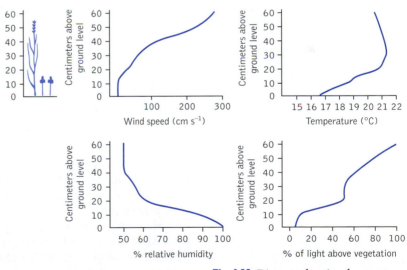

Fig. 4.11 Diagram showing the structure of grassland vegetation and its effect on the microclimate of the habitat.

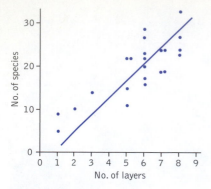

Fig. 4.12 Graph showing the relationship between the number of bird species and the number of layers in the vegetation stratification. Data from Blondell [30].

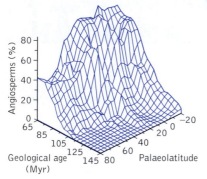

Fig. 4.13 Estimated percentage representation of flowering plants (angiosperms) at different times in geological history and at different latitudes. Angiosperms have always been most abundant in the low-latitude (tropical) regions. From Crane & Lidgard [36].

variations. Ultraviolet radiation is greater in the tropics, and this can increase mutation rates. At the same time, there is a positive feedback in that the new varieties are in competition with one another, leading to an intense degree of selection for the most fit. Together, these two mechanisms would lead to a more rapid rate of evolution in the tropics.

One intriguing proposal is that diversity generates diversity. In other words, evolutionary diversification is itself promoted by high diversity. It is difficult to erect a theoretical model on which such a proposal can be based, but there is circumstantial evidence from a study of island endemism [37] that regions with high diversity exhibit a faster rate of diversification. It is possible that this argument is circular, however. If some external factor, such as conditions promoting productivity as suggested by the metabolic model, results in high diversity it is also likely to create the required environment for enhanced diversification. Thus there may be a link between diversity and diversification that is not directly causal. In biogeography, as in many areas of study, one should never assume that correlation implies causation.

Jonathan Davies and his colleagues from Imperial College London and from Kew Gardens have surveyed data relating to the rate of diversification of plant groups in the tropical regions [25]. Although they found that molecular evolution was indeed more rapid in high-energy environments, they were not convinced that this is the driving force resulting in increased diversity. They found that the accumulation of species was faster in regions of high biomass (and high energy), which could be the result of speciation being more rapid or extinction rates being lower. Either way, it appears that the energy–biomass model remains the most robust explanation currently available for latitudinal gradients in diversity.

There remain, however, additional factors that need to be considered.

The Legacy of Glaciation

Just as a fast rate of speciation in the tropics can be regarded as a factor leading to their increased biodiversity, a slower rate of extinction could also be involved. One possible explanation for greater extinction rates in the higher latitudes is the instability of climatic conditions over the past 2 million years. This subject is covered in greater detail in Chapter 12, but the question of climatic history and its variation with latitude is clearly relevant to a discussion on the latitudinal gradients in diversity.

The Earth's climate has been constantly changing, and it has been considerably colder over the past 2 million years than was the case for the previous 300 million years. As a consequence, the high latitudes have been disrupted by the development of glaciers over the land surface. The effects of these changes on the biogeographical patterns of plants and animals are considered in Chapter 12, but it is evident that the most severe disruption, in the form of ice masses that have spread and destroyed all vegetation over major areas, has occurred largely in the high latitudes. The tropics, apart from the high mountains, have consequently been subjected to less obvious climatic stress. This idea of a climatically stable tropical belt, if it is indeed true, could account for some of the diversity still found in the tropics: the plants and animals could be a relic accumulation of species from a former age. But has the tropical region actually been climatically more stable than the temperate region? The general conclusion that has emerged, particularly from the studies of Paul Colinvaux of the Smithsonian Institute, Panama, and his co-workers [38], is that the tropical lowlands of Amazonia have also been considerably colder (perhaps 5 or 6°C colder than at present) during recent times (the last million years or so) and that the tropical forests could not have remained intact over the whole of their current range. This climatic shift has meant that the tropical rainforest has undoubtedly been restricted in the altitudinal range it was able to cover and has also been at least partially fragmented as a result of cold and drought during the glacial periods of higher latitudes. Forests that remained during these cold episodes are also likely to have changed in their composition, with a higher proportion of trees that are now associated with higher altitudes and hence cooler conditions.

The equatorial forests, however, have still had to endure less disturbance than their temperate counterparts. Some areas have probably maintained themselves in a forested form throughout the period of stress, even though their species composition and architectural structure may well have changed. It has been suggested that, if they were fragmented, this could have actually assisted in the progress of evolution and diversification, for isolated populations, as we see in Chapter 6, may diverge in their evolution and form separate species that fail to interbreed when brought into contact once more. In this case, the impact of climatic change in fragmenting the tropical forests may have added to, rather than subtracted from, their diversity. Evidence from molecular studies of some American songbirds obtained by John Klicka and Robert M. Zink of the University of Minnesota, however, does not support this idea [39]. The time of evolutionary divergence can be estimated from the similarity/dissimilarity of their mitochondrial DNA, and most seem to have diverged from common ancestors up to 5 million years ago, considerably further back in time than can be accounted for by the Ice Ages of the last 2 million years.

It is reasonable to conclude, therefore, that many factors contribute to the species richness of the tropical regions and the lower richness of higher latitudes; no single explanation accounts for all the observed data.

The Mid-Domain Effect

It could be argued that the concentration of species in the low latitudes is precisely what one would expect from a random or null model. Take a landmass stretching roughly from pole to pole, such as the joined continental masses of North, Central, and South America and regard it as a single domain. Then take all the species occurring over that landmass and assess the area occupied by each. Finally, randomize all the patches occupied by the different species, and it emerges that the greatest degree of overlap between organisms lies in the middle of the combined land area, the so-called **mid-domain effect (MDE)**. According to this proposal, the diversity of species in the low latitudes is therefore the outcome of purely random processes [40].

The MDE proposal has proved controversial, however. It operates on the basis of uniformity of the nature of the domain and also of the ecology of the species analyzed. Geographical variations in landmasses, coupled with ecological variation in the limitations of different species, and further compounded by the geological and climatic histories of different parts of the domain, cannot be excluded from a full explanation of patterns of biological richness over the surface of the planet. Tests of the MDE involving specific domain areas (such as North America) and particular groups of organisms (such as birds) have shown that there are serious weaknesses in the hypothesis [41].

Latitude and Species Ranges

A further approach to the question of latitudinal gradients of diversity is based on the observation that high-latitude organisms have broader geographical ranges than those from the low latitudes. This apparently general feature of biogeography was first pointed out by E. H. Rapoport in the 1970s, but rose to prominence as a result of the work of George C. Stevens [42], who coined the term **Rapoport's Rule**. Much work has now been carried out to test the generalization, and the species of high latitudes do, on the whole, display wide geographic ranges, great altitudinal ranges, and broad ecological tolerances in comparison with tropical species. But there remains doubt as to whether this is a local effect that only makes itself felt in more northerly latitudes (above about 40–50°N), or whether it continues to be operative in the equatorial regions. Klaus Rohde [43], of Armidale, Australia, considers that Rapoport's Rule is of local application only and cannot be applied in the tropical regions. The occurrence of broad-range species in the high latitudes could itself be a consequence of the impact of successive glaciations, leaving only the most adaptable and flexible species behind. It can also be argued that the greater seasonal fluctuations of the high latitudes will select for wide-tolerance organisms. Detailed testing of the "rule" using a whole range of statistical techniques has failed to support a general global relationship between latitude or elevation and range extent among species [44].

An unusual extrapolation of Rapoport's Rule has been illustrated by Katherine Smith and James Brown of the University of New Mexico, who have examined the diversity of fish as one proceeds deeper into the ocean [45]. Overall diversity peaks in the upper 200 m (650 ft) of water and then declines with depth, and the species of greater depth have wider tolerance, being able to tolerate shallow waters as well. Narrow-range species were restricted to the upper layers of the ocean.

The use of Rapoport's Rule (perhaps better termed "the Rapoport effect" [46] since its general application is now questioned) as an explanation for the cause of latitudinal gradients of diversity is now virtually dismissed. Given the lower species richness of the high latitudes, it is only to be expected that there will be less competition for resources and that ecological and geographical ranges of species will therefore be more extensive than in the species-dense tropics. The Rapoport effect is likely to be a consequence of latitudinal gradients in species richness rather than its cause.

Although the Rapoport effect has proved more restricted in its application than was originally anticipated, it provides an excellent example of the kind of question that biogeographers are now asking. Stepping beyond the framework of maps and patterns of distribution [47], many biogeographers are examining the mechanisms that underlie their observations. In many respects they are asking ecological questions within a much larger scale framework of space and time. James Brown of the University of New Mexico has coined the term **macroecology** to cover this approach to biogeographical and ecological research [48].

Diversity and Altitude

As discussed in Chapter 2, patterns of vegetation and hence of biomes are generally related to latitude, and these are broadly repeated with respect to altitude, higher altitudes often bearing biome types that are more typical of higher latitudes. One might therefore, expect that the global patterns of biodiversity, declining from equator to poles, will be reflected with increasing altitude. Many studies have been conducted on different groups of organisms investigating the changes in species richness with altitude. Most of these studies have concentrated on tropical mountains, where the full range of climatic variation is found. But the outcome has not generally conformed with expectation. Many studies have demonstrated that the richness of species, particularly plants, increases with altitude, reaches a peak, and then declines again at very high altitudes. Some of the results are summarized in Fig. 4.14.

Vascular plants (those with a water-conducting system, including ferns, conifers, and flowering plants) are among the least difficult to survey in tropical regions, so it is not surprising that more information is available for these than most other groups of organisms. Surveys from various parts of the tropical and subtropical world indicate an increase in overall richness with altitude to a peak at about 1,500 to 2,000 m (5,000–6,500 ft), depending on location, followed by a decline at higher altitudes. On Mount Kinabalu in Borneo (Fig. 4.14a) there is a very clear and distinct peak [49], while in the Himalayan Mountains of Nepal (Fig. 4.14b) the peak is rather more spread and diffuse [50]. There are some problems in interpreting these data, partly because the lower altitudes tend to be more extensive in area than higher ones (leading to inflated diversity), and the lower altitudes are also often subject to more intensive disturbance by human settlement and agriculture than higher ones (which could decrease or increase diversity, depending on the nature of the human impact).

Tropical mountains create complex local climates, however, and these may in part explain the humped pattern of plant richness with altitude. Moderate altitudes often receive more rainfall than either low or very high altitudes in the tropics and subtropics. In the case of the Himalayas, for example, the lower elevations lie in the monsoon rain belt, so there is a distinct dry season alternating with a period of heavy rainfall. Cloud formation in the cooler air of the midaltitudes leads to high humidity and more generally distributed rainfall, so that the forests are permanently moist. Under these conditions, there is often an abundance of **epiphytes**—

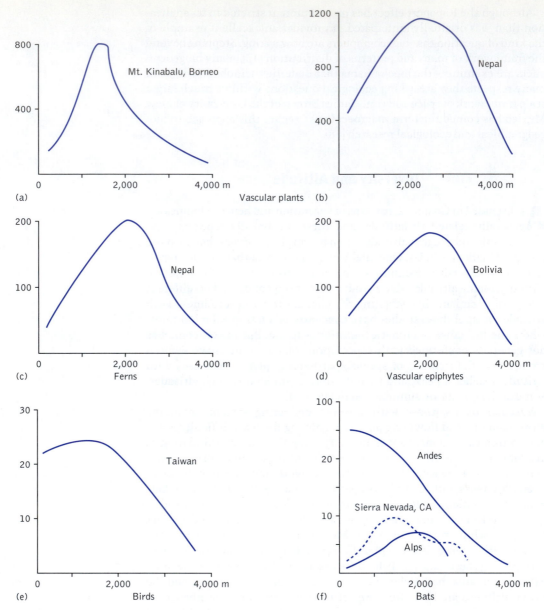

Fig. 4.14 Altitudinal gradients of diversity in certain plant and animal groups. Simplified summaries from a range of sources.

plants that use other vegetation for support, rooting upon trunks, branches, and even the leaves of more robust species, but limiting their demands to support rather than any form of direct parasitism. Mosses and lichens often live as epiphytes in such a forest, as do ferns and certain flowering plant families, such as orchids and bromeliads.

Studies concentrating on ferns and epiphytes in general, such as those shown in Fig. 4.14c and d, from Nepal [51] and Bolivia [52] respectively, demonstrate that these groups have very strong peaks in richness around the 2,000-m altitude, where the mountains are clothed in "cloud forest," a highly humid forest type, in which plants living entirely in the canopy with no roots reaching the ground are unlikely to experience desiccation. The midaltitude peak in plant richness owes much to this abundance of epiphytes in the cloud forest zone [53]. At higher altitudes, conditions for growth become more difficult as average temperatures fall and frost becomes increasingly likely. The alpine zone of mountains, like the polar tundra regions, is a zone in which few plants can survive and the extinction rate of immigrant species is likely to be high.

The study of altitudinal gradients of diversity in animal groups has proved more challenging than similar studies of plants. This is because it is much easier to survey static plants rather than mobile, often inconspicuous animal species. Survey methods for animals are consequently more complex, often involving trapping. Birds and bats have proved the most amenable to survey, and some general outcomes are shown in Fig. 4.14e and f. Studies of the birds of Taiwan, Southeast Asia [54], showed that bird richness exhibited a slight tendency to a hump-shaped relationship with respect to altitude. More detailed analysis of its causes demonstrated a strong positive correlation between the productivity of vegetation and bird diversity. In the case of birds, therefore, it seems that the energy hypothesis, which has been widely employed as an explanation for latitudinal diversity, is also appropriate for the interpretation of variations in bird diversity with altitude. Any midaltitudinal peak is a consequence of altitudinal variations in plant productivity as determined by local climate. Bird studies in the Andes of Colombia [55], however, show a more complex picture. The overall change in bird richness with altitude shows a steady decline from highest values at low altitude. Analysis of the data shows that the bulk of the birds at low altitude are widely distributed. Birds associated particularly with the Andes Mountains peak in diversity at around 2,000 m (6,500 ft). In this region the lowland occupies a large area and is continuous with the lowland forests of the interior. The midaltitudes have experienced a long-term process of speciation leading to a richness of endemics.

Bats have been studied in a very wide range of mountain regions of the world, and the results have been surveyed in detail by Christy McCain of the University of California, Santa Barbara [56]. A few samples are shown simplified in Fig. 4.14f. Some locations in the world show a general decline in species with altitude (such as Peru), while other locations (such as the European Alps and Yosemite, California) show a midaltitude peak in richness. The general conclusion from these studies is that bat diversity, like plant diversity, is largely controlled by the availability of water and the associated temperature regime. Good conditions for primary production of vegetation are also good for bats, presumably because their insect food is more abundant under such circumstances.

The overall conclusion regarding patterns of diversity with altitude, therefore, is that the energy model used in latitudinal studies probably holds good for many types of organism in relation to altitude. The complications found in altitude patterns of diversity probably relate largely to the peculiarities of local climates in mountain regions. It is also undoubtedly true that many such patterns have been obscured by a long history of human modification of montane environments, especially in the temperate regions of the world [57].

Biodiversity Hotspots

Although species richness does generally follow a latitudinal gradient, its pattern often proves more complex when we examine the picture in detail. For example, the Amazon Basin contains approximately 90,000 flowering plant species, whereas equivalent areas in Africa and in Southeast Asia contain only about 40,000 each. Certain areas that appear to be exceptionally rich in species have been termed **biodiversity hotspots** by the conservationist Norman Myers [58]. He originally proposed 10 hotspots, largely identified on the basis of plant diversity, for his argument was based on the idea that if vegetation is diverse all else will follow. But this is not always entirely true, as we have already seen in the case of aphids. Figure 4.15 shows the areas of high plant diversity in Africa compared with areas of high bird diversity, and it can be seen that there is relatively little overlap. Thus one cannot assume parallelism in trends of biodiversity between groups of organisms.

Myers's original work has been developed and expanded. Figure 4.16 shows the location of 25 of the hottest spots for biodiversity on Earth based on a consideration of many of the better known groups of organisms (plants, mammals, birds, reptiles, amphibians, etc.). When we add all the land-surface area of the hotspots together, they comprise only about 1.4% of the Earth's terrestrial total, yet they contain about 44% of the world's vascular plants and 35% of the vertebrates from the four main groups [59]. Stuart Pimm and Peter Raven [60], research conservationists from Columbia University, New York, and the Missouri Botanical Garden,

(a)

(b)

Fig. 4.15 The areas of Africa that are particularly rich in (a) plant species compared with those areas that are rich in (b) endemic birds. As can be seen, the two do not always correspond.

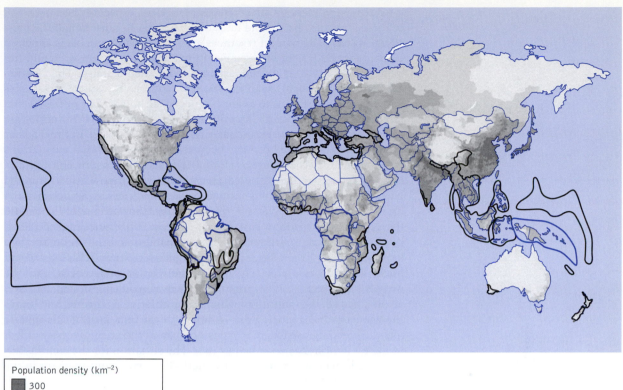

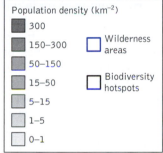

Population density (km^{-2})

- 300
- 150–300
- 50–150
- 15–50
- 5–15
- 1–5
- 0–1

Wilderness areas

Biodiversity hotspots

Fig. 4.16 The location of areas of exceptionally high biological richness (biological hotspots) in relation to human population density. From Cincotta et al. [61].

respectively, have calculated that even if the 25 hotspots were given protection, the likely extinction rate of species would be about 18%. If protection is delayed, then the extinction rate could be as high as 40%. Obviously, these areas must form a focus for global conservation activity, but it would be a mistake to confine attention to them because more than half of the world's species are located elsewhere. There are also regions of the world, such as the polar regions, where diversity is not high but the organisms present are very distinctive and often restricted in distribution.

The map displayed in Fig. 4.16 also shows the pattern of human settlement density around the world, and it can be seen that many of the biodiversity hotspots lie in areas of high human population density [61]. Demographic analyses have shown that over a billion people (about 20% of the world's population) live within the biodiversity hotspots. This must represent a serious threat to habitat survival, especially because some of these areas also have rapid rates of population expansion and the pressures on space and natural resources are intense. Human beings are often blamed for the loss of biodiversity in many parts of the world, and frequently our species is indeed responsible for such damage. The destruction of habitats, such as rainforests, is undoubtedly resulting in species loss, but this is difficult to quantify. Indeed, although 90% of the coastal rainforest of eastern South America has been cleared, no bird species are known to have become extinct as a result [62]. This does not mean that other groups of organisms have been as fortunate, nor that extinctions will not occur in the future as populations settle in to new equilibria. It is possible that the constricted areas of forest are now supersaturated with species and that rapid extinction rates will ensue. Sometimes the human impact on biodiversity can be subtle, as in the case of the cichlid fishes of Lake Victoria in East Africa. At least 500 species of cichlid fish live in this lake, and, although many are capable of hybridization, they are kept separate by their brightly colored patterns; they refuse to accept mates who do not precisely match their color preferences. But the increasing human activity around the lake is resulting in constant soil erosion and sewage pollution, which is cumulative over time, and leaves the waters turbid, reducing the visibility for the fish. Lost in this aquatic fog, the breeding barriers may break down, and the great diversity of fish in Lake Victoria may soon be lost [63]. The situation for the cichlids has been made even worse by the introduction of a voracious predator in the form of the Nile perch.

Hotspots, however, are not necessarily those areas that have suffered least from human impact. In the case of the Mediterranean region of Europe, for example, there is high diversity accompanied by a very long history of human activity, which has often proved very destructive. The island of Crete is only 245 km (150 miles) long by 50 km (30 miles) wide and has been isolated as an island for about 5.5 million years. It has supported human populations since at least the arrival of Neolithic peoples about 8,000 years ago. Since then, climatic changes have resulted in the development of very dry conditions in summer, and additional disturbance by earthquake and volcano activity has been experienced. The increase in human populations, their need for agricultural land, and their intensive pastoralism has resulted in the stripping of much of the original vegetation [64]. Nonetheless, Crete has 1,650 species of plant, 10% of which are endemic to the island. The fossils of Crete tell us that many species have become extinct during its recent history, yet it still remains a remarkably species-rich island. Indeed, the Mediterranean climate areas generally are very rich in plant species, perhaps as a result of the high intensity of habitat patterns and the severity of the local impacts of drought and fire [65]. It is possible that moderate human pressures can increase biodiversity as a result of diversifying habitats. There is certainly some evidence that biological diversity and human population density are positively related over the less extreme density range [66]. One must be careful before assuming that such a correlation implies causation, however.

Explaining why hotspots are so rich is even more difficult than solving the latitudinal gradient question. Indeed, one can detect a general low-latitude concentration in the hotspots, so the problem of gradients is clearly confused with that of hotspots. But additional mechanisms are also at work. Perhaps these hotspots are centers of evolution; or perhaps they are relict fragments of former diverse communities. If these explanations are correct, then one might expect hotspots to be rich in endemic organisms, confined to the region either because of their recent evolution or their extinction in surrounding areas. Several attempts have been made to test the correlation of hotspots with endemism, with some conflicting results. Analysis of the global richness pattern of bird species alone showed no significant correlation between overall richness and the richness of endemic species [67]. But a wider analysis using data from amphibians, reptiles, birds, and mammals [68] showed some degree of correlation between diversity and endemicity. Endemics have long been the subject of special attention for conservation simply because of their rarity or their narrow geographical range. It is reassuring, therefore, to know that the establishment of protection for those areas where endemic vertebrates are abundant also provides a safeguard for a wide range of other species and is good overall policy for biodiversity conservation. Work on marine fish diversity in the South Pacific [69] also suggests that fish biodiversity hotspots act as centers of evolution and that diversity in surrounding areas is related to distance from such hotspots. This supplies hotspots with another good reason for conservation, as they may prove to be the focal points for future species generation.

Diversity in Time

The study of changing species diversity in time may also provide clues to understanding why diversity is concentrated in certain parts of the world. Studies over long periods of time (tens or hundreds of millions of years) can be very informative, but such studies suffer from the disadvantage that it is often difficult to determine how many species there were within a fossil group. One can, however, more easily examine changes in the species composition of a habitat over a short period (decades or centuries). Change over periods of time of this order is termed **succession**, especially if such change follows a predictable and directional course of development [70].

A simple example of succession is the invasion of vegetation following the retreat of a glacier, as has been illustrated by studies in Alaska [71]. Warmer conditions cause the melting of ice, and the ice front gradually recedes, leaving bare rock surfaces and crushed rock fragments in sheltered pockets and crevices. Such primitive soils may be rich in some of the elements needed for plant growth, such as potassium and calcium, but are poor in organic matter. These soils usually have a very limited capacity for water retention, poor microbial populations, little structure, and low levels of nitrogen. A plant that can grow even under these stressed conditions is the Sitka alder (*Alnus sinuata*). This is a low-growing bushy tree that owes its success in part to its association with a bacterium that grows in association with its roots. This microbe forms colonies in swollen nodules on the alder's roots and is able to take nitrogen from the atmosphere and convert it to ammonium compounds that

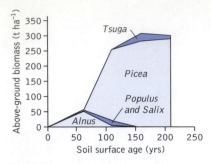

Fig. 4.17 Change in major species composition (expressed in terms of biomass, i.e., above-ground dry weight) during the development of forest following ice retreat in Alaska. The dominance of one species, spruce (*Picea*), is established as the biomass increases with successional development of the plant community. From Bormann & Sidle [71].

can subsequently be used (together with materials derived from alder photosynthesis) to build up proteins. Therefore, the alder manages perfectly well despite the low levels of nitrate in the soil. Because of the gradual death of roots and the return of litter to the soil from the alder, the growth of the tree increases the amount of nitrate in the soil and thus fertilizes it. But this very process of modifying the soil environment eventually proves the downfall of the alder, because it permits the invasion of other, less highly adapted plants, among them the Sitka spruce (*Picea sitchensis*). After about 80 years, the Sitka spruce trees, which are more robust and faster-growing than the alder shrubs, assume dominance in the vegetation and begin to shade out the pioneer alders. Thus, by their very existence at the site, the alders have effectively sealed their own fate and made the next step in the succession inevitable. This driving mechanism that underlies the successional process is termed **facilitation**, and it ensures a progressive and reasonably predictable development within the vegetation.

The course of succession also leads to an accumulation of biomass during the course of time. This is shown in Fig. 4.17, in which the biomass of the major tree species can be seen increasing over the course of 200 years of a succession in Glacier Bay, Alaska. Alder, poplar (*Populus*), and willow (*Salix*) are replaced by Sitka spruce and hemlock (*Tsuga*), and, while alder only achieved a maximum biomass of about 50t/ha, the spruce/hemlock forest grows to a biomass of over 300t/ha. Such an increase in biomass naturally involves the development of a more complex canopy structure, and, as we have seen, the diversity of animal species often follows the increase in structural complexity in vegetation.

The colonization of land exposed by glacial retreat is an example of **primary succession**, which effectively begins with nonliving components. A gap in a tropical rainforest created by the death of an old tree or by wind damage, on the other hand, already has a soil containing seeds, invertebrate animals, and plant nutrients. The further colonization and development of such a gap is termed **secondary succession**. The opening in the canopy gradually becomes filled by the invasion and growth of new vegetation, as shown in Fig. 4.18. The development in structural complexity can be seen in Fig. 4.19, in which plants gradually replacing one another and growing in stature produce successively more complex patterns of canopy cover. This in turn leads to an increasing diversity of microhabitats and hence species in the course of this secondary succession.

The word "diversity" is often used rather loosely by ecologists and conservationists, but it is not easy to define. The term is often used to convey the idea of the number of species per unit area of ground, but this is not really an adequate definition of diversity. One should really call this **species richness**. So the latitudinal gradients that we examined earlier in this chapter are actually gradients of richness. The term *diversity* also contains an implication regarding the way in which the number of animals or plants (or their biomass) is allocated among the species present. A community, for example, that contains 100 individuals belonging to 10 different species could have 91 individuals belonging to one species and only one each for all the others. This is a less diverse community than one in which there are 10 individuals belonging to each of the 10 species. Yet both communities have the same species richness. Many attempts have been made to devise an index of diversity that takes into account degree of **evenness** (or **equitability**) as well as the richness of a community. Perhaps the most widely used such indices is that of Claude Shannon [72].

In the case of successions, both richness and diversity tend to increase with time, especially for the animal component, because this is closely related to architectural complexity. But this may not always be the case with plants. Later stages in succession, as in the case of the alder/spruce sequence, may become dominated by a few large-bodied species, which effectively reduces plant diversity. In the early stages of succession, however, increasing diversity seems to hold generally true for both plants and animals. This can be illustrated by reference to Fig. 4.20, which shows the number of plant species and their area of cover during the course of succession on abandoned pasture in the United States. This is the development of vegetation following the abandonment of agricultural land, and its reversion to woodland is called an **old field succession**. The data shown in the diagram are derived from the work of F. A. Bazzaz [73], who studied an old field succession in Illinois. It covers a period of 40 years and illustrates the increasing number of plant species present (richness). It also shows a general flattening of the assemblage of bars, indicating that fewer species are dominating the community and more species are occupying a more equitable share of the available space. This is a graphic way of expressing the concept of diversity, which can clearly be seen to be increasing through successional time in this instance. There is no strong indication of dominance here, although in later stages one species is beginning to account for a large proportion of the total vegetation cover. (Note that the cover values are expressed as logarithms, which tends to make such dominance less evident.)

The final, mature stage in a succession is termed the **climax** and represents an equilibrium state. But the achievement of overall equilibrium does not mean that the community is static. Individual trees will become senile and die, minor catastrophes such as wind-blow and fire may cause openings to develop within the canopy, and these gaps, as we have seen, become occupied by small-scale secondary successional developments. Some members of pioneer species groups will find new opportunities to survive for a short while and reestablish themselves within such gaps, eventually to be replaced by more persistent species. The final state of vegetation is thus best conceived as an assemblage of individuals of various ages in a constant state of flux, the composition changing in response to the chance arrival of different species and the local extinction of others, all modified by the inevitable but more gradual environmental changes, such as those of climate.

(a)

(b)

(c)

(d)

Fig. 4.18 Gaps in forests, created by the death of old trees or minor catastrophes such as wind-blow or fire, become filled by the regrowth of young trees, often passing through a succession of different species. Habitat heterogeneity and canopy complexity are an outcome of this process.

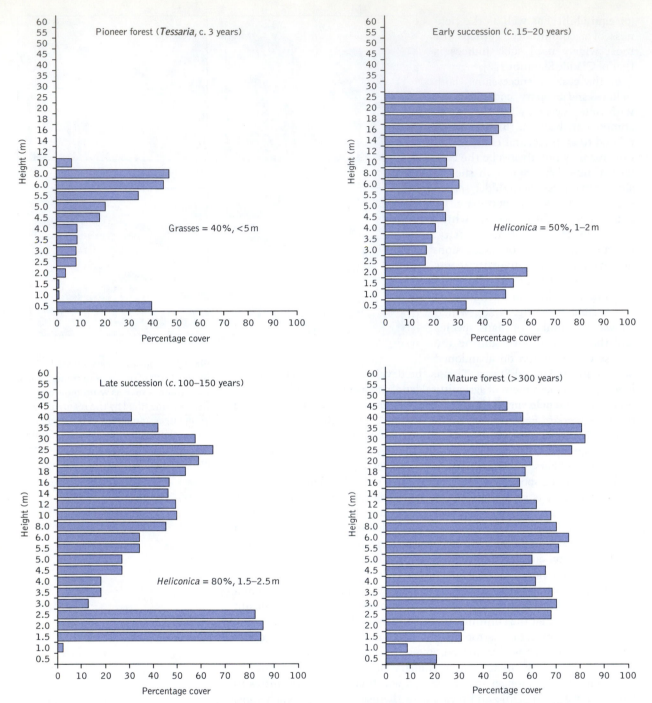

Fig. 4.19 Canopy profiles of a rainforest during the course of successional development over a period of about 300 years. The canopy structure is expressed in terms of leaf percentage cover at different heights above the ground. From Terborgh & Petren [29].

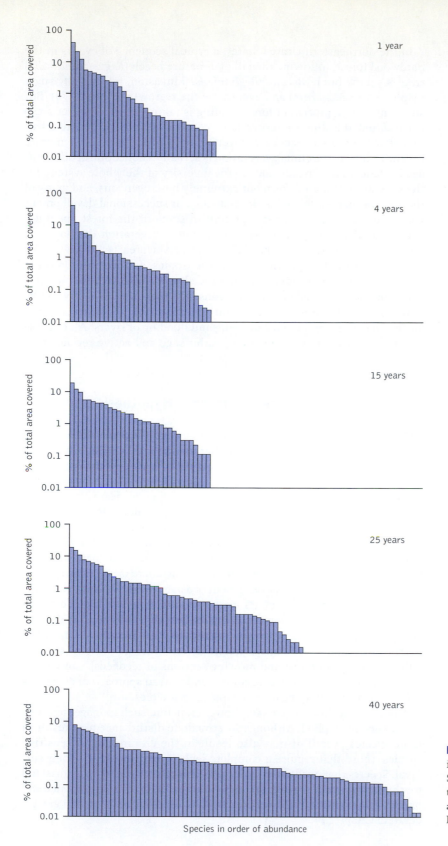

Fig. 4.20 Increasing species diversity in an old field succession in Illinois. Species are ranked in order of abundance, the latter being expressed as percentage area covered on a log scale. Data from Bazzaz [73]. After May [79].

In the northeastern United States, a typical sequence of events in the hardwood forest, following the fall of a mature beech (*Fagus grandifolia*) tree, is yellow birch (*Betula alleghaniensis*) invasion, followed by sugar maple (*Acer saccharum*) and eventually the regrowth of beech [74]. But since this cyclic process of forest healing is taking place wherever a gap has resulted, the climax forest actually consists of a mosaic of patches all in different stages of recovery, together with some patches of mature beech. Therefore the climax vegetation is actually a collection of different-aged patches. This, in fact, adds to the diversity of the whole system, for the vegetation is not uniform but extremely heterogeneous, and many of the species that would be lost from an area, if successional development had effectively ceased, are still present in some of the forest openings. Thus, the complexity of the time element in vegetation development allows even more species to be packed into a given area.

If we return to the question of why the tropics, and tropical forests in particular, are so rich in species, then we have established that successional processes can add a further means by which high diversity can be maintained. The forest is constantly undergoing disturbance from storms, local fire, and the meandering and flooding of rivers. All of these events leave the forest in a state of turbulence and active regeneration that contributes to the diversity of the whole.

Intermediate Disturbance Hypothesis

The fact that areas experiencing some degree of disturbance are often richer has given rise to a proposal that moderate disturbance in general enhances diversity. A site that is totally static, if such an ideal ever exists, would develop a stable and uniform biota. In practice, there is always some degree of change, at least on a local scale, including the death and collapse of individual trees. Even such local changes create niches for species that require more open conditions and might be considered to belong to earlier stages in succession, and their arrival (in the case of plants often from dormant seeds in the soil) increases diversity. Thus, even in a "natural" habitat (i.e., one in which human activity is absent) there is a regular cycle of disturbance, leading to the development of a patchwork of habitats.

Suburban human settlements, parks, and gardens can provide a rich variety of habitats that promote diversity, which is why human density and biodiversity can be positively related. Habitat management, in the form of moderate disturbance, preferably in a mosaic pattern (such as felling clearings in forest and mowing sections of reedbeds), can lead to increased diversity. This has become a widespread approach of conservationists, who are applying biogeographical principles.

A note of caution is necessary, however, before such management can be universally applied. Although intermediate disturbance may result in more species overall, it may also result in the elimination of sensitive species, those that require a complete lack of disturbance for their survival. Species that need extensive ranges of forest, or reedbed, or grassland for the maintenance of populations will not be favored by disturbance and habitat fragmentation. These conservation issues are discussed in greater detail in Chapter 14.

Dynamic Biodiversity and Neutral Theory

It is easy to assume that patterns of biodiversity are static, but this is not so. Although the tropics have always been richer in species than the temperate and polar regions, there is a constant flux of species over geological time. Species are always appearing and disappearing, and are always on the move, so the idea that the natural world has attained some kind of equilibrium state may not be valid. Some biogeographers maintain that the concept that compatible species, each with its distinctive and individual niche, become sorted over the course of time to assemble themselves into a stable equilibrium community, is imposing more order onto nature than is really present. Stephen Hubbell of the University of Georgia has proposed that species vary in their abundance in a random manner. He has set out a **neutral theory of biodiversity** that considers assemblages of species as a collection of randomly selected individuals [75]. The proposal can be modeled using computer simulations and often provides a good match with what is actually found in nature, especially for plants that are all doing very much the same job in the ecosystem, namely, fixing energy from the Sun and carbon dioxide from the atmosphere. The fact that they all have similar functions suggests that it should not matter which species are present in the system. The neutral theory seems to work best when considering species at a selected trophic level. But at such a level, chance, according to Hubbell, plays a great part in determining what species are present in an ecosystem, perhaps a far greater part than has been appreciated in the past. Communities assemble themselves over time as a result of chance arrivals and establishments rather than the shuffling of species with their different niche requirements, ultimately becoming sorted into a collection of coexisting components, each with its distinctive role.

According to neutral theory, therefore, the species present in a community depends more on powers of dispersal and availability for immigration than on the compatibility of niches. Despite the robust nature of the predictions resulting from neutral models of community assemblage, there have been questions and criticisms [76]. There is evidence for stabilizing forces at work in nature that may lead to particular end points (in terms of community composition) being achieved. An analysis of long-term vegetation change using stratified fossil pollen grains preserved in the lakes of North America, for example, shows a great deal of variation in assemblages over time, but there is a tendency for variability to decline with time as a general stability is attained. Such results, based on relatively long-term data (over a period of 10,000 years) indicate that random forces and neutral drift alone cannot account for the nature of communities and their biodiversity [77].

Chance can operate, however, in a number of different ways, apart from the stochastic nature of immigration. One of these is through the initial conditions obtaining when a community begins its development. Slight differences in the original state can strongly affect the outcome of community establishment and therefore its final biodiversity. This idea of minor initial conditions being highly influential is known as **chaos theory**. It can be illustrated by considering a pencil balanced on the tip of its sharpened point. Any minute imperfection in the point, or in the hand of the operator, will determine the direction in which the pencil will eventually fall when it is released. The same is true of ecosystems; very small chance events, such as the occurrence of a certain set of weather conditions, wind direction, or time of year at the start of a succession, can have a great influence on the ultimate outcome.

The fact that species continue to come and go even within apparently stable ecosystems can be observed by monitoring this process of change. The tropical rainforests are generally regarded as both biodiverse and stable in their species content, but observations on the forests of Central America, northern South America, and the Amazon Basin have all demonstrated that continual changes are proceeding within the composition of these ecosystems. In particular, the woody climbing plants (lianas) are becoming more abundant in all of these regions [78]. Neither the causes nor the consequences of these changes are yet clear. It may be a response to global atmospheric or climatic changes, or it may be an example of cyclic or random variation in species composition. The growth of lianas often impairs the performance of some tree species, so the outcome of this widespread vegetation change may prove considerable from the point of view of the future biodiversity of these regions.

The entire question of why some parts of the world are richer in species than others, and what factors influence the assemblage of communities, thus remains one of active debate and research. It is one of the most profound and important questions that is asked of biogeography, for the answer will be extremely valuable in predicting future changes and managing the threatened biodiversity of the planet. Explaining the global patterns of biodiversity is a process that requires a consideration of many factors. Some of these factors have been discussed here, but further understanding of the subject demands a knowledge of many other aspects of biogeography. How is it that many species manage to occupy the same habitat? What factors limit the geographical range of individual species? How have such ranges changed during the course of the Earth's history? How do new species evolve, and why do they evolve in particular ways? These are some of the questions that must be faced if the complex issue of biodiversity is to be further understood.

Summary

1 Biodiversity means the full range of life on Earth, including all the different species found, together with the genetic variation between populations and individuals, and the variety of ecosystems, communities and habitats present on our planet.

2 We are losing species at an unknown, but undoubtedly accelerating, rate. We need to know more about the variety of life on Earth before we can even appreciate how fast we are losing it.

3 Only about 1.8 million of the species that live on Earth have so far been described. This is a very small percentage, perhaps less than 5%, of the likely total of species on the planet.

4 The tropics are generally richer in species than the high latitudes, possibly as a result of high productivity and food availability, high biomass and hence complex structure, past patterns of evolution, survival of fragments of habitats through the cold episodes of the last 2 million years, and also the degree of small-scale disturbance resulting in a mosaic of successional processes.

5 The term *diversity* involves both species number (richness) and pattern of allocation of numbers or biomass between the different species (evenness). It generally increases during the course of succession.

6 The species composition (and therefore the biodiversity) of any community is constantly changing. Chance is likely to play a part in the assemblage of communities and in subsequent changes, but stabilizing forces are also at work.

7 There is a tendency for human populations to be dense in biodiversity hotspots. It is possible that people enhance biodiversity by diversifying habitats, but the destructive tendency of high population densities is a matter for conservation concern.

Further Reading

Bardgett RD, Usher MB, Hopkins DW, eds. *Biological Diversity and Function in Soils*. Cambridge: Cambridge University Press, 2005.

Gaston KJ, Spicer, JI. *Biodiversity: An Introduction*. 2nd ed. Oxford: Blackwell Publishing, 2004.

Groombridge B, Jenkins MD. *World Atlas of Biodiversity: Earth's Living Resources in the 21st Century*. Berkeley: University of California Press, 2002.

Hubbell SP. *The Unified Neutral Theory of Biodiversity and Biogeography*. Princeton, NJ: Princeton University Press, 2001.

Lomolino MV, Riddle BR, Brown JH. *Biogeography*. 3rd ed. Sunderland, MA: Sinauer Associates, 2006.

Perlman DL, Adelson G. *Biodiversity: Exploring Values and Priorities in Conservation*. Oxford: Blackwell Science, 1997.

Walker LR, del Moral R. *Primary Succession and Ecosystem Rehabilitation*. Cambridge: Cambridge University Press, 2003.

References

1 May RM, Nee S. The species alias problem. *Nature* 1995; 378, 447–448.

2 Wilson EO. *Biodiversity*. New York: National Academic Press, 1988.

3 Erwin TL. Beetles and other insects of tropical forest canopies at Manaus, Brazil, sampled by insecticidal fogging. In: Sutton SL, Whitmore TC, Chadwick AC, eds. *Tropical Rain Forest: Ecology and Management*, pp. 59–75. Oxford: Blackwell Scientific Publications, 1983.

4 Stork N, Gaston K. Counting species one by one. *New Scientist* 1990; 127, 43–7.

5 Thomas CD. Fewer species. *Nature* 1990; 347, 237.

6 Pace NR. A molecular view of microbial diversity and the biosphere. *Science* 1997; 276, 734–740.

7 Holms B. Life unlimited. *New Scientist* 1996: 148, 26–29.

8 Fyfe WS. The biosphere is going deep. *Science* 1996; 273, 448.

9 O'Donnell AG, Goodfellow M, Hawksworth DL. Theoretical and practical aspects of the quantification of biodiversity among microorganisms. In: Hawksworth DL, ed. *Biodiversity: Measurement and Estimation*, pp. 65–73. London: Chapman & Hall, 1995.

10 Zettler LAA, Gomez F, Zettler E, Keenan BG, Amils R, Sogin ML. Eukaryotic diversity in Spain's River of Fire. *Nature* 2002; 417, 137.

11 May RM. A fondness for fungi. *Nature* 1991; 352, 475–476.

12 Groombridge B, ed. *Global Biodiversity: Status of the Earth's Living Resources*. London: Chapman & Hall, 1992.

13 Morell V. New mammals discovered by biology's new explorers. *Science* 1996; 273, 1491.

14 Wright MG, Samways MJ. Gall-insect species richness in African fynbos and karoo vegetation: the importance of plant species richness. *Biodiversity Lett* 1996; 3, 151–155.

15 Lawton JH, Bignell DE, Bolton B, Bloemers GF, Eggleton P, Hammond PM, Hodda M, Holt RD, Larsen TB, Mawdsley NA, Stork NE, Srivastava DS, Watt AD. Biodiversity inventories, indicator taxa and effects of habitat modification in tropical forest. *Nature* 1998; 391, 72–76.

16 Vellend M, Lilley PL, Starzomski BM. Using subsets of species in biodiversity surveys. *J. Applied Ecology* 2008; 45, 161–169.

17 Siemann E, Tilman D, Haarstad J. Insect species diversity, abundance and body size relationships. *Nature* 1996; 380, 704–706.

18 Aarssen LW, Schamp BS, Pither J. Why are there so many small plants? Implications for species coexistence. *J. Ecology* 2006; 94, 569–580.

19 da Silva W. On the trail of the lonesome pine. *New Scientist* 1997; 155, 36–39.

20 Chapin FS, Zavaleta ES, Eviner VT, Naylor RL, Vitousek, PM, Reynolds HL, Hooper DU, Lavorel S, Sala OE, Hobbie SE, Mack MC, Diaz S. Consequences of changing biodiversity. *Nature* 2000; 405, 234–242.

21 Gaston KJ. Global patterns in biodiversity. *Nature* 2000; 405, 220–227.

22 Duellman WE. Patterns of species diversity in anuran amphibians in the American tropics. *Ann Missouri Bot Garden* 1988; 75, 70–104.

23 Collins NM, Morris MG. *Threatened Swallowtail Butterflies of the World. IUCN Red Data Book*. Cambridge: IUCN, 1985.

24 Currie DJ, Paquin V. Large-scale biogeographical patterns of species richness of trees. *Nature* 1987; 329, 326–327.

25 Davies JT, Barraclough TG, Savolainen V, Chase MW. Environmental causes for plant biodiversity gradients. *Phil Trans R Soc Lond* B 2004; 359, 1645–1656.

26 Brown JH, Gillooly JF, Allen AP, Savage VM, West GB. Toward a metabolic theory of ecology. *Ecology* 2004; 85, 1771–1789.

27 Algar AC, Kerr JT, Currie DJ. A test of metabolic theory as the mechanism underlying broad-scale species-richness

gradients. *Global Ecology and Biogeography* 2007; 16, 170–178.

28 Whittaker RJ, Nogues-Bravo D, Araujo MB. Geographical gradients of species richness: a test of the water-energy conjecture of Hawkins et al. (2003) using European data for five taxa. *Global Ecology and Biogeography* 2007; 16, 76–89.

29 Terborgh J, Petren K. Development of habitat structure through succession in an Amazonian floodplain forest. In Bell SS, McCoy ED, Mushinsky HR, eds. *Habitat Structure: The Physical Arrangement of Objects in Space*, pp. 28–46. London: Chapman & Hall, 1991.

30 Blondell J. *Biogeographie et Ecologie*. Paris: Masson, 1979.

31 Dixon AFG. *Aphid Ecology: An Optimization Approach*, 2nd ed. London: Chapman & Hall, 1998.

32 Blackburn TM, Gaston KJ. A sideways look at patterns in species richness, or why there are so few species outside the tropics. *Biodiversity Lett* 1996; 3, 44–53.

33 Rosenzweig ML. *Species Diversity in Space and Time*. Cambridge: Cambridge University Press, 1995.

34 Rohde K. The larger area of the tropics does not explain latitudinal gradients in species diversity. *Oikos* 1997; 79, 169–172.

35 Rohde K. Latitudinal gradients in species-diversity—the search for the primary cause. *Oikos* 1992; 65, 514–527.

36 Crane PR, Lidgard S. Angiosperm diversification and paleolatitudinal gradients in Cretaceous floristic diversity. *Science* 1989; 246, 675–678.

37 Emerson BC, Kolm N. Species diversity can drive speciation. *Nature* 2005; 1015–1017.

38 Colinvaux PA, De Oliveira PE, Moreno JE, Miller MC, Bush MB. A long pollen record from lowland Amazonia: forest and cooling in glacial times. *Science* 1996; 274, 85–88.

39 Klicka J, Zink RM. The importance of recent ice ages in speciation: a failed paradigm. *Science* 1997; 277, 1666–1669.

40 Colwell RK, Lees DC. The mid-domain effect: geometric constraints on the geography of species richness. *Trends in Ecology and Evolution* 2000; 15, 70–76.

41 Hawkins BA, Diniz-Filho JAF. The mid-domain effect cannot explain the diversity gradient of Nearctic birds. *Global Ecology and Biogeography* 2002; 11, 419–426.

42 Stevens GC. The latitudinal gradient in geographical range: how so many species coexist in the tropics. *Am Naturalist* 1989; 133, 240–256.

43 Rhode K. Rapoport's rule is a local phenomenon and cannot explain latitudinal gradients in species diversity. *Biodiversity Lett* 1996; 3, 10–13.

44 Ribas CR, Schoereder JH. Is the Rapoport effect widespread? Null models revisited. *Global Ecology and Biogeography* 2006; 15, 614–624.

45 Smith, KF, Brown, JH. Patterns of diversity, depth range and body size among pelagic fishes along a gradient of depth. *Global Ecology and Biogeography* 2002: 11, 313–322.

46 Gaston KJ, Blackburn TM, Spicer JI. Rapoport's rule: time for an epitaph? *Trends Ecol Evol* 1998; 13, 70–74.

47 Blackburn TM, Gaston KJ. There's more to macro ecology than meets the eye. *Global Ecology and Biogeography* 2006; 15, 537–540.

48 Brown JH. *Macroecology*. Chicago: University of Chicago Press, 1995.

49 Grytnes JA, Beaman JH. Elevational species richness patterns for vascular plants on Mount Kinabalu, Borneo. *J. Biogeography* 2006; 33, 1838–1849.

50 Vetaas OR, Grytnes JA. Distribution of vascular plant species richness and endemic richness along the Himalayan elevation gradient in Nepal. *Global Ecology and Biogeography* 2002; 11, 291–301.

51 Bhattarai KR, Vetaas OR, Grytnes JA. Fern species richness along a central Himalayan elevational gradient, Nepal. *J. Biogeography* 2004; 31, 389–400.

52 Kroemer T, Kessler M, Gradstein SR, Acebey A. Diversity patterns of vascular epiphytes along an elevational gradient in the Andes. *J. Biogeography* 2005; 32, 1799–1809.

53 Kueper W, Kreft H, Nieder J, Koester N, Barthlott W. Large-scale diversity patterns of vascular epiphytes in Neotropical montane rain forests. *J. Biogeography* 2004; 31, 1477–1487.

54 Ding T-S, Yuan H-W, Geng S, Lin Y-S, Lee P-F. Energy flux, body size and density in relation to bird species richness along an elevational gradient in Taiwan. *Gobal Ecology and Biogeography* 2005; 14, 299–306.

55 Kattan G, Franco P. Bird diversity along elevational gradients in the Andes of Colombia: area and mass effects. *Global Ecology and Biogeography* 2004; 13, 451–458.

56 McCain CM. Could temperature and water availability drive elevational species richness patterns? A global case study for bats. *Global Ecology and Biogeography* 2007; 16, 1–13.

57 Nogues-Bravo D, Araujo MB, Romdal T, Rahbek C. Scale effects and human impact on the elevational species richness gradients. *Nature* 2008; 453, 216–219.

58 Myers N. The biodiversity challenge: expanded hot-spots analysis. *The Environment* 1990; 10, 243–256.

59 Myers N, Mittermeier RA, Mittermeier CG, da Fonseca GAB, Kent J. Biodiversity hotspots for conservation priorities. *Nature* 2000; 403, 853–858.

60 Pimm SL, Raven P. Extinction by numbers. *Nature* 2000; 403, 843–845.

61 Cincotta RP, Wisnewski J, Engelman R. Human population in the biodiversity hotspots. *Nature* 2000: 404, 990–992.

62 Brooks T, Balmford A. Atlantic forest extinctions. *Nature* 1996; 380, 115.

63 Seehausen O, van Alpen JJM, Witte F. Cichlid fish diversity threatened by eutrophication that curbs sexual selection. *Science* 1997; 277, 1808–1811.

64 Rackham O, Moody J. *The Making of the Cretan Landscape*. Manchester: Manchester University Press, 1996.

65 Cowling RM, Rundel PW, Lamont BB, Arroyo MK, Arlanoutsou M. Plant diversity in mediterranean-climate regions. *Trends Ecol Evol* 1996; 11, 362–366.

66 Araujo, MB. The coincidence of people and biodiversity in Europe. *Global Ecology and Biogeography* 2003; 12, 5–12.

67 Orme CDL *et al.* Global hotspots of species richness are not congruent with endemism or threat. *Nature* 2005; 436, 1016–1019.

68 Lamoreux JF, Morrison JC, Ricketts TH, Olson DM, Dinerstein E, McKnight MW, Shugart HH. Global tests of biodiversity concordance and the importance of endemism. *Nature* 2006; 440, 212–214.

69 Mora C, Chittaro PM, Sale PF, Kritzer JP, Ludsin SA. Patterns and processes in reef fish diversity. *Nature* 2003; 933–936.

70 Moore PD. A never-ending story. *Nature* 2001; 409, 565.

71 Bormann BT, Sidle RC. Changes in productivity and distribution of nutrients in a chronosequence at Glacier Bay National Park, Alaska. *J Ecol* 1990; 78, 561–578.

72 Spellerberg IF, Fedor PJ. A tribute to Claude Shannon (1916–2001) and a plea for more rigorous use of species richness, species diversity and the "Shannon-Wiener" Index. *Global Ecology and Biogeography* 2003: 12; 177–179.

73 Bazzaz FA. Plant species diversity in old field successional ecosystems in southern Illinois. *Ecology* 1975; 56, 485–488.

74 Forcier LK. Reproductive strategies in the co-occurrence of climax tree species. *Science* 1975; 189, 808–810.

75 Hubbell, SP. *The Unified Neutral Theory of Biodiversity and Biogeography*. Princeton, NJ: Princeton University Press, 2001.

76 Ostling A. Neutral theory tested by birds. *Nature* 2005; 436, 635–636.

77 Clark JS, McLachlan JS. Stability of forest biodiversity. *Nature* 2003; 423, 635–638.

78 Phillips OL. et al. Increasing dominance of large lianas in Amazonian forests. *Nature* 2002, 418, 770–774.

79 May RM. The evolution of ecological systems. *Scientific American* 1978; 238 (3), 118–133.

The Engines of the Planet I: Plate Tectonics

This chapter first explains the evidence for plate tectonics. It then describes how that process affects the patterns of life on the continents in two ways. First, it changes them directly, by changing the patterns of interconnection of the continents. Second, it causes changes in the patterns of the continents, oceans, shallow seas, mountains, and ocean currents, which have indirect effects on biogeography by altering climate patterns. It also produces different types of island, which may show different biotic histories. The engine of plate tectonics thus continually provides new challenges and opportunities for living organisms, to which they respond via the engine of evolutionary change, as explained in the next chapter.

The Evidence for Plate Tectonics

As explained in Chapter 1, the idea that continents could fragment and move across the face of the planet was first suggested by the German meteorologist Alfred Wegener in 1912, but was rejected by scientists because he could not suggest any mechanism for such a phenomenon. It was only in the 1960s that new discoveries vindicated Wegener and revealed the motive force. The first breakthrough came with the invention of techniques that used the phenomenon of **palaeomagnetism**. This uses the presence of magnetized particles in many rocks to trace the movements of the rocks, and therefore also of the landmasses in which they lie. Obviously, if the continents had never moved, these "fossil compasses" should all point to the present magnetic poles—but they do not [1]. Instead, if a series of rocks of different ages from one continent are studied, and the positions of one of the magnetic poles at these different times are plotted on a map, it looks as though the pole has gradually moved across the Earth's surface (Fig. 5.1). Of course, it is instead the continents that have moved across the poles. Furthermore, if similar paths of "polar wandering" are constructed for each of today's continents, they also show that these have moved relative to one another. Finally, if we plot these paths on a globe and move the continents back along the paths that they have followed through time, we find that they gradually come together in a pattern very similar to that which Wegener first proposed. As he noted, other evidence for the positions of the continents can be

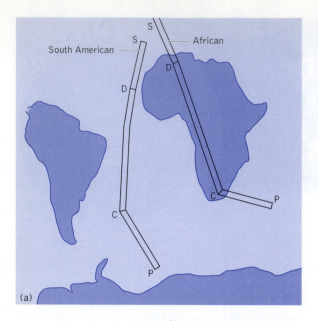

(a)

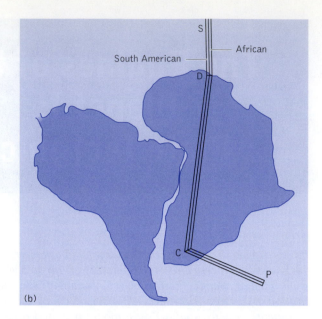

(b)

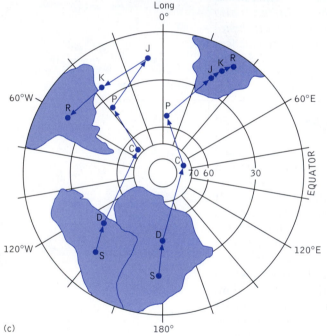

(c)

Fig. 5.1 Polar wandering patterns of South America and Africa, to illustrate the concept.
(a) The position of the South Magnetic Pole relative to each continent in its present position is shown for the Silurian (S), Devonian (D), Carboniferous (C) and Permian (P).
(b) The continents are moved together until their polar wandering patterns overlap, proving that they moved as a single landmass during this period of time.
(c) The palaeomagnetic data suggest that the continents moved across the South Magnetic Pole as shown here; their paths only diverge from one another during the Jurassic (J), to reach position K in the Cretaceous and R today, as the widening South Atlantic Ocean separated them.

gained from the types of rocks that were laid down within them (e.g., desert sandstones or glacial deposits).

The pattern of continental movements suggested by this palaeomagnetic research was supported and confirmed by study of the floor of the oceans. This study revealed a system of great submarine chains of volcanic mountains, and another system of deep troughs or **trenches** around the edges of the Pacific Ocean. In 1962, the American geophysicist Harry Hess [2] suggested that the volcanic chains were **spreading ridges**, where

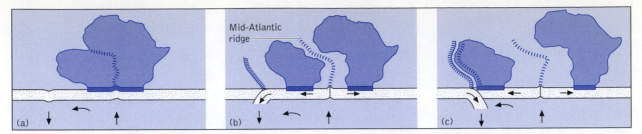

Fig. 5.2 How and why South America and Africa drifted apart.
(a) The two continents were originally part of a single continent, Gondwana (the rest of which is not shown). An upward convection current from the deeper layers of the Earth appeared under them, with the corresponding downward current further to the west, in the Pacific Ocean.
(b) The two continents move apart, separated by the new South Atlantic Ocean. Down the center of this runs the mid-Atlantic spreading ridge, on either side of which new ocean crust is continually created. This extension movement is balanced by the appearance, in the Pacific, of an ocean trench where old ocean crust disappears into the Earth.
(c) South America has moved westwards until it is adjacent to the ocean trench. To continue to balance the still-widening South Atlantic Ocean, the ocean crust that disappears into the trench is now derived from the west. This old crust now disappears below western South America, causing earthquakes and the rise of the volcanic Andes mountains.

new seafloor is being formed as the regions on either side move apart, and that the **trenches**, in contrast, are where old ocean-floor is consumed, disappearing downward into the Earth. He theorized that all this activity is the result of great convection currents that bring heated material to the surface from the hot interior of the Earth. The spreading ridges, which extend for 72,000 km (45,000 mi), mark the positions where these upward currents reach the surface, while the trenches indicate where the corresponding downward currents return cooler material to the depths of the Earth. The different plates may also move past one another at regions known as **transform faults**, which are regions of active earthquake activity; perhaps the best known of these is the San Andreas fault in California, where the eastern edge of the Pacific plate is rotating northward past the western edge of the North American plate.

Where spreading ridges lie within the oceans, their activity will cause continents to move apart by what is known as **seafloor spreading**, and ultimately may lead them to collide with one another. Sometimes a ridge extends under a continent; its activity will then cause the gradual drifting apart of those regions of the continent that lie on either side of the ridge (Fig. 5.2). As these move apart, they become separated by a new, widening ocean, the floor of which is similarly expanding to one side or another, away from the spreading ridge that runs along the center of the new ocean. As a result, the surface of the Earth, known as the **lithosphere**, is occupied by a number of areas known as **tectonic plates**, which may contain continents and parts of oceans, or may consist only of ocean floor (see Fig. 5.3). Because the moving elements therefore include the ocean floors as well as the continents, the study of their movements is known as **plate tectonics** rather than continental drift.

When it first appears from the depths of the Earth, the new ocean crust is still hot and rich in iron minerals, which are sensitive to the prevailing direction of the Earth's magnetic field. Due to changes in the flows of material within the mantle of the Earth, this magnetic field

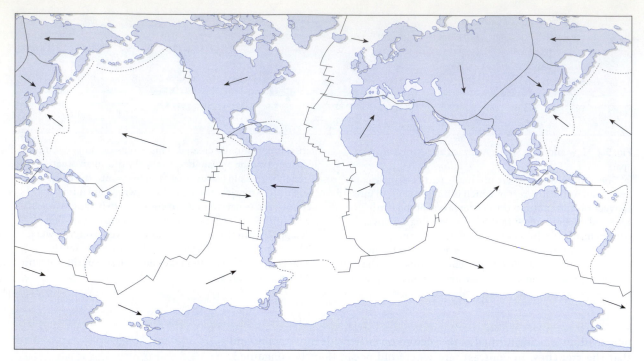

Fig. 5.3 The major tectonic plates. Lines within the oceans show the positions of spreading ridges: dotted lines indicate the positions of trenches. Lines within the continents show the divisions between the different plates. Arrows indicate the directions and proportionate speeds of movement of the plates. The Antarctic plate is rotating clockwise.

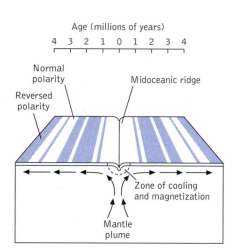

Fig. 5.4 Diagram of a portion of seafloor across a midoceanic ridge, showing the symmetrical pattern of bands of varying width but of alternating polarity. After Stanley [11].

reverses direction every 10^4 to 10^6 years. In 1963 the American geologists Fred Vine and Drummond Matthews [3] found that, as a result, there are strips of varying width, magnetized in opposite directions, in the new ocean crust on either side of the spreading ridge. The symmetry of these bands on either side of the ridge provides striking evidence of the reality of seafloor spreading (Fig. 5.4). Because the ages of these stripes can easily be determined, these also provide direct evidence of the past positions of the continents. By removing from the map any ocean floor younger than, for example, 65 my, one can return the continents to their positions at that time. However, all ocean floor older than 180 my has disappeared by being returned to the Earth's interior at the great submarine trenches, and the positions of the continents before that time therefore have to be deduced from palaeomagnetism. The former patterns of the union of continents can also be deduced from matching the sequences of rock types or rock ages and from dating the times of the rise of mountain chains that mark their collision. (For the geological timescale, see Fig. 5.5 on the next page.)

All this evidence for the theory of plate tectonics was so overwhelmingly convincing that it quickly gained general acceptance by biogeographers during the 1960s.

Palaeobiogeography was particularly important in the formulation of Wegener's original theory of continental drift. Nevertheless, the degree of detail provided by the geophysical data is now on the whole more precise than that of the palaeobiogeographical evidence, so that this biological evidence often has only a confirmatory role. The geophysical data, by establishing the times at which landmasses split or united, have also identified the units of time and geography within which it is appropriate to make palaeobiogeographical analyses [4]. Until then, such analyses often made little sense, for they frequently combined units of time within which major changes of geography had taken place, or combined geographical areas in inappropriate patterns.

Today, we can see that the distribution patterns of plants 350–250 my ago (Fig. 1.3) and of vertebrates 300–270 my ago (Fig. 1.7) reflect the geography that appears when the continents are placed in the locations suggested by the geophysical data, and the outlines of shallow seas are added. Fossil marine faunas provide evidence on the development of faunal provinces on either side of widening oceans (see p. 291). The shapes of the fossil leaves of flowering plants indicate the climatic regime of the areas they inhabited, as do the types of plant themselves, and their pollen also provides information on climatic change during the Ice Ages (see Chapter 12).

Although palaeobiogeography therefore often plays a subordinate role, in some situations its data provide more direct, detailed evidence than the geophysical record. For example, both the fossil marine faunas on either side of the Panama Isthmus (see p. 292) and the fossil mammals of North and South America (see p. 359) provide clearer and more reliable evidence of the date of the final linking of those two continents than do any geophysical data from the area.

For reasons explained later in this chapter, sea levels have varied over time, and they have therefore covered varying amounts of the lower-lying margins of the continents; this submerged area is known as the **continental shelf**. In identifying the outlines of landmasses through time, it is important to remember that it is the edge of the continental shelf, and not the coastline, that marks the true edge of the continent. Between the continental shelves, the deep oceans separate the continental plates. Sometimes the whole of these plates has been above sea level. At other times comparatively shallow "epicontinental seas" have covered the edges of the continents (e.g., the North Sea today) or formed seas within the continents (such as Hudson Bay today). Because seafloor spreading

Era	Period	Epoch	Approximate duration in millions of years	Approximate date of commencement in millions of years BP
Cenozoic	Quaternary	Pleistocene	2.4	2.4
Cenozoic	Tertiary	Pliocene	2.6	5
Cenozoic	Tertiary	Miocene	20	25
Cenozoic	Tertiary	Oligocene	11	36
Cenozoic	Tertiary	Eocene	18	54
Cenozoic	Tertiary	Paleocene	11	65
Mesozoic	Cretaceous		81	146
Mesozoic	Jurassic		62	208
Mesozoic	Triassic		37	245
Palaeozoic	Permian		45	290
Palaeozoic	Carboniferous		72	362
Palaeozoic	Devonian		46	408
Palaeozoic	Silurian		32	440
Palaeozoic	Ordovician		75	510
Palaeozoic	Cambrian		60	570
Proterozoic			4000	

Millions of years ago: 50, 100, 150, 200, 250, 300, 350, 400, 450, 500, 550, 4600

Formation of the Earth's crust about 4600 million years ago

Fig. 5.5 The geological timescale.

does not provide data on the presence or spread of these seas, palaeobiogeography also provides crucial evidence on the times during which these subdivided areas of land.

Changing Patterns of Continents

About 340 my ago there were four separate northern landmasses (see Fig. 10.1, p. 204). The largest of these, **Euramerica**, formed of conjoined North America plus Europe, united about 340 my ago with the northern edge of the great supercontinent known as **Gondwana**; this huge area of land included what is now five of today's landmasses—Antarctica (which was itself made up of two originally separate landmasses), South America, Africa, Australia, and India. Later, about 295 my ago, this supercontinent was joined by two other Northern Hemisphere continents, Siberia and Kazakhstan, the collision causing the rise of the Ural Mountains. Finally, about 260 my ago, this was joined by a number of smaller fragments (including north China, south China, Tibet, Indochina and Southeast Asia) that had split off from the northern edge of Gondwana and moved northward. The result was a single world continent we call **Pangaea** (Plate 3a). However, it was not long before Pangaea started to become divided. About 160 my ago, Gondwana separated from the northern landmass, now made up of North America and Eurasia, that we call **Laurasia**.

Over the last 440 million years, the histories of what we see today as the Northern and the Southern Hemisphere continents were entirely different. That of the southern continents was one of continual fragmentation. From about 135 my ago, Gondwana became progressively broken up into a number of separate tectonic plates, each bearing a separate continent. Plate 3 shows how Africa, India/Madagascar, New Zealand, Australia, and South America in turn separated from Antarctica, and Fig. 5.6 shows this sequence in diagrammatic form. As explained in Chapters 10 and 11, some of these continents moved considerable distances northward, bearing their faunas and floras across zones of latitude with differing climates, to which they had to adapt. The final collision of India with Asia added a new element to the floras and faunas of Asia (see p. 346), and the approach of Australia to Southeast Asia allowed a complex interchange between the biotas of those two areas (see p. 347). Even the comparatively small northern movement of South America led to its connection with North America about 3 my ago, and an even more complex interchange between their faunas (see p. 359).

In contrast to this complex geographical history of the Southern Hemisphere, that of the Northern Hemisphere has been comparatively uniform. Its two continents, North America and Eurasia, have never been far apart, so that the dispersal of organisms between them has usually been fairly easy. However, three factors have, over the past 180 my, subdivided or connected them in several different patterns (Fig. 5.7): continental movements, the expansion or contraction of shallow seas, and the rise or erosion of mountain chains. As explained in Chapter 11, this led to changes in the relationships between their floras and faunas. For most of this time, until about 30 my ago, much of Europe was covered by a shallow

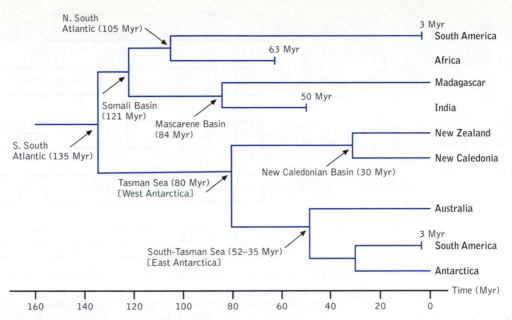

Fig. 5.6 Geological diagram to show the progressive fragmentation of Gondwana. Cross-bars on the lines indicate the times of collision of Africa and India with the southern margin of Eurasia, and of the establishment of the Panama Isthmus linking South America to North America. South America is shown twice, to show its times of separation from Africa and from Antarctica. After Sanmartin, Enghoff, & Ronquist [12], modified.

135 my. Split between Africa and Madagascar/India. *105 my*. This split continues clockwise round Africa to form the South Atlantic between South America and Africa.

110 my. Separation of Madagascar/India from the rest of East Gondwana.

84 my. Separation of Madagascar from India.

80 my. Separation of New Caledonia/New Zealand from Australia/Antarctica.

50–35 my. Separation of Australia from East Antarctica, at first (*50 my*) by a shallow sea and later (*35 my*) by deep ocean.

30 my. Separation of South America from West Antarctica.

30 my. Separation of New Caledonia from New Zealand.

epicontinental sea, whose precise limits varied according to sea levels. The area was therefore rather like the East Indies today—a great archipelago of islands of varying size, which formed only a partial barrier to the dispersal of animals and very little barrier to the dispersal of plants.

To begin with, in the Middle Jurassic, the whole of the Northern Hemisphere, from western North America through Eurasia to eastern Siberia, was made up of a single landmass, Laurasia (Fig. 5.7, A; Plate 3b), though the two ends of this chain were separated by the equivalent of the Bering Sea. By the start of the Late Cretaceous, 90–80 my ago, the shallow Mid-Continental Seaway in North America and Turgai Sea (also known as the Obik Sea) in Eurasia had subdivided both of these continents into western and eastern sections, but Siberia and Alaska had become connected to one another. The result was the establishment of two continents, Euramerica and Asiamerica (Fig. 5.7, B; Plate 3c). The Norwegian Sea had started to open and separate North America/Greenland from Europe, but two different routes interconnected the two continents until the Eocene (Fig. 11.19). The Thulean route was along a ridge of land that is now submerged, while the de Geer route lay further north (see p. 363).

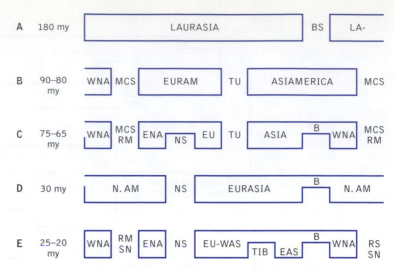

Fig. 5.7 The interrelationships between the land areas of the Northern Hemisphere over the last 180 my.

A. Early/Mid-Jurassic, 180 my ago (cf. Plate 3B). A single landmass, Laurasia, extends around nearly the whole of the Northern Hemisphere; the Bering Sea (BS) between Asia and Alaska is the only break in the near-circle of land.

B. Early Late Cretaceous, 90–80 my ago (cf. Plate 3C). The Mid-Continental Seaway (MCS) and the Turgai Sea (Tu) have divided this into two landmasses, Euramerica (EURAM) and Asiamerica.

C. End of the Late Cretaceous, 75–65 my ago (cf. Plate 3D). The Mid-Continental Seaway or the Rocky Mountains (RM) divide North America into two parts. A western part (WNA) is linked to Asia by the Bering region (B) where the former Bering Sea has dried up to to create a land bridge. The eastern part (ENA) is linked to Europe. The Turgai Sea is still in place, but the expanding Norwegian Sea (NS) now separates the southern parts of Europe from North America/Greenland.

D. Early Oligocene, 30 my ago (cf. Plate 3E). Now the continents are only divided by the completed Norwegian Sea, and their northern parts are linked by the Bering connection.

E. Late Miocene, 26–20 my ago. The Rocky Mountains and Sierra Nevada (RM, SN) subdivide North America, while the Tibetan Plateau (TIB) and adjacent Gobi Desert separate southern China and Southeast Asia from Europe.

Although the Mid-Continental Seaway became smaller and finally disappeared near the end of the Cretaceous, this was due to the beginning of the uplift of the Rocky Mountains, so one or the other formed a barrier between western and eastern North America for most of this time, but the Turgai Sea still separated Europe from Asia (Fig. 5.7, C; Plate 3d). The Norwegian Sea, expanding northward between Greenland and Scandinavia, separated southern Europe from North America. From then on, Siberia and Alaska were from time to time connected across the Bering region, but the effectiveness of this as a barrier depended on sea levels and (because of the high northern latitude of its position) on climate (cf. Chapter 11, p. 364).

By the Early Oligocene, 30 my ago, erosion had leveled the early Rocky Mountains to a plain, uniting the western and eastern parts of North America, the Turgai Sea had disappeared, and the Norwegian Sea had completely separated North America/Greenland from Europe. The continents of North America and Eurasia that we see today had therefore appeared for the first time (Fig. 5.7, D; Plate 3e). A new phase of mountain-building that began in the Late Oligocene, 25 my ago, and continued into the Miocene, gave rise to the Sierra Nevada Mountains and to a renewed Rocky Mountain chain, so that the biotas of the Pacific margin of North America were now largely separate from those of the eastern part of the continent (Fig. 5.7, E). In Asia, mountain-building caused the uplift of the Tibet Mountains that, together with the associated deserts to the north (such as the Gobi Desert) isolated China from the rest of southern Eurasia.

How Plate Tectonics Affects the Living World

Plate tectonics has overwhelmingly been the most important factor in causing major, long-term changes in the patterns of distribution of organisms. The movements of the plates are usually quite slow (only about 5–10 cm year[1], 2–4 in—about the same rate as the growth of our fingernails), so that any resulting changes must have been extremely gradual. The most obvious effect has been the direct one, the splitting and collision of landmasses altering the patterns of land within which new types of living organism could evolve and spread.

But the changing positions of the continents indirectly affected life in several other ways by changing the patterns of climate. The most obvious of these changes was the result of their movement across the latitudinal bands of climate, thereby causing different areas of land to lie in cold polar regions, in cool, damp temperate regions, in dry subtropical regions, or in the hot, wet equatorial regions. The distribution of land relative to the poles is also an important factor; it is noteworthy that during the two periods of time when there were great ice caps at the poles (first about 350–250 my ago, as well as the more recent period that caused the Ice Ages), the poles were surrounded by land, not water. Because of this, any ice and snow that fell on the land formed a white surface that reflected back the light and heat of the sun, so that the land became progressively cooler. If, instead, the pole is surrounded by sea-water, the snow often melts, leaving the sea's dark surface free to absorb the Sun's rays.

Another indirect effect of plate tectonics on living things arises from the fact that new mountains, oceans, or land barriers deflect the atmospheric and oceanic circulations, thus changing the climatic patterns on the landmasses. There are several ways in which this takes place, as we shall see.

Between the continents lie the deep oceans, with their spreading ridges (Fig. 5.3). The amount of activity and the length of these ridges have varied over time. The ridges form chains of huge, undersea volcanic mountains. The more active and extensive these are, the more volume they occupy within the oceans. During periods of increased tectonic activity, sea levels therefore rise, and comparatively shallow **epicontinental seas** cover the lower-lying regions of the continents. These areas, such as the North Sea or Hudson Bay today, are known as **continental shelves**. Even though they are much shallower than the oceans, these seas form just as effective a barrier to the spread of terrestrial organisms. As shown in Fig. 5.7, they were particularly extensive in the Mesozoic, when such seas covered much of North America and Eurasia. But these shallow seas not only affect life by forming barriers to distribution, they also affect the climate of the surrounding areas of land. The climate of any area largely depends on its distance from the sea, which is the ultimate source of rainfall. The central part of great supercontinental landmasses, such as Eurasia today or Gondwana in the past, is therefore inevitably dry, and the climate of such areas experiences great daily and seasonal changes of temperature. The breakup of a supercontinent, or the spread of shallow epicontinental seas into the interior of continents, would have brought moister, less extreme climates to these regions.

Mountain chains are another major influence on climate, and their appearance, location, and orientation are all the result of plate tectonics. Mountain chains appear when two continents collide, as when Asia collided with Euramerica in the Permo-Carboniferous to form the Ural Mountains. Another example is the collision between India and Asia in the Eocene, 55–45 my ago, leading to the uplift of the huge, high Tibetan Plateau, which covers over a million square kilometers and also, later, causing the rise of the Himalayan Mountains. As a result, air that had cooled during the Central Asian winter could no longer escape southward, and Central Asia also became isolated from any seas that could have been the source of rain-bearing winds, so that its climate became both cooler and drier. Later, toward the end of the Miocene, a large sea that had covered much of western Central Asia gradually shrank. This caused a further increase in summer temperatures there and increased seasonality in the Indian rainfall, so that savanna replaced the old tropical vegetation in parts of India.

Mountain chains also appear if a continent comes to lie next to an ocean trench, where the descent of lighter ocean-crust material below the edge of the continent causes the rise of volcanic mountains along its margin; this is the cause of the appearance of the Andes (Fig. 5.2). Finally, if a continent comes to lie across the position of a spreading ridge, the presence of this heated area below the continent causes the appearance of a mountain range. This is the reason for the appearance of the Rocky Mountains in North America, which started to rise in the latest Cretaceous to Middle Eocene; by 10–15 million years ago, they had reached half their present height. The uplift of the Sierra Nevada Range began in the latest Oligocene, but most of it took place over the last 10 million years, while the Cascade and Coast Ranges only rose over the last 6 million years. All these mountains will have increased the seasonality of the climate of North America. It has also been suggested that, by diverting the westerly winds into a more northerly track, they may have played a part in initiating the climatic cooling that eventually led to the Ice Ages. The effects of the rise of the Andes in South America during the Late Cenozoic were less severe because the continent as a whole is narrower and is mainly at lower latitude.

New mountain ranges inevitably affect climate patterns, especially if they arise across the paths of the prevailing moisture-bearing winds, since areas in the lee of the mountains then become desert. Such deserts can be seen today in the Andes, to the east of the mountain chain in southern Argentina, and to the west along the coast from northern Argentina to Peru—the prevailing winds in these two regions blow in opposite directions.

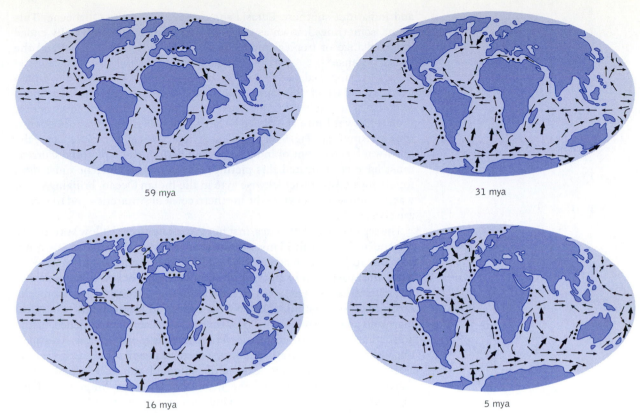

59 mya

31 mya

16 mya

5 mya

Fig. 5.8 Reconstruction of the distribution of the continental landmasses and the inferred pattern of circulation of the ocean currents over the last 59 million years. The large dots indicate regions of upwelling, and the large arrows indicate possible regions of bottom-water formation. (In many cases, the formation of bottom water is by the sinking of warm saline water.) Modified from Angel [13], with permission; after Haq [14].

The splitting and movement of the continents will also have affected their climates by altering the patterns of water circulation in the oceans (Fig. 5.8). When the continents were united in Pangaea and the Atlantic oceans had not begun to open, there was only a single huge ocean, sometimes called **Panthalassa** ("all oceans"), which was the ancestor of today's Pacific Ocean. This ocean must also have had a simple pattern of circulation of the waters as huge northern and southern **gyres**, and with a strong westward Equatorial Current directed into the embayment between what later became Laurasia and Gondwana (Figs. 10.1, 10.2, pp. 307, 309). (A gyyre is a large, horizontally rotating mass of water that fills almost all of an ocean basin, see p. 273.)

Laurasia and Gondwana gradually separated from one another during the Jurassic; geologists refer to the seaway between them as the **Tethys Ocean.** So, by the Early Cretaceous a warm, westwardly directed Equatorial Current would have become continuous around the whole world. It would have run along the southern margin of Eurasia and North America and then westward through the gap between North and South America. But as Africa started to move northward during the Cretaceous, the Tethys Ocean progressively became narrower and shallower, and finally closed when Africa and India met southern Eurasia some time during the Oligocene. This event, sometimes known as the Terminal Tethyan Event, finally ended any exchange of tropical marine faunas between the Atlantic and the Indian oceans.

As this happened, the waters of the Equatorial Current had to find a new path. Part of the current was probably diverted into the South Atlantic, where it started to drive a gyre system in that widening ocean, then northward into the North Atlantic, and finally westward between the two Americas (Fig. 5.8). In the Late Cretaceous and the Paleocene, the northward movement of India across the path of the Equatorial Current must have complicated this picture. However, the Current must have finally formed an anticlockwise gyre in the Indian Ocean, bringing warm waters southward, down to the northern coast of Antarctica and the western coast of Australia.

The separation of Antarctica, first from Australia (by a shallow sea c. 50 my ago and by deep ocean 35 my ago) and then from South America 30–28 my ago, was to have far-reaching effects. First, the northward movement of Australia progressively reduced the westward flow of warm Pacific water into the Indian Ocean and instead directed it anticlockwise along the north coast of Australia. This put an end to any movement of tropical marine faunas between the Atlantic and the Pacific, and added to the power of the South Atlantic gyre system. But the more important result of the separation of Australia and South America from Antarctica was the establishment of the cold, eastwardly directed Antarctic Circumpolar Current, which runs around the whole periphery of Antarctica. This, together with associated strong winds, separated the weather system of Antarctica from that of its warmer, more northern, neighbor and led to fundamental changes in the weather of the Southern Hemisphere. The major result was the progressive cooling of Antarctica: glaciers started to form there c. 45 my ago, had grown to at least 40% of their present size by 35 my ago, and covered nearly all of the continent by 29 my ago [5]. There was some warming in the Late Oligocene 25 my ago, so that less than 50% of Antarctica was ice-covered from 26 to 14 my ago. Later, the collision of northward-moving Australia/New Guinea with Southeast Asia 20 my ago formed a new blockage of the westward Pacific current, forcing some of its waters to flow southward along the eastern margin of Australia and then eastward along Antarctica. This reinforced the Antarctic Circumpolar Current and led to a Mid-Miocene expansion of the eastern Antarctic ice sheet.

The separation of Antarctica from Australia would also have led to a reduction in the amount of evaporation from the now cooler seas around Australia, reducing the rainfall on that continent and causing the arid climate that we see there today. The new continental pattern also added to the power of the gyres in the South Atlantic, including the cold Benguela Current up the western side of southern Africa and the warm clockwise Gulf Stream of the North Atlantic.

Through this period of time, the northward movement of Australia toward Southeast Asia will also have gradually weakened the power of the Pacific Equatorial Current as it was forced to pass through the narrowing pattern of interisland straits in the East Indies, which today is the only low-latitude connection between the world's oceans. In this area, known as the Indonesian Gateway or Indonesian Throughflow [6], the waters at more superficial levels are different from those that lie deeper. The surface water comes from the North Pacific, whose waters are cooler and less saline than those of the South Pacific. They provide most of the surface water just north of Australia, thereby reducing the amount of evaporation of water in the region and therefore also the rainfall in northern Australia. The deeper water in the Indonesian Gateway, on the other hand, is warmer and carries westward a substantial fraction of the heat absorbed by the equatorial Pacific. It is less well understood than the surface flow, but computer models suggest that variations in its power have major effects on the climates of both the Indian and the Pacific oceans. Such variations may well have been caused by the sea-level changes of the later Cenozoic, which united or subdivided the islands on the Sunda shelf (cf. Fig. 11.9).

The meeting between Africa and Eurasia during the Oligocene turned the eastern part of the Tethys Ocean into the ancestor of the Mediterranean Sea. Within this landlocked sea, the rate of water loss by evaporation is greater than its replacement by freshwater from the rivers, which led to an increase in its salinity. When its connection to the Atlantic was briefly lost at the end of the Miocene, this culminated in the drying up of part of the Mediterranean Sea (cf. Chapter 11 p. 342). The closure of the Tethys Seaway also reduced the strength of the Equatorial Current through the Panama gap between the Americas, allowing an increase in the eastward flow of water from the Pacific into the northern Caribbean. This led to a cooling of the water there and may have been the reason for the decline in its coral reefs in the Miocene.

The final major change in the pattern of intercontinental connections was the closure of the Panama Seaway between North and South America by the Panama Isthmus, which finally separated the Pacific Ocean from any communication with the Atlantic Ocean. This took place gradually from 3.5 to 3.1 my ago as both a narrowing and a shallowing of the seaway between the two continents, though there may have been a breakdown in the Isthmus 2.4–2.0 my ago. It is thought that the formation of the Isthmus led to the appearance of the larger ice sheets in West Antarctica and the Northern Hemisphere. Cold, deep Arctic water has been able to enter the North Atlantic since the opening of the Norwegian Sea between Greenland and Scandinavia in the Late Eocene. However, the lands surrounding the North Atlantic have been affected by the warm surface waters of the Gulf Stream, channeled northward by the east coast of North America. The Kuroshio Current in the western North Pacific produces a similar effect, warming Japan.

As explained earlier, these changes in the configuration of the world's continents and oceans have had profound effects on the amount of glaciation of the planet, which in turn affects its climate. Ice reflects the Sun's rays back into space, so the appearance of extensive ice caps in the Eocene is the likely cause of the Terminal Eocene cooling of the planet (Fig. 5.9a). Furthermore, the removal of a large volume of water from the sea causes its level to drop—in the Mid-Oligocene, 30 million years ago, sea levels dropped by at least 120 m (390 ft) (Fig. 5.9b), and the seas therefore

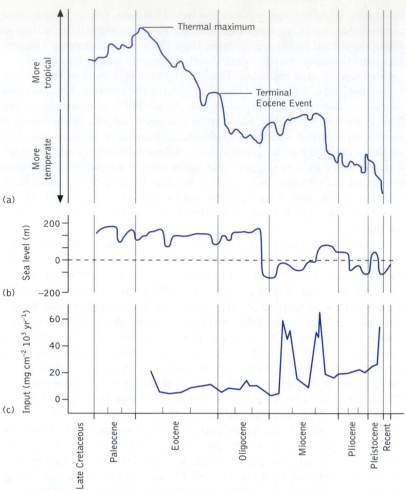

Fig. 5.9 Climatic record of the Late Cretaceous and Cenozoic.
(a) Temperature, as deduced from oxygen isotope analysis of benthic foraminifera from Atlantic deep-sea drilling sites. After Miller et al. [15].
(b) Generalized history of sea-level changes, relative to the present day. After Tallis [16].
(c) Amount of wind-borne dust in deep-sea sediments. After Tallis [16].

retreated from the continental lowlands. But land absorbs less heat than does water and releases it more rapidly. So the draining of seawater from the continents increases the intensity of their seasonal cycles, and their climates become less equable, with hotter, drier summers and colder, wetter winters. Because the oceans themselves were also becoming cooler at this time, less water evaporated to provide rainfall on the land. All these changes caused the continents to become drier, especially in midlatitudes; this is shown by an increase in the amount of wind-blown dust deposited in deep-sea sediments (Fig. 5.9c).

As already shown, the movements of the tectonic plates produced a variety of changes in the world's environmental patterns, in a variety of interacting ways. We can now recognize these factors and can produce complex computer-generated models of the effects of the changes. However, it is still not possible to construct models that accurately match the high temperatures of the Cretaceous and Early Cenozoic, especially in the continental interiors. This at present can only be explained by making additional ad hoc assumptions, for example, of higher solar output of radiation or higher levels of atmospheric carbon dioxide in these earlier periods (some estimates place this at 10 times above current levels!), although little evidence of this has yet been found.

As explained in Chapters 10 and 11, all the processes that have been noted above can have extensive effects on biogeography. As a result of the environmental changes that they caused, organisms had to change their area of distribution or evolve new adaptations. New ecological barriers delimited new areas in which new endemic organisms might appear, or linked them so that the organisms could spread into areas that were previously not available. It should also not be forgotten that the appearance of a new link between two land areas (such as the Panama Isthmus between the Americas) at the same time produces a new barrier between the marine faunas on either side.

The complex interplay between plate tectonics and biogeography in the area between Southeast Asia and Australia, known as Wallacea, is described in Box 5.1

Plate Tectonics and Wallacea

Box 5.1 — Guest Author

by Professor Robert Hall, Earth Sciences Department, Royal Holloway College, University of London

In the nongeological literature, one still occasionally reads interpretations of Earth history that dismiss plate tectonics, relying on supposed land bridges or an expanding planet to account for biogeographic patterns. These suggestions give the impression that plate tectonics is little more than a hypothesis that fits a few facts but overlooks others. Nothing could be further from the truth. Today we have many more techniques of investigation into the structure of the Earth than were available to those who first suggested the theory of plate tectonics. For example, the present spreading centers, where new crust is forming, can now be mapped in great detail throughout the world's oceans, using satellite gravity measurements, as can the transform faults that segment the ridge crests. This helps us to trace the relative movements of plates and so provide increasingly complete models of past plate motions [7]. Above subducting plates there is an almost continuous chain of volcanoes resulting from melting as fluids rise in the mantle. The movements of plates relative to one another also produce earthquakes, and the resulting **seismic** or shock waves can be used to map the sinking plates. Small-velocity variations in the mantle detected with these waves, using methods known as *seismic tomography*, identify slabs carried deep into the lower mantle,

which provide evidence of former subduction episodes. There remain disagreements, but these are not about plate tectonics in general, but merely about details of plate movements, the nature of plate boundaries and similar limited issues.

Different plate tectonic models reflect problems such as difficulties in accurately dating rocks, the incomplete preservation of the rock record due to erosion, deformation and subduction, and the uneven knowledge of global geology. For example, North America and Europe have been studied for longer and by more people than Southeast Asia, Australia, and Antarctica. The oldest ocean crust on the Earth is about 160 million years old. The ocean-floor record of magnetic stripes and transform faults which provides the basis for reconstructing plate movements is therefore missing for the major part of the Earth's history. Models of the arrangements and movements of the plates in the early Earth therefore rely more on interpretations of an incomplete rock record and are consequently less detailed. Plate movements have caused the assembly and breakup of continents several times in the last 4,000 million years. Wegener's interpretation of Pangaea, Laurasia, and Gondwana has stood the test of time, although reconstructions of earlier supercontinents are more controversial.

Early plate tectonic thinking emphasized the rigidity of plates and the narrow zones of deformation between them.

This concept works well for the oceans, but the continents are generally weaker and have a more complex structure than oceans. We now know that continents can deform over large areas, that they have grown by addition of material added during subduction (see the later section in this chapter on Terranes), and that continental crust may be subducted deep into the mantle—although it is buoyant and tends to return to the surface. Some rocks in mountain belts, containing diamonds and other minerals that form only under the high pressures of the Earth's interior, record deep subduction followed by subsequent reappearance at the surface during collision between tectonic plates. Furthermore, movements of subduction zones may lead to extension, and even creation of new oceans, within convergent settings. Subduction may not cease when continents collide, as suggested in early plate tectonic models, and mountain belts are much more complex than originally thought. As in other sciences, a simple paradigm has become more complex with time, as our knowledge has increased. Nonetheless, the plate tectonic framework is at the heart of geological thinking, and some of these ideas are well illustrated by the history of East and Southeast Asia [8].

One of the first people to recognize the fundamental importance of geology in influencing biogeographic patterns was Alfred Russel Wallace, who interpreted the great differences in the distribution of animals across the Malay Archipelago as reflecting differences in geological history. According to Wallace, the islands of Sumatra, Java, and Borneo in the west once formed part of a single continent, which has since been divided into separate islands by changes in sea level, while to the east there was a wide ocean that included remnants of a former Australian and Pacific continent. Wallace, like most other scientists of the 19th and early 20th century, thought in terms of fixed continents, but now all Earth scientists recognize that the surface geography of the planet has changed with time, due to plate movements. No area better illustrates the complex way in which geology has influenced biogeography, and the importance of plate tectonics, than Wallace's Malay Archipelago, which broadly corresponds to today's Indonesia.

Asia has grown by closure of oceans between the continents of Gondwana and the Asian part of Laurasia. Continental blocks were rifted from the northern Gondwana margins, so that oceans south of the rifted blocks widened at spreading centers. To compensate for this action, other oceans further north narrowed by subduction beneath Asia and finally closed, so that the continental blocks they carried collided with the southern margin of Asia. Asia is therefore a mosaic of continental fragments, separated by sutures containing the remnants of the oceans and the active volcanic margins above the subducting plates. This general process was repeated several times during the last 400 million years and will no doubt continue, resulting in the separation of further fragments from Gondwana and their incorporation into a growing Asian continent, which may be the core of a supercontinent of the future. The most recent stages in this process have been the collision of India with Asia and the collision of Australia with Southeast Asia. Both of these major events include numerous smaller events such as the formation of volcanic arcs, accretion of buoyant features on oceanic plates, island arc-continent collisions, and fragmentation of larger blocks by faulting at active margins.

In Asia, from the Alpine-Himalayan mountain chain to Southeast Asia, the mantle structure recorded by seismic tomography reveals this history of subduction and provides a test of tectonic reconstructions. Linear high-velocity anomalies in the lower mantle beneath India and north of India are the result of the closure of different oceans between India and Asia during the Mesozoic and Cenozoic. A broad high-velocity anomaly beneath Indonesia is evidence of a different subduction history north of Australia. Here, subduction ceased in the Mid-Cretaceous after the collision of continental fragments in Sumatra and Java, and resumed about 45 million years ago from the north, beneath Borneo, and from the south, beneath Sumatra, Java, and the Sunda Arc.

As Wallace suggested, geology provides the basis for understanding the distributions of faunas and floras in Southeast Asia, but only via a complex interplay of plate tectonic movements, palaeogeography, ocean circulation, and climate. Plate movements and collisions were intimately linked to changing topography, ocean depths, and land/sea distributions, which in turn influenced oceanic circulation and climate. The convergence of Australia and Southeast Asia has almost closed a former deep, wide ocean. The remaining Indonesian Gateway between the Pacific and Indian oceans is the only low-latitude oceanic passage on Earth, has an important influence on local and probably global climate, and is likely to have been just as significant in the past. As one ocean closed, smaller deep marine basins opened, volcanic arcs formed, and mountains rose. The distribution of land and sea was altered, and changes in sea level further contributed to a complex palaeogeography. Understanding first the geology and then the palaeogeography, and their oceanic and climatic consequences, are vital steps on the way to interpreting the present distributions of plants and animals. But the plate tectonic engine is the driving force for this whole sequence of changes.

Islands and Plate Tectonics

The nature of islands and the biotas that they contain are also affected by the way in which the direct or indirect effects of plate tectonics created them. These islands may be of three different types (Fig. 5.10). As explained in Chapter 8, p. 224, as a result of their different histories, different types of island may contain biotas with different characteristics.

The first type of island was originally a part of a nearby continent, but became separated from it by rising sea levels (e.g., Britain, Newfoundland, Sri Lanka, Sumatra, Java, Borneo, New Guinea, Tasmania), or by tectonic processes that split them away from an adjacent continent (e.g., the larger islands of the Mediterranean, Madagascar, New Zealand, and New Caledonia).

A second type of island is part of a volcanic **island-arc**. The most obvious of these are the Kurile and Aleutian island-arcs that lie along the edge of the Pacific. Here, old ocean crust is being forced into the depths of the crust, the resulting stresses causing the appearance of volcanic islands. The Lesser Sunda Islands of the East Indies have similarly formed where the northward-moving Australian plate is undercutting Southeast Asia.

Other islands form as a result of the activity of what geologists call **hotspots** (Fig. 5.11). These are scattered, but fixed, locations over 700 km (435 mi) deep within the Earth, from which plumes of hot material rise to form volcanoes at the Earth's surface. Where this volcano lies within an ocean rather than within a continent, it may either remain submerged as a **seamount** or **guyot**, or grow to rise above the surface as a volcanic island. However, because the seafloor is part of a moving tectonic plate, the island is gradually carried away from the plume of hot material. Volcanic activity in the island then ceases, and the surrounding seafloor cools and contracts, while the island itself is subject to erosion. As a result, over thousands of years the island gradually disappears below the surface of the ocean. Meanwhile, a new volcano is developing in that part of the seafloor that now lies above the hotspot. Over millions of years, repetition of this process causes the appearance of a chain of volcanoes (extinct, except for the youngest), the orientation of which clearly shows the direction of movement of the underlying seafloor. There are several such chains of islands in the Pacific Ocean, as shown in Fig. 5.10—the bend partway along each of these chains was caused by a change in the direction

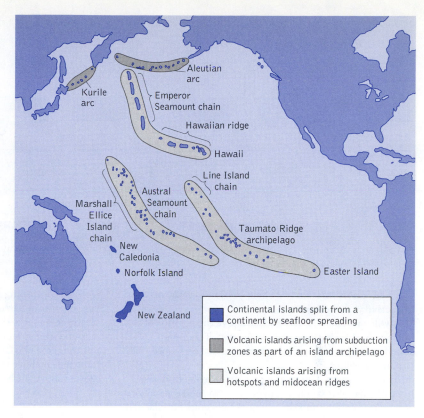

Fig. 5.10 Map of the Pacific Ocean, showing examples of the three types of island. After Mielke [17].

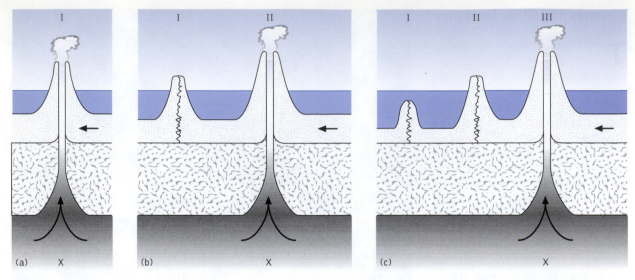

Fig. 5.11 The formation of a chain of islands as the result of the activity of a hotspot.

(a) A volcanic island, I, has formed above the position of a hotspot, X, deep within the Earth's crust. The more superficial layer of the crust is being carried westward, due to plate tectonics.

(b) This movement carries the first island away from the hotspot, so that it is starting to erode away, and the seafloor around it is cooling and shrinking. A new volcanic island, II, now forms above the hotspot.

(c) The continuing movement of the superficial layer of the crust has now carried the second island away from the hotspot, and it is therefore starting to erode. The seafloor around islands I and II is cooling and shrinking, so that the further eroded island I is now merely a seamount that does not appear above the surface. Yet another island, III, now forms above the hotspot.

of motion of the Pacific plate about 43 my ago. The longest is that caused by the activity of the hotspot that today lies under the Hawaiian Islands (see Fig. 8.2).

Terranes

Where volcanic islands, or their eroded, submerged remains reach the edge of a trench within the ocean, they are merely recycled back into the interior of the Earth, leaving no trace—the oldest member of the Emperor seamount chain (Fig. 8.2) is about to disappear into the great trench that lies just east of the Kamchatka Peninsula of Asia. But where the trench lies adjacent to a continent, though the ocean crust will simply be subducted, any superficial islands, seamounts, reefs, or other masses of volcanic material are scraped off against the edge of the continent. There they form individual patches known as **terranes**—small regions within which the rocks are quite different from those of the surrounding area. These are best known from the eastern margin of the Pacific, where a complex of 42 terranes makes up a strip 80–450 km (50–280 mi) wide along the western margin of the United States and Canada; another 48 terranes make up much of Alaska (Fig. 5.12). Others lie along the western edge of South America—especially at the northern end, where their arrival from the Early Cretaceous onward caused the rise of the eastern, central, and western ranges of the northern Andes.

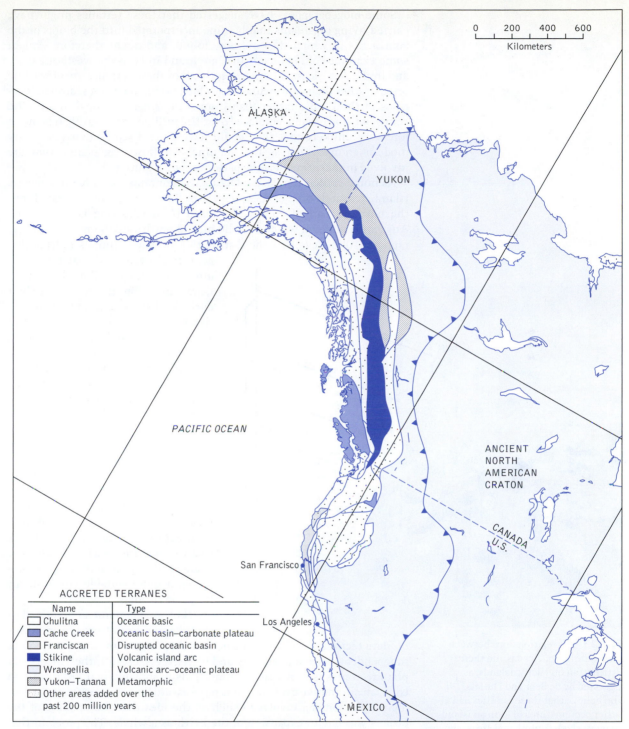

Fig. 5.12 Western North America, to show the terranes that lie along its western edge. Over the past 200 my, islands, seamounts, and other superficial features on the seafloor of the Pacific Ocean have been scraped off and added to the western edge of the North American plate as the seafloor itself slipped beneath it. After Jones et al. [18].

Map labels: ALASKA, YUKON, PACIFIC OCEAN, ANCIENT NORTH AMERICAN CRATON, CANADA, U.S., San Francisco, Los Angeles, MEXICO

Scale: 0 200 400 600 Kilometers

ACCRETED TERRANES

Name	Type
Chulitna	Oceanic basic
Cache Creek	Oceanic basin–carbonate plateau
Franciscan	Disrupted oceanic basin
Stikine	Volcanic island arc
Wrangellia	Volcanic arc–oceanic plateau
Yukon–Tanana	Metamorphic
Other areas added over the past 200 million years	

Some biogeographers have suggested that these terranes might have carried living organisms that became incorporated into the biotas of the continents in which they are now found, and might therefore explain some examples of organisms that are found in both the western Pacific and in the Americas. However, nearly all of these terranes are of oceanic origin, rather than continental fragments, originated from scattered locations in the Pacific rather than from some single notional mid-Pacific continent, and became emplaced over 100 million years ago [9]. It is therefore very unlikely that they could have carried representatives of any modern groups, such as the mammals or flowering plants, that are involved in these trans-Pacific patterns of distribution.

As noted earlier, the Pacific Ocean also contains a number of volcanic island arcs, and some of these appear to have become incorporated into the northern part of New Guinea and its neighboring islands as the Australian plate on which they lie moved northward. Some panbiogeographers [10] believe that the distribution of some animals reflects this geological event and that these had already become distributed along the island arc before the collision. Their current distribution, according to this theory, is therefore the result of such a "Noah's Ark" event, in which living organisms were carried as passengers on moving islands (Fig. 5.13). However, these volcanic islands were mostly small and short-lived, and it seems very unlikely that such animals as cicadas and birds of paradise, which are very poor at dispersing, would have been able to disperse along such an oceanic chain. A more straightforward explanation is that their distribution reflects their ecological preferences, the soils and environment in the areas of New Guinea that were contributed by these former islands differing from the rest and providing a more suitable environment for these animals.

All of the phenomena described in this chapter result from the processes of plate tectonics, and the evidence for them is drawn from many quite independent aspects of geology and biology. As a result, the "theory" of plate tectonics is immensely strong. This is shown when, from time to time, an alternative explanation is put forward for some aspect of the geological phenomena involved—such as the idea that the opening of the oceans was due to expansion of the Earth as a whole. The problem that arises for such theories is that they do not provide satisfactory solutions for other phenomena that are easily explained by the theory of plate tectonics. In this case, it is difficult to identify the origin of the vast quantity of water needed to fill these enlarging oceans, and the suggested changing

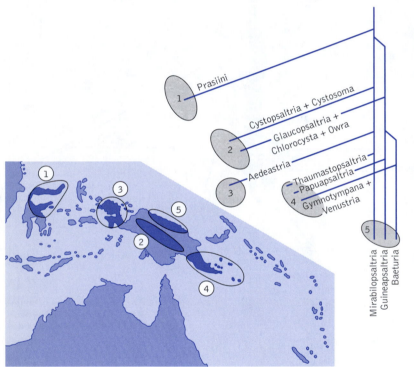

Fig. 5.13 The relationships between some families of cicadas and their distribution on Pacific islands, as suggested by de Boer and Duffels [19], who believe that these families had earlier occupied areas 1–5 in an island-arc (gray-shaded numbered areas, above right), which then collided with parts of New Guinea and neighboring islands to form areas 1–5 (white circles) where these families are found today.

diameter of the Earth is incompatible with the coherent pattern of palaeo-magnetic results [9]. Such ideas may also raise problems with other aspects of our planet's history—for example, an expanding Earth would have caused a rapid reduction in its rate of rotation.

The theory of plate tectonics is based on a great variety of independent lines of evidence. Such a theory is known as a paradigm, and the theory of plate tectonics is the central paradigm of the earth sciences. The theory of evolution by natural selection, which is considered in the next chapter, is similarly based on a great variety of lines of evidence and is the central paradigm of the biological sciences.

Summary

1 The fact that continents had moved across the face of the Earth was first proved in the 1960s by data from the magnetized particles preserved in their rocks. This was soon followed by data from the ocean beds, which showed that the Earth's surface is covered by a pattern of moving "plates," bounded by spreading ridges where new material appears from the Earth's interior and by deep trenches where old material returns to its depths.

2 As a result of this process of plate tectonics, the pattern of the continents has changed greatly over the last 350 my. At first there were three northern continents, plus a huge southern continent that has been called Gondwana. Later all these landmasses joined to form a single supercontinent, Pangaea, which later broke up into the continents that we see today. Palaeontologists have been able to show how these geographical changes affected the evolving faunas and floras that inhabited them.

3 The most direct effect of the movements of the tectonic plates on the climates of the different continents was caused by their movements across the latitudinal zones of climate. But the movements also affected the climate in other ways, for example, by changing sea levels, by the appearance of mountain chains, and by changing the patterns of ocean currents, and also resulted in the Ice Ages. All of these phenomena affect the biota of the continents and islands.

4 Plate tectonics is also the cause of the appearance of volcanic islands, either as chains above geological hotspots deep in the Earth or as island arcs where ocean crust is disappearing. These islands provide interesting biogeographic studies and are home to many endemic species.

5 Small volcanic islands, reefs, and other volcanic masses may also become attached to the edges of continental masses, where they are known as terranes, but there is no evidence that they contributed any living organisms to the continents into which they eventually became absorbed.

Further Reading

Edwards J. *Plate Tectonics and Continental Drift.* North Mankato, MN: Smart Apple Media, 2005.

References

1 Runcorn SK. Paleomagnetic comparisons between Europe and North America. *Proc Geol Assoc Canada* 1956; 8: 77–85.

2 Hess HH. History of ocean basins. In: Engel AE, James HL, Leonard BF, eds. *Petrologic Studies: A Volume in Honour of A.F. Buddington*, pp. 599–620. Denver, CO: Geological Society of America, 1962.

3 Vine FD, Matthews DH. Magnetic anomalies over a young ocean ridge. *Nature* 1963; 199: 947–949.

4 Cox CB. Vertebrate palaeodistributional patterns and continental drift. *J Biogeogr* 1974; 1: 75–94.

5 Crowley, TJ, North GR. Palaeoclimatology. *Oxford Monogr. Geol. Geophys* 1991; 8: 1–330.

6 Schneider N. The Indonesian Throughflow and the global climate system. *J Climate* 1998; 11: 676–689.

7 Hall R. Cenozoic geological and plate tectonic evolution of SE Asia and the SW Pacific: computer-based reconstructions and animations. *J Asian Earth Sciences.* 2002; 20: 353–434.

8 Hall R. SE Asia's changing palaeogeography. *Blumea* 2009; 54: 148–161.

9 Cox CB. New geological theories and old biogeographical problems. *J Biogeogr* 1990; 17: 117–130.

10 Heads M. Regional patterns of diversity in New Guinea animals. *J Biogeogr* 2002; 29: 285–294.

11 Stanley S. *Earth and Life through Time*. New York; W. H. Freeman, 1986.

12 Sanmartin I, Ronquist F. Southern Hemisphere biogeography inferred by event-based models: plant versus animal patterns. *Systematic Biology* 2004; 53: 216–243.

13 Angel MV. Spatial distribution of marine organisms: patterns and processes. In: Edwards PJR, May NR, Webb NR, eds. *Large Scale Ecology and Conservation Biology*, pp. 59–109. British Ecological Society Symposium no. 35. Oxford: Blackwell Science, 1994.

14 Haq BU. Paleoceanography: a synoptic overview of 200 million years of ocean history. In: Haq BU, Milliman HD, eds. *Marine Geology and Oceanography of Arabian Sea and Coastal Pakistan*, pp. 201–231. New York: Van Nostrand Reinhold, 1984.

15 Miller KG, Fairbanks RG, Mountain GS. Tertiary oxygen isotope synthesis, sea level history, and continental margin erosion. *Paleoceanography* 1987; 2; 1–19.

16 Tallis JH. *Plant Community History: Long-Term Changes in Plant Distribution and Diversity*. London: Chapman & Hall, 1991.

17 Mielke HW. *Patterns of Life. Biogeography in a Changing World*. Boston: Unwin Hyman, 1989.

18 Jones DL, Cox A, Coney P, Beck M. The growth of western North America. *Scientific American* 1982; 247 (5): 70–84.

19 de Boer AJ, Duffels JP. Historical biogeography of the cicadas of Wallacea, New Guinea and the West Pacific. *Palaeogeogr Palaeoclimatol Palaeoecol* 1996; 124: 153–177.

The Engines of the Planet II: Evolution, the Source of Novelty

This chapter explains how evolution by natural selection works and presents the evidence for it. The genetic mechanism that leads to natural variations in the characteristics of organisms is explained. Also discussed are the definition of the species and the way in which new species appear, as is the rôle of isolation in enabling this process of speciation. Studies of "Darwin's finches" in the Galápagos Islands have shown the effectiveness of natural selection. The technique known as cladistics provides a reliable method of discovering the patterns of evolution and relationship between different species.

Each of the myriad organisms in the living world belongs to a particular species, and each species is adapted to a particular way of life and lives in a particular environment. Every aspect of this environment makes its demands on the structure or the physiology of the organism: the average state of the physical conditions, together with their daily and annual ranges of variation; the changing patterns of supply and abundance of food; and the occasional increased losses due to disease, predators, or increased competition from other organisms. Every species must be adapted to all this, so that it can tolerate and survive the hostile aspects of its environment and yet take advantage of the opportunities it offers.

In the early 19th century, it was accepted that each species had always existed precisely as we now see it. God was thought to have created each one, with all its detailed adaptations, and these had remained unchanged. Fossils were considered to be merely the remains of other types of animal or plant, each equally unchanging during its span of existence, that God had destroyed in a catastrophe (or a number of catastrophes) such as the Biblical Flood. But in his journey around the world in the ship HMS *Beagle* from 1831 to 1836, Darwin made two observations that eventually led him to reconsider these assumptions. The first resulted from his 5 weeks' stay in the Galápagos Islands in the eastern Pacific, isolated from South America by 960 km (600 mi) of sea. There, it seemed as though the mocking birds and tortoises were variable, rather than immutable, and this was also suggested by the fact that the fossil mammals that he had collected in South America were similar to the living forms but much larger. However, both of these observations contradicted the accepted view that every species was a special creation and had no blood relationship with any other species.

The Reality of Evolution

The realization that evolution is a process that underlies most of the phenomena of the biological sciences is no more a theory than is the proposal that the force of gravity is responsible for the patterns of interactions between the planets and the sun, and for many of the phenomena that we experience on our own planet. The reality of evolution is proved by the great range of different areas of scientific inquiry in which its action is the only scientifically plausible cause of the patterns that we see.

The patterns of biogeography are strong evidence for evolution, for they show that each type of organism that appears is restricted to the part of the world in which it appeared. For example, the great variety of Australian marsupials is restricted to that continent, in which they radiated, and the unique and strange mammals that evolved in South America remained confined to that continent until the Panama land bridge formed (see p. 359). This is also true of the relationship between the patterns of distribution of extinct animals and the geography of the times in which they lived, such as the way in which the distribution of dinosaurs conforms to the patterns of landmasses during the Cretaceous Period.

The sequence in which different organisms appear in the geological record is also strong evidence for evolution. If each new type evolved from other, preexisting organisms, we should see them appearing in turn, the simpler before the more advanced. The palaeontological record provides varied and detailed evidence of this, for fish first appeared over 450 million years ago, followed by amphibians over 360 million years ago, the first reptiles about 40 million years later, and the first mammals about 200 million years ago. The great radiation of the mammals did not start until 65 million years ago, and the first primitive types of human being only appeared about 6 million years ago.

If this sequence in time is because each new lineage has evolved from another, we would expect that the gradual diversification of successful new lineages should be reflected in a hierarchy of characteristics. This in turn should make it possible to construct a hierarchy of groupings into which the organisms can be placed. We see this in our classification of animals and plants. For example, all mammals have hair and produce milk to feed their young. There are many types of mammal but one group, known as the perissodactyls, has limbs in which the main axis runs through the third digit; they include the tapirs, with their short, fleshy trunk, the horned rhinoceros, and the equids (horses and zebras). So the equids have all the preceding characters, but all also have their own specializations, including high-crowned teeth and limbs ending in a single hoofed digit. They have an excellent fossil record, in which we can trace the gradual appearance of these characteristics (see p. 335), which evolved in response to the spread of grasslands.

Similarly, if these similarities in adult structure are the result of evolutionary relationship, we would expect to be able to find similar relationships in the embryology of the groups. Again, that is precisely what we find. For example, in the early stages of their development, mammals still show traces of the gill clefts of fish. They also show relics of the system of membranes that reptiles employ, while they are in the egg, in order to use their yolky food supply and to obtain oxygen. Similarly, the early larval stages of many marine organisms betray evidence of their relationships. These are sometimes surprising—for example, the larvae of barnacles show clearly that they are crustaceans, related to crabs, rather than being limpet-like molluscs that their adults closely resemble.

So, every one of the above areas of biological knowledge is evidence for the action of evolution. It follows that any alternative that purported to be an alternative explanation for any one of these would also have to explain all the others.

The Mechanism of Evolution—Natural Selection

The explanation that Darwin eventually deduced and published in 1858 in his great book *On the Origin of Species* is now an almost universally accepted part of the basic philosophy of biological science. Darwin realized that any pair of animals or plants produces far more offspring than would be needed simply to replace that pair. For example, many fish produce millions of eggs each year, and many plants produce millions of seeds. Nevertheless, the number of individuals normally remains almost unchanged from generation to generation. It follows that they must be competing with one another in order to survive. They must survive disease, parasites, predators, shortages of food and nutriments, or hostile weather conditions, simply in order to reach the age of reproduction. Then they must succeed in the competition to attract a mate, or the complete uncertainty as to which pollen grain will pollinate and fertilize the plant ovum, so that they may reproduce in their turn. To do so, they must employ the multitude of characteristics that they possess and have inherited from their parents. But the offspring of any single pair of parents are not identical to one another. Instead, they vary slightly in their inherited characteristics. Inevitably, some of these variations will prove to be better suited than others to the hazards of life that the individual experiences. The offspring that have these favorable characteristics will then have a natural advantage in the competition of life, and will tend to survive at the expense of their less fortunate relatives. By their survival, and eventual mating, this will lead to the persistence of the favorable characteristics into the next generation, while the less advantageous characteristics will gradually disappear. This process of differential survival is known as **natural selection**.

Evolution is therefore possible because of competition between individuals that differ slightly from one another. But why should these differences exist, and why should each species not be able to evolve a single, perfect answer to the demands that the environment makes on it? All the flowers of a particular species of plant would then, for example, be of exactly the same color, and every sparrow would have a beak of precisely the same size and shape. Such a simple solution is not possible because the demands of the environment are neither stable nor uniform. Conditions vary from place to place, from day to day, from season to season. No single type can be the best possible adaptation to all these varying conditions. Instead, one particular size of beak might be the best for the winter diet of a sparrow, while another, slightly different size might be better adapted to its summer food. Since, during the lifetimes of two sparrows differing in this way, each type of beak is slightly better adapted at one time and slightly worse adapted at another, natural selection will not favor one at the expense of the other. Both types will therefore continue to exist in the population as a whole, as in the case of the oystercatchers described on p. 79.

Because we do not normally examine sparrows very closely, we are not aware of the many ways in which the individual birds may differ from one another. In reality, of course, they vary in as many ways as do different individual human beings. In our own species we are accustomed to the multitude of trivial variations that make each individual recognizably unique: the precise shape and size of the nose, ears, eyes, chin, mouth and teeth, the color of the eyes and hair, the type of complexion, the texture and waviness of the hair, the height and build, and the pitch of voice. We know of other, less obvious characteristics in which individuals also differ, such as their fingerprints, their degree of resistance to different diseases, and their blood group. All of these variations are, then, the material on which natural selection can act. In each generation, those individuals with the greatest number of advantageous characteristics will be more likely to survive and breed, at the expense of those with fewer.

The Genetic System

The mechanism that controls the characteristics of each organism and their transmission to the next generation lies within the cell, in a rather opaque object called the nucleus. Inside this lie a number of thread-like bodies called **chromosomes**, made of a complex molecule called deoxyribonucleic acid, or **DNA** (see Box 6.1). Each characteristic of the organism is the result of the activity of a particular part of the DNA, which is referred to as a **gene**. It is the biochemical activity of these genes that is responsible for the characteristics of every cell of an individual, and thus for the characteristics of the organism as a whole. There might, then, be a particular gene that determined the color of an individual's hair, while another might be responsible for the texture of the hair and another for its waviness. (There are just over 20,000 genes in human beings.) Each gene exists in a number of slightly different versions, or **alleles**. Taking the gene responsible for hair color as an example, one allele might cause the hair to be brown while another might cause it to be red. Many different alleles of each gene may exist, and this is the main reason for much of the variation in structure that Darwin noted. The total of all the genes, which makes up the total genetic inheritance of an organism, is known as its **genotype**. The activity of the genotype produces the characteristics of the individual (its morphology, physiology, behavior etc.); this is known as the **phenotype**. But in some cases the environment can modify the phenotypic expression of the genes. For example, identical twins develop from the splitting of a single developing egg and therefore have exactly the same genotype. But they nevertheless may come to differ from each other if they are brought up experiencing different amounts of sunlight or food. This slight plasticity of the genotype is valuable from an evolutionary point of view, for it makes it possible for a single genotype to survive in slightly different habitats.

An individual, of course, inherits characteristics from both its parents. This is because each cell carries not one set of these gene-bearing chromosomes,

(a)

Plate 1 (a) Desert scrub vegetation in eastern Iran. The most conspicuous feature is the deciduous shrub, *Zygophyllum eurypterum*, together with smaller plants of *Artemisia herba-alba*. This vegetation is typical of the drier regions of alluvial flats.

(b)

Plate 1 (b) Desert scrub vegetation in eastern Iran in which the main shrub is *Haloxylon persicum*. This vegetation occupies the moister and more saline regions of the alluvial flats.

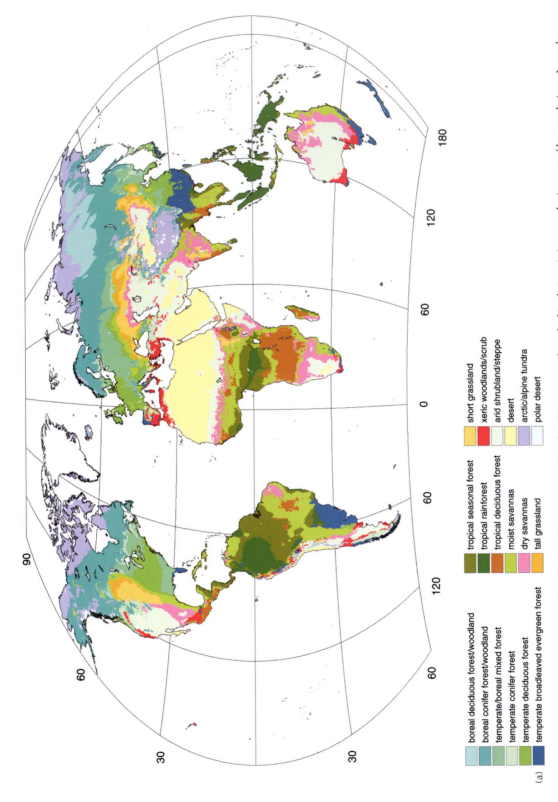

Plate 2 (a) Global map of the present distribution of vegetation on Earth. This is somewhat idealized since it ignores the impact of human beings. It can be regarded as the potential vegetation if climate alone were the operative determinant.

Legend:

- boreal deciduous forest/woodland
- boreal conifer forest/woodland
- temperate/boreal mixed forest
- temperate conifer forest
- temperate deciduous forest
- temperate broadleaved evergreen forest
- tropical seasonal forest
- tropical rainforest
- tropical deciduous forest
- moist savannas
- dry savannas
- tall grassland
- short grassland
- xeric woodlands/scrub
- arid shrubland/steppe
- desert
- arctic/alpine tundra
- polar desert

(a)

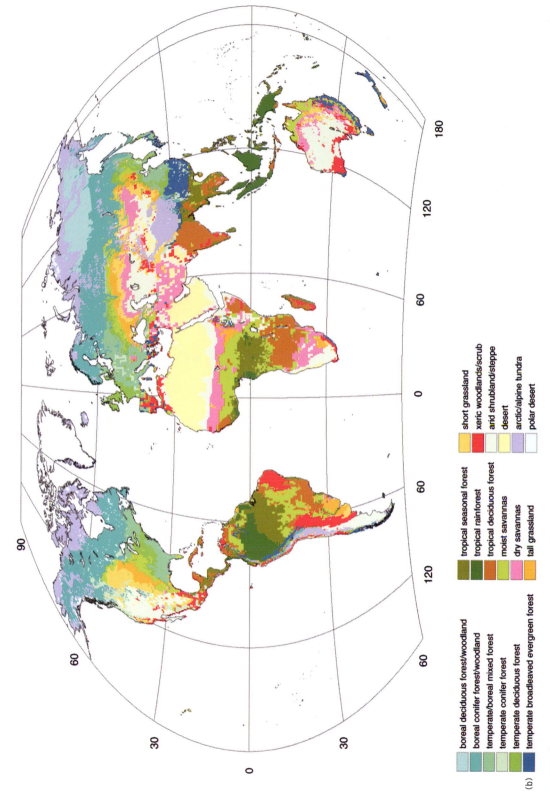

Plate 2 (b) Global map of vegetation derived from the predictions of a computer model, BIOME 3. As in the case of (a), this is a potential vegetation map that is based solely on climatic conditions. It can be seen that the two maps are very similar, which means that this model is a reliable predictor of vegetation under given climatic conditions. This close correspondence suggests that the model could be used to predict reliably the consequences of future climatic change for vegetation distribution patterns. Reproduced from Haxeltine & Prentice (Chapter 3, [30]), with permission.

boreal deciduous forest/woodland
boreal conifer forest/woodland
temperate/boreal mixed forest
temperate conifer forest
temperate deciduous forest
temperate broadleaved evergreen forest

(b)

tropical seasonal forest
tropical rainforest
tropical deciduous forest
moist savannas
dry savannas
tall grassland

short grassland
xeric woodlands/scrub
arid shrubland/steppe
desert
arctic/alpine tundra
polar desert

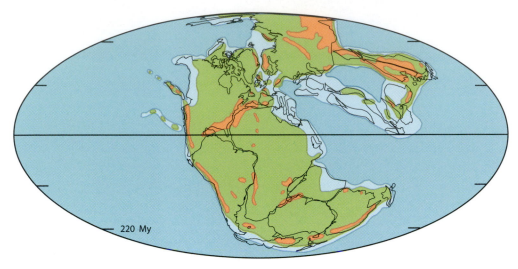

Plate 3A Middle/Late Triassic

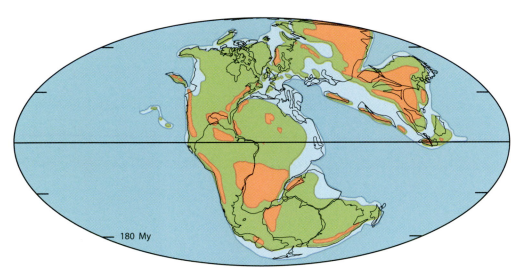

Plate 3B Early/Middle Jurassic

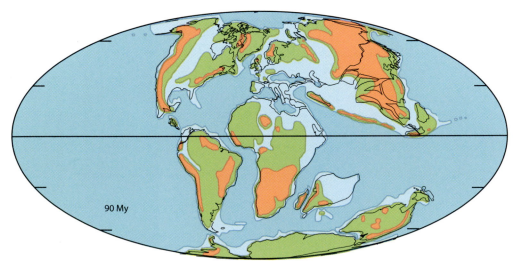

Plate 3C Early Late Cretaceous

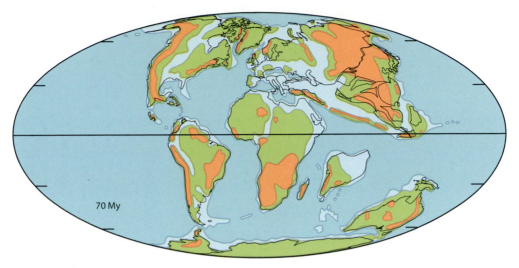

Plate 3D End Cretaceous

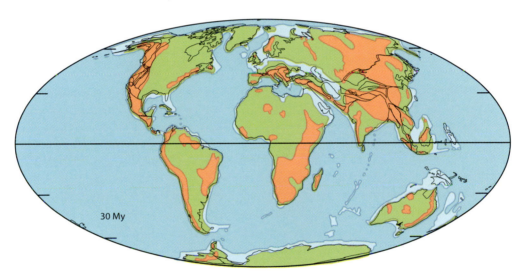

Plate 3E Early Oligocene

Plate 3 Palaeocontinental reconstructions at five periods of time in the past. Tan areas indicate highlands (>1500 m), green areas indicate lower parts of the continents. Light blue areas indicate shallow seas (<200 m), darker blue areas indicate ocean basins. (Copyright ©2004 by C.R. Scotese, with permission.)

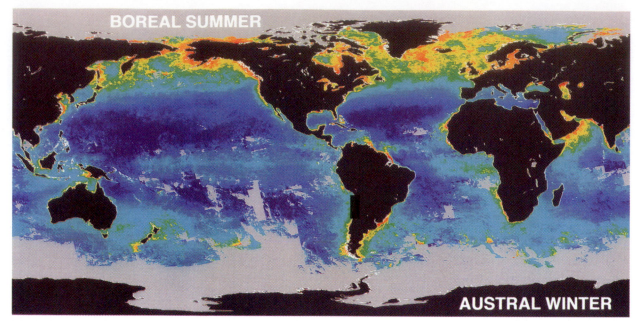

(a)

Plate 4 Density of plankton, as measured by chlorophyll concentrations from the sea surface to a depth of 25 m, averaged over 1978–86. Colors indicate concentrations of chlorophyll on a log scale: extremes are purple ($< 0.06\,\mathrm{mg\,m^{-3}}$), via dark blue, light blue, and green to orange-red ($1–10\,\mathrm{mg\,m^{-3}}$). Reproduced from Longhurst (Chapter 9, [5]), by permission of NASA/Goddard Space Flight Center.
(a) Northern Hemisphere summer, June–August.

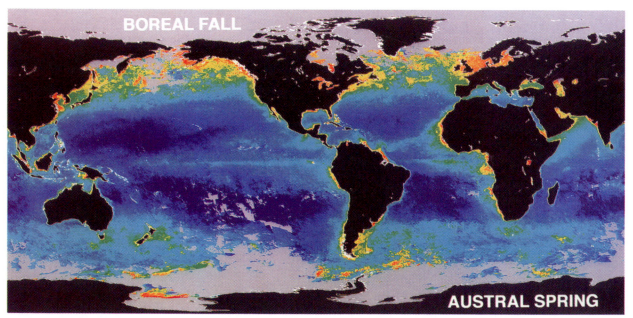

(b)

Plate 4 (b) Northern Hemisphere autumn, September–November.

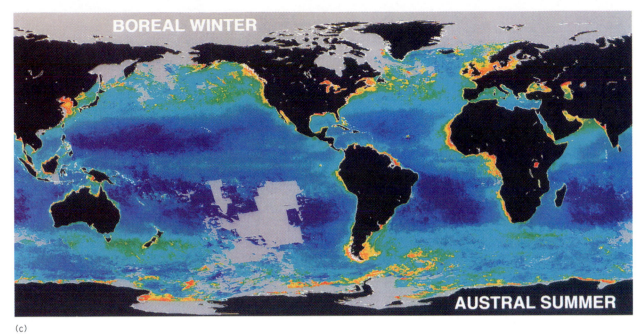

(c)

Plate 4 (c) Northern Hemisphere winter, December–February.

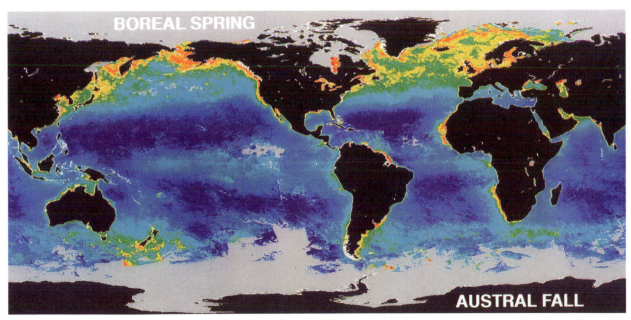

(d)

Plate 4 (d) Northern Hemisphere spring, March–May.

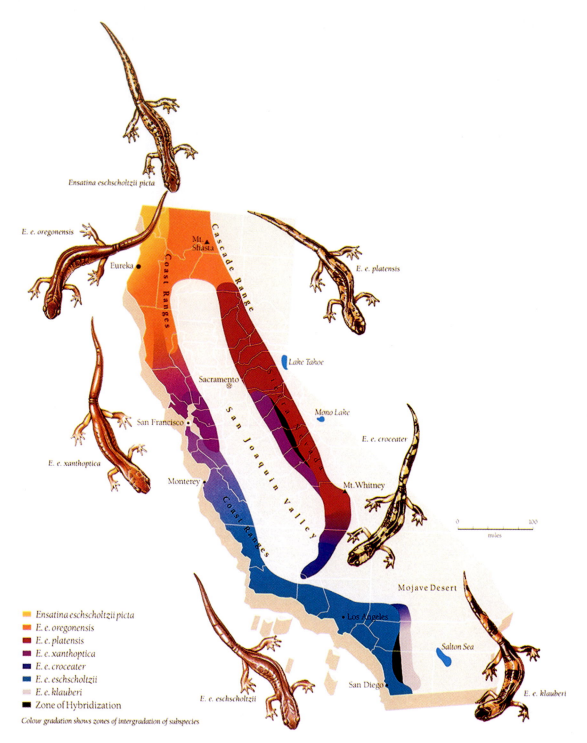

Ensatina eschscholtzii picta

E. e. oregonensis

E. e. platensis

E. e. croceater

E. e. xanthoptica

E. e. eschscholtzii

E. e. klauberi

Mt. Shasta

Cascade Range

Coast Ranges

Eureka

Lake Tahoe

Sacramento

Sierra Nevada

Mono Lake

San Francisco

San Joaquin Valley

Monterey

Mt. Whitney

Coast Ranges

Mojave Desert

Los Angeles

Salton Sea

San Diego

0 _____ 100
miles

■ Ensatina eschscholtzii picta
■ E. e. oregonensis
■ E. e. platensis
■ E. e. xanthoptica
■ E. e. croceater
■ E. e. eschscholtzii
■ E. e. klauberi
■ Zone of Hybridization
Colour gradation shows zones of intergradation of subspecies

Plate 5 Ring species of the salamander *Ensatina* in the western United States. One species (*E. oregonenis*) is found in the north and up into Oregon and Washington. It then divides in northern California and forms a more or less continuous ring around the San Joaquin Valley. The salamanders vary in form from place to place, and consequently they have been given a number of taxonomic names. Where the coastal and island sides of the ring meet in southern California, the salamanders behave as good species at some sites (black zones on the map).

Reproduced from Thelander CG *Life on the Edge*. Berkeley, CA: Biosystems Books 1994, p. 233, by permission of Ten Speed Press, California.

but two: one set derived from the individual's mother, and the other derived from its father. Both parents may possess exactly the same allele of a particular gene. For example, both may have the allele for brown hair, in which case their offspring would also have brown hair. But very often they may hand down different alleles to their offspring; for example, one might provide a brown-hair allele, while the other provided a red-hair allele. In such a case, the result is not a mixing or blurring of the action of the two alleles to produce an intermediate such as reddish-brown hair. Instead, only one of the two alleles goes into action, and the other appears to remain inert. The active allele is known as the **dominant** allele, and the inert one as the **recessive** allele. Which allele is dominant and which is recessive normally is firmly fixed and unvarying; in the above example, the brown-hair allele is usually dominant, and the red-hair allele is usually recessive.

This, then, is the genetic system that provides two vital properties of the organism. First, it provides the stability that ensures that its complex systems will function and be adapted to the demands of the environment. Second, it provides the plasticity that allows it to respond to minor changes in that environment. But how do modifications of its characteristics take place?

The genes themselves are highly complex in their biochemical structure. Although normally each is precisely and accurately duplicated each time a cell divides, it is not surprising that from time to time—due to the incredible complexity of the molecules involved—there is a slight error in this process. This may happen in the cell divisions that lead to the production of the sexual gametes (the male sperm or pollen, the female ovum or egg). If so, the individual resulting from that sexual union may show a completely new character, unlike that of either parent. In the example referred to earlier, such an individual might have completely colorless hair. Such sudden alterations in the genes are known as **mutations**.

This genetic system can lead to changes in the characteristics of an isolated population, in two ways. First, new mutations may appear and, if they are advantageous, spread through the population. Second, since each individual carries several thousand genes, and each may be present in any one of its several different alleles, no two individuals (unless they are identical twins) carry exactly the same genetic constitution. Even if no mutations have taken place, so that they carry the same sets of characteristics, these may be present in different combinations. Inevitably, therefore, each isolated population will come to differ from the others in its genetic content, some alleles being rarer or, perhaps, being absent altogether. As mating goes on in different populations, new combinations of alleles will appear haphazardly in each, and this will lead to further differences between them.

Whether they are new mutations, or merely new recombinations of existing alleles, new characteristics will therefore appear within an isolated population. Any of these that confer an advantage on the organism are likely to spread gradually through the population and so change its genetic constitution. However, it is important to realize that chance, as well as its genetic constitution, plays a role in determining whether a particular individual survives and breeds. Even if a new, favorable genetic change appears in a particular individual, it may by chance die before it

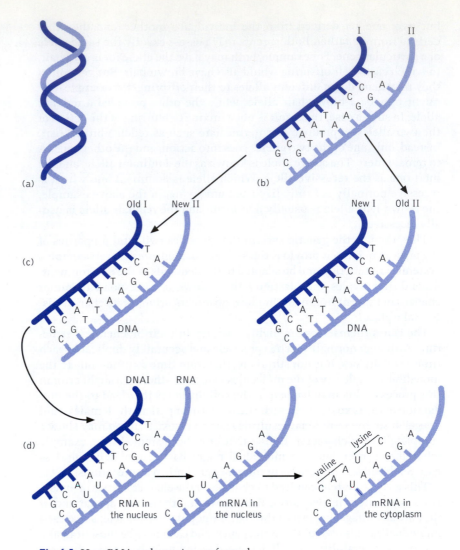

Fig. 6.1 How RNA and proteins are formed.
(a) The DNA thread is made up of two strands that spiral around one another in a double helix.
(b) The two strands are connected to one another via their nucleotides, which bind together in pairs: A (adenosine) and T (thymine), or C (cytosine) and G (guanine).
(c) At cell division, the two strands separate, and each goes to one of the new cells, where it builds a new partner identical to the old partner.
(d) During the manufacture of proteins, a thread of DNA first builds a thread of RNA, in which U (uracil) has replaced the T (thymine). This thread of RNA detaches itself to form messenger RNA (mRNA), which travels from the nucleus to the ribosome in the cytoplasm. There, the nucleotides in the mRNA act as triplets, each of which forms a template for the formation of a particular amino acid—here, valine and lysine.

Each chromosome is composed of a pair of strands that spiral around one another (Fig. 6.1a) to form a double helix. Each strand is made up of a string of molecules called **nucleotides**, of which there are just four types present in the string: adenine (A), cytosine (C), guanine (G), and thymine (T). The nucleotides of one strand link with those of its partner in the double helix but only in pairs: A with T, C with G (Fig. 6.1b). When a cell divides to produce two new body cells, each thread of DNA separates from its partner, and each of the daughter cells receives one of these two threads. Within each new cell, the thread of DNA then builds a new partner, but because of the unique pairing system of A and T versus C and G, each new partner is an exact replica of the original partner before cell division took place (Fig. 6.1c). This precision of replication is essential if the new body cells are to retain the nature of the parent cell. The chromosomes themselves are also paired, one chromosome in each pair having originated from the mother and one from the father; this is known as the **diploid** condition.

In the process of sexual reproduction, each sperm or ovum contains only one of each pair of chromosomes; this is known as the **haploid** condition. During the process of production of these sperm and ova, there is some exchange of material between the maternal and the paternal chromosomes, a process known as **recombination**. This is an important part of the genetic system because it allows the next generation to have some characteristics of each of its parents, rather than being a simple replica of one or other of them.

As already mentioned, DNA is also responsible for all the characteristics of the individual. But to do this, it has to pass through several intermediate stages. First, it is modified within the nucleus into a slightly different molecule called ribonucleic acid, or RNA (Fig. 6.1d) This, in a form known as messenger RNA, then travels out of the nucleus into the **cytoplasm** that surrounds the nucleus, where it manufactures proteins (Fig. 6.1d). These are large molecules made up of units called **amino acids**, which are responsible for controlling all the metabolic processes within the cell. In manufacturing proteins, the nucleotides of the RNA act as triplets: each successive group of three nucleotides forms a template to which particular amino acids can attach, thus creating a chain of linked amino acids. In the example shown in Fig. 6.1d, the amino acids valine and lysine are forming.

There are about 20 different common amino acids, and each type of protein is composed of a unique sequence of them, which may be thousands of amino acids long. The precise sequence of amino acids in each protein controls its properties. Some proteins are structural—for example, muscle tissue is made up of a protein called myosin. But the most important function of proteins lies in their activity as catalysts, or **enzymes**, in which they moderate and control all cell processes. The expression of the DNA code, therefore, is ultimately effected through the activities of these enzymes.

reproduces, or all of its offspring may similarly die, so that the new mutation or recombination disappears. Nevertheless, however rare each genetic change may be, each is likely to reappear in a certain percentage of the population as a whole. In a larger population, each mutation or recombination will therefore reappear sufficiently often that the effects of random chance are reduced. The underlying advantages or disadvantages that they confer will then eventually show themselves as increased or decreased reproductive success. For this reason, it is the population, and not the individual, that is the real unit of evolutionary change. In smaller populations, however, chance will play a greater role in controlling whether a particular allele becomes common or rare or disappears; this effect is known as **genetic drift** because it is not controlled by selective pressures. Smaller populations therefore contain less genetic variability and are less closely adapted to their environment; they are therefore more likely to become extinct than larger populations. (This can be a particular problem in island populations; see p. 226.)

All that is now required for the appearance of a new species is that some of the new characteristics of this isolated population fit it for a way of life that is in some fashion different from that of the ancestral population from which it became isolated.

However great may be the area of land (or water) within which a particular species is found, it is not present everywhere there. Any area is a patchwork quilt of differing environments, of meadow, pond and woodland, of dense forest or of regrown forest—and even the meadow or woodland is itself made up of a myriad of differing habitats, as we saw in Chapter 4. As a result, the species is broken up into many individual populations that are separate from one another. Furthermore, no two patches of woodland, no two freshwater ponds, will be absolutely identical, even if they lie in the same area of country. They may differ in the precise nature of their soil or water, in their range of temperature, or their average temperature, or in the particular species of animal or plant that may become unusually rare or unusually common in that locality. Each population independently responds to the particular environmental changes that take place in its own location. The response of each population is also dependent on the particular pattern of new mutations and of new genetic combinations that have taken place within it. Each population will therefore gradually come to differ from the others in its genetic adaptations.

Provided that the barriers between the two populations are great enough to prohibit genetic exchange between them, the foundations for the appearance of a new species have now been laid. If two such divergent populations should meet again when the process of divergent adaptive change has not gone very far, they may simply interbreed and merge with one another. If, however, they have become significantly different in their adaptations, though they may still be able to mate and have fertile offspring, known as **hybrids**, these are likely to have a mixture of the characters of their two parents. Because each of the parents had already become adapted to its individual environment, these hybrid offspring, not particularly adapted to either of the environments, will not be favored by natural selection. From the point of view of each of the well-adapted parent populations, this hybridization is disadvantageous because it merely leads to the production of poorly adapted individuals that will not survive. So, evolution will then favor the appearance of any characteristics that reduce the likelihood of hybridization. These are known as **isolating mechanisms**, and they can take two different forms—systems that prevent mating between related species from taking place and systems that lead to reduced fertility if such mating *does* take place.

Premating isolating mechanisms are common in animals such as birds and insects that have a complicated courtship and mating behavior. This is because small differences in these rituals may in themselves effectively prevent interbreeding. In the case of Darwin's finches related species recognize one another because they have different songs, and they do not mate with an individual that sings the "wrong" song [1]. Sometimes the preference for the mating site may differ slightly. For example, the North American toads *Bufo fowleri* and *B. americanus* live in the same areas but breed in different places [2]. *B. fowleri* breeds in large, still bodies of water such as ponds, large rainpools, and quiet streams, whereas *B. americanus* prefers shallow puddles or brook pools. Interbreeding between species is also hindered by the fact that *B. americanus* breeds in early spring and *B. fowleri* in the late spring, although there is some mid-spring overlap. However, where the group concerned is rapidly evolving, such premating barriers may not have had sufficient time to become effective. For example in ducks, where color pattern and behavior are the premating barriers, 75% of the British species are known to hybridize.

Many flowering plants are pollinated by animals that are attracted to the flowers by their nectar or pollen. Hybridization may then be prevented by the adaptation of the flowers to different pollinators. For example, differences in the size, shape, and color of the flowers of related species of the North American beard-tongue (*Penstemon*) adapt them to pollination by different insects or, in one species, by a hummingbird (Fig. 6.2). In other plants, related species have come to differ in the time at which they shed their pollen, thus making hybridization impossible. Even if pollen of another species does reach the stigma of a flower, in many cases it is unable even to form a pollen tube because the biochemical environment in which it finds itself is too alien. It cannot therefore grow down to fertilize the ovum. Similarly, in many animals the spermatozoa of a different species cause an allergic reaction in the walls of the female genital passage, and the spermatozoa die before fertilization.

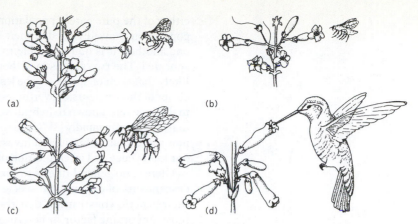

Fig. 6.2 Four species of the beard-tongue (*Penstemon*) found in California, together with their pollinators. Species (a) and (b) are pollinated by solitary wasps, species (c) by carpenter bees, and species (d) by hummingbirds. After Stebbins [20].

Postmating isolating mechanisms do not prevent mating and fertilization, but instead ensure that the union is sterile. In some cases, the two species may mate together, but they will have no offspring—the mating is sterile. In other cases, the two species may mate and have offspring, but these descendants are themselves sterile. An example is the case of the horse and the ass. Although these are separate species, they do sometimes breed together, but the resulting mule (male) or hinny (female) is sterile. A third category of cases comprises those in which the two species may breed together and have offspring, but the fertility of the offspring is reduced. The hybrids therefore soon become extinct in competition with the more fertile results of mating within each species.

The basic cause of all these incompatibilities is usually to be found in the genetic system. Sometimes the structure and arrangement of the genes on the chromosomes are so different that the normal processes of chromosome splitting and pairing that accompany cell division are disrupted. Other genetic differences may disrupt the normal processes of embryological development, or the growth and maturation of the hybrid individual. Whatever it may be, the final result is the same—the hybrid mating is sterile or, if offspring are produced, they are themselves sterile or have reduced fertility.

The final biogeographical step toward the appearance of a new species occurs when hybrids between the two independent populations are found only along a narrow zone where the two populations meet. Such a situation suggests that, though continued interbreeding within this zone can produce a population of hybrids, these hybrids cannot compete elsewhere with either of the pure parent populations. But, of course, if two closely related populations have diverged in their adaptations but are not in contact with one another, it is quite possible that isolating mechanisms will not have appeared. The two groups may then be able to mate together, though it is likely that their offspring will be less well adapted to the environment than either of their parents. Nevertheless, 25% of plant species and 10% of animal species are known to hybridize, and the number of cases known in animals is growing rapidly [3]. This seems to be an area of biology in which the new molecular techniques of investigation may lead to marked changes in our knowledge and interpretation of the evolutionary process.

There is no general rule as to the length of time that it will take for the descendants of one original species to diverge so far from one another in their genetic constitution that they have become separate species. The most important factor in determining the rate of genetic change is the speed at which the environment changes. If it changes rapidly, the organism must also change rapidly, or else become liable to extinction. But the rate at which an organism can respond is also dependent on population size. In a small population, the random effect of genetic drift may by chance produce a new mixture of genetic characteristics that match the new requirements of the environment. This is less likely to happen in a larger population, where the sheer size of the gene pool makes rapid evolutionary change of this kind less likely.

The fastest well-documented example of speciation is that of some species of cichlid fish in a small lake in Africa. This lake became separated from the large Lake Victoria by a strip of land that has been dated by radiocarbon analysis at 4,000 years old [4], and the species of fish in the small lake therefore cannot be older than that. Many of the species of the fruitfly *Drosophila* that are found in the narrow valleys between the lava flows on the slopes of the volcanic mountains in the Hawaiian Islands (see p. 234) are probably also only a few thousand years old. But even this comparatively short period of time is longer than biologists have studied any species, so we have never been able to watch while a new species emerges (though see the study of Darwin's finches, below). However, where a species has been extending its range and, in the process, has had to adapt to new environments, we can see the resulting pattern of evolutionary change laid out on the landscape as a pattern of biogeography. In a few cases, the end products of that process have come into contact and demonstrated the extent of the genetic change by refusing to mate with one another: they have evolved into separate species, known as **ring species**.

One of the best examples of a ring species is the pattern of distribution of the salamander *Ensatina* in the western United States [5]. The story appears to have begun with the species *E. oregonensis* living in Washington State and Oregon, which spread into northern California, where it formed the new species *E. eschscholtzii*. As this species continued to spread southward, encircling the hot lowlands of the San Joaquin Valley, it developed populations with different genetic constitutions. One result of this was that the populations had different color patterns (see Plate 5): the populations on the western side of the valley became lightly pigmented, while those on the eastern side developed a blotchier pattern. Biologists therefore gave them different names but, because these populations were able to interbreed with one another, so that hybrids were found, they were recognized as merely separate subspecies of the single species *E. eschscholtzii* (*E. e. picta*, *E. e. platensis*, etc.).

In two areas, these two sets of subspecies came into contact. At some time in the past, the subspecies that lives in the San Francisco area, *E. e. xanthoptica*, colonized the eastern side of the valley, where it met one of the "blotched" subspecies, *E. e. platensis*. Here, the amount of genetic divergence between these two subspecies is not enough to prevent them from interbreeding (Plate 5). The other meeting point is below the southern end of the San Joaquin Valley, in Southern California, where *E. e. eschscholtzii* met *E. e. croceater* and *E. e. klauberi*. (There is nowadays a gap in the ring of subspecies in this area, perhaps because of climatic change, but populations of *E. e. croceater* are found both northwest and east of Los Angeles, showing that the chain was once complete [6].) Over much of this area of overlap, the two subspecies hybridize to a limited extent; approximately 8% of the salamanders of the area are hybrids. However, at the extreme southerly end of the chain, the length of time since the two types of salamander diverged from their common northern ancestors is at its maximum. As a result, the genetic differences between them are so great that they behave as completely separate species, *E. eschscholtzii* and *E. klauberi*.

Sympatry vs. Allopatry

In all the examples considered so far, it has been assumed that the divergence between the populations as they evolved into separate species took place in isolation from one another, a situation known as **allopatric speciation**. Until comparatively recently, it was generally accepted that, with the exception of polyploidy (see Concept Box 6.2), this was the almost unvarying way in which speciation takes place. But there is now a lively discussion as to whether new species can also arise by **sympatric speciation**, *within* the area of distribution of the ancestral species. There is even evidence for this in Darwin's finches (see p. 190), where two types of *Geospiza fortis*, with different beak sizes, live alongside one another. They prefer to mate with individuals with a similarly sized beak, and genetic analyses show that there is reduced genetic flow between the populations of the two types, which also have different songs. This could well be an example of sympatric speciation.

A more detailed example concerns two different races of the fruitfly *Rhagoletis pomonella* that are found in southern North America, one infesting the fruits of the native hawthorn and the other infesting those of the introduced apple trees. It was suggested that the appearance of a second race had taken place sympatrically by the fruitfly adapting to a new host within its original area. However, recent work [7] implies that the original genetic difference between the two races took place about 1.57 my ago, when there were two separate populations of *R. pomonella*—one in Mexico and the other in the United States. A genetic change took place in the northern race that shortened the length of the development period of the overwintering pupa. This allowed it to infest the apple trees, which fruit 3 to 4 weeks before the hawthorn. It was only later that the two populations,

Polyploidy

Concept Box 6.2

A quite different method by which new species can appear is by **polyploidy**—the doubling of the whole set of chromosomes in the nucleus of a fertilized egg cell. The result is that each chromosome automatically has an identical partner, with which the normal processes of pairing and cell division can take place. This may occur in the development of a hybrid individual (in which case it can overcome any genetic isolating mechanisms), or in the development of an otherwise normal offspring of parents from a single species. In either case, the new polyploid individual will be unlikely to find another similar individual with which to mate, and the origin of new species by polyploidy has therefore been important only in groups in which self-fertilization is common. Only a few animal groups fall into this category (e.g., flatworms, lumbricid earthworms, and weevils), but in these groups an appreciable proportion of the species probably arose in this way. In plants, however, in which self-fertilization is common, polyploidy is an important mechanism of speciation. Perhaps three-quarters of all plant species have probably arisen in this way, including many valuable crop plants such as wheat (see p. 422), oats, cotton, potatoes, bananas, coffee, and sugarcane. Polyploid species are often larger than the original parent type, and also hardier and more vigorous; many weeds are polyploids.

An example of a pest species resulting from such polyploidy is the cord-grass, a robust plant of coastal mudflats around the world. There are several species of this plant, but none of them was a serious pest until two species met in the waters around the port of Southampton, UK, in the latter half of the 19th century. An American species of cord-grass, *Spartina alterniflora*, was brought into the area, probably carried in mud on a boat, and was able to hybridize with the native English species, *S. maritima*. The hybrid was first found in the area in 1870 and was named *Spartina × townsendii*. It contained 62 chromosomes in its nucleus but, because these chromosomes were derived from two different parent species, they were unable to join together in compatible pairs before gamete formation; the hybrid therefore did not produce fertile pollen grains or egg cells. It was nevertheless able to reproduce vegetatively, and it is still found along the coasts of western Europe. But in 1892 a new fertile cord-grass appeared near Southampton and was named *S. anglica*. This has 124 chromosomes. As a result of this doubling of the number found in the sterile hybrid, the chromosomes could once again form compatible pairs and fertile gametes could be produced. This new species, formed by natural polyploidy, has been extremely successful and has spread around the world, often creating problems for shipping by forming mats of vegetation within which sediments are deposited, and so contributing to the silting-up of estuaries.

Polyploidy can also be artificially produced, for example, by the use of colchicum, an extract of the meadow saffron plant, *Colchicum autumnale*. Techniques of this kind have been used to provide new strains of commercially valuable plants, such as cereals, sugar beet, tomatoes, and roses.

now infesting different host plants, came into contact and so provided an apparent example of sympatry. It may be significant that this new interpretation has only become possible as a result of the new techniques of investigation of genetic differences at the molecular level (see p. 215) that allow estimation of the dates of divergence. This suggests that similar examples of apparent sympatry should be accepted only after similar detailed investigation.

It is more difficult to use the possibility of an earlier phase of allopatric speciation as a possible explanation in other cases, such as where two or more distinct forms of a fish are found in a single lake. For example, two forms of the arctic charr *Salvelinus alpinus* are to be found in Lake Rannoch in Scotland [8]. As shown in Fig. 6.3, the main difference is in the mouth structure. The upper fish in the illustration has a relatively small mouth with small teeth and mainly eats the zooplankton found in the upper layers of the lake. The lower fish has a larger, rounder mouth and bigger teeth, and eats a range of invertebrates and even fish, feeding mainly near the bottom. It is difficult to imagine any way in which different populations of these fish could exist in the lake, sufficiently separate from one another to be able to evolve into separate species. There are similar examples of such a situation in mountain lakes.

The Great Lakes of East Africa show far more dramatic examples of this same phenomenon, for they are large enough to provide a great diversity of environments and old enough for these ecological opportunities to have become realized through evolutionary change. The cichlid fishes, in particular, have been able to take advantage of this (Fig. 6.4) and are the most rapidly speciating organisms known. These fish are extremely good at changing their diet, and the different jaw shapes and diets shown in Figure 6.4 have evolved repeatedly and independently in several of these lakes. The result is the presence of hundreds of different types of cichlid in each of the Great Lakes Tanganyika, Victoria, and Malawi [6].

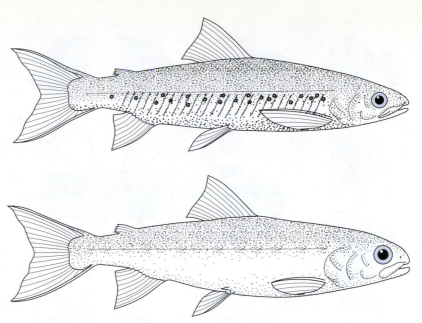

Fig. 6.3 Two different forms of the arctic charr from Loch Rannoch in Scotland. The upper fish has narrow jaws and small teeth, and feeds mainly on plankton. The lower fish has heavier jaws and larger teeth, and feeds on a wider range of prey, including invertebrates and other fish. From Walker et al. [8].

Recent research on the cichlids of Lake Victoria [9] has thrown considerable light on the mechanism of this rapid speciation. Genetic mechanisms for postmating isolation are frequently weak or absent in these fishes. However, they are mainly brightly colored, show obvious differences in characteristics that allow the two sexes to recognize one another, and depend on vision for their selection of mates. The waters of the lake contain suspended particles that absorb particular wavelengths of light. As a result, blue colors that are clearly visible in the surface waters become increasingly unclear at greater depths, where red colors instead are more visible. The female cichlids have mating preferences for conspicuously colored males, and their color preference is correlated with a genetic difference, for they have evolved different alleles of the visual pigments in their eyes. So, in fish that live close to the surface, where blue is more visible, the fish themselves are blue and their visual pigment is more sensitive to that color. In those that live at greater depths, where red is more visible, the fish are red and their visual pigment is more sensitive to that color. As long as the rate of change in the relative visibility of the colors changes gradually with depth, these differences in the environment and in the genetics and mating preferences of the fish are sufficient to keep the different populations genetically separate, so that they are potential new species. However, where there is a more rapid change in relative visibility, so that the two populations encounter each other more frequently, they interbreed with one another, forming a single population within which both types of visual pigment are to be found. This type of evolutionary change, associated with differences in the sensory systems and behavior, has been called **sensory drive**.

There is an unhappy ending to this tale. The fish need good visibility in order for the females to be able to see the color of male that they prefer,

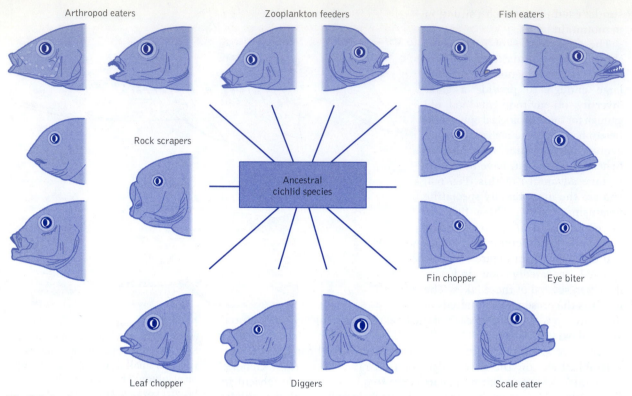

Arthropod eaters

Zooplankton feeders

Fish eaters

Rock scrapers

Ancestral
cichlid species

Fin chopper

Eye biter

Leaf chopper

Diggers

Scale eater

Fig. 6.4 A selection of the different shapes of head and mouth that have evolved for different feeding habits in the cichlid fishes in Lake Malawi. Redrawn, after Fryer & Iles [4].

but recent changes in their environments are making this more and more difficult. Because of increasing human activity around Lake Victoria, nutrients and silt are draining into the lake from surrounding farmland, together with sewage from urban settlements, causing turbidity and poor light penetration in the waters. Human activity could thus bring the diversification of fish in this lake to a halt [10].

Now that we have established how new species arise, we can turn to the methods we use to define just what a species is, how different species are related to one another, and where the species "lives."

Defining the Species

There is a great variety of definitions of the **species**, which focus on a variety of different aspects of the organisms, so let's first of all consider what type of data is *normally* the concern of the biogeographer.

Practical constraints usually lead the scientist, in analyzing the distribution of species, to rely on those that are easy to observe and collect, and on morphological characteristics that can be measured. This in turn normally results in the use of data from such groups as the vertebrates, the larger invertebrates, and macroscopic plants. (Studies on other groups, such as mosses, lichens, or worms show that they often exhibit similar biogeographical patterns.)

As we have seen, the most important factor in preserving the identity and nature of a species is isolation, and this aspect is emphasized in the **biological species concept**, which focuses on its reproductive isolation: "the species is a group of natural populations whose members can all breed together to produce offspring that are fully fertile, but that in the wild do not do so with other such groups." (So the horse and the ass belong to separate species because, although they can breed together to produce a mule, the mule is always sterile.)

In deciding whether a particular group of populations should be recognized as a separate species, biologists in practice usually start by finding that they can recognize the group by particular aspects of their appearance, that is, by their physical characteristics. In defining the species, the biologist will then try to use characteristics that appear to be important in its way of life, perhaps because they are obviously used by the members of the species when they mate—as is the case in the cichlid fishes described earlier. This may be fairly easy to detect if the habits of the species are not too different from ours (i.e., it is a terrestrial vertebrate and uses its eyes in social interaction), but it becomes progressively more difficult as one examines groups that are more and more different from us in structure and environment. Species recognition must be quite different in, for example, a blind burrowing worm, and in all groups that merely release their eggs and sperm into the water, or that release their pollen into the air. Here, one can only try to identify the characteristics of the species that are connected to its ecology.

This takes us to the second type of definition of the species, the **ecological species concept**, which arises from the organism's need to find its own place in the natural world. Because it must do so in the face of competition from other organisms, it must develop a set of characteristics (morphological, behavioral, physiological, etc.) that give it an advantage over those competitors. (These features are, of course, genetically based, so we are also describing the species as a distinct genotype.) These specializations define its ecological niche, which may be quite narrow if competition is intense, as in the case of the tanagers and monkeys described in Chapter 2 (p. 80). If there is less competition, or if conditions are so changeable that the species must remain more variable in order to be able to survive during these changing environmental demands, then the species may show more variation in its own characteristics. (A good example is found in Darwin's finches on the next page.) Here, as always, the organism has adapted to the demands of the environment.

It should never be forgotten however that evolution is a continuing, dynamic process. So, from time to time, detailed research will reveal related populations that are in the process of becoming separate species. They have therefore not yet reached the point at which their differing adaptations, and the necessary isolating mechanisms, have become completely established. This evidence of evolution in progress provides support for our concept of speciation by natural selection, rather than posing a problem for it.

These definitions obviously cannot be used in the case of fossil species, where we have no knowledge of their abilities to interbreed with one another. Here, morphological data are the only type of evidence available. However, again, it is best to try to find those features that are connected to the way of life or reproduction of the species, and to try to find an assemblage of characteristics that appear to be reliably associated with one another. Where we have an extremely good fossil record, as in the case of some invertebrates and our own species (see p. 411), we may find a series of linked forms, evolving and changing over a long period of time. We can then only subdivide them into a sequence of intergrading "species."

Any scientific definition reflects the nature and detail of the information available at the time. That is why definitions of the species have changed over time, as we know more and more about the genetic and ecological nature, as well as the population structure, of the species we observe.

A Case Study: Darwin's Finches

Until recently, it was thought that evolution took place too slowly for it to be detectable over the timescale of scientific studies of living organisms, so that it could only be detected in the fossil record. However, it is now clear that this perception was wrong. One of the most detailed and fruitful studies of evolution in action has been carried out precisely where Darwin first noted evidence of it—in the Galápagos Islands in the eastern Pacific Ocean, 960 km (600 mi) from the nearest land (Fig. 6.5), and on the birds now known as **Darwin's finches**, which are found only in those islands. Modern evidence suggests that these finches are descended from a flock of South American finches that entered the Islands 2–3 million years ago and evolved into 14 different species belonging to five different genera. A number of them feed on buds or insects, cactus seeds, flowers, and pollen, while six species, belonging to the genus *Geospiza*, feed on the seeds of some two dozen species of plant, which they find on the ground.

For many years, the "Finch Unit" led by the British workers Peter and Rosemary Grant at Princeton University have studied the geospizid finches on the tiny Galápagos island called Daphne Major, only 34 ha (84 acres) in area and 8 km (5 mi) from the nearest large island. The largest of them is *Geospiza magnirostris* and the smallest is *G. fuliginosa*, while *G. fortis* is of intermediate size (Fig. 6.6). The size of the beak is the only difference in the appearance of these three species, and it is a vital aspect of their lives, for it determines the size of the seed that the bird is best suited to eat. Both the size of the finch's body and that of its beak are strongly dependent on those of its parents; that is, there is a strong inherited factor in their size.

The work of the Finch Unit has been vividly described in Jonathan Weiner's excellent book, *The Beak of the Finch* [11]. Since 1973, they have

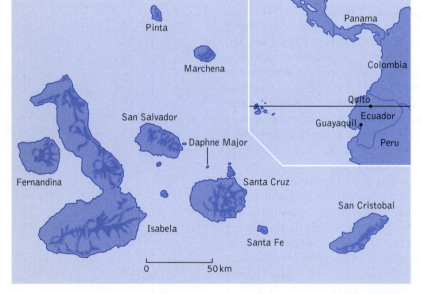

Fig. 6.5 Map of the Galápagos Islands.

Fig. 6.6 Galápagos ground finches with beaks of different size: from left to right, *Geospiza magnirostris*, *G. fortis*, *G. fuliginosa*, and *G. difficilis*.

measured and weighed some 20,000 of these birds, belonging to 24 generations. Their populations have ranged from fewer than 300 birds in the hardest, driest year, to approximately 1,000 in the best year. Since 1977, the Finch Unit has recognized, measured, and ringed every single bird, and recorded with which others it has mated, how many offspring they had, how many survived, in turn mated, and so on. They have recorded the abundance of each type of food plant and its seeds, the hardness of the seeds, and the pattern of temperature and rainfall. Analysis of their results showed with startling clarity the extent of the differences in the environment from year to year, and the immediacy of the impact of these changes on the populations of the finches.

When food is plentiful, all three species prefer to eat the softest types of seed. In drier seasons, there are fewer of these favorite seeds because they are produced by smaller plants, which tend to wither and die during a drought. As a result, each species has then to become more specialized, spending more of its time feeding on those seeds to which its beak size is best adapted. For example, *G. magnirostris* has the biggest and strongest beak, and is the best at cracking the hardest seeds, which belong to a little creeping plant called *Tribulus cistoides*. The fruit of this plant (Fig. 6.7) breaks up into hard, spiny "mericarps," each of which is c. 7 mm long and contains up to six seeds. *G. magnirostris* can crack two mericarps in less than a minute and extract at least four seeds, while *G. fortis* obtains only three seeds in over 1.5 min. Therefore, in this particular, tiny example of competition, *G. magnirostris* is gathering food at 2.5 times the rate of *G. fortis*. Furthermore, not all the *G. fortis* birds can even try to compete for this food, for only those with beaks at least 11 mm long can crack a mericarp; those whose beaks are only 10.5 mm long cannot do so. Meanwhile, little *G. fuliginosa* has to feed on smaller, softer seeds. This provides a very precise example of the nature and results of the link between morphology (beak size) and ecology (availability of different types of food).

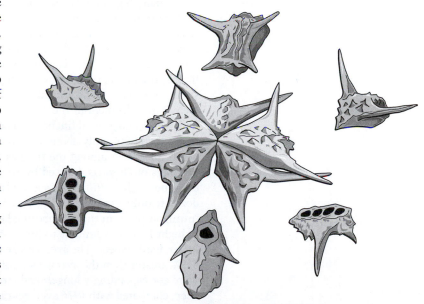

Fig. 6.7 The fruit of *Tribulus*, to show how it breaks up into mericarps. Some of these mericarps have been opened by finches, which have removed the seeds, leaving holes.

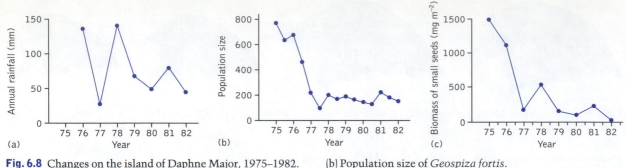

Fig. 6.8 Changes on the island of Daphne Major, 1975–1982. After Grant [21], with permission. (a) Annual rainfall (that of 1975 is unknown).

(b) Population size of *Geospiza fortis*. (c) Biomass of small seeds.

But evolution is not merely about relative ease of existence; natural selection is about life and death. Lying in the Pacific Ocean, the climate of the Galápagos Islands is greatly affected by the cyclical changes in ocean temperature known as El Niño (during which there is heavy rainfall) and La Niña (during which the rainfall is much reduced [see Chapter 12]). The Finch Unit was fortunate that their studies covered a period when the adaptations of the finches were tested to their fullest, for it included two of the most extreme years in the century—both the driest year, 1977, and the wettest year, 1983.

The drought of 1977 (Fig. 6.8a) first affected the amount of food available for all the ground finches. During the wet season of 1976, there had been more than 10 g of seeds per square meter of ground. Through 1977, as the drought struck harder and harder, the plants failed to flower and set the new season's seeds. The finches therefore had to continue to eat the seeds produced in 1976: by June, there were only 6 g/m² of seeds, and by December only 3 g/m² (Fig. 6.8c). The food shortage hit first the new generation of finches because of the shortage of the small seeds on which all the youngest, smallest birds feed. As a result, all the fledglings of 1977 died before they were 3 months old. But the drought also hit the adults. In June 1977 there had been 1,300 ground finches on Daphne Major; in December, there were fewer than 300—only about a quarter of the population had survived. But death hit harder among the finches that could only eat softer seeds. The numbers of *G. fortis* dropped by 85%, from 1,200 birds down to 180 (Fig. 6.8b), but little *G. fuliginosa* suffered most, for its population dropped from a dozen to only a single bird [12], so that the population could only recover by immigration from the neighboring larger island of Santa Cruz. Finally, the drought had been kindest to birds with big beaks, who could feed on the larger, harder seeds. The average size of the *G. fortis* birds that survived was 5.6% greater than the average size of the 1976 population, and their beaks were correspondingly longer (and stronger)—11.07 mm long and 9.96 mm deep, compared with 1976 averages of 10.68 and 9.42 mm.

As if that opportunity to watch and document the harsh workings of natural selection were not enough, 1983 allowed the Finch Unit to see a total reversal in the demands of the environment of Daphne Major. That was the year of the strongest 20th-century El Niño event; the rainfall was 10 times the previously known maximum. The island was drenched, and its plants grew rampantly—by June, the total mass of seeds was almost 12 times greater than in the previous year. This time, small soft seeds predominated—they formed up to 80% of the total mass of seeds, up to 10 times more than the previous maximum. That was partly because the smaller plants grew luxuriously and produced many more seeds, and partly because the growth of *Tribulus* had been hampered by smothering grasses.

As a result of these lush conditions, the finch population spiralled upward. By June, there were more than 2,000 ground finches on the island, the numbers of *G. fortis* having increased by more than four times. But in 1984 there was only 53 mm of rain, and in 1985 only 4 mm (2 and 0.15 in., respectively). Now there was a new episode of drastic selection, but in the opposite direction from that which had followed the drought of 1973. Now it was selecting smaller birds, with smaller beaks, more suited to eating the plentiful smaller seeds [13].

The moral of this second part of the finch story is that conditions, and selection, can oscillate violently. The Finch Unit had witnessed and documented a total reversal of the selection pressures within six years—a timespan that would have been totally invisible in the fossil record, had there been one.

But there was still another aspect of the action of evolution on Daphne Major to be found in the Finch Unit's records. Normally, any species has its own niche, and hybridization between species is therefore disadvantageous, for the hybrid young are less well adapted than either of their parents and therefore cannot compete successfully with either of them. That had indeed been true on Daphne Major before the flood year. Interspecies mating was rare, and the occasional hybrid was unsuccessful and did not find a mate. But when there is so much food available that competition has for the moment disappeared, hybridization is no longer necessarily a disadvantage—and that was the situation on Daphne Major in the Year of Abundance, 1983. After that, hybrids between *G. fortis* and *D. fuliginosa* in fact did better than pure-bred members of either species and, by 1993, about 10% of the finches of the island were hybrids [14]. It would now seem that the three supposed species are in reality merely members of one species that shows an unusual degree of variation. But such a flexibility is advantageous in an environment that shows the immense oscillations in climate and environment that the El Niño system brings to the Galápagos Islands. So, the process and results of evolution are closely tailored to the situation in which they take place.

The Finch Unit has therefore seen evolution in action. Their records show precisely what Darwin's explanation had predicted and what biologists have long accepted. Different species show adaptations (here, of beak size and power) that fit them to live on different resources in the environment, thereby reducing competition with other species living alongside them. But that adaptation is not immutable: it cannot afford to be, for the environment itself is fluid and changeable. Thus, as from year to year the environment makes new demands and provides new opportunities, so the adaptations of the species will change in harmony with those demands and opportunities. It is changes of this kind in the characteristics of a species that have led to the evolution of strains of bacteria that are resistant to commonly used antibiotics, to strains of insects that are resistant to DDT and to other chemicals designed to control pests, and to strains of malarial parasite that are resistant to modern drugs. Adaptation by evolution is never far behind our efforts to control the biological world.

Controversies and Evolution

There is now a vast amount of evidence for Darwin's explanation of evolution by natural selection. Nevertheless, controversy still exists about some details of the circumstances in which new species evolve or the rate at which this happens. For example, some biologists believe that evolutionary change normally takes place at a steady, gradual rate—a concept known as **gradualistic evolution**. Others instead believe that, even if genetic alterations gradually accumulate within a population, this may not be reflected in detectable morphological or physiological changes until they are so numerous as to shift the balance of the whole genotype. At this point, a comparatively large number of changes are seen to take place at the same time; this is known as the **punctuated equilibrium** model of evolutionary change. Each group of theorists provides examples that may support their view—and, at times, each interprets the same example as supporting their own view!

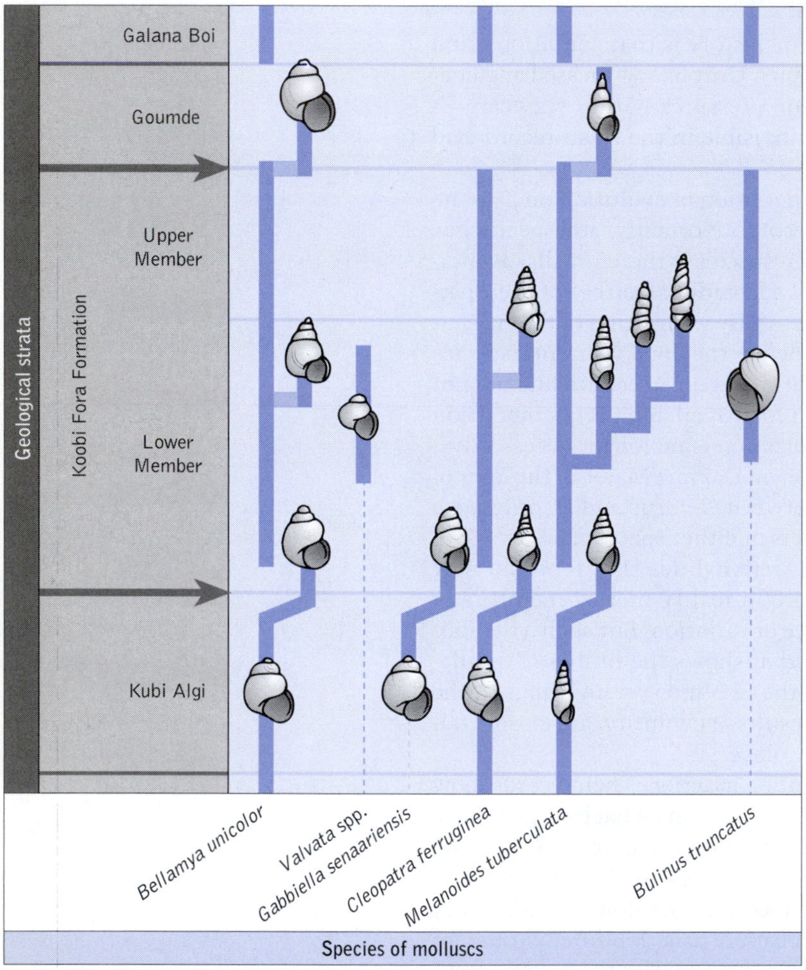

It is also difficult to isolate such underlying patterns from the more direct effects of the environment. For example, a study of the fossil shells of gastropod (spiral-shelled) molluscs that lived in northern Kenya over the last few million years shows long periods during which their structure and size remained unchanged, interrupted by shorter periods (5,000–50 000 years) during which they changed rapidly (Fig. 6.9). This was interpreted as an example of punctuated equilibrium [15]. However, the fact that the periods of change took place in several lineages at about the same time suggests that they were the result of external events that affected all of them, rather than resulting from some inherent evolutionary mechanism.

Fig. 6.9 Evolutionary changes in fossil gastropod molluscs in northern Kenya. The arrows indicate the levels at which sudden evolutionary changes took place simultaneously in several different species. From Dowdeswell [22].

The main difficulty with such studies is that, in general, the fossil record is not sufficiently detailed for us to be able to be certain whether gradualistic evolution or punctuated evolution was involved. In any case, we have no reason to believe that either style of evolution systematically prevails over the other. Therefore the real point of interest should instead be to identify the circumstances under which one or other type of evolution would be more likely to take place.

These controversies, and others like them, are common in any area of science, as new observations provoke new theories or suggest modifications of existing theories. But the protagonists in these disputes are only arguing about details of evolution by natural selection: all accept that the explanation itself is correct and is indeed the only one that makes sense of the phenomena of the living world. It is particularly important to realize this, as some groups within society are basically opposed to the idea of evolution, which they view as conflicting with their belief that the human race was created by divine action. One such group is the **Creation Science** organization, who believe an interpretation of the Bible that holds that everything on the planet was created in a short burst of divine activity a few thousand years ago, and insist that all the "supposed evidence for evolution" is spurious [16]. Another group, who believe in **Intelligent Design**, consider that the process of evolution by natural selection is incapable of producing the degree of adaptation that they see in the organic world, and therefore that it must have appeared by divine action. But these aspects of the biological world are the result of millions of years of continual improvement, as each organism competes with its neighbors. In any case, to the observant scientist, many organisms are less than perfect: even our own species would be better off without wisdom teeth or an appendix! Such groups try to present these academic controversies as symptoms of widespread and fundamental skepticism of the validity of Darwin's explanation. They are, of course, nothing of the kind. (Darwin's explanation is in any case not incompatible with religion. It is perfectly possible that the biological phenomena that scientists have discovered are merely documenting the gradual way in which the world and all its fauna and flora were created—a view that is accepted by many biologists.)

These groups also sometimes suggest that Darwin's explanation is founded on a circular argument, as follows: "Darwin suggests that it is the fittest that survive. But how do we know which is the fittest? Why, because those are the ones that survive, of course!" But the catch-phrase "survival of the fittest" was not Darwin's, and in any case, as noted earlier, the central points of his reasoning are straightforward, as follows:

1 All organisms normally produce many more offspring than are required to replace them in the population, but the population remains approximately constant in its numbers—so only a small proportion of the offspring survive.

2 The individuals are not identical in their characteristics; many of these differences are inherited from one generation to the next, and some of these differences will affect how well the individual is adapted to the environment.

3 It is therefore likely that the offspring that survive will be those whose characteristics make them better adapted.

4 It is thus more likely that these individuals will reproduce and that their characteristics will be more frequent than the others in the next generation. This relative increase in numbers lies at the heart of the concept of natural selection, and it does not involve any circular argument.

Another misunderstanding concerns the appearance of "new" characters. Some critics object that natural selection can only affect existing characters and does not explain the appearance of new characters. However, to the biologist, there are no "new" characters, but merely the modification of previously existing characters. The bony shell of a tortoise, the scales of a reptile, the feathers of birds, and the hair of mammals are all merely modifications of the scaly skin of fishes, which contain layers of bone covered by horn. The five-fingered limbs of land vertebrates are merely an elaboration of the skeleton of the fin of early bony fishes. This can be demonstrated by the fossil record, which documents the history of such changes through geological time, and by comparative embryology, which shows how these evolutionary changes were achieved by the process of development. So, for example, the appearance of such a "new" type of mammal as the horse is shown in the fossil record to be merely a gradual change of the limbs, skull, and teeth of a little dog-like creature that lived over 50 million years ago. The embryology of the horse similarly shows how the skeleton of its limbs gradually change in proportions and structure to produce the elongate, simplified limbs that we see today.

Similarly, the theory of evolution does not predict that such elaborate structures as the eyes will suddenly appear, in all the complexity that produces perfect vision. Instead, one can see in the animal kingdom a complete spectrum of different kinds of eye, from those that merely provide an indication of the direction from which light is coming, to the complicated vertebrate eye that provides an accurate, color image of the world. All that natural selection requires in order to improve any system is that it is slightly better than the other systems around it.

The idea that natural selection of genetic characteristics is the mechanism of evolution is, therefore, deeply rooted in biological science today and, like the idea of evolution itself is supported by a great deal of evidence (see Concept Box 6.3). This support from a wide variety of independent areas of scientific inquiry makes natural selection an immensely strong explanation of the method by which evolution takes place. It is therefore the central paradigm (see p. 3) of the biological sciences, just as the theory of plate tectonics is the central paradigm of the earth sciences. The two systems are based on quite independent and different sets of evidence—biological evidence in the case of evolution and physical evidence in the case of plate tectonics. The fact that the two systems provide similar dates for origins of the individual islands in groups such as the Hawaiian Islands (see p. 232) and for the organisms that are found on these islands therefore gives the two paradigms even greater strength as explanations of the workings of the world.

Charting the Course of Evolution

When we observe a number of related species or genera, it at first seems a hopeless task to try to establish precisely how the different taxa are related to one another. (Whether the group in question is a species, a genus, or some larger unit, it is known as a **taxon**, (plural **taxa**), whose interrelationships are studied in **taxonomy**.) How can we start to put some order into the puzzle and simplify the task of understanding the course that their evolution had taken? The solution to all this was

The techniques of modern genetics allow us to document the details of the genome of any organism, to compare it with that of others, and to deduce the pattern of relationship between them and the times of their divergence from one another. It also allows us to identify the precise genes that are responsible for each characteristic and process in the organism. As a result, we can now see the genetic basis for each of the varied lines of evidence for evolution.

So, to take the evidence from geographical distribution, we can identify individual genes that link together the apparently very different mammalian lineages that evolved from a common ancestor in Africa, known as the Afrotheria (see below, p. 199). (This was a complete surprise, for there had not previously been any strong indication of some of these relationships.) Similarly, a particular stretch of DNA is found only in the group of mammalian lineages found only in North America and Eurasia (Fig. 10.5, p. 317). We also find that the sequence of dates of divergence of related groups suggested by the fossil record are reflected in the degree of difference of their DNA. So, the evidence from human genetics shows that the first divergence of our lineage took place when our ancestors migrated from Africa into Eurasia, and also documents our gradual expansion across Asia, into North America, and down into South America, and from Asia across the island groups of the Pacific Ocean. Genetics also supports the classification of organisms that has been built up by taxonomists. For example, it confirms that the different types of anteater found in South America, Africa, and Australia are, despite many similarities in their structure, related to other lineages of mammal in those continents rather than to one another.

The genes are also intimately involved in the processes of embryology, in which a single-celled fertilized egg is transformed into a complex organism composed of many types of tissue and a variety of organs of different function. Some studies of the involvement of genes in this have provided a very great surprise. It turns out that the same linked groups of genes, known as *homeoboxes*, are responsible for the segmentation of the body in such diverse and ancient groups as the annelid worms, crustaceans, insects, and vertebrates. So these groups of genes have been in existence over all the hundreds of millions of years since these groups diverged from one another. At a very different level, genetics also shows us the DNA basis for the different embryological processes that led to the quite different structural features found in the various breeds of dog or pigeon, whose domestication by human selection Darwin had cited as one of his proofs of similar selection in nature.

In addition to the evidence from genetics, modern biological science has provided convincing examples of natural selection in action. We now know many examples in which different populations of a single species live under slightly different circumstances and show particular adaptations to these circumstances. If natural selection works, we would expect that, if we exchange populations of these species, each of them would gradually evolve the adaptations needed in their new environment. For example, some populations of the fish known as the guppy (*Poecilia reticulata*) live in localities in which predation pressure is intense, while others live where that pressure is lower. Those in the high-predation localities mature and breed at a smaller size and earlier than those in low-predation localities, so as to minimize the chances of being eaten before they have reproduced. In one experiment, populations from high-predation localities were introduced to low-predation sites—and, within a few years, each had evolved the new strategy appropriate to its new environment [17].

worked out in 1950 by the German taxonomist Willi Hennig [18], and it is called the cladistic method (see Concept Box 6.4).

Hennig's method seems clear and simple, but the problem arises when deciding which characters to use. The most readily available are aspects of the morphology of the organism, but the danger is that similar ways of life can lead different groups that are not closely related to show similar characteristics. There are many instances of this in the fossil record, but usually a careful study of the complete organism will reveal other features showing that the similarity is the result of convergent evolution. A good example is found in the extinct litoptern mammal herbivores of South America, whose limb adaptations to rapid movement on the open, grassy pampas show a remarkable similarity to the limbs of horse ancestors—but many features of their skulls show that the litopterns are really part of a great radiation of South American herbivores, and not at all closely related to horses. But where the taxa are more closely related, as when one is investigating the

First of all, Hennig tried to identify a group of taxa that were all related to one another, sharing a common ancestor and including all its descendants. Such a lineage is known as a **clade**. He then treated the process of evolutionary change in this clade as a series of branching events, or "dichotomies," at each of which a single group divides into two daughter groups. At each dichotomy, known as a node, one or more of the characteristics of the group changes from the original ancestral or **plesiomorphic** state into a derived or **apomorphic** state. The plesiomorphic characters are recognized by comparison with an "outgroup," which is closely related to the lineage being studied, but is not a part of it. The evolutionary history of the group can then be portrayed as a branching **cladogram**. Thus, in Fig. 6.10, characters a-g evolved after the divergence between the group 1, which is the outgroup, and groups 2–5. They are therefore derived, or apomorphic, relative to the characters of group 1 (in which these characters have remained primitive, or plesiomorphic), but plesiomorphic for groups 2–5. Other new, apomorphic characters then evolved at different points within the evolutionary history of groups 2–6 and can therefore be used to analyze their patterns of relationship.

In constructing a cladogram, the characters shown by the different taxa are listed, and the taxa are then arranged so that those that show a similar set of characters are placed in adjacent positions on the branching "tree." As far as possible, it is assumed that each apomorphic evolutionary event only occurred once in the history of each group of related taxa (a concept known as economy of hypothesis, or **parsimony**), and the taxa are arranged on the cladogram in such a way as to minimize the number of parallelisms. For example, in Fig. 6.10 it is most parsimonious to believe that character h, which is absent in group 5, has evolved twice, because that involves the assumption of only that single additional evolutionary event (Fig. 6.10a). The alternative is to transfer the origin of group 2 to near the base of groups 3/4, with the consequent need to assume that characters i–k had been lost in the evolution of group 2 (shown in brackets)—an assumption of three additional evolutionary events, instead of only one (Fig. 6.10b).

Of course, species show an enormous variety of characteristics, some of which are linked to one another by such aspects as function, development, or behavior. In constructing the cladogram, the taxonomist therefore has to be careful to try to avoid selecting more than one of each set of linked characters. Even then, it is still possible to select different arrays of characteristics that may give rise to different cladograms. Though the use of a greater number of characteristics may minimize the importance of these difficulties, it becomes progressively more difficult to arrange the taxa and analyze the results. This has led to the introduction of computer programs such as PAUP (Phylogenetic Analysis Using Parsimony) that calculate the cladogram showing the most parsimonious arrangement of the taxa. Even then, it is not rare for more than one cladogram to show equally parsimonious, and therefore equally likely, possible patterns of evolutionary relationship. The relationship between a particular set of taxa may also be so uncertain that they have to be shown as a converging to a single point; this is known as a *polychotomy*, and the cladogram is said to be not fully "resolved." An example of these problems comes from the study of the Hawaiian plants known as silverswords (Fig. 6.11).

interrelationships of species or genera, and the characteristics being used are much finer points of detail, the results of convergent evolution may be far more difficult to detect. In any case, the problem is not confined to the fossil record. Until the advent of molecular methods, some "groups" of living organisms that biologists had assembled on the basis of what seemed to be quite reasonable features were in fact not closely related at all. For example, a number of lineages of songbird, such as wrens, warblers, nuthatches, treecreepers, robins, shrikes, crows, magpies and jays, seemed to have representatives in Australia, and it was thought that they had originated in the Old World and spread from Southeast Asia through the East Indies to Australia. But molecular studies [19] showed that in fact the songbirds originated in Australia. Only after their dispersal out of Australia in the Eocene did they produce the huge worldwide radiation that today includes nearly half of all living species of bird.

But the new revelations of molecular methods can also work the other way around by showing that taxa that had previously been thought to be quite unrelated are in fact part of a single radiation. The most spectacular example is the demonstration that the huge elephants, the tiny elephant-shrews, the insectivorous tenrecs and long-snouted aardvarks, the rabbit-like conies, the Cape golden mole, and the aquatic sea-cows and manatees are all the descendants of a single early radiation of mammals that took place in Africa some 55 my ago. As a result, they are now placed in a single group known as the Afrotheria (see p. 318), and provide a completely new example of endemic biogeography, instead of being a rather puzzling array of diverse mammals whose history and biogeography had all to be analyzed separately.

These new insights into relationships depend on analysis of the DNA of the organisms (see Concept Box 6.1). The constitution of the DNA of any individual is virtually unique, depending on the combination of molecular structures received from its parents, together with any changes that have taken place in the process. The more closely related two individuals are, however, the more similar their DNA profiles are likely to be.

DNA is also found in other parts of the cell as well as the nucleus, such as in the **mitochondria** (which are responsible for the control of respiration in cells) and the **chloroplasts** (the green structures in plant cells where the conversion of energy from sunlight takes place); these types of DNA are not involved in cell reproduction. They are therefore not subjected to the recombination process (during which some genetic material is exchanged between the maternal and paternal chromosomes), seen at the commencement of each new generation, and they are consequently more stable. They are also haploid and therefore contain a smaller variety of genetic information than the chromosomes in a diploid cell. A further great advantage of the use of extranuclear DNA in such research is that the rate of change of its molecule is far greater than that of nuclear DNA, so that differences in this show up in a much smaller number of generations. It also survives better than nuclear DNA in dead tissues, so it has proved more useful in analyzing preserved and dried specimens from museum collections. Studies of extranuclear DNA found in the mitochondria, known

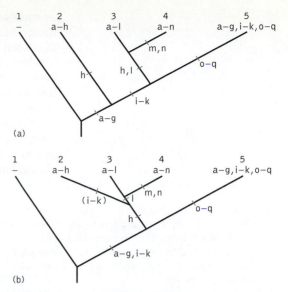

Fig. 6.10 Cladogram of the relationships between five groups, using characteristics a to q. The positions at which characters were lost are shown in brackets.

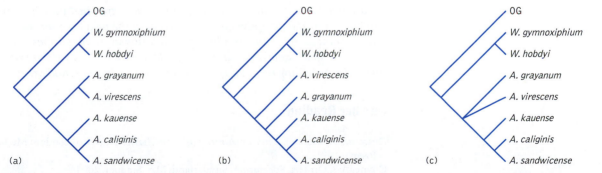

Fig. 6.11 Three equally parsimonious cladograms of the interrelationships of species of the Hawaiian silversword genera *Argyroxiphium* and *Wilkesia*, showing different interpretations of the relationships between *A. virescens* and *A. grayanum*. OG is the outgroup used in the analysis. From Funk & Wagner [23].

as mtDNA, have therefore become increasingly important in phylogenetic research while that found in the chloroplasts of plants is also proving useful in the study of their interrelationships.

When taxonomists study the evolutionary relationships between organisms, either within a species or comparing species, the DNA profile is therefore an ideal source of information, since it is not confused by superficial resemblances of structure caused by such phenomena as convergent evolution. Techniques are now available for the direct study of DNA sequences of living organisms, so relationships can be worked out using such molecular methods. The application of these methods to fossil material is limited by the gradual breakdown and eventual loss of DNA in the process of fossilization.

Summary

1 The organisms that make up the living world are separated into distinct species, each of which has a particular assemblage of adaptations—characteristics that enable it to survive in a particular environment, utilizing a particular source of nourishment.

2 Evolution, the gradual modification of one type of organism or structure into another, is the fundamental phenomenon that underlies most of the processes of the living world. It was Darwin who first put forward convincing evidence for its occurrence, and suggested that natural selection is the mechanism of evolution.

3 Natural selection arises from the fact that each pair of individuals produces far more descendants than are needed to replace them in the population, so that there is competition for survival between these descendants. This leads to the natural selection of those individuals that have the greatest number of favorable characters. In the short term, this ensures that the species remains adapted to its existing environment. In the longer term, if the environment changes, natural selection then acts on the population to change its characteristics, so that it remains adapted.

4 These characteristics are controlled by a genetic mechanism, so they are inherited from generation to generation. This mechanism involves the continual production of slight variations in the characteristics of the species, due to the recombination of existing characteristics, and also the appearance of new features by mutation.

5 Natural selection takes place independently in each population of a species, adapting it to local conditions. Continued isolation between these populations can lead them to gradually become so different from one another that they have become new species. However, under some circumstances, a new species can arise in the same area as its parent species.

6 Cladistics, a method that presents the pattern of relationship between species as a series of bifurcations, provides a clear way of presenting those relationships. Though it can be misleading if based only on morphological characteristics, in which convergent evolution due to similar ways of life may take place, this danger can now be avoided by the use of molecular methods.

Starting with an explanation of the discovery and mechanism of evolution, we have now followed the process of speciation, explained the role of isolation, defined the nature of the species, and explained how its course can be simply and unequivocally portrayed. In the next chapter, we show how advances in understanding the relationships of species to one another in space and time have at long last provided us with a reliable method of revealing the history of lineages, biota, biomes, and areas.

Further Reading

Desmond A, Moore J. *Darwin (A Biography of Charles Darwin).* London: Michael Joseph, 1991.

Coyne JA & Orr HA. *Speciation.* Sunderland, MA: Sinauer, 2004.

Schluter D. *The Ecology of Adaptive Radiation.* Oxford: Oxford University Press, 2000.

Weiner J. *The Beak of the Finch.* London: Vintage Books, 1994.

References

1 Grant PR, Grant BR. Species before speciation is complete. *An. Miss Bot Gdn* 2006; 93: 94–102.

2 Blair AP. Isolating mechanisms in a complex of four species of toads. *Biol Symp* 1942; 6: 235–249.

3 Mallet J. Hybrid speciation. *Science* 2007; 446: 279–283.

4 Fryer G, Iles TD. *The Cichlid Fishes of the Great Lakes of Africa; Their Biology and Evolution.* Edinburgh: Oliver & Boyd, 1972.

5 Kuchta SR, Parks DS, Mueller RL, Wake DB. Closing the ring: historical biogeography of the salamander ring species *Ensatina eschscholtzii. J. Biogeogr* 2009; 36: 982–995.

6 Jackman TR, Wake DB. Evolutionary and historical analysis of protein variation in the blotched forms of salamanders of the *Ensatina* complex (Amphibia: Plethodontidae). *Evolution* 1994; 48: 876–897.

7 Feder JL. *et al.* Allopatric genetic origins for sympatric host-plant shifts and race formation in *Rhagoletis. Proc Nat Acad Sci USA* 2003; 100: 10314–10319.

8 Walker AF, Greer RB, Gardner AS. Two ecologically distinct forms of Arctic Charr *Salvelinus alpinus* (L.) in Loch Rannoch. *Scot Biol Conserv* 1988;43: 43–61.

9 Seehausen O, *et al.* Speciation through sensory drive in cichlid fish. *Nature* 2008; 4552: 620–626.

10 Seehausen O, van Alphen JJM, Witte F. Cichlid fish diversity threatened by eutrophication that curbs sexual selection. *Science* 1997; 277: 1808–1811.

11 Weiner J. *The Beak of the Finch.* London: Vintage Books, 1994.

12 Grant PR, Boag PT. Rainfall on the Galápagos and the demography of Darwin's finches. *Auk* 1980; 97: 227–244.

13 Gibbs HL, Grant PR. Ecological consequences of an exceptionally strong El Niño event on Darwin's finches. *Ecology* 1987; 68: 1735–1746.

14 Grant PR. Hybridization of Darwin's finches on Isla Daphne, Galápagos. *Phil Trans R Soc London B* 1993; 340: 127–139.

15 Williamson PG. Palaeontological documentation of speciation in Cenozoic molluscs from Turkana Basin. *Nature* 1981; 293: 437–443.

16 Scott EC. *Evolution vs. Creationism.* Berkeley; University of California Press, 2004.

17 Reznick DN, *et al.* Evaluation of the rate of evolution in natural populations of guppies (*Poecilia reticulata*). *Science* 1997; 275: 1934–1937.

18 Hennig W. *Grunzüge einer Theorie der phylogenetischen Systematik.* Berlin: Deutscher Zentralverlag. 1950. English translation of 3rd edition: *Phylogenetic Systematics,* trans. DD Davis & R Zanderl. Urbana, University of Illinois Press, 1966.

19 Barker KF, Cibois A, Schikler P, Feinstein J, Cracraft J. Phylogeny and diversification of the largest avian radiation. *Proc Nat Acad Sci USA* 2004; 101: 11040–11045.

20 Stebbins GL. *Variation and Evolution in Plants.* New York: Columbia University Press, 1950.

21 Grant PR. *Ecology and Evolution of Darwin's Finches.* Princeton, NJ: Princeton University Press, 1986.

22 Dowdeswell WH. *Evolution: A Modern Synthesis.* London: Heinemann, 1984.

23 Funk VA, Wagner WI. Biogeographic patterns in the Hawaiian Islands. In: WI Wagner, VA Funk, eds. *Hawaiian Biogeography: Evolution on a Hotspot Archipelago,* Washington, DC: Smithsonian Institution Press, 1995.

From Evolution to Patterns of Life

The previous chapter explained how the process of evolution leads to the appearance of new species. As time goes by, that may lead to the appearance of a radiation of related species. We now need to explain how modern methods allow us to establish precisely how the different species are related to one another, when each of them diverged from its relatives, and how they came to spread from their place of origin to other locations. Only then shall we be able to use that understanding to trace the biogeographic history of lineages of organisms, of biotas, of biomes, and of the areas in which they are found.

If the evolutionary process continues after a new species has evolved, the many descendants of the original species may eventually spread over large areas of the planet. Sometimes the areas in which related species are found today may be widely separated from one another, a situation known as a *disjunct* distribution. Trying to understand how and why these patterns appeared and changed makes us confront a series of questions—and the more of these questions we can answer, the more confidence we can have in the correctness of our analysis of the reasons for these patterns of life.

1 On the geographical/geological side, when did two areas in which we see the organisms living today become separate from one another?

2 Was this the result of geological processes such as plate tectonics, which might have caused the areas to split apart or to become separated by mountain-building, or was it the result of environmental events such as changes in sea level or climate?

3 On the biological side, how are the taxa related to one another?

4 Did their appearance in separate areas take place before or after the appearance of a barrier between them?

5 When did they start to become separate due to evolutionary change?

Until comparatively recent years, we were unable to answer many of these questions. As explained in Chapter 1, it was only in the 1960s that understanding of the processes of plate tectonics enabled us to answer questions 1 and 2. Only a few years after that, in the late 1960s, the advent of molecular analysis of the proteins of living organisms, together with Hennig's cladistic method allowed us to obtain reliable answers to question 3. Most methods of analysis in historical biogeography therefore start with data that provide answers to these three questions, so we can now turn to the problems posed by questions 4 and 5.

Dispersal, Vicariance, and Endemism

Let us imagine a species that has recently evolved. It is likely, to begin with, to extend its area of distribution or **range** until it meets barriers of one kind or another to its further spread; this is known as **range extension** (Fig. 7.1, left). Sometimes it is eventually able to disperse across this barrier, which now provides the necessary isolation for the population on either side to differentiate into separate species (Fig. 7.1, center). (This is sometimes referred to as **jump dispersal**.) But the barrier could also have appeared within the area of a distribution of an existing species, subdividing it into separate populations, which could then diverge from one another into separate species (Fig. 7.1, right), a process known as **vicariant speciation**. After it has arisen, the new species will gradually extend its area of distribution until it reaches barriers of one kind or another (physical, ecological, etc.) beyond which it cannot readily spread. It is said to be **endemic** to that particular area, being found there and nowhere else.

Various definitions of **areas of endemicity** have been proposed, the criteria required ranging from the biological (relatively extensive sympatry of the taxa involved) to the physical (areas delimited by barriers). No two organisms live in *exactly* the same area; even those that live in the same lake or island will have at least slightly different ecological preferences and will therefore not live in precisely the same set of locations. This is a particular problem at smaller scales of study, where local ecology is important. At the other end of the scale, where major continental or subcontinental areas are the subjects of study, though the areas are easy to define, problems of tectonic subdivision or fusion are more likely to arise. This is then more likely to lead to successive patterns of subdivision and subsequent reunion into different patterns—a phenomenon known as a reticulate pattern (see below, p. 211).

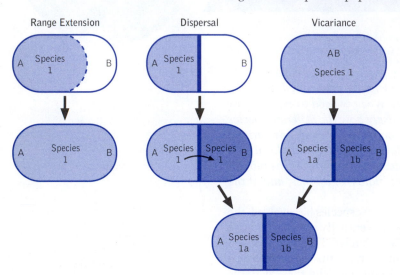

Fig. 7.1 Range extension (left): the species is at first found only in area A. Later, it gradually extends its range of distribution into the neighboring area. However, in the absence of any barrier between the two areas, it cannot differentiate into a new separate species.

Dispersal (center): the species is at first restricted to area A by a barrier that separates it from area B, and only later disperses across the barrier. The two populations of the species now each differentiate into separate species.

Vicariant speciation (right): a species originally occupies the whole of area A and B, but these two areas become separated from one another by a barrier. The original species then differentiates, by gradual genetic change, into two separate species separated by the barrier.

N.B. The results of dispersal and of vicariant speciation are identical.

Phylogenetic Biogeography and Cladistic Biogeography

The analysis of the significance of the endemism patterns found in related species lies at the heart of much of modern biogeography, but there are two different approaches to it. The first starts with the selection of a group of biological taxa that are assumed to have diverged from a common ancestor, and then uses their pattern of distribution to study the implications for the history of the areas to which they are endemic. The earliest and clearest example of this method is the 1966 work of the Swedish entomologist Lars Brundin, who was the first to realize the potential of cladistics (see p. 198) as a tool for analyzing the distribution patterns of areas of endemism [1]. He referred to his method as **phylogenetic biogeography**. Studying the distribution of three subfamilies of chironomid midges in the Southern Hemisphere, he first produced a cladogram of the evolutionary relationships of all the species. In place of the name of each species in the cladogram, he then instead inserted the name of the continent on which it is found, and where it is therefore endemic, transforming the phyletic cladogram into a **taxon–area cladogram** (Fig. 7.2). The result was a consistent pattern in which the African species appeared to have diverged first, followed in turn by those of New Zealand, South America, and Australia. So, cladistics plus information on endemicity produces a biological taxon-area cladogram in which information on a number of lineages is now also providing information on their sequence of association into the biotas of the different continents.

This sequence, based on the evolutionary relationships of the midges, was independently supported by geophysical data on the sequence of breakup of the Gondwana supercontinent. Again, Africa was the first to break away, followed in turn by New Zealand, South America, and Australia. (India and Antarctica do not appear in this analysis because they do not contain these midges.) So, the biological taxon-area cladogram was paralleled by a **geological area cladogram**. This in turn had useful implications as to the apparent geological ages of the different groups of midge, because the dates of separation of the continents were known from the geophysical data.

The second type of approach to the relationship between patterns of endemism and the history of the areas concerned starts with the identification of areas that are assumed to have had a simple history of successive subdivision from an original single unit. So these did not subsequently fuse together, nor were they colonized more than once by any individual taxon, while any geological subdivision was always accompanied by speciation. The presumed history of the subdivision is shown as a geological area cladogram, and its accuracy is then tested by seeing the extent to which the biological cladograms of the organisms that are found in these areas conform with the geological area cladogram. Because it starts with the areas, this type of approach is known as **area biogeography**, but most of the examples of its use are usually known as **cladistic biogeography**.

A good example of this approach is the work of the British palaeontologist Colin Patterson [2], who included the locations of fossils of a number of **monophyletic** groups (i.e., groups all of whose members are descended from a single common ancestor), and showed interesting parallels and differences between their taxon-area cladograms. The geological area cladogram (Fig. 7.3a) reflects the plate tectonic history of the areas, which began with the division of Pangaea into Laurasia and Gondwana. Laurasia then

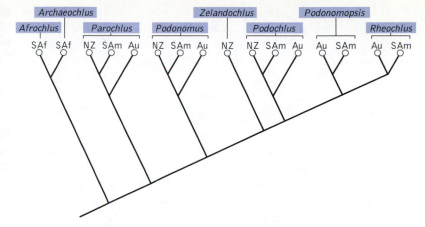

Fig. 7.2 Simplified taxon–area cladogram of some of the Gondwana genera of podonominine chironomid midges studied by Brundin [1]. The names in italics are those of the genera involved, while the circles represent individual species. The initials indicate the continent in which each species is found:

Au, Australia
NZ, New Zealand
SAf, South Africa
SAm, South America

The African genera appear to have diverged first. In each of the other genera, the divergence of the New Zealand species preceded the divergence between the South American and the Australian species.

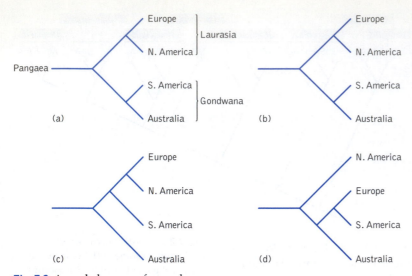

Fig. 7.3 Area cladograms of several vertebrate groups. After Patterson [2].
(a) Geological area cladogram of the breakup of the continents.
(b) Consensus taxon-area cladogram of beech trees, chironomid midges, xylotine flies, etc.
(c) Taxon-area cladogram of hylid frogs.
(d) Taxon-area cladogram of syrphid flies.

split into North America and Europe, while South America and Australia were two of the later fragments of Gondwana. The biological taxon-area cladograms of beech trees, chironomid midges, xylotine flies, and many other taxa (Fig. 7.3b) are similar to this geological area cladogram, suggesting that their histories were all responding to the sequence of events shown in the geological area cladogram. The taxon-area cladogram of hylid frogs (Fig. 7.3c), however, shows South America in an intermediate position, between North America and Australia. This could be explained by the fact that South America later became linked to North America via the Panama Isthmus (cf. Fig. 9.13) and that this was the event that allowed hylid frogs to disperse northward. The taxon-area cladogram of syrphid flies (Fig. 7.3d) is unlike any of the others, suggesting that it may be the result of incorrect taxonomy. (Conflicting results such as this can be useful in science, for they may reveal areas in which further research is needed.)

It is clearly valuable to have two sets of data, one on geological history and the other on biological history, for it enables us to compare the patterns of each, which may show interesting parallels or differences. When the underlying assumptions of the cladistic biogeography method are fulfilled, there is a neat concordance between the geological and the biological cladograms (Fig. 7.4a). However, there are many possible exceptions. For example, two areas may unite, their two biotas then merging into a single common biota, as happened after the disappearance of the Turgai Strait in the Oligocene allowed the European and Asian biotas to unite in a common Eurasian biota by range extension (Fig. 7.4b). (This has also been called **geodispersal** because it results from a geological rather than from a biological event.) Another example is when the splitting of a single area into two parts is *not* accompanied by biological speciation, so that the same taxon is present in more than one area (Fig. 7.4c). The same result is found if one of the areas resulting from the geographical split is colonized by recent dispersal (Fig. 7.4d). In contrast, sympatric speciation may take place within a single area, which therefore contains more than one related taxon (Fig. 7.4e). Alternatively, one of the taxa may become extinct, which will also alter the apparent taxon-area cladogram (Fig. 7.4f). Finally, the same geological event may have quite different implications for the distributions of biological groups of different ecologies.

Interpreting the Biogeography of the Past

The further we go back in time to examine the patterns of life in the past, the more uncertain we become about the reliability of what we know. This is because of the limitations of the fossil record and of our ability to identify the taxa concerned from their fossil remains. Biogeographers have implicitly recognized this by focusing on larger taxonomic units such as genera or families rather than species, and by similarly analyzing larger geographical units such as continents rather than individual regions within them. This has had the additional advantage that large-scale geological or geographical events, such as plate tectonic movements or changes in sea level, are well documented. In the case of ancient, extinct groups, this gave enough information for us to be able to correlate their distribution with the former patterns of land and sea that geological information revealed in the 1970s. This provided satisfying explanations, for example, of the patterns of Permo-Carboniferous plants or land vertebrates, or of Late Cretaceous dinosaurs, early mammals, and plant spores (p. 21).

Attempts to explain the more ancient patterns have therefore tended to accept and use this increasingly detailed geological information, and to then try to fit the biological information to it in order to identify the sequence of patterns of distribution that are most likely to have led to that which we see today. That pattern may have arisen from the operation of four different biological processes: vicariance, **duplication**, dispersal, or extinction. (Where we are dealing with large areas such as continents in the distant past, it is impossible to estimate whether new species that appear within them arose by dispersal or vicariance at the local level. Their appearance can therefore only be ascribed to a process of "duplication" that does not make any assumptions as to how the taxa reached another area.) However, these four processes differ in the extent to which the results provide useful information on the history of the areas concerned.

In the case of both vicariance and duplication, the new species is still found in the area in which it originated. Its geographical relationships to related species, especially to the ancestral species, have therefore not changed, and consequently these provide firm data on biogeographical relationships. By contrast, where a new species arises by dispersal, it is now found in an area where its ancestor did not occur. Similarly, in the case of extinction, part of the ancestral area of endemism has been lost. In both of these cases, the pattern of ancestor–descendant geographical relationships has been broken. These phenomena therefore do not provide information on the biogeographical history of the groups involved and do not help us in trying to choose between alternative hypotheses. To try to avoid this uncertainty, modern techniques therefore try to discover systems that minimize the extent to which we have to invoke dispersal or extinction in order to produce the final biogeographic patterns. It is only if

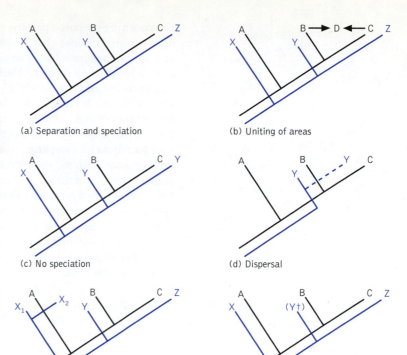

(a) Separation and speciation

(b) Uniting of areas

(c) No speciation

(d) Dispersal

(e) Sympatry

(f) Extinction

Fig. 7.4 The possible relationships between a cladogram of the history of the separation or fusion of areas (in black) and the corresponding cladogram of the evolution of the taxa (in blue).
(a) Areas A, B, and C have separated from one another, and this has been accompanied by the evolution of separate taxa (X, Y, and Z).
(b) Areas B and C have become united, so that the resulting area D contains taxa Y and Z.
(c) Areas B and C have separated, but taxon Y has not separated into two species, and is therefore present in both areas.
(d) Taxon Y has dispersed from area B to area C, and is therefore present in both areas.
(e) The results of sympatric evolution of two taxa, X_1 and X_2, in area A.
(f) The results of the extinction of taxon Y in area B.

the resulting proposed pattern of sequential biogeographical changes still does not make sense that we see whether the occurrence of dispersal or extinction may have been the cause.

As a result of all these problems, cladistic biogeographers had to develop increasingly complicated statistical methods for analyzing the history of biotas, which have been well reviewed and explained by Crisci et al. [3]. Two different types of method have been recognized **pattern-based** and **event-based**. Pattern-based methods such as **Brooks parsimony analysis** (BPA) and **parsimony analysis for comparing trees** (PACT) [4] first try to find a common pattern of relationships, or **general area cladogram** (GAC) that shows the physical history of the areas of endemism involved. Any data that do not fit this GAC are then assumed to have been the result of dispersal, extinction, or speciation, and the computer program then attempts to find the explanation that involves the smallest number of such events. However, no attempt is made to identify which of these processes is involved in each event, so that this is left to the investigator to evaluate; nor is any estimation made of the times at which these events may have taken place. This vagueness makes it very difficult to evaluate and compare the results of these techniques [5].

Event-based methods instead specify which biological process was involved at each point in the branching cladogram. It is able to do this because each type of biological event (vicariance, duplication, dispersal, and extinction) is given a cost, so that the computer algorhythmic program can then search for the set of events that can best explain the GAC by involving the lowest cost—that is, the most "parsimonious" explanation. This is then compared with the cost that is expected by chance under the null hypothesis that no relationship exists between the taxon cladogram and the area cladogram. (A **null hypothesis** is one that assumes that the statistical relationship between two phenomena is due to chance alone.) As long as the suggested GAC costs less than that of the null hypothesis, it is reasonable to conclude that it is the most likely to be correct. A further important advantage of event-based methods is that they can also incorporate estimates of the dates of the events, using geological information.

A good example of such a method is **parsimony-based tree fitting**. In this method, "duplication" (where a new species arises in the same area as its closest relative) and vicariance are given a low cost, 0.01, while, for reasons explained earlier, higher costs are given to dispersal (2.00) and extinction (1.00). (Randomization trials have shown that this pattern of costs provides the greatest likelihood of obtaining a result that provides the best fit of the historical taxon/area relationships, as opposed to random patterns.) A computer program known as **TreeFitter** is then used to find the lowest-cost explanation of the biogeographic history of the biota in question.

The Spanish biogeographer Isabel Sanmartin has given a good example of the use of this method to explain the distribution patterns of the species of the southern beech tree *Nothofagus* in the Southern Hemisphere [6]. Over the past 25 years, attempts to explain this have perhaps provided a greater number of research papers in historical biogeography than any other problem. Analysis of the distribution of 23 of its 35 species allows us to compare the advantages and disadvantages of parsimony-based tree fitting and BPA. In Fig. 7.5, the numbers shown on the trees in Fig. 7.5a and 7.5d represent characters, or sets of characters, that define the different ancestral stages in the phylogeny. Some characters (e.g., no. 45) were present in the common area from which all the later areas were derived by subdivision, while others (e.g., nos. 37 and 41) only arose later, in more restricted areas that had appeared later in this process of subdivision. Fig. 7.5a shows the general area cladogram that results from a BPA analysis of the distribution of the characters. Some characters are found to be present in all the species, and others in subordinate groups. The analysis suggests that the Southern South America (SSA) clade is closest to an Australia (AUS)–New Zealand (NZ) clade, the New Guinea (NG)–New Caledonia clade being separate from these. The TreeFitter program finds that the total "cost" of this

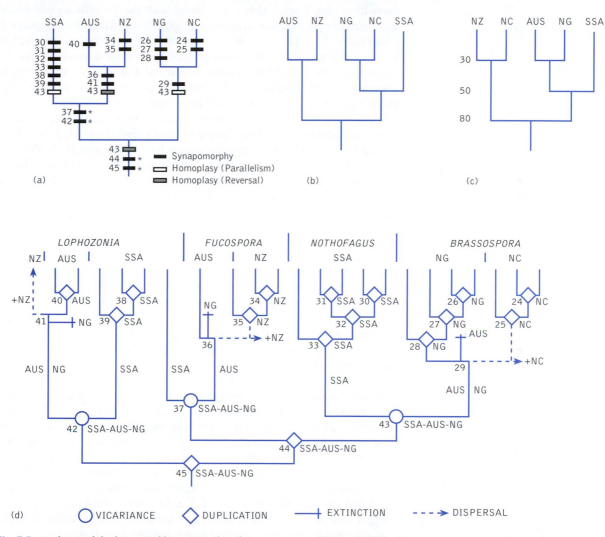

Fig. 7.5 Analyses of the historical biogeography of *Nothofagus*, modified from Sanmartin [7].
(a) General area cladogram suggested by Brooks parsimony analysis.
(b) Alternative general area cladogram suggested by TreeFitter.
(c) Geological area cladogram (the figures indicate the age of the events in millions of years).

(d) Detailed TreeFitter solution mapped onto the geological area cladogram.
The numbers on (a) and (c) indicate where in the phylogeny the characters, or sets of characters, that define the different groups made their appearance. AUS, Australia; NC, New Caledonia; NG, New Guinea; NZ, New Zealand; SSA, Southern South America.

area cladogram, which involves one extinction and one dispersal, is 3.21. However, the TreeFitter program also finds an alternative area cladogram (Fig. 7.5b) with the same cost, but in which New Guinea and New Caledonia are shown as closest to Southern South America, while Australia and New Zealand are now the more isolated clade.

This shows the extent to which these methods can provide significantly different results from the same data. However, in the case of problems involving the landmasses of the Southern Hemisphere, geological data can provide crucial diagnostic help. This is because the sequence of breakup of the various areas and the dating of these events are known with considerable confidence (Fig. 7.5c). If this is combined with the TreeFitter event-based reconstruction of the historical biogeography of the genus *Nothofagus*, the result is Fig. 7.5d. This suggests that the ancestor of *Nothofagus* evolved in Gondwana before that supercontinent split up, and also diversified into the subgenera *Lophozonia* and *Fucospora* and the ancestor of the subgenera *Brassospora* and *Nothofagus*. *Lophozonia* and *Fucospora* later became extinct in New Guinea (cost, 2.00) but dispersed from Australia to New Zealand (cost, 4.00), while *Brassospora* became extinct in Australia (cost, 1.00) but dispersed from New Guinea to New Caledonia (cost, 2.00). There were also three "cheap" vicariance events (cost, 0.03) and 16 duplication events (cost, 0.16). The higher total cost (9.19) shows the extent to which this reconstruction of the biogeographical history of *Nothofagus* differs from the geological history of the breakup of Gondwana—which is not a surprising conclusion. (In fact, palaeontological evidence supports some aspects of this reconstruction because it suggests that the southern beech tree became extinct in New Zealand—probably because of the Oligocene marine flooding of the island. If so, the presence of *Lophozonia* and *Fucospora* there must be the result of dispersal, as the reconstruction suggests. Similarly, fossil evidence shows that *Brassospora* was originally present in Australia, but became extinct there—just as the reconstruction suggests.)

Several observations can be made. First, TreeFitter provides a very clear portrayal of its solutions because these directly show the location and relative timing of the events, as well as the direction of the implied dispersals, unlike the BPA model. Second, use of the reliable geological cladogram provides an input from the real world, for it introduces data that arise from the quite independent phenomena of plate tectonics. Finally, this geology-aided TreeFitter solution also suggests that dispersals and extinctions are considerably more common than is likely to appear in unaided TreeFitter, with its relative heavy cost penalties for these events. Nevertheless, as will be seen (see ch. 11), the frequency with which molecular data indicate that dispersal, rather than vicariance, is the cause of many patterns in historical biogeography makes one suspect that this is the correct interpretation. The fact that TreeFitter would be unlikely to suggest this solution if the geological evidence were not available might lead one to question whether these heavy penalties are realistic.

Isabel Sanmartin and the Swedish biogeographer Fredrick Ronquist [7] have shown that this technique can also be used to infer a general biogeographic pattern for a set of different groups of Southern Hemisphere organisms. They used the phylogenies of 54 animal groups and 19 plant groups, and found interesting differences between the two. The patterns of relationship between the animal groups showed a close relationship between

Australia and South America. This is presumably a result of the long, ancient link between the two areas before their separation—both tectonically, by plate tectonic rifting, and biologically, by the glaciation of Antarctica and consequent extinction of most of its biota. The animal data show a weak relationship between Australia and New Zealand—so low that it is not statistically significant. In contrast, the plant data show a stronger link between Australia and New Zealand. Apparently, the 2,000-km (1240-mi) distance between these two areas is not sufficiently great to exclude long-distance dispersal of the plants, aided by the strong winds of the West Wind Drift. In a recent paper [8], Muellner et al. comment that the resistant properties of seeds make it much easier for plants to get from place to place and therefore plants are less affected by the details of physical geography than are animals.

Sanmartin and Ronquist point out that such data contradict the strong contention of the vicariance biogeographers, who held that dispersal was always a rare, random event, which could never lead to coherent, systematic patterns of distribution, and should therefore be ignored as mere "noise" in the biogeographic record. The Australian biogeographers Lyn Cook and Michael Crisp have suggested methods that can take account of such nonrandom, asymmetric factors in evaluating species trees [9]). Recent phylogeographic work also shows that the distributions of some southern temperate marine macroalgae, starfish, and gastropods are the result of dispersal by the ocean currents driven by the West Wind Drift, rather than being the relics of ancient vicariance [10].

It must not be forgotten, however, that the same area cladogram may have arisen more than once but at different times, as a result of a repetition of the same geographical change—for example, the separation between western North America and eastern North America during periods B, C and E in Fig. 5.7, a phenomenon known as **pseudocongruence**. One advantage of event-based methods of analysis is that they integrate dating information and can therefore detect pseudocongruence, unlike pattern-based methods.

Reticulate Patterns

All of the preceding methods involve the assumption that there is a single area cladogram that describes the relationships between the areas involved. However, in many cases, especially where a long period of time is involved, some of the areas may have had more than one type of relationship with one another, leading to a network-like, or **reticulate pattern** of relationships. This is prevalent in the historical biogeography of the Northern Hemisphere where, unlike the simple, tectonically driven, progressive subdivision and scattering of land areas in the Southern Hemisphere, the areas of land came into several different patterns of relationship due to changes in sea level (see Chapter 5, Fig 5.7). Both BPA and TreeFitter assume that the biogeographic history of the organisms being analyzed was the result of a series of vicariant events that can be portrayed as a general area cladogram. They would therefore not be able to analyze a reticulate pattern, in which a group might appear in a new area that had previously separated from the area in which the organism is found.

To cope with such a situation, Ronquist [11] has suggested a variant of parsimony-based tree fitting, in which no general area cladogram is assumed. In this event-based technique, known as **dispersal-vicariance analysis** (DIVA), vicariance and duplication are given a zero cost, while dispersal and extinction events cost 1.00.

DIVA has also been used [12] to find general patterns in the distribution of 57 monophyletic northern groups, also using geological and molecular information to sort events into four periods of time. The DIVA analysis showed a satisfying degree of correspondence between the events in each period of time and the pattern of intercontinental relationship present at that time. DIVA has also been used to find where and when the elements of a particular biome appeared, and how they spread during the gradual evolution of the biome by its assembly from different elements. It has also been used to estimate the time of origin of particular lineages that are specific to a particular biome, which may have implications in relation to the timing of particular connections, such as that between Greenland and Europe (see p. 363). The American plant biogeographers Michael Donoghue and Stephen Smith used DIVA to analyze patterns of distribution in the temperate forests of the Northern Hemisphere [13]. Their results suggested that many of their plant groups originated and diversified in eastern Asia and later spread into North America across the Bering region (rather than via Europe). Furthermore, these dispersals took place much later, and perhaps more frequently, than the dispersals of animal taxa suggested by Sanmartin, Enghoff, & Ronquist [12]. This implies that the insect fauna already resident in these forests must have had to adapt to the newly immigrant plants by switching hosts or foods. Such a degree of detail also enables investigation of the relationship between the times of arrival of new taxa, the climatic events that took place at that time, and such aspects of the ecology of the migrants as methods of dispersal and whether they were deciduous or evergreen.

DIVA thus holds out the promise of rewarding inquiries into details of the evolution of the community of a biome and of coevolution within it—aspects of historical biogeography that had previously seemed inaccessible. These inquiries may also reveal how, when, and why some of the unusual biomes that we see in the past were transformed into the more familiar ones that we see today.

However, as frequently happens, the now-widespread use of DIVA is revealing some aspects of its use that require caution or modification. In a recent paper [14], Kodandaramaiah has pointed out some limitations in the technique. For example, it does not distinguish between range expansion and dispersal unless molecular-based estimates of the divergence times of the lineages are also taken into account, or range expansion is given a lower cost than dispersal. The technique's option to limit the number of areas allowed in the analysis can also lead to errors, so that it is preferable to run the analysis several times, using different levels of constraint. Kodandaramaiah also discusses other problems with the technique, and some alternative methods, but he remains optimistic that model-based techniques hold the key to future progress.

In the above-mentioned techniques, the principle of parsimony is extremely valuable in ensuring that we find the simplest solution that is compatible with the known data. However, that principle also has fundamental limitations, such as the difficulty of incorporating data from the relative dating of the divergence of lineages, the fossil, record, and palaeogeography. New methods have now addressed these problems (see Guest Author Box 7.1).

Box
7.1

Guest Author

Breaking the Chains of Parsimony: A Parametric Model-Based Approach to Historical Biogeography

by Dr. Isabel Sanmartín, Real Jardín Botánico-CSIC, Madrid

Although they represent a considerable advance over more traditional cladistic methods, event-based approaches still have several limitations. For example, the cost of the events (i.e., the likelihood of an event) cannot be estimated from the data, but must be defined in advance, using ad hoc criteria intended to improve its performance, such as maximizing the conservation of biogeographical patterns along a phylogeny (see main text). But the most important limitation of both types of method is their reliance on the principle of parsimony. This has several implications.

First, as explained in the main text, in the attempt to find biogeographical patterns that remain unchanged from the ancestor to its descendants, event-based methods have to penalize events that may alter the geographic range of a taxon by making them more "expensive." Dispersal and extinction are such events, and the principle of parsimony therefore is likely to underestimate the frequency with which these events take place.

Second, parsimony-based reconstructions ignore two types of error associated with biogeographical inference. The first of these errors is that it requires that only the most parsimonious phylogenetic tree be used for reconstructing the biogeographical history of the group, assuming that this tree represents the true phylogeny. The second is that only "minimum-change" biogeographical reconstructions are considered in the analysis, even though alternative reconstructions that are almost as likely are possible [14].

Third, use of parsimony limits the type of data that can be used in the biogeographical analysis to the topology of the phylogenetic tree and the distribution of the terminal species. Other relevant evidence does exist, such as the times of divergence between lineages, and information from the fossil record or paleogeographic reconstructions. But these can only be considered secondarily, in the interpretation of the inferred events (i.e., cladistic biogeography), or incorporated indirectly in the analysis, as in parsimony-based tree fitting. For example, in analyzing the historical biogeography of *Nothofagus* (Fig. 7.5), a general area cladogram representing the geological sequence of Gondwana fragmentation was used to infer the relative timing and frequency of nonvicariant events such as extinction and transoceanic dispersal. But this type of information cannot be incorporated directly into inference algorithms if the principal of parsimony is employed.

In the last few years, a whole new class of methods has been developed that is not limited by these inherent biases of the parsimony approach. These new methods are termed **model-based** or "parametric" because they are based on models of range evolution whose parameters (variables) are quantifiable biological processes such as extinction, dispersal, or range expansion. They are analogous to the models used in phylogenetic studies to trace the evolution of a character in a phylogeny, but here the character states are the geographic ranges of the terminal species. Change between character states (or, here, alterations in the geographic range) is modeled as a stochastic process that changes along the branches of the phylogenetic tree according to a probabilistic model. This model determines the probability of range evolution from ancestor to descendants along the phylogeny, and it is dependent both on time (the amount of evolutionary change between ancestor and descendants) and on the rates (probability of occurrence) of biogeographical processes such as dispersal, extinction within one area, or range expansion that alter the geographic range of a taxon along the biogeographic tree. Instead of assigning these processes a fixed cost (i.e., likelihood of occurrence) as in event-based methods, model-based approaches allow us to estimate the frequency (i.e., rates) of these biogeographical processes directly from the data.

The result is a model of range evolution (Fig. 7.6), which can then be combined with the phylogenetic tree and the distribution of the terminal species to estimate the geographic range of the ancestors in the phylogeny, and also the probability of change between geographical ranges along the phylogeny. The method uses standard phylogenetic inference algorithms such as maximum likelihood or Bayesian inference. (The latter is a statistical technique that allows us to combine the range of uncertainty associated with data from one source, such as phylogenetic inference, with that arising from another, separate source, such as ancestral range reconstruction, in such a way as to improve the precision of any estimate based on both of them.)

These two statistical methods have the advantage over parsimony in that they allow us to evaluate all possible biogeographical scenarios in estimating the relative probabilities of ancestral areas, rather than considering only the most parsimonious [17]. In addition, Bayesian inference can effectively account for phylogenetic uncertainty, since rates of biogeographical processes in the matrix (e.g., dispersal) are estimated by integrating over all possible topologies and branch lengths in the phylogeny.

Relative to previous approaches, model-based methods offer many other advantages [18]. First, they allow the incorporation into biogeographic inference of estimates of the times of evolutionary divergence between lineages, represented by the length of branches in the phylogeny (Figure 7.6). As a result, unlike parsimony, they can take account of the fact that the likelihood of biogeographical changes such as dispersal is higher along long branches than along shorter ones. If the branch lengths represent the time since the divergence of a new taxon, these methods allow us to integrate this information directly into the inferred biogeographical history, and are therefore better in disentangling pseudocongruence from real shared biogeographic history.

Second, these methods allow us to use data from independent sources of evidence, such as the organism's fossil record, its dispersal capability, ecological preferences, or the availability of land connections [19], and integrate these into the biogeographical model. This can be done either in the form of new factors in the transition matrix, or by using variables that scale the likelihood of dispersal or extinction rate according to geographic distance, area size, and so on. For example, in an island system, rates of change between areas in the transition probability matrix could be made dependent on geographic distance or on the strength of wind currents facilitating dispersal [15].

Finally, model-based approaches provide a more rigorous statistical framework for the testing of alternative biogeographic hypotheses than do event-based methods. For example, two different biogeographic scenarios can be formulated in terms of alternative models and compared on the basis of how well they fit the data. Since each biogeographic scenario is described in terms of biogeographical processes such as dispersal and extinction, we can identify the processes that best explain the data by identifying the best-fitting model.

Recently, two new inference methods that are based on explicit models of quantifiable biogeographical processes have been proposed. The first one is a Bayesian island biogeography model [15] that allows the estimation of rates of dispersal/biotic exchange between isolated geographic areas from DNA sequence data and species distributions. Because this integrates phylogenetic and biogeographic uncertainty, estimates of biogeographical parameters are not dependent on a particular phylogeny. The method can therefore be used across a range of taxonomic groups that differ in their ecological preferences or dispersal capabilities. The second method is a diversification–extinction–cladogenesis (DEC) model [17] that uses maximum likelihood to estimate the ancestral geographic ranges, the rate of change between ranges, and biogeographical scenarios describing how ranges are divided at speciation nodes for a single individual group.

The Bayesian island model restricts geographic ranges in the transition matrix to single areas, and models dispersal as the equivalent to "jump dispersal" (the crossing of a geographic barrier). It would therefore be more appropriate for analyzing biogeographical scenarios in which areas are isolated by geographical barriers and dispersal is immediately followed by speciation; that is, the species that disperses to a new area is not expected to retain its widespread distribution for long. Instead, the DEC model accepts widespread states (ancestral distributions formed by two or more areas) in the matrix and models dispersal as the equivalent of range extension (the ancestor moves into a new area but also retains its original distribution). It is thus more suitable for biogeographical continental scenarios in which lineages are expected to expand their ranges and later speciate by vicariance or range division (see ref. 17 for a more detailed explanation of the two methods).

Despite their potential advantages, model-based approaches are not without their own limitations and challenges. The biggest challenge is computational feasibility and learning how to balance this with increasing realism of biogeographic scenarios [17]. For example, for methods allowing widespread ancestors such as the DEC model, the size of the matrix increases exponentially with the number of areas—so, there would be three possible states for two areas (A, B, AB), six possible states for three areas (A, B, C, AB, AC, BC), 15 for four areas, and so on. To reduce the size of the matrix without decreasing the number of initial areas, one can use models in which transitions between certain ancestral ranges are prohibited according to some biological, geographic, or geological criteria. For example, in biogeographic continental scenarios, one could restrict widespread states (geographic ranges comprising two or more areas) to combinations only of areas that are geographically adjacent to one another. All other combinations would imply an extinction event in the intervening area or the crossing of a barrier between the two areas, so these widespread states are not allowed. For island systems such as Hawaii, in which dispersal proceeds along the island chain, one can limit dispersal to follow a "stepping-stone" model, by making the rate of dispersal between nonadjacent islands in the chain equal to zero [14]. Whether or not these constraints are biologically realistic depends on each particular scenario. Despite these challenges, parametric models of biogeographical evolution represent an exciting new area of research for the integration of patterns, processes, and time within a common historical biogeographic framework.

The Molecular Approach to Historical Biogeography

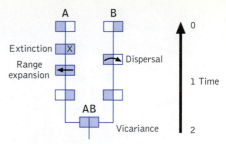

The investigation techniques described in the preceding section have allowed great improvements in our understanding of biogeographical events of the distant past, because the use of geological evidence has enabled us to estimate when the taxa involved started to differentiate from one another. But if we consider the more recent past, our interest turns to a more detailed level of geographical and taxonomic change that such now-distant and large-scale processes as plate tectonic movements and changes in sea level are rarely of any help. Fortunately, advances in our ability to perceive rates of change in biological processes have now made it possible for us to interpret biogeographical changes that took place more recently.

In the late 1960s, the Japanese geneticist Motoo Kimura suggested that the vast majority of the mutations that become permanently established in the DNA molecule are neutral from the point of view of fitness—they neither harm nor help the organism concerned. They are therefore not affected by selective pressures, and so they steadily accumulate through time. As a result, the greater the number of molecular differences between two organisms, the more distant in time was their evolutionary separation from one another. We can thus place a timescale against the phylogenetic tree. Where possible, this can also be calibrated by reference to an outside datum, such as a geological event or a well-dated fossil relevant to the date of separation of two lineages.

In the early days of testing Kimura's theory, relatively few proteins had been subjected to detailed structural analysis. But when these (such as hemoglobin and cytochrome c) were checked, they did indeed display a steady change in the course of time, which provided support for the neutral theory of molecular evolution. A variety of other biochemical molecules, such as the DNA, found in the mitochondria or chloroplasts, messenger RNA, and ribosomal RNA (see Chapter 6), are now used in these studies. However, as ever, it turns out that there are complications in this line of investigation [20]. First, this molecular clock's rate of ticking does vary between evolutionary lineages. Though these are similar within a given group, such as the primates, they may be different for rodents because of their smaller body size, shorter lives, and less stable population sizes—all features that affect the rates of mutation or of the repair of mutation. Studies of different molecules in the same taxa can often add greater precision.

These problems are minimized when we are studying the evolution of closely related groups, such as species or subspecies, and this has given rise to the new approach known as **phylogeography,** developed by the American biogeographer John Avise [21]. Its methodology is similar to that of phylogenetic biogeography because it, too, uses cladistics. However, its analyses are based on data from mitochondrial DNA (mtDNA), which evolves rapidly, is passed down only via maternal inheritance, and does not undergo the complex genetic changes and exchanges of meiosis. As a result, it can reveal relationships at the level of species or species complexes, and can therefore document distributional changes that have taken place in the more recent evolutionary past. By combining such results from clades from within different groups

Fig. 7.6 Simple example of a parametric biogeographic tree. The original range, AB, has been split by vicariance into two branches, A and B. In branch A, the range has changed first by range expansion and then by extinction in part of the new, larger range. In branch B, dispersal has taken place into a new area, where the new taxon is now endemic.

occupying the same areas, comparative phylogeography can also reveal previously unexpected geographical relationships. For example, the American biogeographers Brian Arbogast and Jim Kenagy [22] have studied the patterns of distribution of four different mammals that live in the forests of North America (Fig. 7.7).

Two new insights resulted from this research. First, they all show a genetic difference between a Pacific *coastal* lineage and an interior *continental* lineage. In three of the cases shown in Fig. 7.7, this difference is present within the distribution of what had previously been considered a single species. Only in the case of the tree squirrel *Tamasciurus* had the difference been recognized as distinguishing two species, *T. douglasi* and *T. hudsonicus*. Second, the geographical position of the break between the two lineages is similar in all the groups. This suggests that they all responded similarly, by vicariant evolution, to an episode of fragmentation of their ranges. Both the fossil record and the genetic evidence suggest that this may have been connected with Quaternary episodes of cyclical fragmentation, contraction, and expansion of the boreal forest. The later expansion allowed the coastal and continental lineages to come into renewed contact, which sometimes led to an overlap of their ranges. Neither of these insights was obvious from earlier analyses of these groups. Comparative phylogeography is now playing an increasingly important part in unraveling very detailed aspects of the patterns of biogeography that have resulted from events in the last few million years. In order to interpret the detailed information on the geographical distribution and genetic constitution of minor clades that comparative phylogeography can sometimes provide, it has now become necessary to develop complex computer methods such as **nested clade analysis**, or **NCA**.

The NCA method was developed by Alan Templeton of Washington University, and it illuminates the evolutionary history of a species over space and time, using information from its molecular structure. It first defines a series of haplotypes of a given molecule, first arranging these into groups that differ from one another by only a single mutational difference— the "1-step" clades. These groups are then treated as the units for an identical process that produces "2-step" clades, which similarly differ from one another by only a single mutation. This process is continued until there is only a single clade unit that includes all the haplotypes.

This technique can be illustrated by an investigation into the phylogeography of the Greek populations of the European chub *Leuciscus cephalus*, a fish that lives in rivers [23]. Figure 7.8 shows the area in which different populations of the fish were sampled, together with illustrations (A, B) of two different theories as to how their distributions might have come about. Bianco's hypothesis (Fig. 7.7A) suggests that the Danubian population spread to both eastern and central Greece via the Black Sea, while the western Greek populations were the result of a southward spread down the west coast. Economidis and Banarescu (Fig. 7.7B) instead suggest that the spread of the Danubian chub via the Aegean only reached eastern Greece, while other Danubian chub spread southward by changes in the patterns of drainage of rivers (1–4) first to central Greece and then westward to the coastal regions.

The NCA of the chub populations is shown in Fig. 7.8c and shows that they fall into three groups. The existence of a Danubian/Central Greek

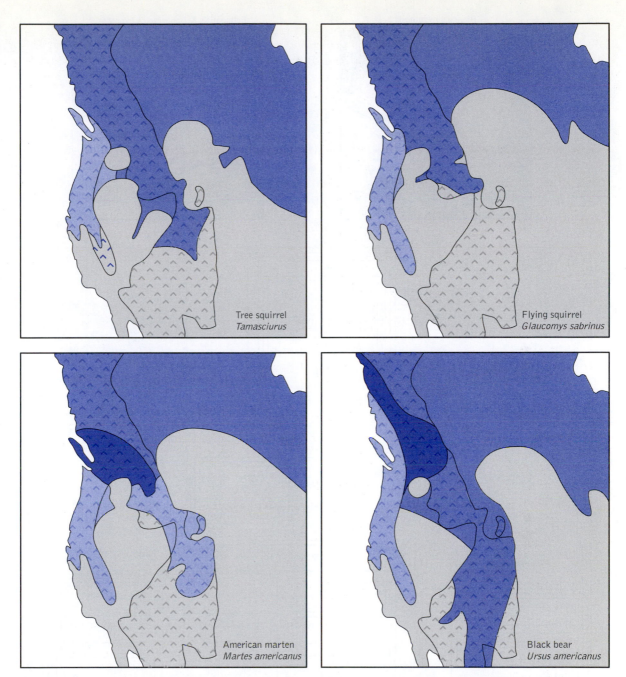

Fig. 7.7 The phylogeographical structure of four mammal taxa from the northern forests of North America, to show the difference between the current taxonomy and the results of molecular phylogeography. The range of the mammals of Pacific Coast lineage is shown in light blue, the range of continental lineage in medium blue, and the area of overlap of two lineages in dark blue. After Arbogast & Kenagy [22].

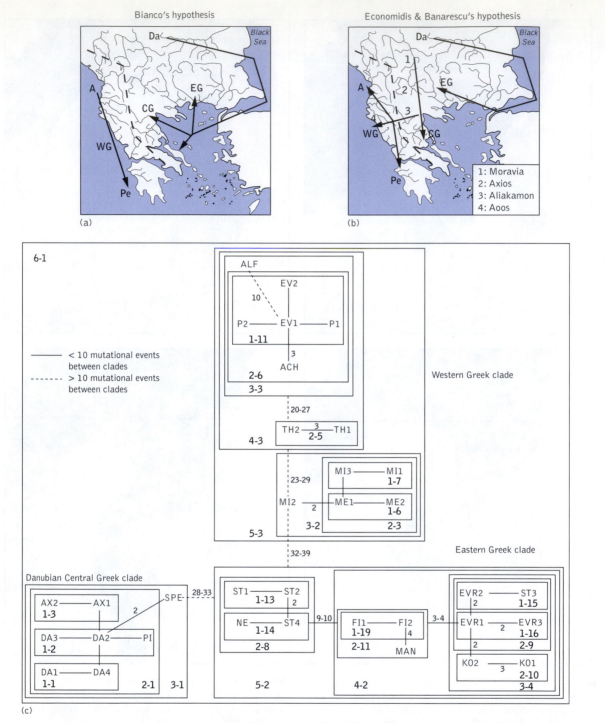

Fig. 7.8 A, B, two hypotheses as to the biogeographical history of populations of the chub, *Leuciscus cephalus* in Greece. From Durand, Templeton et al. [23].

C, Nested clade analysis of the interrelationships of these populations. Letters (e.g., ALF) represent abbreviations of the different locations in which they were found. Numbers indicate the hierarchy of clades: e.g., 3–3 indicates a three-step clade that includes 2–6, a two-step clade that in turn includes a one-step clade 1-11. Each solid line without a number represents a single mutational change from one of the two genotypes to the other. Any numbers beside the lines indicate the number of mutations that took place between the two genotypes linked by the lines; e.g., there were between 32 and 39 mutational differences between the western Greek clade, 5–3, and the eastern Greek clade, 5–2. See text for further explanation.

clade and of a separate eastern Greek clade supports those aspects of Economidis and Banarescu's theory. On the other hand, the presence of a quite separate western Greek clade supports Bianco's hypothesis that the chub of this region originated directly from the north and not from the Danubian populations. Other analyses of the data allow one to estimate the relative times, and therefore sequence, of these events and therefore their relationship to other geological or climatic phenomena.

Comparative phylogeography is now starting to throw light on an increasingly varied range of topics within the biogeography of the last few million years. These range from the expansion of a species out of the refugia to which it became restricted during the Ice Ages [24] and the direction of gene flow between geographically separate populations of a single species [25], to the relative contributions of historical and recent processes in causing the distribution of genes in a population today [26]. Inevitably, it is starting to overlap with investigations into processes that took place earlier in the Tertiary [27]. A recent paper [28] by three biogeographers (Richards, Carstens, & Knowles) working in the University of Michigan has suggested techniques that allow for the integration of genetic data and distributional data with palaeoclimatic modeling, and so testing alternative hypotheses about the timing and pattern of divergence of the populations. John Avise has recently published a paper [29] outlining the new perspectives that phylogeography has enabled. He also suggests some potentially fruitful directions that new research may take, such as comparative phylogeographic appraisals of multispecies regional biotas to identify shared patterns resulting from major historical events.

Investigations of the genetics and distribution of populations today have to cope with a very large amount of data, and this is provoking the development of complex computer programs. The variety, complexity, and rate of appearance and modification of these are such that it is not appropriate to try to explain them in a book such as this, and students are instead advised to follow them up in the current relevant research literature.

Molecules and the More Distant Past

Although the difficulties in the use of molecular data, mentioned earlier, are minimized when studying more recent and more closely related groups, such data are still very valuable in studying events in the more distant past [30]. The molecular data can be used to set up a phyletic tree of relationship with the implied dates of divergence between the different lineages. A similar tree of relationship is set up using data from the fossil record, which also gives dates of divergence based on the ages of the fossils. This in its turn introduces the possibility of errors. The most fundamental and inevitable of these is that the age of any fossil, placed on a given lineage, is a minimum age, because other, undiscovered, older fossils may lie on this same lineage before its divergence from its closest relatives. Other errors may arise from incorrect estimates of relationship, or if the age estimates of the fossil-bearing rocks are wrong. But recent techniques reduce the impact of any such individual error by calculating the effect that each datum has on the overall degree of difference between the molecular tree and the fossil-based tree [31]. Those that produce the greatest degree of inconsistency can then be removed from the data set and/or be reappraised. The results can then also be compared with different models of the rates of substitution in the molecular data tree in order to find a cross-correlated set that minimizes the differences between the two independent systems.

The extent to which plate tectonic, geological, and timescale information has been integrated into event-based methods of analysis of historical biogeography has created major problems for the proponents of pattern-based methods. The response of some of them has been to object to the use of any geological information, claiming that "biogeography should not be subordinate to geology," and to point out that the geological information might be wrong and that there are other theories of the history of the continents, such as the expanding earth theory [32]. However, the value of using this information is not that it makes the biological data dependent on the geological data and theories, but that it provides an independent picture of events, with which the biological picture can be compared and evaluated. This is normal good practice in any part of science and, if an expanding-earth theory seemed likely to provide a better match for the biological scenario, biogeographers would be quite prepared to consider it. At present, that is not the case.

Comparisons with the ages of plate tectonic events are not, however, the only examples of the uses of molecular data in relating biological events to environmental events. For example, ancient patterns of river drainage have been revealed as causative agents in the biogeography of freshwater crabs in the Cape of South Africa [33] and of freshwater molluscs in western North America [34]. Perhaps the most complex set of factors is that being revealed in the history of the biota of the Amazon Basin (cf. Ch. 11. p. 362). Here, molecular phylogeny has not only shown that the accepted taxonomy of such groups as parrots was incorrect, but has also revealed that the diversification patterns were much older than had been thought and had been influenced by the location and timing of episodes of mountain-building. Other variations in the climate of the Basin seem to have been linked to changes in the location of the low-level jet stream, itself affected by the long-term precessional changes in the Earth's orbit [35]. Faced with more detailed biological data, biogeographers are now having to become aware of ideas and information in a greater variety of other branches of science than ever before.

Summary

1 If two closely related species are widely separated from one another, did the original species disperse across a preexisting barrier, or did the barrier arise later, separating them by a process known as vicariance? The two possible answers to this question, as causes of patterns of disjunct patterns of distribution, have been supported by two opposing, disputing schools of biogeographers.

2 Cladistics, together with information on the areas of endemism of the taxa, can produce a taxon-area cladogram. Comparison of this with the plate tectonic history of the areas sometimes shows a relationship between the two processes. Further results may be obtained by comparing the taxon-area cladograms of different groups.

3 The many uncertainties involved in interpreting the results of these methods led to the development of complicated statistical analyses to try to resolve the problems. More recently, event-based methods have been developed. These specify whether dispersal, vicariance, extinction, or duplication of taxa were

In this chapter, we have shown how a species is defined, how it evolved and is related to other species, where it lives, and how we can use our knowledge to interpret its history. That understanding will now help us to reconstruct how the patterns of life on the planet came to appear through time. This, the study of historical biogeography, will make up Chapters 8 to 10 of this book. But first we shall examine the patterns of life on islands, which vividly illustrate evolution in action, and in the oceans that surround them.

Further Reading

Lomolino MV, Heaney LR, eds. *Frontiers of Biogeography.* Sunderland, MA: Sinauer Associates, 2004.

Pennington RT, Cronk QCB, Richardson JA. Plant phylogeny and the origin of major biomes. *Phil. Trans. Roy. Soc. Lond. B* 2004; 359: 1453–1656.

References

1 Brundin LZ. Phylogenetic biogeography. In: Myers AA, Giller PS, eds. *Analytical Biogeography*, pp. 343–369. London: Chapman & Hall, 1988.

2 Patterson C. Methods of paleobiogeography. In: Nelson G, Rosen DE, eds. *Vicariance Biogeography: A Critique*, pp. 446–489. New York: Columbia University Press, 1981.

3 Crisci JV, Katinas L, Posadas P. *Historical Biogeography: An Introduction.* Cambridge, MA; Harvard University Press, 2000.

4 Garzon-Orduña IJ, Miranda-Esquivel DR, Donato M. Parsimony analysis of endemicity describes but does not explain: an illustrated critique. *J Biogeogr* 2008; 35: 903–913.

5 Wojcicki M, Brooks DR. PACT: an efficient and powerful algorithm for generating area cladograms. *J Biogeogr* 2005; 32:755–774.

6 Sanmartin I. Event-based biogeography: integrating patterns, processes and time. In: Ebach MC, Tangney RS, eds. *Biogeography in a Changing World*, pp. 135–159. London: CRC Press, 2007.

7 Sanmartin I, Ronquist F. Southern hemisphere biogeography inferred by event-based models: plants versus animal patterns. *Systematic Biology* 2004; 53: 216–243.

8 Muellner, AN, Pannell CM, Coleman A, Chase MW. The origin and evolution of Indomalesian, Australasian and Pacific island biotas: insights from Aglaieae (Meliaceae, Sapindales). *J Biogeogr* 2008; 35: 1768–1789.

9 Cook LG, Crisp M. Directional asymmetry of long-distance dispersal and colonization could mislead reconstructions of biogeography. *J Biogeogr* 2005; 32: 741–754.

10 Waters JM. Driven by the West Wind Drift? A synthesis of southern temperate marine biogeography, with new directions for dispersalism. *J Biogeogr* 2008; 35: 417–427.

11 Ronquist F. Dispersal-vicariance analysis: a new biogeographic approach to the quantification of historical biogeography. *Systematic Biology* 1997; 46: 195–203.

12 Sanmartin I, Enghoff H, Ronquist F. Patterns of animal dispersal, vicariance and diversification in the Holarctic. *Biol. J. Linn. Soc. Lond.* 2001; 73: 345–390.

13 Donoghue MJ, Smith S. Patterns in the assembly of temperate forests around the Northern Hemisphere. *Phil. Trans. Roy. Soc. Lond. B* 2004; 359: 1633–1644.

14 Kodandaramaiah U. Use of dispersal-vicariance analysis in biogeography—a critique. *J Biogeogr* 2010; 37: 3–11.

15 Sanmartin I, van der Mark P, Ronquist F. Inferring dispersal: a Bayesian approach to phylogeny-based island biogeography, with special reference to the Canary Islands. *J. Biogeog.* 2008; 35: 428–449.

16 Sanmartin I, Wanntorp L, Winkworth R.-C. West Wind Drift revisited: testing for directional dispersal in the Southern Hemisphere using event-based tree fitting. *J Biogeog* 2007; 34: 398–416.

17 Ree RH, Smith SA. Maximum likelihood inference of geographic range evolution by dispersal, local extinction, and cladogenesis. *Syst. Biol.* 2008; 57: 4–14.

18 Ree RH, Sanmartin I. Prospects and challenges for parametric models in historical biogeographical inference. *J Biogeogr* 2009; 36: 1211–1220.

19 Ree RH, Moore BR, Webb CO, Donoghue MJ. A likelihood framework for inferring the evolution of geographic range on phylogenetic trees. *Evolution* 2005; 59: 2299–2311.

20 Pulquério MJF, Nichols RA. Dates from the molecular clock: how wrong can we be? *Trends in Ecology and Evolution* 2006; 22: 180–184.

21 Avise JC. *Evolutionary Pathways in Nature: A Phylogenetic Approach.* Cambridge: Cambridge University Press, 2006.

22 Arbogast B, Kenagy GJ. Comparative phylogeography as an integrative approach to historical biogeography. *J Biogeogr* 2001; 28: 819–825.

23 Durand JD, Templeton AR, et al. Nested clade and phylogeographic analysis of the chub, *Leuciscus cephalus* (Teleostei, Cyprinidae) in Greece: implications for Balkan Peninsula biogeography. *Mol. Phylogen. & Evol.* 1999; 13: 566–580.

24 Hewitt GM. Genetic consequences of climatic oscillations in the Quaternary. *Phil. Trans. Roy. Soc. Lond. B* 2004; 359: 183–195.

25 Zheng XJ, Arbogast BS, Kenagy GJ. Historical demography and genetic structure of sister species: deermice (*Peromyscus*) in North American temperate rain forest. *Molecular Ecology* 2003; 12: 711–724.

26 Templeton AR, Routman E, Phillips CA. Separating population structure from population history: a cladistic analysis of the geographical distribution of mitochondrial DNA haplotypes in the tiger salamander, *Ambystoma tigrinum*. *Genetics* 1995; 140: 767–782.

27 Riddle BR, Hafner DJ. A step-wise approach to integrating phylogeographic and phylogenetic perspectives on the history of a core North American warm desert biota. *J. Arid Environ.* 2006; 65: 435–461.

28 Richards CL, Carstens BC, Knowles LL. Distribution modelling and statistical phylogeography: an integrative framework for generating and testing alternative biogeographical hypotheses. *J Biogeogr* 2007; 34: 1833–1845.

29 Avise JC. Phylogeography: retrospect and prospect. *J Biogeogr* 2009; 36: 3–15.

30 Donoghue PCJ, Benton MJ. Rocks and clocks: calibrating the Tree of Life using fossils and molecules. *Trends in Ecology and Evolution* 2007; 22: 425–431.

31 Near TJ, Sanderson MJ. Assessing the quality of molecular divergence time estimates by fossil calibrations and fossil-based model selection. *Phil. Trans. R. Soc. Lond. B* 2004; 359: 1477–1483.

32 McCarthy, D. Are plate-tectonic explanations for trans-Pacific disjunctions plausible? Empirical tests of radical dispersalist theories. In: Ebach MC, Tangney RS, eds. *Biogeography in a Changing World*, pp. 177–198. New York: CRC Press, 2007.

33 Daniels SR, Gouws G, Crandall KA. Phylogeographic patterning in a freshwater crab species (Decapoda: Potamonautidae: *Potamonautes*) reveals the signature of historical climatic events. *J Biogeogr* 2006; 33: 1538–1549.

34 Liu H-P, Hershler R. A test of the vicariance hypothesis of western North American freshwater biogeography. *J Biogeogr* 2007; 34: 534–548.

35 Bush MB. Of orogeny, precipitation, precession and parrots. *J Biogeogr* 2005; 32: 1301–1302.

Life, Death, and Evolution on Islands

The limited and precise area and biota of islands provide us with three unique fields of biogeographic research. The first of these includes the ways in which their isolation and unusual, unbalanced biota provoke changes in the adaptations of their mainland colonists. Second, their limited area and biota, together with their diversity, allows us to make statistical analyses of the interactions between the processes of colonization, extinction, isolation, and island area; this also has some useful implications for the design of nature reserves. Third, the study of the sequence of colonization of islands that have been devastated of life provides invaluable insights into the ways in which ecosystems develop and change.

Study of the biogeography of the continents is quite different from that of islands. Those great areas of land have, over long periods of time, changed in their positions and interconnections, while new types of organism have been able to evolve, compete with one another, and change their patterns of distribution. Their complex ecosystems, with a great variety of interacting species, makes their interpretation quite difficult. The focus of biogeographical research on the biota of the continents has therefore been mainly on the interpretation of these long-term patterns and on the factors that underlie them.

In contrast, the limited biota, and therefore comparatively simple ecosystems, of oceanic islands provide unique insights into three quite different areas of research. The first is the ways in which the environment affects and controls the evolutionary process. Islands not only provide one of the essential necessities of evolution—isolation; they also ensure that the new colonist, arriving from the mainland environment, is confronted with a very different environment. On the one hand, its former competitors, predators, and parasites may be absent, presenting new opportunities, while on the other hand, the smaller area and differing patterns of climate provide new limitations. This aspect of island biogeography is dealt with in the first part of this chapter, culminating in an examination of how these factors have operated in forming the biota of the Hawaiian Islands.

The second area of research is the relevance of island biogeography to evolution and speciation: what controls the number of species on an island? The huge diversity and number of islands (there are over 20,000 in the Pacific alone) provide the equivalent of an enormous, ongoing natural experiment. How much can we learn from comparisons between the biotas of islands of different age, history, climate, size, or topography, or of islands that lie at different latitudes or at different distances from their source of colonists? Is it possible to deduce any general, underlying regularities that may control the diversity of the island biota, and perhaps enable us to predict this? In particular, this research attempts to identify and quantify the factors that control three phenomena: the rate at which new species reach an island, the rate at which species become extinct on an island, and the number of species that an island can support.

The third area is the role of biogeography in the evolution of island ecosystems. The great volcanic explosion of Krakatau in 1883 created islands such as Rakata, which were initially bare of life. Research on the colonization of Rakata and on the appearance of different ecosystems in different parts of the island and in similar islands provides unique opportunities to study the processes of ecological change and increasing complexity within a comparatively limited arena.

These are the interlocking, fascinating themes explored in this chapter.

Types of Island

As we have seen (Chapter 5), our planet shows islands of three different types, and their differing origins lead to their having biotas of different characteristics. Those of former continental fragments would originally have been a subset of the biota of the continent itself, but will have changed because of independent evolution and extinction within the new island. Some of that evolutionary change will have been a response to the different conditions of life on an island compared with those on the mainland (see below). In addition, if the fragment were to gradually move further and further away from the parent mainland, its biota would gradually become dominated by the results of transoceanic dispersal.

All of the biota of island arcs and of island chains caused by geological hotspots (see p. 169), on the other hand, originally arrived by transoceanic dispersal. Like that of the continental fragments, the biota of these islands will subsequently have become changed by evolution, but they may also show a progressive ecological change as the island ecosystem changes and offers new opportunities.

The islands of an arc all appear more or less simultaneously, while those of a hotspot chain appear (and disappear) in turn, but in both cases a number of islands exist at the same time. This provides the potential for interisland dispersal and for a more complex pattern of evolutionary change—a pattern that is sometimes called **archipelago speciation**.

Getting There: Problems of Access

Oceans are the most effective barriers to the distribution of land animals. Because few terrestrial or freshwater organisms can survive for any length of time in seawater, organisms can only reach an island if they possess special adaptations for transport by air or water. Dispersal to oceanic islands is therefore by a sweepstakes route (see p. 42), the successful organisms sharing adaptations for crossing the intervening region rather than for living within it. This greatly restricts the diversity of life that is capable of dispersing to an island.

Some flying animals, such as birds and bats, may be capable of reaching even the most distant islands unaided, using their own powers of flight, especially if, like water birds, they are able to alight on the surface of the water to rest without becoming waterlogged. Smaller birds and bats and, especially, flying insects may reach islands by being carried passively on high winds. These animals may, in their turn, carry the eggs and resting stages of other animals, as well as the fruits, seeds, and spores of plants.

Most land animals cannot survive in seawater long enough to cross oceans and reach a distant island, but some may occasionally make the journey on masses of drifting debris. Natural rafts of this kind are washed down the rivers in tropical regions after heavy storms, and entire trees may also float for considerable distances. Such a floating island could carry small animals such as frogs, lizards, or rats, the resistant eggs of other animals, and specimens of plants not adapted to oceanic dispersal. It is rare for such raft-aided dispersal to be seen and documented, but there was a good example of this in the West Indies in September 1995 [1]. Soon after two hurricanes had passed through the Caribbean, a mass of logs and uprooted trees, some over 10 m long, was found on the beach of the island of Anguilla, and at least 15 individuals of the green iguana lizard, *Iguana iguana*, were seen on logs offshore and on the beach. They included both males and females in reproductive condition, and specimens were still surviving on the island (where the species was previously unknown) over 2 years later. Judging by the track of the hurricanes, the lizards probably originated on the island of Guadeloupe, some 250 km (160 mi) away, and the journey probably lasted about a month.

A few plants have developed fruits and seeds that can be carried unharmed in the sea. For example, the coconut fruit can survive prolonged immersion, and the coconut palm (*Cocos nucifera*) is widespread on the edges of tropical beaches. However, since the beach is as far as most seaborne fruits or seeds are likely to get, only species that can live on the beach are able to colonize distant islands in this way. The fruits or seeds of plants that live inland would be less likely to reach the sea, and, even if they were able to survive prolonged immersion and were later cast up on a beach and germinated, they would be unable to live in a beach environment. They are therefore only able to reach island habitats by evolving different methods of dispersal, by winds or animals.

It is potentially far easier for a plant to adapt to long-distance dispersal. Very many plants show some adaptation to ensure that the next generation is carried away from the immediate vicinity of the parent [2]. It requires little elaboration of some of these dispersal devices to make it possible for them to traverse even wide stretches of ocean. In addition to this, successful colonization requires only a few fertile spores or seeds, whereas in most animals it requires the dispersal of either a pregnant female or a breeding pair. The spores of most ferns and lower plants are so small (0.01–0.1 mm) that they are readily carried considerable distances by winds; they therefore tend to have very wide patterns of dispersal. Microbial spores can be carried even greater distances. Some plants have seeds that are specially adapted to being carried by the wind. Orchid seeds, for example, are surrounded by light, empty cells, and some have been known to travel over 200 km (125 mi). *Liriodendron* and maple seeds have wings, and the seeds of many members of the Asteraceae (daisies and their relatives) have tufts of fluffy hairs; those of thistles have been carried by the wind for 145 km (90 mi).

Many fruits and seeds have special sticky secretions or hooks to make them adhere to the bodies of animals. Examples are the spiny fruits of burdocks and beggarticks, and the berries of mistletoe, which are filled with a sticky juice so that the seeds that they contain stick to birds' beaks. The seeds of many plants can germinate after passing through a bird's stomach, and those of some (e.g., *Convolvulus, Malva, Rhus*) can germinate after up to 2 weeks there. In the Canary Islands off the western coast of Africa grows the plant *Rubia*, whose seeds are carried in fleshy fruits that are eaten by gulls. One study [3] showed that the seeds remained in the bird's digestive system for 9–17 hours, during which time the gulls could have flown nearly 700 km (435 mi), and over 80% of the seeds germinated after passing through the bird.

Dying There: Problems of Survival

Like any other population, the island population of a species must be able to survive periodic variations in its environment. But island life is more hazardous than that on the mainland, for several reasons. Catastrophe, such as volcanic eruption, has longer-lasting effects in an island situation, for there is little opportunity for a species to leave the area and return subsequently, nor is reinvasion easy should extinction take place. On the mainland, by contrast, chance extinction of a species in a particular area can soon be made good by immigration from elsewhere. An island will therefore contain a smaller number of species than a similar mainland area of similar ecology. For example, study of a 2-hectare plot of moist forest on the mainland of Panama showed that it contained 56 species of bird, while a similar plot of shrubland contained 58 species. The offshore Puercos Island, 70 ha in area and intermediate between the two mainland plots in its ecology, contained only 20 of these species [4].

Since its success and survival are the only measures of an organism's degree of adaptation to its environment, the fact that a species has become extinct also demonstrates that it was not able to adapt to the biotic or climatic stresses to which it was exposed. The island environment is inevitably different from that of the mainland, which was the source of the colonists, and adaptation to it is not easy. First, if the colonists are few in number, they can include only a very small part of the genetic variation that provided the mainland population with the flexibility to cope with environmental change; this is sometimes known as the **founder principle**. Second, small populations are also far more susceptible to random non-adaptive changes in their genetic makeup (see p. 181). Since it is less likely to be closely adapted to its environment, a small population is therefore also more liable to chance extinction.

The extent of this added risk to survival in islands is shown by the fact that, although only 20% of the world's species and subspecies of bird are found only on islands, they contributed 155 (90%) of the 171 taxa that are known to have become extinct since 1600—an extinction rate about 50 times as great as that on the continents. The influence of area on extinction rate is underlined by the fact that 75% of the island extinctions took place on small islands [5].

Because of these obvious risks in island life, it has been assumed that there was a one-way traffic of colonists from the continents outward, where they island-hopped across the ocean. However, here, as in so many fields of biogeography, molecular studies have shown us otherwise. A recent study [6] of the molecular genetics of the Monarch flycatcher birds, which are found in nearly every Pacific archipelago including the Hawaiian Islands, has shown that, after an initial single colonization from the mainland, there was a single radiation into six genera and 21 species wholly within the islands, followed by a reverse colonization of Australia within the last 2 my.

A species that can make use of a wide variety of food is therefore at an advantage on an island, for its maximum possible population size will be greater than that of a species with more restricted food preferences. This advantage will be especially great in a small island, in which the possible population sizes are in any case smaller. This is probably the reason why, for example, although on the larger islands of the Galápagos group in the eastern Pacific both medium-sized and small-sized *Geospiza* finches can coexist, on some of the smaller islands of the group there is only a single type, of intermediate size [7].

Chance extinction is also a particular danger for predators, since their numbers are inevitably always far lower than those of their prey. As a result, island faunas tend to be unbalanced in their composition, containing fewer predators, and fewer varieties of predator, than a similar mainland area. This in turn reinforces an island's fundamental lack of variety of animal and plant life, which has resulted from the hazards involved in entry and colonization. The complex interactions of continental communities containing a rich and varied fauna and flora act as a buffer that can cope with occasional fluctuations in the density of different species, and even with temporary local extinction of a species. This resilience is lacking in the simple island community. As a result, the chance extinction of one species may have serious effects and lead to the extinction of other species. All these factors increase the rate at which island species may become extinct.

There are several different reasons why a particular organism may be absent from a particular island. It may be unable to reach it; it may reach it but be unable to colonize it; it may have colonized it but later have become extinct, or it may simply as yet, by chance, not have reached the island [8]. It is often very difficult to decide which of these possible reasons was the cause in any particular case. On the other hand, there is another side to all the above problems. Once a species has arrived in an island and found a niche in the island biota, it is to some extent protected from many of the problems that may plague its continental relatives, as pointed out by Cronk [9]. The surrounding ocean waters often protect islands from episodes of climatic change experienced by the continents, while the comparatively small number of species on the island reduces competition, and the frequent absence of the parasites and predators also makes life easier for the successful immigrant. Finally, the island species is also protected from the appearance of new, more competent species that have evolved on the continents. All of this explains the fact that species may survive on islands, as what Cronk has called "relict endemics," long after their continental relatives have disappeared. For example, the plant *Dicksonia* is found both as a living plant and as a 9-million-year-old fossil on the little (122 km^2 [47 mi^2]) island of St. Helena, isolated in the South Atlantic, though none of its relatives survive in Africa, and its closest known living relatives are found in New Zealand.

Adapting and Evolving There

Colonists may encounter many difficulties when they first enter an island, but there are rich opportunities for those species that can survive long enough for evolution to adapt them to the new environment. These opportunities exist because of the lack of many of the parasites and predators that elsewhere would prey upon the species, and of many of the other species with which it normally competes. Opportunity for alterations in behavioral habits, diet, and way of life provides in turn opportunity for the organism to become permanently adapted, through evolutionary change, to a new way of life. This process requires a longer period of time and is therefore unlikely to take place except on islands that are large and stable enough to ensure that the evolving species does not become extinct. But if an island does provide these conditions, then remarkable evolutionary changes may occur as colonizing species become modified to fill vacant niches. This can be seen at a whole range of different levels, from the trivial to the comprehensive.

A good example at the trivial level is found in the Dry Tortugas, the islands off the extreme end of the Florida Keys, which only a few species of ant have successfully colonized [10]. One species, *Paratrechina longicornis*, on the mainland normally nests only in open environments under, or in the shelter of, large objects. However, on the Dry Tortugas it also nests in environments such as tree trunks and open soil, which on the mainland are occupied by other species. Not every species, however, is capable of taking advantage of such opportunities in this way. On the Florida mainland, the ant *Pseudomyrmex elongatus* is confined to nesting in red mangrove trees, occupying thin, hollow twigs near the treetop. Although it has managed to colonize the Dry Tortugas, it is still found there only in this very limited nesting habitat.

At a higher level of opportunity, a whole category of niches may be unoccupied. For example, many islands are treeless because the seeds of trees are usually much larger and heavier than those of other plants, and therefore do not often disperse over long distances. The valuable adaptive property of the tree niche is that its height allows it to shade-out its competitors and also to live longer [11]. The modifications needed to produce a tree from a shrub that already possesses strong, woody stems are comparatively slight—merely a change from the many-stemmed, branching habit to concentration on a single, taller trunk. For example, although most members of the Rubiaceae are shrubs, this family has produced on Samoa the tree *Sarcopygme*, 8 m (26 ft) tall and with a terminal palm-like crown of large leaves. Although more comprehensive changes are needed to produce a tree from a herb, many islands show examples of this. In many cases, the plants involved are members of the Asteraceae, perhaps because they have unusually great powers of seed dispersal, are hardy, and often already have partly woody stems. (It has been suggested that the arborescent habit may be primitive in the Asteraceae.) To this family belong both the lettuces, which have evolved into shrubs on many islands, and the sunflowers. Five different types of tree, 4–6 m (13–20 ft) high, found on the isolated island of St. Helena in the South Atlantic have evolved there from four different types of immigrant sunflower (Fig. 8.1). Two of these (*Psiadia* and *Senecio*) are endemic species of more widely distributed genera, while the other three (*Commidendron*, *Melanodendron*, and *Petrobium*) are recognized as completely new genera.

At an even higher level of opportunity, even birds may find it difficult to colonize oceanic islands. As a result, any successful bird colonist, originally adapted to a particular, limited diet and way of life, may find a whole range of other avian niches empty and available to it. The most spectacular example of this is the radiation of the cardueline finches on the Hawaiian Islands, from an original colonist that probably had the standard finch adaptation for cracking seeds using a stout beak to birds

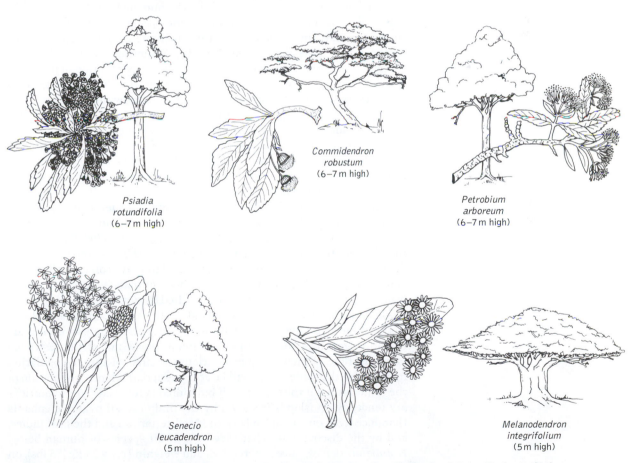

*Psiadia
rotundifolia*
(6–7 m high)

*Commidendron
robustum*
(6–7 m high)

*Petrobium
arboreum*
(6–7 m high)

*Senecio
leucadendron*
(5 m high)

*Melanodendron
integrifolium*
(5 m high)

Fig. 8.1 The varied trees that have evolved from immigrant sunflowers on St. Helena Island. From Carlquist [11].

with an array of beak shapes and feeding methods that parallel many of those of the whole range of songbirds (see below p. 236).

Examples of such evolution in isolation are also provided by the Great Lakes of East Africa ("islands" of water in an expanse of land), large enough to provide a great diversity of environments and old enough for these ecological opportunities to have become realized by evolutionary change (see p. 187, Chapter 6).

Even if they do not find opportunities of this kind, their life in islands can cause other changes in the immigrant species. One tendency is for island species (especially those on small islands) to lose the dispersal mechanisms that originally allowed them to reach their new home. Once on the restricted area of the island, the ability for long-distance dispersal is no longer of value to the species: in fact, it is a disadvantage, for the organism or its seeds is now more likely to be blown out to sea. Seeds tend to lose their "wings" or feathery tufts, and many island insects are wingless—18 of the 20 endemic species of beetle on the island of Tristan da Cunha have reduced wings. The loss of wings by some island birds may also be partly for this reason and partly because there are often no predators from which they need to escape. A few examples are the kiwi and moa of New Zealand, the elephant birds of Madagascar, and the dodo of Mauritius (the last three are extinct, but only because they were killed by humans).

It is also not unusual to find that island species are different in size from their mainland relatives, and this phenomenon has been discussed by the American ecologist Ted Case [12]. Sometimes an island lacks a particular type of predator because the size of the population of its prey is not large enough to provide a reliable source of food. This may decrease the death rate of the prey species and allow it to grow more rapidly, for it can feed at times and places that were previously dangerous. The predator might also have preferred to catch larger specimens of the prey species, so that its absence has removed this former penalty for growth. For all these reasons, the average size of island herbivores, such as some rodents, iguanid lizards, and tortoises, may become greater than that of their mainland relatives.

Island species may also change in size if some of their competitors are absent. For example, if a smaller competitor is absent, then an island species may become smaller to colonize that vacant niche. The converse may be true if its competitor was larger, as in the case of the Komodo dragon (*Varanus komodoensis*), a giant lizard that lives on Komodo Island and nearby Flores Island, in Indonesia. These animals increased in size to occupy niches that on the mainland are filled by much larger animals.

All these organisms evolved in islands to fill habitats normally closed to them. But other evolutionary changes frequently found on islands are the direct result of the island environment itself, not of the restricted fauna and flora. We have seen how serious the effect of a small population may be. But the same island will be able to support a larger population of the same animal if the size of each individual is reduced. This evolutionary tendency on islands is shown by the find of fossil pygmy elephants that once lived on islands in both the Mediterranean and the East Indies, and by the discovery of what may be a dwarf species of human being, *Homo floresiensis*, only 1 meter high and weighing only 25 kg (55 lbs), on the island of Flores in the East Indies [13].

The tendency for small island species to become larger and for large island species to become smaller is sometimes called the **island rule**, and relevant data have been reviewed and analyzed by the New York biogeographer Mark Lomolino [14]. He found that the rule was true for mammals in general, as well as for examples in some birds, snakes, and turtles, though another study [15] suggested that it did not apply to the Carnivora. It has also been suggested that the island rule also applies to deep-sea species—perhaps because, like islands, this environment suffers from decreased total food availability [16].

The Hawaiian Islands

As has been seen, many aspects of island life are unique, though many others differ only in degree from life on the continental landmasses. The result of the action of all these different factors can be seen by examining the flora and fauna of one particular group of islands. The Hawaiian Islands provide an excellent example, for they form an isolated chain, 2,650 km (1,650 mi) long, lying in the middle of the North Pacific, just inside the Tropics (Fig. 8.2). Sherwin Carlquist has provided an interesting account of the islands and of their fauna and flora, pointing out the significance of many of the adaptations found there [17]. A collection of papers on Hawaiian biogeography [18] contains many fascinating contributions, and others are to be found in a collection of papers on Pacific island biotas in general [19].

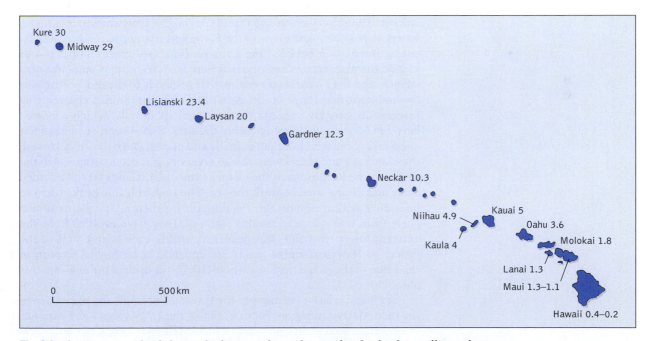

Fig. 8.2 The Hawaiian island chain. The figures indicate the age of each island, in millions of years.

The Hawaiian Islands are part of the most outstanding example of a volcanic island chain generated by a geological hotspot (see p. 169). They rise from a seafloor that is 5,500 m (18,000 ft) deep and that is moving northwestward at 8–9 cm (3 in) per year. The whole chain comprises 129 volcanoes; 104 of these were originally high enough to reach the surface and form islands (or add to an existing island). The youngest and biggest island, Hawaii, is only 700,000 years old and is made up of six volcanoes, two of which are still active. The most westerly island still visible above the sea, Kure, is nearly 30 my old, but the Emperor Seamount chain, consisting of other submerged islands and volcanoes, extends further westward and then northward to near the Kamchatka Peninsula of Siberia. There, the Meiji Seamount is perched on the edge of the Kurile Trench and will eventually disappear back into the depths of the Earth.

Meiji was formed by the hotspot 85 my ago, at the present-day location of Hawaii, so islands and submerged volcanoes have been forming in the central North Pacific Ocean for at least that length of time. However, there have been periods in the past when there were no visible islands, or only low islands that were soon submerged. The most recent of these periods lasted 18 my and ended with the appearance of Kure Island about 30 my ago. Kure is therefore the oldest island that was a potential home for colonists that might be ancestral to the biota of the Hawaiian Islands of today. Although most later islands (until the last 5 my) were small, there continued to be the occasional larger, higher island that might have provided a home for the descendants of the biota of Kure.

The rocks that form the Hawaiian Islands can be accurately dated, and the modern technique of calculating the rates of molecular change in the lineages of animals and plants provides a similar system of dating the times of divergence of different genera and species. As the American workers Hampton Carson and David Clague [20] have shown, it is therefore now possible, in the case of the Hawaiian Islands, to produce a fascinating integration between the gears of those two great engines of our planet: the plate tectonic engine that here leads to the appearance of new islands, and the evolutionary engine that responds to these opportunities by producing new forms of life. As will be seen later in this chapter, this integration strongly suggests that the ancient islands, which appeared between 30 my and 10 my ago, were the sites of evolution of some of the ancestors of today's Hawaiian animals and plants. Once they had arrived, these animals and plants were able to diversify and, often, to spread to the newly formed additions to the chain as the older islands steadily eroded away and disappeared beneath the sea. The high volcanic peaks seem to be islands within islands, for it is difficult for their alpine plants to disperse from one island to another. Instead, they have evolved from the adjacent lower-lying flora independently within each island: 91% of the genera of Hawaiian alpine plants are endemic, a far higher proportion than that of the island flora as a whole (16.5%) or even of the angiosperms alone (20%) [21].

The closest relatives of many of the Hawaiian animals and plants live in the Indo-Malaysian region. For example, of the 1,729 species and varieties of Hawaiian seed plants, 40% are of Indo-Malaysian origin but only 18%

are of American origin. Also, nearly half of the 168 species of Hawaiian ferns have Indo-Malaysian relatives, but only 12% have American affinities. This is not surprising, for the area to the east is almost completely empty, while that to the south and west of the Hawaiian chain contains many islands, which can act as intermediary homes for migrants. Organisms adapted to life in these islands would also be better adapted to life in the Hawaiian Islands than would those from the American mainland. However, about half of the 21 lineages of birds found on the Hawaiian Islands appear to have originated in North America.

Mechanisms of Arrival

The way in which the Hawaiian birds reached the islands is obvious enough. One of the plants that probably came with them is *Bidens*, a member of the Asteraceae, whose seeds are barbed and readily attach themselves to feathers; about 7% of the Hawaiian nonendemic seed plants probably arrived in this way [21]. The Hawaiian insects, too, arrived by air. Most of the Hawaiian insects are those with light bodies, which are most readily blown long distances by the wind, although heavier dragonflies, sphinx moths, and butterflies are also found there. Similarly, because ferns have spores that are much smaller and lighter than the seeds of angiosperms, they provide a greater proportion of the flora in the Hawaiian Islands than in the rest of the world. Of the nonendemic seed plants of the Hawaiian Islands, about 7.5% almost certainly arrived carried by the wind, while another 30.5% have small seeds (up to 3 mm in diameter) and may also have arrived in this way. For example, the unusually tiny seeds of the tree *Metrosideros* have allowed it to become widely dispersed through the Pacific Islands. It is a pioneering tree, able to form forests on lowland lava rubble with virtually no soil—a great advantage on a volcanic island. *Metrosideros* shows great variability in different environments, from a large tree in the wet rainforest, a shrub on wind-swept ridges, to as little as 6 in (15 cm) high in peatlands; it is therefore the dominant tree of the Hawaiian forest. Although these differences are probably at least partially genetically based, the different forms are not distinct species, and intermediates are found where two different types (and habitats) are adjacent to one another.

The carriage of seeds within the digestive system of birds is an important mechanism for the arrival of plants; about 37% of the nonendemic seed plants of the islands (e.g., blueberry, sandalwood) probably arrived in this way. Significantly, many plants that succeeded in reaching the islands are those that, unlike the rest of their families, bear fleshy fruits instead of dry seeds (e.g., the species of mint, lily, and nightshade found in Hawaii). An exception to this appears to be *Viola*, but it is significant that the Hawaiian members of this genus are most closely related to those of the Bering region of the northern Pacific, from which some 50 species of bird regularly overwinter in the Islands.

Dispersal by sea accounts for only about 5% of the nonendemic Hawaiian seed plants. As well as the ubiquitous coconut, the islands also contain *Scaevola toccata*; this shrub has white, buoyant fruits and forms dense hedges along the edge of the beach on Kauai Island. Another seaborne migrant is *Erythrina*; most species of this plant genus have buoyant, bean-like seeds. On Hawaii, after its arrival on the beach, *Erythrina* was unusual in adapting to an island environment, and a new endemic species, the coral tree *E. sandwichensis*, has evolved on the island. Unlike those of its ancestors, the seeds of the coral tree do not float—an example of the loss of its dispersal mechanism often characteristic of an island species.

The successful colonists of the Hawaiian Islands are the exceptions; many groups have failed to reach them. There are no truly freshwater fish and no native amphibians, reptiles, or mammals (except for one species of bat), while 21 orders of insect are completely absent. As might be expected, most of these are types that seem in general to have very limited powers of dispersal. For example, the Formicidae (ants), which are an important part of the insect fauna in other tropical parts of the world, were originally absent. They have, however, since been introduced by humans, and 36 different species have now established themselves and filled their usual dominant role in the insect fauna. This proves that the obstacle was reaching the islands, not the nature of the Hawaiian environment.

Evolutionary Radiations within the Hawaiian Islands

As ever, the absence of some groups has provided greater opportunities for the successful colonists. Several insect families, such as the crickets, fruitflies, and carabid beetles, are represented by an extremely diverse adaptive radiation of species, each radiation derived from only a few original immigrant stocks. For example, the fruitflies, belonging to the closely related genera *Drosophila* and *Scaptomyza*, have undergone an immense radiation in the Hawaiian Islands; over 1,300 species are known worldwide, of which over 500 have already been described from the Hawaiian Islands, where probably another 250–300 species are awaiting description. The abundance of species of fruitflies in the islands is probably due partly to the great variations in climate and vegetation to be found there, and also to the periodic isolation of small islands of vegetation by lava flows, each such island providing an opportunity for the evolution of new species. But another major factor has been that the Hawaiian fruitflies, in the absence of the normal inhabitants of the niche, have been able to use the decaying parts of native plants as a site in which their larvae feed and grow. This may also have been encouraged by the fact that their normal food of yeast-rich fermenting materials is rare in the Hawaiian Islands.

Molecular studies indicate that the common ancestor of all the drosophilids of the Hawaiian Islands diverged from the Asian mainland type about 30 million years ago. Comparison with the ages of the different islands (Fig. 8.2) suggests that this must have happened on Kure Island when that island was young and nearly 900 m (3,000 ft) high. Similarly, these studies indicate that *Scaptomyza* diverged from *Drosophila* 24 my ago, which suggests that this event took place on Lisianski, which was at one time an island about 1,220 m (4,000 ft) high. Detailed studies of the chromosome structure of the Hawaiian "picture-winged" fruitflies are now making it possible to reconstruct the sequence of colonizations that must have taken place [22]. As might have been expected, the older islands to the west in general contain species ancestral to those in the younger islands in the east (Fig. 8.3). The youngest, Hawaii itself, has 19 species descended from species in older, more westerly islands, but none of its species appears to be ancestral to those in the older islands. (The present-day islands of Maui, Molokai, and Lanai together formed a single island until the postglacial rise in sea levels; their *Drosophila* species are therefore treated together as a single fauna.) Within Hawaii itself, molecular studies show that the species on the more southern part of the island, which is up to 200,000 years old, evolved from those on the more northern part, which is 400,000 to 600,000 years old.

The same phenomenon of a great adaptive radiation has taken place in other groups of animals and plants. In general, therefore, although the islands contain comparatively few different families, each contains an unusual variety of species, nearly all of which are unique to the islands— over 90% of the Hawaiian plant species are endemic to the islands.

There are many other examples of Hawaiian adaptive radiations, but three are of particular interest: the silverswords and lobeliads among the plants [22], and the honeycreepers among the birds. The silverswords are descended from the tarweeds of southwest North America (members of the family Asteraceae, which includes sunflowers and daisies), and probably arrived on the islands as sticky seeds attached to the feathers of birds. They have produced only three genera in the islands (*Dubautia*, *Argyroxiphium*, and *Wilkesia*), but these have colonized a variety of habitats. For example, on the bare cinders and lava of the peak of Haleakala 3,050 m (1,000 ft) high on the island of Maui, two of the few plant species that can survive there are silverswords. *Dubautia menziesii* is adapted to this arid environment by its tall stem and stubby succulent leaves, while *Argyroxiphium sandwichense* is covered by silvery hairs that reflect the light and heat. A few hundred meters below the bare volcanic peaks, conditions are at the other extreme, because most of the rain falls at heights of from 900–1,800 m (2,950–5,900 ft); these regions receive 250–750 cm

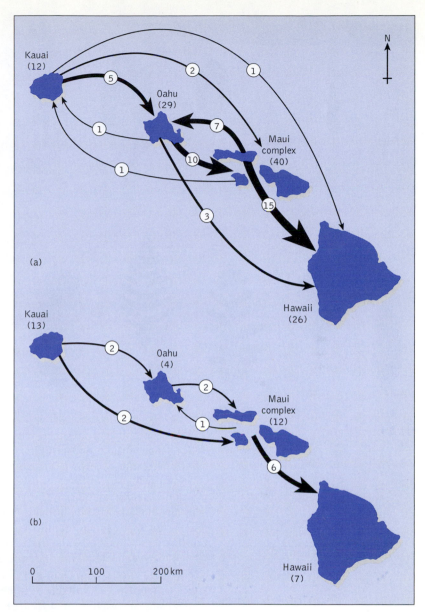

Fig. 8.3 The dispersal events within the Hawaiian Islands that are suggested by the interrelationships of the species of "picture-winged" *Drosophila* flies (above) and of silversword plants (below). The width of the arrows is proportional to the number of dispersal events implied, and the number of species in each island is shown in parentheses. From Carr et al. [22], by permission of Oxford University Press.

Fig. 8.4 The leaves of different species of the lobeliad genus *Cyanea*.

C. leptostegia adult

C. linearifolia

C. stictophylla

C. angustiflora

C. tritomantha

C. lobata

C. leptostegia juvenile

C. solanacea

C. shipmanii

C. asplenifolia

C. grimesiana

C. solanacea quercifolia

(10–30 in) of rain per year. The upper regions of Mount Puu Kukiu, 1,770 m (5,800 ft) high, on Maui are covered by mire in which thrives another silversword, *A. caliginii*. On the island of Kauai the heavy rainfall has led to the development of dense rainforest, in which *Dubautia* has evolved a tree-like species, *D. knudsenii*, with a trunk 0.3 m (1 ft) thick, and large leaves to gather the maximum of sunlight in the dim forest. Kauai bears another silversword that shows the tendency for island plants to become trees. In the drier parts of this island grows *Wilkesia gymnoxiphium*, with a long stem which carries it above the shrubs that compete with it for light and living space. This species also shows another example of the loss of the dispersal mechanism that first brought the ancestral stock to the island: the seeds of *Wilkesia* are heavy and lack the fluffy parachutes usually found among the Asteraceae. The dispersal pattern of the 28 species in the three silversword genera within the Hawaiian Islands is rather similar to that of the drosophilid flies (Fig. 8.3) [22]. Molecular evidence suggests that these three genera had a common ancestor that originally arrived on Kauai.

Lobeliads are found in all parts of the world, but they have undergone an unusual adaptive radiation in the Hawaiian Islands because their normal competitors, the orchids, are rare. The Hawaiian lobeliads include 150 endemic species and varieties, making up six or seven endemic genera. Over 60 species of the endemic genus *Cyanea* are known, showing an incredible diversity of leaf form (Fig. 8.4) [17]. The plants range from the tree *C. leptostegia*, 9 m (30 ft) tall (similar in appearance to the tarweed *Wilkesia*) to soft-stemmed *C. atra*, only 0.9 m (3 ft) tall. The species of another genus, *Clermontia*, are less varied in overall size, but are very varied in the size, shape, and color of their flowers. These are mainly tubular and brightly colored, a type of flower that is often associated with pollination by birds. On isolated islands such as Hawaii, the adaptation of larger flowers to bird pollination may have taken place because the large insects that would normally pollinate such flowers on the mainland are absent. It is no coincidence that the adaptive radiation of the Hawaiian lobeliads has been accompanied by the adaptive radiation of a nectar-eating type of bird, the honeycreepers [23].

The ancestors of these birds were probably a species of finch-like immigrant from Asia [24] that fed on insects and nectar. From that original immigrant species, adaptive radiation has produced 11 endemic genera comprising the endemic family Drepanididae (Fig. 8.5), but another c. 30 extinct species are known. Many of the genera, such as *Himatione, Vestiaria, Palmeria, Drepanidis,* many species of *Loxops,* and one species of *Hemignathus* (*H. obscurus*) are still nectar eaters, feeding from the flowers of the tree *Metrosideros* and the lobeliad *Clermontia.* Since insects, too, are attracted to the nectar, it is not surprising to find that many nectar-eating birds are also insect-eaters, and it is a short step from this to a diet of insects alone. *Hemignathus wilsoni* uses its mandible, which is slightly shorter than the upper half of its bill, to probe into crevices in bark for insects, and *Pseudonestor xanthophrys* uses its heavier bill to rip open twigs and branches in search of insects. Other types have heavy, powerful beaks, which they use for cracking open seeds, nuts, or beans. The light bill of the recently extinct *Ciridops* was used for eating the soft flesh of the fruits of the Hawaiian palm *Pritchardia.*

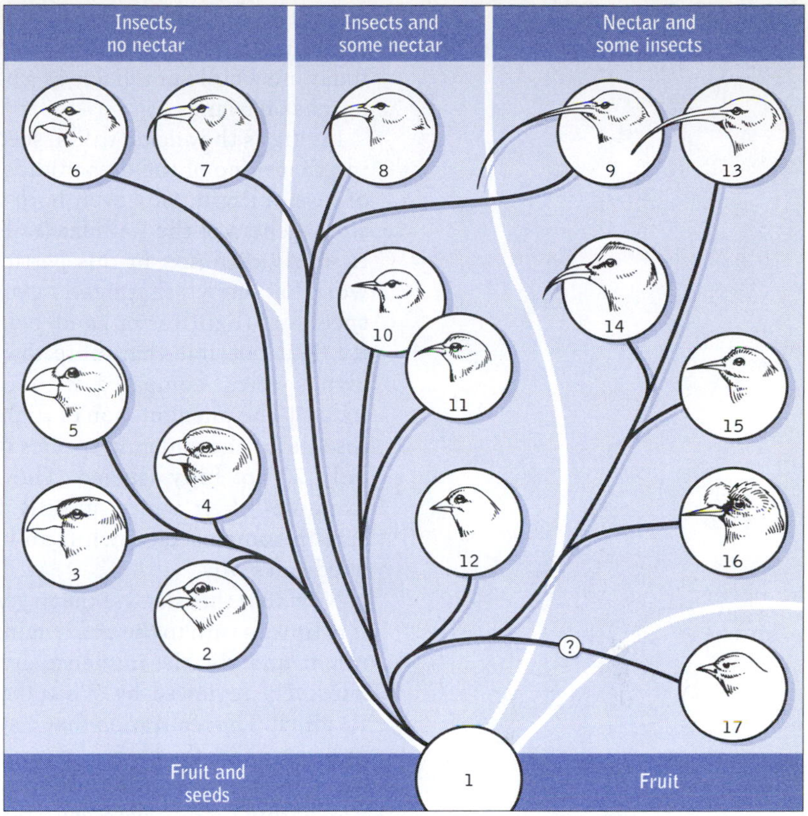

Fig. 8.5 The evolution of dietary adaptations in the beaks of Hawaiian honey-creepers. 1, Unknown finch-like colonist from Asia; 2, *Psittirostra psittacea*; 3, *Chloridops kona*; 4, *Loxioides bailleui*; 5, *Telespyza cantans*; 6, *Pseudonestor xanthophrys*; 7-11, *Hemignathus munroi, H. lucidus, H. obscurus, H. parvus, H. virens*; 12, *Loxops coccineus*; 13, *Drepanis pacifica*; 14, *Vestiaria coccinea*; 15, *Himatione sanguinea*; 16, *Palmeria dolei*; 17, *Ciridops anna*. Taxonomy after Pratt et al. [81].

It is interesting to speculate why these Hawaiian birds should have radiated into so many more species than did Darwin's finches on the Galápagos Islands (p. 190). Possible reasons may be that the Hawaiian Islands are older, contain a far greater variety of environments, lie further apart than the islands in the Galápagos archipelago, and are not subject to the extreme climatic changes of the el Niño/la Niña cycle (see pp. 404–406).

Studies of the Hawaiian avifauna also show the unreliability of the modern biota as a basis for estimating rates of biotic change or of the relationship between island area and number of species. It has long been known that about a dozen Hawaiian bird species became extinct after the arrival of Europeans and the animals that they introduced. But studies by the American biologists Storrs Olson and Helen James [25, 26] revealed at least 50 now-extinct species of Hawaiian bird—more than the entire avifauna of the islands today. These included flightless types of ibis, rail and goose-like ducks, six hawks and goshawk-like owls, and nearly two dozen species of drepanidid finches, mostly of insectivorous type. Most of these species were still alive when the Polynesians arrived in the Islands in about AD 1500, but had become extinct before Europeans arrived 300 years later. Similar evidence for bird extinctions has been reported from many other Pacific islands, and it is clear that the patterns of distribution, endemism, and numbers of species on individual islands that are seen today are wholly unreliable as a basis for generalizations about bird populations on islands.

Living as they do on fruits, seeds, nectar, and insects, it is not surprising that none of the drepanidids shows the island fauna characteristic of loss of flight. However, both on Hawaii and on Laysan to the west, some genera of the Rallidae (rails) have become flightless. (This is particularly common in this particular family of birds. Palaeontological work has shown that nearly every Pacific island once had at least one species of flightless rail and, before extinction caused by human activity, it is possible that more than 800 of the islands may have had its own species, compared with the 27 flightless species that survive today!) The phenomenon of flightlessness is also common in Hawaiian insects: of the endemic species of carabid beetle, 184 are flightless and only 20 are fully winged. The Neuroptera or lacewings are another example—their wings, usually large and translucent, are reduced in size in some species, while in other species they have become thickened and spiny.

Now that the island scene in general has been set in the above sections, it is time to turn to the challenging task of trying to find general rules that might underlie the immense diversity of island biota. This is comprehensively reviewed by Whittaker and Fernández-Palacios (see Further Reading). The realization that dispersal, rather than vicariance, is the normal source for the biota of oceanic islands (see Chapter 7), together with the demonstration by molecular studies of the extent of allopatric speciation within them, have given a renewed impetus to the analytical study of island biotas [27, 28].

Integrating the Data: The Theory of Island Biogeography

One of the most obvious characteristics of the biota of islands is that it is strongly affected by the degree of isolation of the island. However diverse the habitats that it offers, the variety of the island life depends, in the short term, very much on the rate at which colonizing animals and plants arrive. This, in turn, depends largely on how far the island is from the source of its colonizers, and on the richness of that source. If the source is close, and if its biota is rich, then the island in its turn will have a richer biota than another, similar island, which is more isolated or which depends on a source with a more restricted variety of animals and plants. Each sea barrier further reduces the biota of the next island, which in turn becomes a poorer source for the next. For example, the data provided by Van Balgooy [29] make it possible to map the diversity of conifer and flowering plant genera in the Pacific island groups (Fig. 8.6). This clearly shows that diversity is much lower in the more isolated island groups of the central and eastern Pacific.

In several of the more westerly island groups, however, the diversity is much higher than their geographical position alone would lead one to predict. A logarithmic graph of the relationship between the number of genera

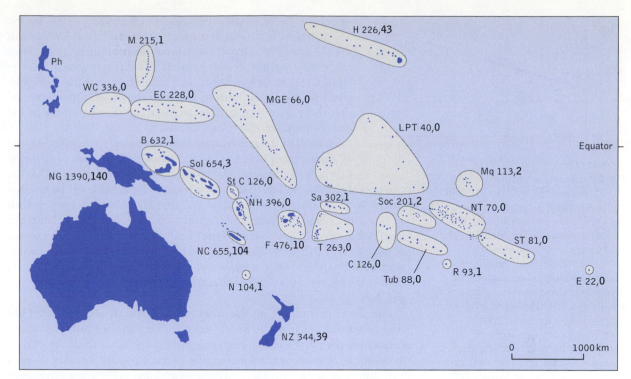

Fig. 8.6 The distribution of conifers and flowering plants in the Pacific islands. The first number beside each island group is the total number of genera found there; the second is the number of endemic genera found there. B, Bismarck Archipelago; C, Cook Islands; E, Easter Island; EC, East Carolines; F, Fiji Islands; H, Hawaiian Islands; LPT, Line, Phoenix, and Tokelau Island groups; M, Marianas; MGE, Marshall, Gilbert, and Ellis Islands; Mq, Marquesas; N, Norfolk Island; NC, New Caledonia; NG, New Guinea; NH, New Hebrides; NT, Northern Tuamotu Islands; NZ, New Zealand; Ph, Philippines; R, Rapa Island; Sa, Samoa group; Soc, Society Islands; Sol, Solomon Islands; ST, Southern Tuamotu Islands; StC, Santa Cruz Islands; T, Tonga group; Tub, Tubai group; WC, West Carolines. Data from Van Balgooy [29].

and the area of the islands (Fig. 8.7) clearly shows that, in most cases, the generic diversity is simply dependent on island area. (The fact that some islands therefore have more genera than do the islands from which most of their flora is derived proves that other genera were once also present in these latter islands, but have since become extinct there.) Nearly all of the more isolated island groups (shown as triangles in Fig. 8.7) have, as would be expected, a much lower diversity than would be predicted from their areas alone. The number of land and freshwater bird species in each island shows a similar relationship to island area (Table 8.1), but this is probably due, not to island area directly, but to the resulting higher floral diversity.

As we have seen, the number of species found on an island depends on a number of factors: not only on its area and topography, its diversity of habitats, its accessibility from the source of its colonists, and the richness of that source, but also on the equilibrium between the rate of colonization by new species and the rate of extinction of existing species. Many individual observations and analyses of such phenomena have been made over the past 150 years. As explained in Chapter 1, a quantitative theory

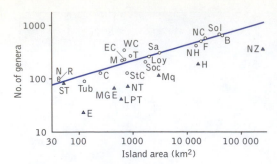

Fig. 8.7 The relationship between island area and the diversity of conifer and flowering plant genera in the Pacific islands. The more isolated islands are indicated by triangles. The data from the other islands lie very close to a straight line (the regression coefficient), suggesting that generic diversity in these islands is almost wholly controlled by island area—the correlation coefficient is 0.94, indicating a very high degree of correlation. For abbreviations, see legend of Fig. 8.6, plus Loy, Loyalty Islands. Data from Van Balgooy [29].

to explain these relationships was put forward in 1967 by the American ecologists Robert MacArthur and Edward Wilson in their book, *The Theory of Island Biogeography* [10]. They made two main suggestions: that the changing, and interrelated, rates of colonization and immigration would eventually lead to an equilibrium between these two processes, and that there is a strong correlation between the area of the island and the number of species it contains.

To explain the relationship between rates of colonization and of immigration, MacArthur and Wilson took the case of an island that is newly available for colonization. They pointed out that the rate of colonization will at first be high because the island will be reached quickly by those species that are adept at dispersal and because these will all be new to the island. As time passes, immigrants will increasingly belong to species that have already colonized the island, so that the rate of appearance of new species will drop (Fig. 8.8). The rate of immigration will also be affected by the position of the island, for it will be higher for islands that are close to the source of their colonists and lower for those that lie further away (Fig. 8.9).

The rate of extinction, on the other hand, will start at a low level but gradually rise. This is partly because, since every species runs the risk of extinction, the more that have arrived, the more species there are at risk. In addition, as more species arrive, the average population size of each will diminish as competition increases—and a smaller population is at greater risk of extinction than a larger population.

The theory of island biogeography (or TIB) suggests that the two curves that represent these two conflicting processes will intersect at a point where the rates of immigration and extinction are equal, known as the "turnover rate," so that the number of species is constant at this equilibrium number.

At first, the few species present can occupy a greater variety of ecological niches than would be possible on the mainland, where they are competing with many other species. For example, in the comparison mentioned earlier between the Panama mainland and Puercos Island, the smaller number of bird species in the island were able, because of reduced competition, to be far more abundant: there were 1.35 pairs per species per hectare (2.5 acres) in Puercos Island, compared with only 0.33 and 0.28, respectively, for the two mainland areas [30]. This effect of release from competition was especially noticeable in the antshrike (*Thamnophilus doliatus*). On the mainland, where it competed with

Table 8.1 The relationships between island area and the diversity of bird genera and nonendemic flowering plant genera in some Pacific islands. Data from Van Balgooy [29]; Mayr [79]; MacArthur & Wilson [80].

	Area (km²)	Angiosperm Genera	Bird Genera
Solomon Islands	40,000	654	126
New Caledonia	22,000	655	64
Fiji Islands	18,500	476	54
New Hebrides	15,000	396	59
Samoa group	3,100	302	33
Society Islands	1,700	201	17
Tonga group	1,000	263	18
Cook Islands	250	126	10

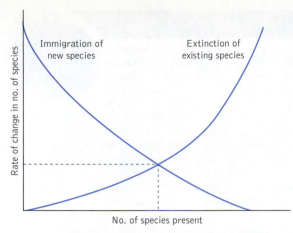

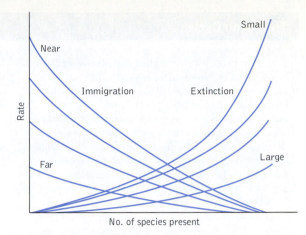

Fig. 8.8 Equilibrium model of the biota of an island. The curve of the rate of immigration of new species and the curve of the rate of extinction of species already on the islands intersect at an equilibrium point. The interrupted line drawn vertically from this point indicates the number of species that will then be present on the island, while that drawn horizontally indicates the rate of change (or "turnover rate") of species in the biota when it is at equilibrium. After MacArthur & Wilson [80].

Fig. 8.9 The interrelationship between isolation and area in determining the equilibrium point of biotic diversity. Increasing distance of the island from its source of colonists lowers the rate of immigration (left). Increasing area lowers the rate of extinction (right). After MacArthur & Wilson [80].

over 20 other species of ant-eating bird, there were only eight pairs of antshrike per 40 ha (100 acres); on Puercos Island, where there was only one such competitor, there were 112 pairs per 40 ha (100 acres).

This effect of release from competition will be reversed if the island is later colonized by a new species whose diet overlaps with that of one of the earlier immigrants. This may result in the extinction of one of them. This result may be because they compete too closely with one another, so that they cannot coexist. Alternatively, it may be because the competition between them leads to a reduction in the population size of each—because each species has to become more specialized in its ecological requirements. This in turn renders the species more vulnerable to extinction. In either case, the rate of extinction on the island will have increased.

In all these theoretical cases, the number of species present in the biota will obviously be the result of the balance between the rate of immigration and the rate of extinction. MacArthur and Wilson suggested that the biota will eventually reach an equilibrium, at which the rates of immigration and of extinction are approximately equal, and that this equilibrium level is comparatively stable.

Many immigrant island species, now with less competition than they had on the mainland, will be able to expand into new habitats. But if other competing new colonists now arrive, they may later find their distribution and evolutionary expansion reduced. This concept of alternating expansion and construction gave rise to the theory of the taxon cycle (see Concept Box 8.1).

The Theory of Island Biogeography was widely welcomed, for it gave biogeographers a theoretical background they could use to compare their own individual results, and therefore encouraged a more structured and

In addition to his attempt to establish general principles in the theory of island biogeography, Edward Wilson had earlier [31] suggested that the distribution and ranges of individual species in island communities went through stages of expansion and contraction, which he named the **taxon cycle.** The methodology of this theory is to establish categories for island species that have different ecological and distributional characteristics, and then infer that these differences are the result of their having colonized the island at different times and interacted with one another.

The concept is best understood by visualizing a time sequence, imagining the history of a species from its first dispersal from the mainland and arrival in an island. The species is likely to arrive in one of the ecologically marginal habitats, such as shore communities, grassland, or lowland forest. Here, they were at first generalists, with a wide and continuous range of distribution. However, taking advantage of the lack of its normal competitors, predators, or parasites the species may extend its ecological range into other environments, such as inland forests or montane rainforest. The species is in the expansion phase of the taxon cycle.

The next event is the arrival of other species, whose competition expels the original species from its original marginal habitat, so that its distribution is restricted to the more specialized, inland habitats. It is now in a contraction phase of the taxon cycle, with a more patchy, less continuous pattern of distribution. The original species will have become genetically adapted to life in these habitats and so may have become recognizable as a new, endemic species. This process may take place independently on more than one island but, if so, each of these new species will be most closely related to the original colonist from the mainland, rather than to each other.

The taxon cycle theory has attracted a great deal of criticism. At the most fundamental, theoretical level, the problem is that the diversity of patterns of distribution to be found within and between island faunas is so varied that it is easy to find examples that will fit within almost any set of categorizations. Furthermore, it is impossible to prove that the suggested linkages between distribution, adaptation, and relationship are cause-and-effect, or that they form a sequence in time. However, the American biogeographers Robert Ricklefs and Eldredge Bermingham, in a review of the taxon cycle concept [32], have shown that molecular phylogenetic analyses of the times of divergence of 20 lineages of West Indian birds conform to the assumptions of the hypothesis, and phylogenetic analyses of lineages of West Indian anolid lizards [33] have similarly supported it.

less ad hoc approach to biogeographical studies. Its methodology was also extended into types of isolation other than that of dry land surrounded by water, such as mountain peaks (Fig. 8.10), caves, coral reefs, and individual plants, and the theory was even extended to evolutionary time, with individual host–plant species being regarded as islands as far as their "immigrant" insect fauna was concerned.

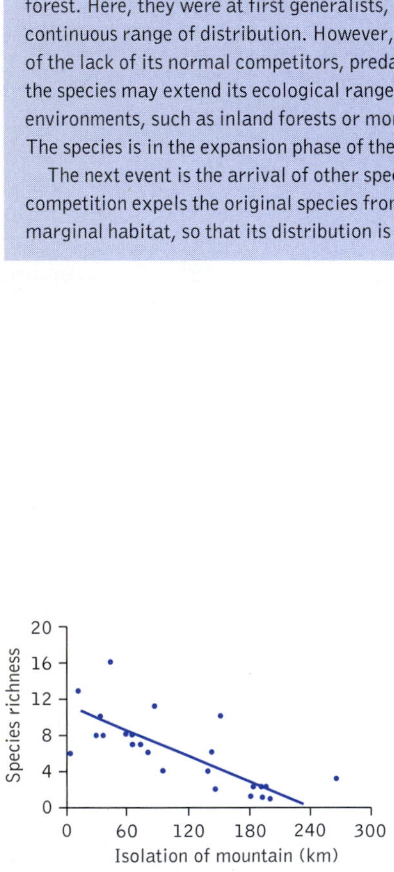

Modifying the Theory

The story of how the Theory of Island Biogeography came to dominate this whole area of research has been outlined in Chapter 1 and will not be repeated here. Even though this almost total acceptance of the validity of the Theory delayed its critical evaluation, its shortcomings were eventually realized. Many studies that had been extensively quoted as supporting the TIB were really far too imprecise, as pointed out by the American ecologist Dan Simberloff [34] and in a review by the British biologist Francis Gilbert [35], while the statistical procedures of many earlier studies were criticized by the American ecologists Edward Connor and Earl McCoy [36]. It was also pointed out that the Theory treats species as simple numerical units, of equal value to one another, so that their possible biological interactions, such as competitive or coevolutionary effects, are therefore ignored. However, this was an essential part of the methodology

Fig. 8.10 The relationship between the degree of isolation of mountain peaks in the southwestern United States and the number of species of mammal found in each. From Lomolino, Brown, & Davis [82].

of the TIB, for it was expressly designed to rise above the inevitable complexity that results from the analysis of individual species in order to see whether this might reveal general rules against which individual species or circumstances might be judged. The general point that arises here is the difficulty of transforming the essentially gradual nature of many biological phenomena into the individual points of defined, quantified data that mathematical treatments require.

There has been considerable criticism of work that had seemed to support the TIB's prediction that the number of species will come to, and remain at, equilibrium, as long as the environment remains constant. Jared Diamond [37], comparing the numbers of bird species found in the Californian Channel Islands in a 1968 survey with those recorded in a 1917 review, concluded that there had been an equilibrium. However, Lynch and Johnson pointed out [38] that fundamental changes in the environment of the Channel Islands had taken place between 1917 and 1968, rendering the example worthless. Lynch and Johnson found similar flaws in other studies of bird faunas. Similarly, Simberloff [39] analyzed the records of the bird faunas of two islands and three inland areas over 26–33 years and found that none of them showed any evidence of regulation toward equilibrium.

Apart from these specific studies, can we be sure that *any* biota that we see today is in a state of stable equilibrium? The level of that equilibrium will alter if the environment changes—and the environment has changed a very great deal over both the distant and recent past. These changes include alterations in climate and in sea level, that may in turn result in changes in island area or in the union or subdivision of islands. We ourselves have caused the extinction of endemic island species and introduced new species to them. It has been estimated that human activities over the last 30,000 years have led to the extinction of thousands of populations of birds and as many as 2,000 bird species [40]. Under these circumstances, it is difficult to be confident that any island biotas are at stable equilibrium. If so, they cannot provide a database either for estimates of the actual numerical equilibrium level in any existing situation or for any prediction of where such an equilibrium might emerge in the future. In any case, the lag-time of many biological phenomena may mean that the situation observed today is merely ephemeral, inasmuch as the biological characteristics are still in the process of adjusting to an earlier event of climatic or sea-level change. As Shafer [41] has observed, no aspect of the concept of an equilibrium level in species numbers can now be considered as established, and this would in any case be of limited value in an environment subject to both cyclical and occasional change.

When considering the rate of turnover of species, the longevity of the dominant species is also important. Case and Cody [42] have pointed out that the life spans of forest trees are so great that turnover is inevitably very slow. This is shown by the fact that the species structure of the forest at Angkor in Cambodia, which started to grow when the ancient Khmer capital was abandoned 560 years ago, has still not become identical to that of the surrounding older forest.

Other research has been directed at trying to modify and improve the TIB. MacArthur and Wilson's first point, that larger islands contain more species, has met general acceptance. But considerable research has been done to identify the factors that produce this effect, as well as their relative

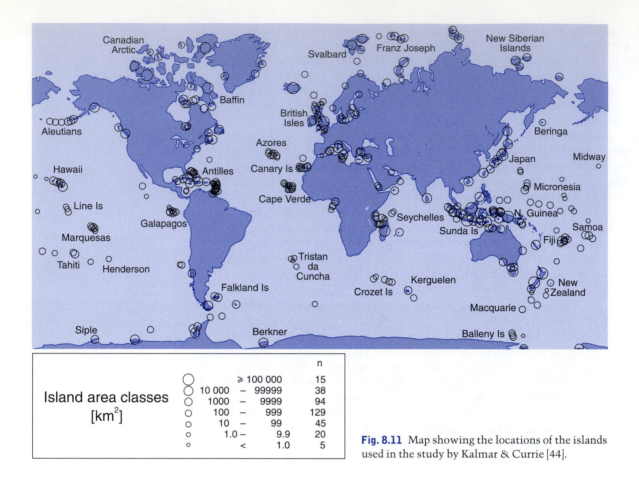

Island area classes [km²]		n
⬯ ≥ 100 000		15
○ 10 000 – 99999		38
○ 1000 – 9999		94
○ 100 – 999		129
○ 10 – 99		45
○ 1.0 – 9.9		20
○ < 1.0		5

Fig. 8.11 Map showing the locations of the islands used in the study by Kalmar & Currie [44].

importance, and to examine the impact of other factors, such as climate and evolutionary history, that the Theory ignores. For example, the American biogeographer David Wright suggested that the fundamental aspect of enlarging the island area is that it increases the amount of energy that falls on the island—a concept known as species-energy theory [43]. Wright proposed that islands are essentially energy collectors, but that the amount of energy that an island can collect will also vary according to its climate, those with a warm, wet climate being more productive and therefore having more species than those that are cool and dry. Wright showed that the species richness of flowering plants on 24 islands world wide, and of land birds on 28 islands of the East Indies, were better explained by his theory than by the Theory of Island Biogeography. More recently, the Canadian biogeographer David Currie and his Hungarian student Attila Kalmar have extended Wright's work to find out whether the degree of isolation of islands affected their species richness [44]. Using data on nonmarine birds from 346 islands (Fig. 8.11), they found that the species richness of an island is correlated to its distance from the nearest island, but far more strongly correlated to its distance from the nearest mainland. Impressively, they showed that a combination of average annual temperature, total annual rainfall, and distance from the nearest continent together explained 87.5% of the bird species richness of these islands. However, in contrast to the TIB, their analysis showed that the slope of the species richness vs. island-area curve depends on climate, not on isolation.

The General Dynamic Model for Oceanic Islands

Islands such as the Hawaiian chain, which result from the activities of an oceanic hotspot, have their own cycle of birth, growth, subsidence, and disappearance (see p. 169). Robert Whittaker and his colleagues in the Biodiversity Research Group at Oxford University have recently combined the patterns of this geological life history with some aspects of the Theory of Island Biogeography to produce a new general dynamic model (GDM) that predicts the patterns of biodiversity, endemism, and diversification through the lifetime of such an island [45].

Whittaker and colleagues provide a graph that illustrates the general pattern of the changes in the area, altitude, and topographic complexity of a hotspot island. Its area and altitude at first steadily increase, but then these begin to be reduced by erosion, which, together with subsidence of the cooling ocean floor as the island moves away from the center of the hotspot, and major landslips resulting from the erosion of the steep volcanic slopes, will also lead to the eventual disappearance of the island below the sea (Fig. 8.12a). Erosion also produces a more complex topography—though this complexity reaches a maximum a little after the time of maximum elevation. This in turn increases the habitat diversity and therefore the number of vacant ecological niches on the island (Fig. 8.12b), and so also increases the potential carrying capacity of the island (K in Fig. 8.12c). These vacant niches may be filled either by new colonists or by the radiation of existing colonists, which together increase the extent to which the potential species-carrying capacity is realized (R in Fig. 8.12c). While the gap between these two figures remains high, the ecological isolation between existing species will be high, and opportunities for the appearance of such new species will be correspondingly high (Fig. 8.12b). As these niches are filled (R gradually coming closer to K) and competition between the species increases, the rate of appearance of new species will decrease (S in Fig. 8.12c).

To begin with, nearly all the vacant niches will be filled by immigrants from neighboring islands. But as the habitat diversity and vacant niche space increases, more and more of these niches are likely to be filled by the evolution of new species from existing immigrants. Over time, an increasing proportion of the biota will be single-island endemics (SIEs) resulting from such within-island evolution, and the fact that these clades will themselves become increasingly species-rich. Though later the number of SIEs is likely to be reduced by, for example, their dispersal to other islands or their extinction as competition increases, Whittaker and his coworkers [45] point out that the number of these SIEs should vary in a predictable manner through the lifetime of the island. The chain of islands resulting from a hotspot, similar in their location and the nature of the animals and plants available, should provide a particularly suitable test of this. Accordingly, the authors use 10 sets of data for organisms from the Hawaiian, Canary,

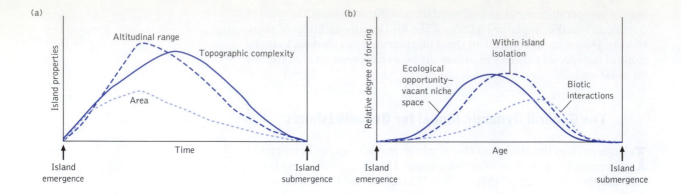

(a)

Island properties

Altitudinal range

Topographic complexity

Area

Time

Island emergence

Island submergence

(b)

Relative degree of forcing

Ecological opportunity~ vacant niche space

Within island isolation

Biotic interactions

Age

Island emergence

Island submergence

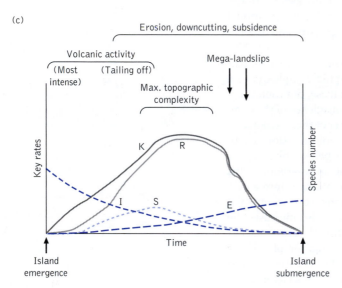

(c)

Erosion, downcutting, subsidence

Volcanic activity

(Most intense)   (Tailing off)

Mega-landslips

Max. topographic complexity

Key rates

K   R

I   S

E

Time

Species number

Island emergence

Island submergence

Fig. 8.12 (a) The geology of the GDM: how the area, height, and topographical complexity of an island may change over its lifetime.
(b) The biology of the GDM: how the number of vacant ecological niches, within-island speciation, and competition may change over its lifetime.
(c) Integrating the geology and the biology of the island's lifetime. The rise in the carrying capacity of the island (K) is followed by an increase in the speciation rate (S) and, after a delay, by a consequent rise in its realized species richness, R. Also shown are the rates of immigration (I) and extinction (E). The kinks in the K and R curves represent mega-landslips, which would lead to an increase in the rate of extinction. After Whittaker et al. [45].

Galápagos, Azores, and Marquesas chains to test their model, and find that they do support it. They comment that their model should in principle apply to other island archipelagos, such as island-arcs and those that were affected by Pleistocene changes in climate and sea level. It will be very interesting to see how their work is received and expanded.

Living Together: Nests, Tramps, and Assemblies

As we have seen, the Theory of Island Biogeography, in its attempt to find general rules governing the number of species to be found in islands, inevitably had to treat the species as identical units, ignoring their history and interactions. Other research has now attempted to find general rules and principles that affect the ways in which individual species associate with one another, avoid one another, or are distributed among the patterns of islands.

If different island faunas are compared with one another, one frequent pattern that emerges is known as **nestedness** (see Guest Author Box 8.2).

By Dr. Richard Field, University of Nottingham

The American biologists Bruce Patterson and Wirt Atmar [46] coined the term 'nestedness', though the concept had previously been proposed independently by Hultén in 1937, Darlington in 1957 and Daubenmire in 1975 (47).

Nestedness is a measurable property of a set of islands or patches (referred to below as 'island systems'), which is commonly, but not always, observed. An island system is perfectly 'nested' (or exhibits perfect nestedness, or comprises perfectly 'nested species subsets') if all species found in less species-rich islands are also found in all more species-rich islands. In other words, all larger species lists contain all species found in all smaller species lists. (An analogy is often drawn with Russian dolls or concentric circles on Venn diagrams—smaller ones fitting entirely within larger ones.) Table 8.2 shows three hypothetical archipelagos with differing degrees of nestedness, ranging from perfect nestedness A to perfect anti-nestedness C.

Much of the literature on this subject has been concerned with the question of whether nestedness is more common in natural systems than expected by chance alone. There is no definitive answer to this question though, by most measures, the majority of archipelago datasets are significantly nested—leading to the obvious question of why this is so. However, some workers argue that observed nestedness in most island systems is not significant because the random or null expectation (the 'null model') is not the sort of random distribution of species occurrences across islands shown in Table 8.2 B. Indeed, passive sampling from log–normal species–abundance distributions, illustrated in Figure 8.13 D, tends to produce strong nestedness in model communities [48], possibly stronger than that typically observed in the field [49]. This has led some to suggest that nestedness is not really typical in natural insular systems [47]. If this is so, the key question about nestedness becomes inverted: why are natural systems less nested than expected from passive sampling? As is often the case with debates about null models (other examples include debates about null models in spatial species richness patterns and in island assembly rules), what is regarded as 'null' depends on both the person doing the research and the aims of that research. For example, what one person considers to be lack of departure from a null model (e.g. passive sampling), another may consider to be a good ecological or biogeographical explanation for the observed nestedness.

Perhaps a more appropriate perspective is that such debates are not addressing the right questions. Rather than asking whether the insular system of interest is 'significantly nested' (which requires some form of null model for testing significance, leading to an intellectual quagmire), we can recognise that nestedness is a measurable parameter or property of our dataset, which provides useful summary information for an island system. Other system-wide parameters include the species richness of the archipelago (the gamma diversity) and the overall beta diversity. These measures interlink well: the archipelago species richness cannot be derived from the species richness values of the individual islands and the overall slope of the species–area or species–isolation relationship, without knowing the degree of overlap in species composition between the islands [see Further Reading]. Both nestedness and beta diversity measure that degree of overlap. Greater overlap of composition means greater nestedness (Table 8.2) but lower beta diversity, and the two measures are opposites in some respects [50]. However, there are many ways of measuring beta diversity, and measured beta diversity may not always be related to measured nestedness [51]. Yet there is no debate about how to determine whether beta diversity is 'significant', nor is there one for species richness. Beta diversity is a parameter that is often measured, typically accompanied by attempts to account for why that level of beta diversity is observed; this seems an appropriate way of analysing nestedness, too [52].

Given the strong conceptual links between nestedness and beta diversity, it is odd that the two literatures are largely separate; indeed, the two terms are rarely used in the same publication. To illustrate this point, as well as the great current interest in nestedness, a search of topic = 'nestedness' on ISI Web of Knowledge on 11th August 2009 produced 401 papers, of which 313 were published since January 2000, but only 22 of these were retrieved by a search for 'nestedness and beta diversity'. Further, some of these 22 barely mention beta diversity—for example, the highly influential nestedness review by several of the leading scientists working on the subject [49], which lists beta diversity as a 'key word' but never mentions it in the paper! Similarly, given the frequent use of nestedness analysis on patches within fragmented ecosystems, it is surprising that there is also minimal overlap between the nestedness and the metapopulation and metacommunity literatures. Such overlap may rapidly increase in the near future.

There are various ways of measuring nestedness, or 'metrics': a useful recent "consumer's guide to nestedness analysis" [47] reviewed eleven nestedness metrics. Two common ones are the C-value [50] and various derivatives of it, and the nestedness 'temperature' [53] (Table 8.2). 'Temperature', which is a measure of disorder or unexpectedness in the species presence–absence matrix, is by far the most commonly used of the metrics. The 'nestedness temperature calculator' of Atmar & Patterson [53] has the operational advantage of calculating the extent to which individual species and islands deviate from the overall nestedness of the island system. This allows the identification of 'idiosyncratic' species and islands —the ones whose patterns are the most different from the overall nestedness of the island system (usually the least nested species or islands are picked out). The computer program published by Wright & Reeves [50] to calculate the C-measure of nestedness has the advantage of producing different values according to the order of entry of the data. So, for example, to calculate the overall nestedness, the islands are ordered from the most to the least species rich. But islands can also be ordered by area, isolation or any other variable of interest, the results indicating the degree of nestedness associated with the variables. Both the temperature measure and the C-value therefore allow investigation of what might be causing the observed level of nestedness. None of the metrics is perfect for all purposes, and various influential reviews have drawn attention to issues such as sensitivity to matrix size and degree of fill [e.g., 49].

Once we have measured nestedness, we can use it in our study system. Specifically, we can try to account for the observed degree of nestedness, in the process learning about the processes at work in our study system, and we may also be able to apply our knowledge of the nestedness of the system in the field, often for conservation. Strongly anti-nested insular systems may be associated with checkerboard species distributions or allopatric speciation events. Strongly nested systems may reflect structure in extinction (often seen as the commonest and strongest cause), colonisation or habitat availability (Figure 8.13). Finding an intermediate situation, with no clear nestedness or anti-nestedness, may be informative because it suggests a lack of such strong, ordered influences on species composition. Recently, nestedness analysis has also been usefully applied beyond typical island or habitat island systems, particularly for interaction networks such as plant–pollinator and host–parasite webs.

For nestedness to be useful for conservation, it should contain information that goes beyond basic biogeographical principles, such as species–area relationships, that are better measured in other ways [54]. Atmar & Patterson, the instigators of nestedness research [53], made perhaps the most elaborate, indepth case for the practical applications of nestedness measurements. They argued that their temperature calculator, as well as measuring nestedness, also measures relative population stabilities and uncertainty in species extinction order, as well as providing both a way of identifying minimally sustainable population sizes, and a way of estimating the historical coherence of the species assemblage. They also argued that information about these features of ecological systems would be very difficult to obtain in any other way. Similarly, Fleishman and her colleagues [55] argued that nestedness analysis, which uses readily available data, is an effective and low-cost tool for resource management and conservation planning.

On the other hand, as with all summary indices, we must be careful not to over-interpret nestedness statistics [56]. Indeed, recent work has found nested structure sometimes to be due to relatively small numbers of accidental occurrences [57]. We should recognise that some conditions make the development of nestedness much more likely, especially the existence of a common species pool for all the islands and some form of hierarchical organisation of species' ecologies [58]. We must also recognise that the causes of nestedness can be many, varied and interacting. For example, continued immigration affects extinction probabilities via the 'rescue effect' [59], while larger areas provide larger 'targets' for immigration [49] so that area and isolation (Figure 8.13) tend to interact as causes of nestedness [60]. Similarly, larger islands typically contain more habitats, and this is typically structured (as in Fig. 8.13), meaning that area is confounded with habitat nestedness. There is enough complexity to keep biogeographers occupied for some time to come!

Table 8.2 Three archipelagos representing different degrees of nestedness. All have six species and six islands and all have the same 'fill': the same total number of species presences (21 or 58.3%). (A) Perfect nestedness, in which all larger species lists contain all species in all smaller species lists. (B) Random species presences, with no nesting or anti-nesting, generated simply by each cell in the table having a 58.3% probability of containing a 1. (C) As non-nested a set of species occurrences as is possible within this archipelago, if the fill of 58.3% is maintained. All matrices shown are 'maximally packed,' meaning that they display the presences towards the top-left corner and the absences towards the bottom-right as much as possible. The 'temperature' measures the amount of nestedness in the archipelago and was calculated by the 'nestedness calculator' of Atmar and Patterson—available at http://aics-research.com/nestedness/tempcalc.html and see Atmar and Patterson [53].

	A Perfect nestedness						**B Random, no nestedness**						**C Perfect anti-nestedness**					
	Sp	Sp	Sp	Sp	Sp	Sp	Sp	Sp	Sp	Sp	Sp	Sp		Sp	Sp	Sp	Sp	Sp
Island	1	2	3	4	5	6	Sp1	2	3	4	5	6	Sp1	2	3	4	5	6
Island 1	1	1	1	1	1	1	1	1	1	1	1	0	1	1	0	1	0	1
Island 2	1	1	1	1	1	0	1	1	1	0	1	0	1	1	1	0	1	0
Island 3	1	1	1	1	0	0	1	1	1	1	0	0	0	1	1	1	0	1
Island 4	1	1	1	0	0	0	1	0	0	1	1	0	1	0	1	0	1	0
Island 5	1	1	0	0	0	0	0	0	1	1	0	1	0	1	0	1	0	1
Island 6	1	0	0	0	0	0	0	1	0	1	0	0	1	0	1	0	1	0
Temperature							31.						80.					
:	0.0						9						3					

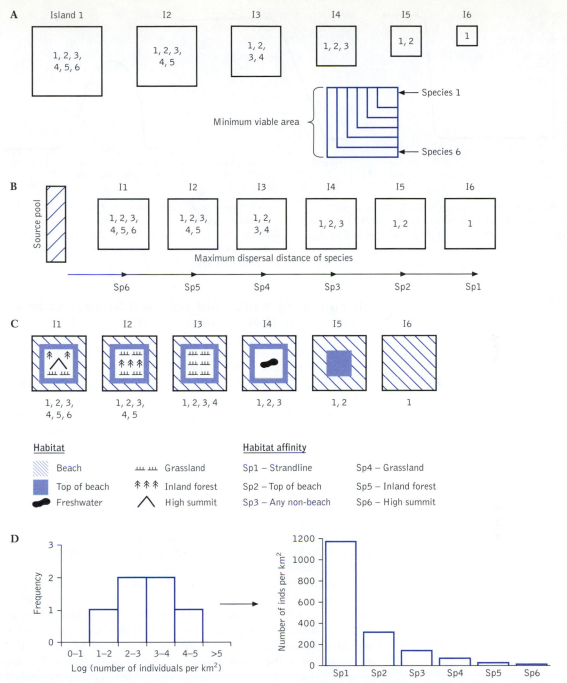

Fig. 8.13 Figure 8.2. Possible causes of nestedness, based on Wright et al. [49]: A area; B isolation; C habitat nesting; D passive sampling. An archipelago comprising six species in six islands and showing perfect nestedness is shown (the same as in Table 8.2 A). In A to C the squares represent the islands, labelled Island 1–6, except for the squares representing minimum viable area for the six species in A. The species found in each island are listed within A-B or below C the island. In B dispersal is assumed only to occur from the mainland source pool to any of the islands. In C the nesting is caused by a combination of habitat nesting (mostly) and habitat specificity or generalism of the species. D represents a log-normal species-abundance relationship in the species pool; the right-hand graph illustrates the strong skew of individuals per species that this represents. Randomly drawing individuals of any species from this species pool is likely to generate the nestedness pattern shown in A to C if the number of individuals found on the six islands varies from low to high.

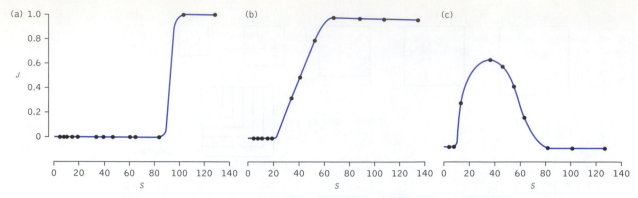

Fig. 8.14 Examples of Diamond's incidence rules. Each graph shows the percentage of 50 islands within which each species is found, plotted against the number of bird species found on each island. Each point represents grouped data for between 3 and 13 islands, except that the two largest values each represent only one island. A, the distribution of a sedentary species, the cuckoo *Centropus violaceus*. B, the distribution of an intermediate tramp, the pigeon *Ptilinopus superbus*. C, the distribution of a supertramp, the pigeon *Macropygia mckinlayi*. After Diamond [61].

In his work on Pacific island birds, Jared Diamond documented the occurrence, or **incidence**, of 513 species on thousands of islands near New Guinea (Fig. 8.14) [61]. He found that, on the one hand, some species (which he called sedentary species) are present only on the largest, most species-rich islands. A few others (which he called supertramps) are absent from these and are usually found in the smaller or most remote and species-poor islands (Fig. 8.13). In between these two extremes, Diamond arbitrarily defined four other categories of lesser tramps. The data suggested that the combinations of bird species to be found on the different islands were not random, but depended on many factors, including size of the island, number of species, availability of suitable habitats, and extent of the required habitat in area or duration in time.

Analyzing the different assemblages of species found in the islands, Diamond drew up a number of generalizations that, he believed, fitted and explained the data; he called these generalizations **assembly rules**. Concentrating on groups of species that occupy similar ecological niches, he discovered that some pairs of species were never found together, apparently because they competed directly with one another, so that the first to arrive was able to exclude the other. This appears to be the case in the colonization of central Pacific islands by wind-dispersed spiders [62]. Some combinations therefore appear to be "permissible", while others are "forbidden." Other such pairs can coexist, but only if they are part of a larger assembly. But precisely what combination of species is permissible varied according to the size and species richness of the island. The data also suggested that potential invaders were unable to colonize islands if their presence there would produce a "forbidden" combination (presumably because the species that were already there did not leave a suitable niche vacant for the invader). However, the invader might be able to enter such a combination on a larger or more species-rich island.

Over the last 30 years, Diamond's suggestions have stimulated a wide range of both critical and supportive comments, which in turn have provoked responses by himself and Gilpin [63]. Some of these comments related to his acceptance of the equilibrium aspect of the TIB. Another basic problem, of course, is the multitude of factors that Diamond invokes in his incidence functions and assembly rules. These include distribution vs. island area, species richness, pattern of habitats, history of changes in climate, area, connection with the mainland, human intervention, and interaction between the species. It could therefore fairly be claimed that some explanation could be found for any island biota, but that one could not rule out the possibility that it was merely the result of chance (the null hypothesis), an approach recently supported in a study on woody flowering plants in 27 islands west of Vancouver Island, Canada [64]. However, as Robert Colwell and David Winkler have shown [65], it is in fact extremely difficult to design a null model that does not itself suffer from serious biases. It is equally difficult to use a real-world mainland biota as a comparison because its composition has been conditioned by competition within a much more varied biota. Given the variety and complexity of island biotas, areas, and history, it is unlikely that any single theory will explain them all, and it is pragmatic to retain a variety of approaches and to see which is the most helpful in each case.

Another line of research has been to try to identify what characteristics are found in earlier or in later colonists. The earliest are found to be **r-selected**: highly mobile, capable of rapid growth to early maturity, and therefore with a high potential rate of population increase. They contrast with the later arrivals, known as **K-selected**: slower at dispersing and reproducing, but with a greater ability to sustain their population when this is getting close to the carrying capacity of the island.

The Theory of Island Biogeography and the Design of Nature Reserves

It is not surprising that the Theory of Island Biogeography was warmly welcomed by those concerned with the management or design of nature reserves, for they could be analyzed as islands amidst the sea of surrounding unprotected land. It seemed to promise an almost magical prescription for ensuring the ideal balance between effectiveness in terms of retaining the maximum number of species and economy in their area (and therefore in their cost of acquisition and management). However, it is now clear that it is no such remedy and reliable forecaster. As the island biogeographers Robert Whittaker and José María Fernández-Palacios comment in their textbook (see Further Reading), which contains an extended coverage of this topic, "The grand hopes for single unifying principles have proven to be as illusory as the end of the rainbow." Perhaps the most fundamental reason for this is that the heart of the TIB lies in generalizations that ignore the interactions between species, but are designed to provide general rules against which individual species or circumstances can be judged. In contrast, the problems of conservation and the design of nature reserves are concerned with the needs and interactions of the whole ecosystem that includes the threatened species.

As we have already seen, only one of the theory's predictions—that the number of species in an island will decrease directly with reduction in its area—now seems reliable. Its usefulness in the context of nature reserves was confirmed by the work of the American ecologist William Newmark [66], who studied the records of the species of large mammal found in 14 National Parks of the western United States. He found that 42 separate populations of the species of lagomorphs, carnivores, and artiodactyls had disappeared from the parks since their foundation. Some of these losses had been the result of deliberate human intervention (e.g., the culling of all the gray wolves from Bryce and Yellowstone), but the patterns of disappearance of other species were correlated closely with the size of the park. Thus, the smallest (Bryce, Lassen, and Zion) have each lost four to six species, while great Yellowstone, 20 times larger than Zion, has lost none. Furthermore, it seemed that area alone was the best predictor of the diversity of large mammals in each park, rather than any other factor such as the range of elevation found in the park, its latitude, or the diversity of plants that it contained. A study of the effects of the fragmentation of tropical forests in West Africa by the British biologists Jennifer Hill and Paul Curran came to similar conclusions [67].

On the other hand, the diversity of the habitats in the park also has an important effect on the maintenance of diversity in the assemblages of species it contains, and may even override the effects of area alone [68]. Shafer [41] comments that it is probably impossible to separate the effects of the area of a nature reserve from other factors, even using sophisticated statistical techniques of multivariate analysis.

Large nature reserves should therefore retain more species and suffer fewer extinctions. But how large need they be? Very little is known about the area requirements of most species, and these requirements will depend on the degree of ecological specialization of the species and on their breeding habits. The larger mammals and birds that are a major preoccupation of many tropical game reserves require a very large home range to meet their energy needs. However, within the size range of the world's nature reserves, 97.9% of which are less than 10,000 km^2 (3,860 mi^2) in area, the variability of prediction of the Theory of Island Biogeography is so great as to be of no practical value when considering the needs of such large animals [69].

Another important consideration is the size of the population that is required in order to ensure it is large enough to preserve enough genetic variability to be able to persist through the occasional environmental disaster and avoid excessive inbreeding. This is not simply a question of counting the number of adult individuals in the population, for normally only a proportion of them are involved in breeding. The required population size may therefore be substantially smaller than the total population. Shafer [41] showed, in the case of the grizzly bears in Yellowstone National Park, that their population needed to be as high as 220 individuals in order that it would contain the minimum required breeding population of 50. Of course, another consideration here is the size of the organism concerned. Other things being equal, a reserve that is far too small to safeguard its grizzly bears might well be quite adequate for the conservation of cold-blooded organisms such as salamanders, whose size and demands on the environment are very different.

It is tempting to assume that a large nature reserve is simply better than a small one, but each in fact has its own set of advantages and disadvantages. This has led to a lively, and continuing debate, sometimes known as the

SLOSS debate—Single Large, Or Several Small, reserves? For example, a larger reserve is likely to support larger populations of the species that it contains, and therefore render them less vulnerable to chance extinction. Similarly, local ecological changes are also less likely to affect all the areas occupied by a particular species within a large park than within a small park.

Another possible advantage of a large reserve is what is known as the **edge effect**. The peripheral zones of a reserve will have an environment that is a transition between that of its interior, or core, and that of the surrounding ecosystem. This may have the advantage of harboring some species that are not found in the core, and so increase the diversity of the reserve. But this may be at the expense of the more important core species, which are likely to be less successful in the transition. The presence of the transition zone will also reduce the effective area of the reserve itself; the smaller the reserve, the greater will be the proportion of its area that is taken up by a transition zone.

In such cases a large reserve therefore has an advantage over several small reserves. At the same time such random occurrences as fire, disease, or the appearance of an overcompetent predator or competitor might exterminate all of the individuals of a given species in one large park, but are unlikely to affect all of a number of small parks. Several small parks are also likely to have a greater number of different habitats.

Small reserves also suffer the disadvantage of being isolated from one another or from any neighboring larger reserve. However, it has been suggested that establishing corridors to interconnect the reserves can reduce these disadvantageous effects. The value of such corridors has been shown in a study of the Australian Carnaby's cockatoo, *Calyptorhynchus funereus latirostris* [70]. Land clearance for annual crops had led to the widespread disappearance of the vegetation on which it feeds. As a result, once the flock of cockatoos had run out of food in one remaining area of such vegetation, it found it difficult to cross the intervening areas of agricultural crops to find the next, leading to a marked reduction in their population size and area of distribution. The provision of wide strips of native vegetation along the sides of the roads, which the birds can follow to find another, larger patch of their preferred food, has led to an improvement in their flock sizes and area of distribution.

However, size alone is not everything. Even a large park will not safeguard the future of all of the species that live within it, if it does not contain environments that are vital for some of them. This is shown by the work of the American zoologists Barbara Zimmerman and Richard Bierregaard [71], who studied the breeding requirements of 39 species of Amazon frog. Some of the frogs gathered to breed in large streams, some in permanent pools of water, and others in temporary pools. The presence of the pig-like peccaries was also important, for they created small wallows in which some of the frog species bred. The obvious lesson here is that we need to know a lot about the ecology of any species that we wish to conserve and that this may be more important than simple calculations of required area. A large park that omitted any of these crucial environments would be of less value in conserving a frog species than a smaller park that contained the appropriate breeding environment.

It is now clear that decisions on the size and location of nature reserves must include an ecological analysis: which species do we wish to preserve, and what are their requirements? Furthermore, we need to be able to estimate the size of the minimum population that is likely to provide a reasonable chance of long-term survival for the species in question, together with an estimate of the area that is needed to support a population of that size. Finally, in some communities there are **keystone species** (see p. 97), that are crucial to the survival of many other species and whose disappearance would therefore fundamentally change the nature of the community. In the case of the frogs mentioned earlier, peccaries are important to their breeding. Any conservation area must therefore be large enough to support the minimum viable population of that species. Considerations of this kind, rather than inferences from abstract curves of the numbers of species arriving, surviving, or becoming extinct, must lie at the heart of the design of nature reserves. There are no shortcuts. Furthermore, the vast range of types of species, problems, and environments suggests that, in any particular case, we should consider a range of the different approaches and models that are now available and consider which may be the most appropriate, rather than relying on any single type of analysis.

Regardless of the usefulness of the Theory of Island Biogeography in the design of nature reserves, islands are of great value in conservation, as their isolation provides a useful insulation from the factors that may be producing extinctions on the mainland. For example, the Australian conservationists Andrew Burridge, Matthew Williams, and Ian Abbott [72] have shown that 35 islands around Australia protect 18 taxa of threatened mammals, and they emphasize the importance of preventing the introduction of exotic mammals, especially foxes. On the other hand, the low population sizes characteristic of islands make their biota more susceptible to extinction, as is shown by studies of human impact on them. For example, of the original mammal species that were autochthonous to islands (i.e., that were not only endemic to them but had actually evolved there), 27% became extinct owing to human activities—and this figure rises to 35% if flying mammals are excluded from the calculation [73].

Building an Ecosystem: The History of Rakata

In most cases, we have little knowledge of the history that lies behind the complex assemblage of animals and plants that inhabit a particular island. We may attempt to compare different islands and to place their biota in a series, or several series, that might represent an historical process, but such an enterprise is fraught with the difficulties of subjective interpretation. We are on surer ground only where the history of the biota of a single island has been documented over an extended period of time. We are now fortunate in having such documentation for one island, Rakata, whose biota is in the process of being reassembled after total destruction. It has been studied since 1979 by the Oxford biogeographer Rob Whittaker and his associates, and since 1983 by the Australian zoologist Ian Thornton and his co-workers. Much of the information in this section is taken from Thornton's book *Krakatau* [74], plus other data from Whittaker et al. [75]. The data accumulated (and still being accumulated by such research programs) allows us to analyze the sequence of colonization of the island and how different methods of colonization contribute to that sequence. Another valuable aspect of these studies, as we shall see, is that it has been possible to extend them to other neighboring new islands, giving us a rare opportunity to make comparative studies.

Rakata is in Indonesia and lies between the major islands of Java (56 km/35 mi away) and Sumatra (50 km/31 mi away), which act as the main sources of its colonists (Fig. 8.15). With an area of 17 km^2 (65 m^2) and approximately 735 m (2,410 ft) high, Rakata is the largest remaining fragment of the island of Krakatau, which was destroyed by an enormous volcanic explosion in 1883. Two other islands, Sertung 13 km^2 (5 mi^2), 182 m (600 ft) and Panjang 3 km^2 (1 mi^2), 147 m (482 ft), are fragments of an older, larger version of Krakatau, and a new island, Anak Krakatau, appeared in 1930. All life on the islands was extinguished by the eruption, which covered them with a layer of hot ash 60–80 m (200–260 ft) deep on average and up to 150 m (490 ft) deep in places. Surveys of the biota of Rakata were made intermittently from 1886, with a gap between 1934 and 1978 (apart from a little work in 1951), and intensive work has been carried out since the centenary of the eruption in 1983. (In the following account, the dates of these surveys are indicated as the length of time that had elapsed since the 1883 eruption; thus 1908 = E + 25. See also the scale at the base of Fig. 8.16.) These surveys show that the patterns of colonization and extinction are not smooth, but are heavily influenced by the times of emergence of new ecosystems and by the linkage between plants and animals due to food requirements or mechanisms of dispersal.

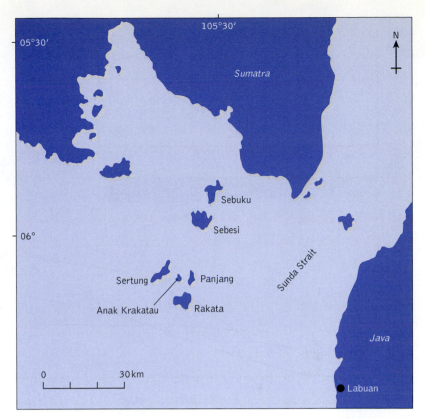

Fig. 8.15 Location of Rakata and neighboring islands. After Whittaker et al. [83].

The Coastal Environment

The development of the biota of Rakata is best understood by analyzing the beach and near-beach environment separately from the inland environment. There are two reasons for this separation. First, the beach environment is largely unchanged by the establishment of living organisms there. It is also itself ceaselessly being destroyed and re-created as ocean currents and storms erode one part of the beach and redeposit its materials elsewhere. This causes frequent local extinction of the biota and simultaneous recolonization elsewhere. Second, these environments

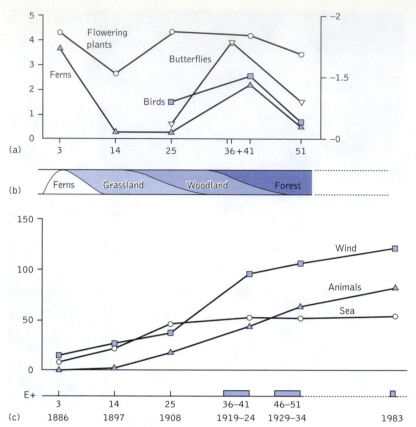

Fig. 8.16 The changing biota of Rakata, 1883–1934. The lower axis shows the dates, and the periods elapsed, since the eruption (E + x); the blocks indicate the periods at which collections were made.
(a) Changes in the immigration rates of ferns and flowering plants (left-side scale) and of butterflies and resident land birds (right-side scale; these were not censused before 1908). After Thornton [74].
(b) A subjective interpretation of the rate and nature of the environmental changes, deduced from the descriptions given by scientific investigators.
(c) Methods of initial colonization by ferns and flowering plants. Beyond the figures for 1934, the level has been extrapolated toward the figures from the 1983 census. After Bush & Whittaker [84].

are primarily colonized by plants that have evolved methods of dispersal by sea, and for whom the ceaseless tides provide daily opportunities to colonize the beach.

There were already nine species of flowering plant on the beaches of Rakata by E + 3 (including two species of shrub and four species of tree). Eleven years later (E + 14), this had risen to 23 species of flowering plant, including 3 species of shrub and 10 species of tree. The flora already included three distinct communities. Along the beach itself lay the strand-line creepers such as *Ipomoea pes-caprae*. The trees and shrubs grew a short distance inland and were made up of stands of the Indian or sea almond tree *Terminalia catappa*, or of the tree *Casuarina equisetifolia*. All of the species in these woodlands are widely distributed on the beaches of Southeast Asia and the western Pacific, showing that they are good at dispersal by sea. The number of trees in these early beach-arrival figures may at first seem surprising, but the larger size of the fruits or nuts of trees makes it easier for them to have flotation devices and not be overwhelmed by the waves. (Not all arrivals are merely individual specimens; in 1986, the beach on Anak Krakatau bore a mass of vegetation 20 m^2 (215 ft^2), including complete palm trees 3–4 m (9–13 ft) tall.) Once they had grown, the early trees and shrubs also provided food in the form of fruit, as well as perching places, for birds and bats whose droppings were the probable source of other trees, such as two species of fig. Early colonists themselves therefore also provide a beachhead for other arrivals. The seeds and fruits of trees (such as *Terminalia*) that had arrived on the beach may also have been taken further inland by both fruit bats and crabs; the lack of mammals other than bats and rats on Rakata may explain why this inland spread seems to have been comparatively slow on the island.

By E + 25, 46 species of plant had arrived in Rakata by sea (Fig. 8.16c), but thereafter the seaborne component of its flora started to level off. Potential colonists are continually arriving by sea—a 2-month-long survey of the beach of Anak Krakatau on two successive years found the fruits, seeds, or seedlings of 66 species of plant. Because it is so easy for these species to colonize the beach, most of them soon do so, and thereafter there will be few new arrivals of seaborne species. So, for example, between E + 41 and E + 106, the diversity of the beach flora increased by only six new species (from 53 to 59 species), and the beach community is now relatively stable, with few new gains or losses.

Life Inland

The history of the colonization of the inland areas of Rakata is more complex because the environment did not remain constant. Instead, the presence and activities of each wave of colonists not only changed the environment, but also produced a greater variety of habitats, some of which were suited to a new selection of colonists. Both the complexity and the variety of the inland ecosystems of Rakata therefore steadily increased. Figure 8.16b is an attempt to give an impression of the timing and rate of these ecological changes, using the descriptions given in the early surveys, so that they can be compared with the changes in some of the animals and plants that were colonizing the island during this period of time.

At first, by E + 3, the devastated, ash-covered inland areas bore a gelatinous film of blue-green algae, or cyanobacteria. This had provided a moist environment in which the spores of 10 species of fern and 2 species of moss had already germinated and grown, as well as four species of herb. (This early preponderance of ferns may be because their spores are lighter than the seeds of flowering plants, but it is interesting to note that there was a similar sudden increase in ferns after the devastation of the planet's plants by the meteorite at the end of the Cretaceous, see p. 313). By E + 14, many new species of flowering plant had arrived, so that the ferns were less dominant, although they still covered much of the upland areas. By that time, definite plant associations could be distinguished at different levels in the island. The interior hills and valleys were covered by a "grass steppe," up to 3 m (10 ft) high, consisting of the wild sugarcane *Saccharum spontaneum*, together with other grasses and scattered trees. Interestingly, some of the creeping plants and a shrub that are normally confined to the beach had been able to extend inland in this still-impoverished Rakata flora.

Grassland also covered the higher slopes of Rakata, but here it was dominated by *Imperata cylindrica*, a grass that is usually the first to colonize fire-cleared areas in the East Indies, together with the bamboo grass *Poganotherum*. Although ferns were still present in these grassland floras, they contributed only 14 species, compared with 42 species of flowering plants, and dominated only the higher regions.

By E + 25, new species of animal-dispersed tree such as figs and *Macaranga* were arriving in the lowlands. As a result, the interior grasslands were being replaced by mixed woodland and forest, and they had almost completely disappeared by E + 45. The forest grew denser, and its canopy gradually closed between E + 36 and E + 51. This caused a progressive change in the physical habitat and microclimate of the forest floor: wind velocity, light intensity, and temperature all decreased, while humidity increased. The graphs of the immigration rates of ferns, flowering plants, butterflies, and birds (Fig. 8.16a) show interesting changes that appear to be results of these ecological changes. (The immigration rate is the rate of addition of new species to the biota, per year. The addition of 10 new species over the space of five years between one survey and the next would therefore be an immigration rate of two species per year.)

For both ferns and flowering plants, the grasslands that replaced the early fern phase do not seem to have provided an environment that encouraged a diversity of new colonists, and the immigration rate of both groups fell. The fall was greater for the ferns, because their immigration rate had earlier been particularly high during the formation of the fern phase. The immigration rate of the flowering plants had improved when they were next sampled, at E + 25, but had fallen by E + 51. This may be because the increasing woodland at first provided a greater variety of habitats for them, but that the later closure of the forest canopy restricted the light. The immigration rate of the ferns, in contrast, continued to increase even during the early stages of canopy formation, probably because the moist forest provided an ideal environment for a second group of ferns. These were mainly shade-demanding species, many of which were epiphytes, living on the trunk and branches of the forest trees. In both butterflies and resident land birds, the immigration rate rose as the forest started to form, but had fallen by the time the canopy had closed. Thus, closure of the canopy took place at the same time as a reduction in the immigration rate of all these groups. Nevertheless, because extinction rates still remained comparatively low, the total number of species increased slightly in the case of the flowering plants and remained approximately constant in the other groups. (The fact that extinction rates remained low during the period of closure of the canopy suggests that patches of open ground or woodland must have remained, either around the edges of the forests or within them. This may have been where trees had fallen and provided a continuing opportunity for the survival of species that preferred an open habitat.) Analysis of the species that were lost suggests that some of them had never become properly established, that some had lived in habitats that had disappeared or had been transformed, and that others had had a very restricted distribution. The higher plant flora of the interior is still gaining new species, so that the forest succession is still continuing and the balance of species in the canopy is still changing.

Rob Whittaker, together with the British biogeographer Richard Field and the Indonesian botanist Tukirin Partomihardjo, have analyzed the data on the floras of Rakata and its neighboring islands, with interesting and important results [76]. Though earlier analyses suggested that many extinctions had been due to the interactions of a complex of variables, their new analysis instead shows that this was the result of sampling errors. Instead, a relatively high proportion of the extinctions were the inevitable results of the loss or transformation of habitats as part of processes of change that affected whole islands or communities.

As noted earlier, the biota of the interior of Rakata also differs from that of the coast in that nearly all of it arrived by air, not by sea. That was not a very difficult journey, for winds from Java and Sumatra have an average speed of 20–22 kph (12–14 mph), so that wind-blown seeds could arrive in Rakata in about 2 hours. However, although some of the airborne arrivals came on the wind, others came in or upon the bodies of animals. To begin with, the ash-covered interior of Rakata was totally uninviting to animals. Thus, as can be seen from Fig. 8.16c, animal-aided dispersal only became a significant contributor to the biota from the time of the E + 25 survey. From then on, the graphs of the arrival of wind-dispersed species and of animal-dispersed species move in parallel. But they also rise more steeply because of a positive-feedback effect between the plants and the animals. The woodlands that had developed by E + 25 provided an environment that other species of flowering plant could colonize (Fig. 8.16b). The increasing diversity of these plants in turn provided food for an increasing diversity of animals, as can be seen from the fact that the immigration rates of both butterflies and birds increased at that time. This effect became especially evident as the forest canopy formed. But the increasing numbers and diversity of animals arriving in the growing forests also brought the seeds of other new species of plant, either within their alimentary canal or adhering to their bodies. The resulting increase in plant diversity in turn encouraged more animal diversity, and so on.

The animal-dispersed component has been the most important one ecologically because the seeds of nearly all the tree species of the inland forests arrived in this way. Wind dispersal, on the other hand, was particularly important in the addition of other plant species, providing all of the forest ferns and many of the herbs and shrubs. Of these, 17% belong to the Asteraceae and 13% to the Asclepiadaceae (milkweeds, whose seeds bear silky hairs rather like those of the Asteraceae and are easily carried in the wind), while over 50% of these species belong to the Orchidaceae, some of which are dispersed by wind and others by animals. (Orchid seeds have no food reserve and need root fungi in order to germinate and grow, which suggests that the original colonists, at least, may have arrived on the legs of birds, in mud that also contained the fungus.) For the flora as a whole, species with small, wind-dispersed seeds form a much higher proportion of the flora of Rakata than they do on neighboring Java.

Rakata is the largest and highest of the three islands that were the surviving, initially lifeless, fragments of Krakatau (see Fig. 8.15). It might have been expected that the forests that eventually appeared on these three islands would be similar to one another and to those on neighboring Java and Sumatra—but they are not. The lowland forests of Rakata are unique because they are dominated by the wind-dispersed tree *Neonauclea* (Rubiaceae), which grows up to 30 m (100 ft) high. The forests of Panjang and Sertung, in contrast, are instead dominated by the animal-dispersed trees *Dysoxylum* (Meliaceae) and *Timonius* (Rubiaceae). Why should the forests of these islands be unique and also different from one another? Various theories have been put forward.

One difference between the trees is that *Neonauclea* is less tolerant of shade than are the other two species. It may therefore be "shaded out" if all three species arrive on an island at about the same time. However, *Neonauclea* arrived on Rakata in 1905, nearly 25 years before the others, and this may have given it a head start and allowed it to become dominant on that island. It also seems that it requires overturned soil or fresh ash in order to establish itself. Few scientific visits were made to the other two islands, so the history of their forests is not well known—only that all three species were present by 1929. The Japanese ecologist Hideo Tagawa and his colleagues [77] have suggested that, if they all arrived at about the same time, the shade produced by the growing *Dysoxylum* and *Timonius* trees might have made it more difficult for *Neonauclea* to flourish. Rob Whittaker and his colleagues initially pointed out that, unlike Rakata, the other islands had been partially covered by up to 1 m (3 ft) of ash during the

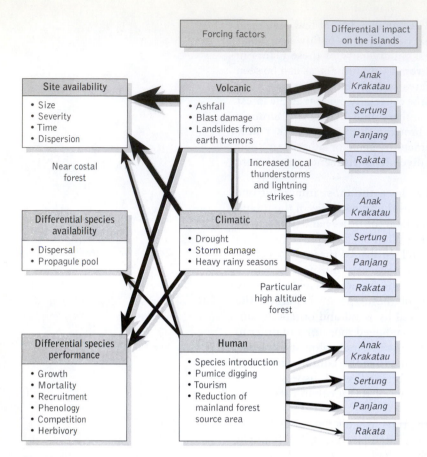

Forcing factors

Differential impact on the islands

Site availability
- Size
- Severity
- Time
- Dispersion

Near costal forest

Differential species availability
- Dispersal
- Propagule pool

Differential species performance
- Growth
- Mortality
- Recruitment
- Phenology
- Competition
- Herbivory

Volcanic
- Ashfall
- Blast damage
- Landslides from earth tremors

Increased local thunderstorms and lightning strikes

Climatic
- Drought
- Storm damage
- Heavy rainy seasons

Particular high altitude forest

Human
- Species introduction
- Pumice digging
- Tourism
- Reduction of mainland forest source area

Anak Krakatau
Sertung
Panjang
Rakata

Anak Krakatau
Sertung
Panjang
Rakata

Anak Krakatau
Sertung
Panjang
Rakata

Fig. 8.17 The relative importance, and hierarchy, of the different factors affecting plant succession in the Krakatau Islands. From Schmitt & Whittaker with permission [85].

eruption of Anak Krakatau in 1930 and that this might well have affected the floras of the islands. They later [76] suggested that there is a very strong chance element determining which species happens to form the dominant element, its success depending on such factors as time of year, prevailing climate and wind direction, what elements of the previous vegetation had survived and which of them were fruiting, which dispersal agents were available, and so on (Fig. 8.17).

The recolonization of Anak Krakatau following occasional eruptions from 1932 to 1973 showed a similar pattern to that following the main Krakatau eruption of 1883. In particular, the coastal and pioneer inland colonists were very similar, and it will be very interesting to see what light the interior forest colonists, only now starting to appear, will throw on the variations shown by the forests of Rakata, Panjang, and Sertung.

All of this is a good illustration of how difficult it is to interpret ecological biogeography, even in the apparently simple situation of the colonization of a lifeless island environment.

Fig trees, which are a particularly important part of the flora of tropical forests, provide another interesting problem in the colonization of Rakata. They are a major component of the forests; by E + 40, the 17 fig species found on Rakata, Panjang, and Sertung made up nearly two-thirds of the total number of tree species there. (Figs are a very important component of the flora because they fruit throughout the year and so sustain the populations of frugivore species through periods of time when other sources of their food are not available. They are therefore used as food by many animals and are keystone species in the ecosystem. In Malaya, one individual fig tree was visited by 32 species of vertebrate, and 29 species of fig tree were used by 60 species of bird and 17 species of mammal.) Just as important, however, is the fact that few of these animals eat only figs, and therefore those fig-eating animals (especially bats) arriving in Rakata were quite likely to bring, within their digestive system, the seeds of other trees. (Bats can retain viable seeds in their gut for over 12 hours, which gives them plenty of time to feed on the mainland and then fly to Rakata.) But figs also provide a problem in colonization, for each species of fig requires the services of its own species of pollinating wasp in order to produce fertile seeds—and the fig is similarly necessary for the life cycle of the wasp. Therefore, for this symbiotic system to become effective and established, there has to be a sufficiently large population of both the fig tree and the wasp, each component having arrived in Rakata independently, the figs having been borne by animals and the wasps having arrived by air. (However, some fig species have solved this problem by hybridizing with one another [78], thus making up for the absence of one of the required species of wasp.)

Finally, one can learn something about the processes and difficulties of island colonization by noting what types of animal or plant have so far not been able to colonize Rakata. Small, nonflying mammals such as cats, monkeys, and most rodents are absent; they are incapable of crossing sea barriers of more than 15 km (9 mi). The exception is the country rat *Rattus tiomanicus*, which has been known to swim 35 km (22 mi) and which had colonized Rakata by E + 45. Because there is no supply of running or standing freshwater on the island, there are no mangrove trees, no freshwater birds, no insects that have aquatic larvae, and no freshwater molluscs. There is as yet no mature forest on Rakata, and therefore none of the birds that require that environment, such as trogons, parrots, nuthatches, hornbills, pittas, and leafbirds. Finally, although some trees (e.g., dipterocarps, which are usually the dominant trees in Southeast Asian rainforests) and bushes have winged seeds, these do not normally travel in the wind for more than a mile (1.6 km), and these species are absent from Rakata.

There are also some interesting interactions at a more detailed level. For example, the bird fauna of Rakata includes the flowerpecker *Dicaeum*, which distributes the seeds of the plant family Loranthaceae—epiphytic plant parasites of the trees of the forest canopy. However, the Rakata forest is not yet old enough to contain the mature and dying trees that the parasite can attack. As a result, the Loranthaceae are absent, together with the butterfly *Delias* that feeds on these plants, even though the butterfly itself is highly migratory and a competent potential colonist.

The complexity of all these ecological and successional changes shows that the colonization history of an island will not follow the simple path predicted by the theory of island biogeography. MacArthur and Wilson themselves pointed this out in their book, in which they gave an example of a single-peaked colonization curve and suggested that the sequence of invasion might affect the nature of colonization and the equilibrium number [9]. The replacement of one plant community by another causes pronounced irregularities in the graphs of immigration and extinction, not only of the plants themselves but also of the associated fauna. The integrated nature of the successional ecosystems therefore makes it likely that the colonization history of an island like Rakata will show pronounced waves of change. The simple monotonic curves predicted by the Theory of Island Biogeography therefore apply only to the situation after the pattern of communities in an island has already become firmly established. There can be no doubt that the detailed studies of Whittaker and his colleagues will continue to provide fundamental data and insights into the varied processes of the establishment of island communities, and of ecological interactions between the different species of the island biota.

Islands, with their limited biota and correspondingly simple ecosystems, provide unique information on three areas of biogeographical research.

1 The islands' impoverished faunas and floras are the ideal situation for rapid evolutionary modification and adaptive radiation of colonists. The events of colonization and subsequent adaptation to the new environment, sometimes taking advantage of major new ecological opportunities, provide many fascinating examples of evolution in action. These processes are illustrated with a detailed study of the biogeography of the Hawaiian Islands.

2 Island life is unusually hazardous, so that there is a complex interaction between the processes of immigration, colonization, and extinction. However, attempts to construct a predictive theory of the numbers of species that would be found on islands of different sizes and locations have proved to be unreliable, and these theories also provide only limited help in the design of nature reserves.

3 Research on the recolonization of the once-lifeless island of Rakata in the East Indies is providing unique insights into the interaction between environmental factors and the arrival of new animals and plants in a developing ecosystem. Comparison between the results of these processes in Rakata and two other neighboring islands pose interesting questions on the role that chance plays in these ecological developments.

Summary

We now turn from the islands to study the oceans that surround them, whose biogeography involves a new dimension, depth, but whose organisms provide fundamental problems in the identification of species and their areas of endemicity.

Further Reading

Whittaker RJ, Fernández-Palacios JM. *Island Biogeography: Ecology, Evolution and Conservation.* 2nd ed. Oxford: Oxford University Press, 2007.

Quammen D. *The Song of the Dodo—Island Biogeography in an Age of Extinctio*n. London: Pimlico/Random House, paperback, 1996.

Grant P. ed. *Evolution on Islands,* Oxford: Oxford University Press. 1988.

References

1 Censky EJ, Hodge K, Dudley J. Over-water dispersal of lizards due to hurricanes. *Nature* 1998; 395: 556.

2 van der Pijl L. *Principles of Dispersal in Higher Plants*, 3rd ed. Berlin: Springer-Verlag, 1982.

3 Nogales M, Medina FM, Quilis V, González-Rodríguez M. Ecological and biogeographical implications of Yellow-Legged Gulls (*Larus cachinnans* Pallas) as seed dispersers of *Rubia fruticosa* Ait. (Rubiaceae) in the Canary Islands. *J Biogeogr.* 2001; 28: 1137–1145.

4 MacArthur RH, Diamond JM, Karr J. Density compensation in island faunas. *Ecology* 1972; 53: 330–342.

5 Diamond JM. Historic extinctions: A Rosetta Stone for understanding prehistoric extinctions. In: Martin PS, Klein RG, eds. *Quaternary Extinctions: A Prehistoric Revolution.* Tucson: University of Arizona Press, 1984.

6 Filardi CE, Moyle RG. Single origin of a pan-Pacific bird group and upstream colonization of Australia. *Nature* 2005; 438: 216–219.

7 Lack D. Subspecies and sympatry in Darwin's finches. *Evolution* 1969; 23: 252–263.

8 Simberloff DS. Using island biogeographic distributions to determine if colonization is stochastic. *Am Naturalist* 1978; 112: 713–726.

9 Cronk QCB. Islands: stability, diversity, conservation. *Biodiversity and Conservation* 1997; 6: 477–493.

10 MacArthur RH, Wilson EO. *The Theory of Island Biogeography*. Princeton, NJ: Princeton University Press, 1967.

11 Carlquist S. *Island Life*. New York: Natural History Museum Press, 1965.

12 Case TJ. A general explanation for insular body size trends in terrestrial vertebrates. *Ecology* 1978; 59: 1–18.

13 Brown P. et al. A new small-bodied hominin from the Late Pleistocene of Flores, Indonesia. *Nature* 2004; 231: 1055–1061.

14 Lomolino MV. Body size evolution in insular vertebrates: generality of the island rule. *J Biogeogr* 2005; 32: 1683–1699.

15 Meiri S, Dayan T, Simberloff D. The generality of the island rule reexamined. *J Biogeogr* 2006; 33:1571–1577.

16 McClain CR, Boyer AG, Rosenberg G. The island rule and the evolution of body size in the deep sea. *J Biogeogr* 2006; 33: 1578–1584.

17 Carlquist S. *Hawaii, a Natural History*. New York: Natural History Press, 1970.

18 Wagner WL, Funk VA (eds.). *Hawaiian Biogeography: Evolution on a Hot-Spot Archipelago*. Washington, DC: Smithsonian Institution Press, 1995.

19 Keast A, Miller SE (eds.). *The Origin and Evolution of Pacific Biotas, New Guinea to Eastern Polynesia: Patterns and Processes*. Amsterdam: SPB Academic Publishing, 1996.

20 Carson HL, Clague DA. Geology and biogeography of the Hawaiian Islands. In: Wagner WL, Funk VA, eds. *Hawaiian Biogeography: Evolution on a Hot-Spot Archipelago*. Washington, DC: Smithsonian Institution Press, 1995.

21 Stone BS. A review of the endemic genera of Hawaiian plants. *Bot Rev* 1967; 33: 216–259.

22 Carr GD, et al. Adaptive radiation of the Hawaiian silversword alliance (Compositae—Madiinae): a comparison with Hawaiian picture-winged *Drosophila*. In: Giddings LY, Kaneshiro KY, Anderson WW, eds. *Genetics, Speciation and the Founder Principle*, pp. 79–95. New York: Oxford University Press, 1989.

23 Raikow RJ. The origin and evolution of the Hawaiian honey-creepers (Drepanididae). *Living Bird* 1976; 15: 95–117.

24 Sibley CG, Ahlquist JE. The relationships of the Hawaiian honeycreepers (Drepaninini) as indicated by DNA-DNA hybridisation. *Auk* 1982; 99: 130–140.

25 Olson SL, James HF. Descriptions of thirty-two new species of birds from the Hawaiian Islands: Part I. Non-Passeriformes. *Ornithol Monogr* 1991; 45: 1–88.

26 James. HF, Olson SL. Description of thirty-two new species of birds from the Hawaiian Islands: Part II. Passeriformes. *Ornithol Monogr* 1991; 46: 1–88.

27 Cowie RH, Holland BS. Dispersal is fundamental to biogeography and the evolution of biodiversity on oceanic islands. *J Biogeogr* 2006; 33: 193–198.

28 Heaney LR. Is a new paradigm emerging for oceanic island biogeography? *J Biogeogr* 2007; 34: 753–757.

29 Van Balgooy MMJ. Plant-geography of the Pacific as based on a census of phanerogam genera. *Blumea* 1971; Suppl. 6: 1–222.

30 MacArthur RH, Diamond JM, Karr J. Density compensation in island faunas. *Ecology* 1972; 53: 330–342.

31 Wilson EO. Adaptive shift and dispersal in a tropical ant fauna. *Evolution* 1959; 13: 122–144.

32 Ricklefs RE, Bermingham E. The concept of the taxon cycle in biogeography. *Global Ecol. Biogeogr.* 2002; 11: 353–361.

33 Losos JB Phylogenetic perspectives on community ecology. *Ecology* 1996; 77: 1344–1354.

34 Simberloff DS. Species turnover and equilibrium island biogeography. *Science* 1976; 194: 572–578.

35 Gilbert FS. The equilibrium theory of island biogeography: fact or fiction? *J Biogeogr* 1980; 7: 209–235.

36 Connor EF, McCoy ED. The statistics and biology of the species-area relationship. *Am Naturalist* 1979; 113: 791–833.

37 Diamond JM. Avifaunal equilibria and species turnover on the Channel Islands of California. *Proc Natl Acad Sci USA* 1969; 64: 57–63.

38 Lynch JF, Johnson NK. Turnover and equilibria in insular avifaunas, with special reference to the California Channel Islands. *Condor* 1974; 76: 370–384.

39 Simberloff DS. When is an island community in equilibrium? *Science* 1983; 220: 1275–1277.

40 Steadman DW. *Extinction and Biogeography of Pacific Island Birds*. Chicago: Chicago University Press, 2006.

41 Shafer CL. *Nature Reserves. Island Theory and Conservation Practice*. Washington, DC: Smithsonian Institution Press, 1990.

42 Case TJ, Cody ML. Testing theories of island biogeography. *Am Scientist* 1987; 75: 402–411.

43 Wright DH. Species-energy theory: an extension of species-area theory. *Oikos* 1983; 41:496–506.

44 Kalmar A, Currie DJ. A global model of island biogeography. *Global Ecol. Biogeogr.* 2006; 15: 72–81.

45 Whittaker RJ, Triantis KA, Ladle RJ. A general dynamic theory of oceanic island biogeography. *J Biogeogr* 2008; 35: 977–994.

46 Patterson BD, Atmar W. Nested subsets and the structure of insular mammalian faunas and archipelagos. *Biological Journal of the Linnean Society* 1986; 28: 65–82.

47 Ulrich W, Almeida-Neto M, Gotelli NJ. A consumer's guide to nestedness analysis. *Oikos* 2009; 118: 3–17.

48 Cutler AH. Nested biotas and biological conservation: metrics, mechanisms, and the meaning of nestedness. *Landscape and Urban Planning* 1994; 28: 73–82.

49 Wright DH, Patterson BD, Mlkkelson GM, Cutler A, Atmar W. A comparative analysis of nested subset patterns of species composition. *Oecologia* 1998; 113: 1–20.

50 Wright DH, Reeves J. On the meaning and measurement of nestedness of species assemblages. *Oecologia* 1992; 92: 416-28.

51 Sfenthourakis S, Giokas S, Tzanatos E. From sampling stations to archipelagos: investigating aspects of the assemblage of insular biota. *Global Ecology and Biogeography* 2004; 13: 23–35.

52 Martínez-Morales MA Nested species assemblages as a tool to detect sensitivity to forest fragmentation: the case of cloud forest birds. *Oikos* 2005; 108: 634–42.

53 Atmar W, Patterson BD. The measure of order and disorder in the distribution of species in fragmented habitat. *Oecologia* 1993; 96: 373–82.

54 Carstensen DW, Olesen JM. Wallacea and its nectarivorous birds: nestedness and modules. *J Biogeogr* 2009; 36: 1540–1550.

55 Fleishman E, Donnelly R, Fay JP, Reeves R. Applications of nestedness analyses to biodiversity conservation in developing landscapes. *Landscape and Urban Planning* 2007; 81: 271–81.

56 Simberloff D, Martin J-L. Nestedness of insular avifaunas: simple summary statistics masking complex species patterns. *Ornis Fennica* 1991; 68: 187–92.

57 Patterson BD, Dick CW, Dittmar K. Nested distributions of bat flies (Diptera: Streblidae) on Neotropical bats: artifact and specificity in host–parasite studies. *Ecography* 2009; 32: 481–7.

58 Patterson BD, Brown JH. Regionally nested patterns of species composition in granivorous rodent assemblages. *J Biogeogr* 1991; 18: 395–402.

59 Brown JH, Kodric-Brown A. Turnover rates in insular biogeography—effect of immigration on extinction. *Ecology* 1977; 58: 445–9.

60 Lomolino MV. Investigating causality of nestedness of insular communities: selective immigrations or extinctions? *Journal of Biogeography* 1996; 23: 699–703.

61 Diamond JM. Assembly of species communities. In: Cody MI & Diamond JM, eds. *Ecology and Evolution of Communities*, pp. 342–444. Cambridge, MA: Harvard University Press, 1975.

62 Garb JE, Gillespie RG. Island hopping across the central Pacific: mitochondrial DNA detects sequential colonization of the Austral Islands by crab spiders (Araneae: Thomisidae). *J Biogeogr* 2006; 33: 201–220.

63 Diamond JM, Gilpin ME. Biogeographical umbilici and the origin of the Philippine avifauna. *Oikos* 1983; 41: 307–321.

64 Burns KC. Patterns in the assembly of an island plant community. *J Biogeogr* 2007; 34: 760–768.

65 Colwell RK, Winkler DW. A null model for null models. In: Strong DR et al., eds. *Ecological Communities and the Evidence*, pp. 344–359. Princeton, NJ: Princeton University Press, 1984.

66 Newmark WD. A land-bridge island perspective on mammalian extinctions in western North American parks. *Nature* 1987; 325, 430–432.

67 Hill JL, Curran PJ. Area, shape and isolation of tropical forest fragments: effects on tree species diversity and implications for conservation. *J Biogeogr* 2003; 30: 1391–1403.

68 Báldi A. Habitat heterogeneity overrides the species-area relationship. *J Biogeogr* 2008; 35: 675–681.

69 Western D, Ssemakula S. The future of the savannah ecosystems: ecological islands or faunal enclaves? *S Afr J Ecol* 1981; 19: 7–19.

70 Saunders DA, Hobbs RJ. Corridors for conservation. *New Scientist* 1989; 121 (1648), 28 January: 63–68.

71 Zimmerman BL, Bierregaard RO. Relevance of the theory of island biogeography and species-area relations to conservation with a case from Amazonia. *J Biogeogr* 1986; 13: 133–143.

72 Burridge AA, Williams MR, Abbott I. Mammals of Australian islands: factors influencing species richness. *J Biogeogr* 1997; 24: 703–715.

73 Alcover JA, Sans A, Palmer M. The extent of extinctions of mammals on islands. *J Biogeogr* 1998; 25: 913–918.

74 Thornton I. *Krakatau—the Destruction and Reassembly of an Island Ecosystem*. Cambridge, MA: Harvard University Press, 1996.

75 Whittaker RJ, Bush MB, Asquith NM, Richards K. Ecological aspects of plant colonization of the Krakatau Islands. *Geo J* 1992; 28: 201–211.

76 Whittaker, RJ, Field R, Partomihardjo T. How to go extinct: lessons from the lost plants of Krakatau. *J Biogeogr* 2000; 27: 1049–1064.

77 Tagawa H, Suzuki E, Partomihardjo T, Suriadarma A. Vegetation and succession on the Krakatau Islands, Indonesia. *Vegetatio* 1985; 60: 131–145.

78 Parrish TL, Koelewijn HP, van Dijk PJ. Genetic evidence for natural hybridization between species of dioecious *Ficus* on island populations. *Biotropica*, 2003; 35: 333–343.

79 Mayr E. Die Vogelwelt Polynesians. *Mitt Zool Mus Berlin* 1933; 19: 306–323.

80 MacArthur RH, Wilson EO. An equilibrium theory of island biogeography. *Evolution* 1963; 17: 373–387.

81 Pratt HD, Bruner PL, Berrett DG. *A Field Guide to the Birds of Hawaii and the Tropical Pacific*. Princeton, NJ: Princeton University Press, 1987.

82 Lomolino MV, Brown JH, Davis R. Island biogeography of montane forest mammals in the American Southwest. *Ecology* 1989; 70:180–194.

83 Whittaker RJ, Jones SH, Partomihardjo T. The rebuilding of an isolated rain forest assemblage: how disharmonic is the flora of Krakatau? *Biodiversity Conserv* 1997; 6: 1671–1696.

84 Bush MB, Whittaker RJ. Krakatau: colonization patterns and hierarchies. *J Biogeogr* 1991; 18: 341–356.

85 Schmitt SF, Whittaker RJ. Disturbance and succession on the Krakatau Islands, Indonesia. *Symp Br Ecol Soc* 1998; 37: 515–548.

Drawing Lines
in the Water

Life clothes the land; it merely stains the seas.

The oceans and seas of the world contain three very different types of habitat: the vast volumes of the open oceans, the floor of the deep ocean far from the light of the surface, and the far richer life of the shallow seas around the continents and oceans. Our understanding of the biogeography of these environments has been limited by two factors. First, it is far more difficult for us, as air-breathing creatures, to survey and sample them—particularly in those cases where they are far from land and at great depth. But it has also become clear that, unlike terrestrial species, marine species often cannot be distinguished by differences in morphology, so that their taxonomy has to be based on more subtle genetic comparison. It is also more difficult to recognize the barriers that lie between the areas of distribution of marine species. As a result, our understanding of marine biogeography is still at an early stage of development.

The biogeography of the continental masses and that of the sea are similar in one way: both involve analysis of the biotas of vast areas of the surface of the globe. However, because their environments are very different, the marine biotas are much more difficult to study. As a result, we know far less about the composition and ecology of marine organisms than we do about those of the land. In many areas of marine research, we are therefore still at the stage of constructing and evaluating hypotheses at a comparatively basic level. This fact is of more than merely academic and technical importance, in view of our desire and need to conserve the world's present diversity of organisms. To do this, we must first understand the fundamental patterns of distribution, both of the ecosystems and of the organisms they contain. Only then can we identify those that are threatened because of their rarity or because of their vulnerability to ecological change—whether natural or the result of human activities.

The terrestrial biogeographical regions are, effectively, the different continents. The interplay of the topography of the land and of the seasonal cycles in climate within each region also produces a considerable variety of physical environments. Those regions are usually separated from one another by barriers of ocean, mountain or desert that make it difficult for organisms to disperse from one to another. The geographical boundaries of the regions are therefore easy to define. Their inhabitants live in air, which has a very low density. As a result, it is impossible for terrestrial organisms to be permanently airborne, and it does not provide much help in their long-range dispersal. On the other hand, it allows plants to become structurally complex. In most parts of the continents (excluding the tundra, steppe, and desert), the plants therefore dominate the environment. In addition to the variety of physical environments, the plants therefore add their own living architecture (grassland, woodland, forest, etc.) within which the animals exist. These habitats provide a framework for biogeographical analysis at a finer level of detail, while our investigations are helped by the fact that we, too, are terrestrial.

The liquid world of the oceans is quite different. The major oceans are all interconnected, so that their geographical boundaries are less clear than those of the continents. As a result, their biotas cannot show such clear differences as those on land. The oceans are also far larger than the continents, for they make up 71% of the surface of our planet and, because of their depth, they account for 97% of its habitable volume. The oceans themselves are continually moving because the water within each ocean basin slowly rotates. These moving waters carry marine organisms from place to place, and also help the dispersal of their young or larvae. Furthermore, the gradients between the environments (and therefore between the different faunas) of different areas of seafloor or of ocean water mass are very gradual and often extend over wide areas that are inhabited by a great variety of organisms of differing ecological tolerances. There are no firm boundaries within the open oceans.

Because of the density and power of the waters, however, there are no large, complex plants to provide the equivalent of the terrestrial biomes. Photosynthesis in the sea is carried out mainly by tiny, single-celled organisms known as **phytoplankton**. The density of plant life in the ocean is therefore far less than that on land, and the primary productivity per unit area is only one-fifteenth of that in tropical rainforest, so that much less solar energy is being fixed in the system.

Nevertheless, there is one way in which the oceans are more complex than the land: they have an important extra dimension, that of depth. The physical conditions of light, temperature, density, and pressure—and often also the concentrations of nutrients and oxygen—change much more rapidly with depth in the seas than they do with altitude on land. These lead to corresponding changes in the biotas. In fact, quite unlike the situation on land, the most important environmental gradients and discontinuities take place in the vertical dimension, not the horizontal, and are far more abrupt. These resulting vertical patterns of distribution also interact with the horizontal patterns.

Because we ourselves are terrestrial and air-breathing, it is difficult for us to study and to census the life of the sea, even in the near-shore or surface-water regions, and even more difficult in its depths. Our knowledge of the fauna of the deep-ocean seabed, which covers an area of 270 million km^2 (10 mi^2), is derived from cores totaling only about 500 m^2 (600 yds^2), together with the areas sampled from dragging a number of trawls and deep-sea sleds over the bottom!

All the oceans interconnect at high southern latitudes, and the Panama Isthmus was only completed in comparatively recent times, about 3 million years ago. Because of this lack of physical barriers in the oceans and their lower productivity, there may have been less opportunity for evolutionary diversification. As presently described, marine families contain fewer genera than terrestrial families, and these genera contain fewer species, so that many fewer species are known from the sea: only about 210,000 species of marine organisms have been described, compared with about 1.8 million from the land (but see p. 280). Similarly, there are probably over 250,000 species of land plant, but only 3,500–4,500 species of phytoplankton. However, even the thus far very limited sampling of the seafloor shows a quite surprising variety of life, which has led the American oceanographers Frederick Grassle and Nancy Maciolek to estimate that the deep sea may contain between 1 million and 10 million species [1]. We have explored less than 5% of the deep sea, and we know less about it than about the dark side of the moon.

As we shall see later, although our understanding of the distribution of marine organisms is still limited, it would appear that, with the exception of those that live in the intertidal or shallow-water regions, they usually have a much wider distribution than those on land—at least at the level of the family or genus. Thus, while most families of mammal are found in a single zoogeographical region, most families of marine organisms are cosmopolitan or widespread throughout the world's oceans. As a result, marine faunas differ from one another in containing different genera or species rather than different families. These genera or species are not known by different English-language names, so we can only refer to them by their Latin names. Thus, although it is easy to explain that, for example, the South American zoogeographical region contains the endemic families of armadillo, anteater, and sloth, one can only explain the differences between the biotas of marine regions by giving lists of Latin names.

The great ocean basins are similar to the continents as the major biogeographical units. Even though their patterns are more diffuse, they exhibit the same phenomena and raise the same problems of explanation by dispersal or vicariance, or of the extent to which differences are due to different histories of enlargement, fusion, or subdivision, or to different evolutionary events. But since the marine faunas are still much less well known, we need to be much more cautious in coming to conclusions, or in assuming that particular deductions and generalizations associated with continental biogeography are necessarily valid for marine biogeography. Nevertheless, a study of the biogeography of the hake *Merluccius* by the South African marine zoologists Stewart Grant and Rob Leslie [2] has shown some interesting parallels with the more familiar maps of dispersal on the landmasses (Fig. 9.1). Fossils and the phylogeny suggested by molecular data make it likely that the genus originated in the Early Oligocene in the shallow epicontinental seas of the northeastern Atlantic-Arctic. From here, Old World and New World lineages diverged down the continental shelves of the eastern and western coasts of the Atlantics. The Old World lineage penetrated beyond the Cape of Good Hope into the Indian Ocean, with subsequent episodes of back-dispersal to and from West Africa. The New World lineage spread through the still-open gap between North and South America into the Pacific Ocean, within which it spread northward, southward, and westward, with an episode of back-dispersal around Cape Horn into the South Atlantic.

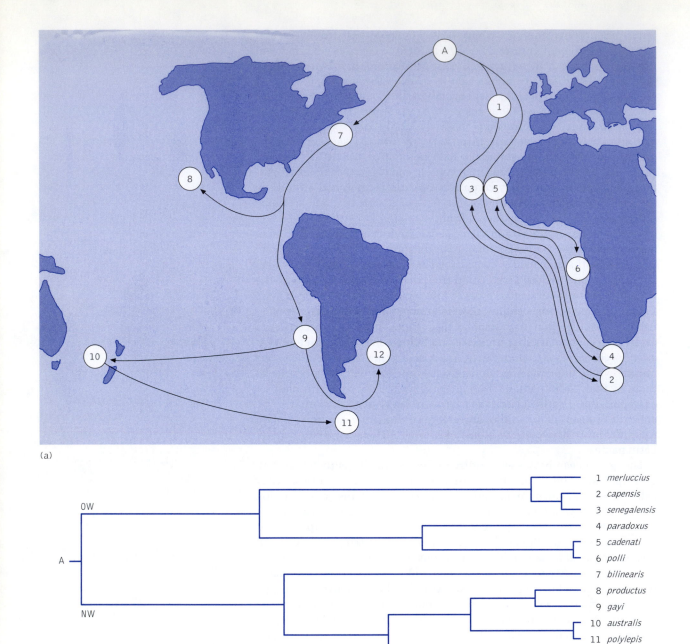

(a)

(b)

		1	*merluccius*
		2	*capensis*
		3	*senegalensis*
		4	*paradoxus*
		5	*cadenati*
		6	*polli*
		7	*bilinearis*
		8	*productus*
		9	*gayi*
		10	*australis*
		11	*polylepis*
		12	*hubbsi*

15 14 13 12 11 10 9 8 7 6 5 4 3 2 1 0

Divergence dates (Myr)

Fig. 9.1 The dispersal of 12 species of hake, *Merluccius*, as suggested by fossil and molecular data. (a) Map showing distributions and dispersal. The numbers refer to the individual species shown in the cladogram of phylogenetic relationships, (b), which also shows the dates of divergence between the different lineages, in millions of years. The numbers refer to the individual species as shown in the cladogram. A, ancestral *Merluccius*; N.W., New World lineage, O.W., Old World lineage. Data from Grant & Leslie [2].

Zones in the Ocean and on the Seafloor

Around the deep ocean that makes up just over 50% of the marine environment are areas of shallow sea adjacent to the continents. The distinction between the two is a result of the basic structure of our planet. The continents rise above the level of the ocean floor because the rocks of which they are made are less dense and therefore lighter than those of the ocean floor. Seas cover the lower-lying parts of the continents, which are known as the continental shelves (Fig. 9.2b). The depth of the water in these seas, which are known as "epicontinental seas" or shallow seas, varies according to how much of it has gone to form ice sheets and glaciers on the continents. At present, they have a maximum depth of 200 m (650 ft) (except in the Antarctic, where the great weight of ice has depressed the continent, so that the edge of the continental shelf lies at about 500 m [1,600 ft]). At this point, known as the **shelf break**, the slight gradient of the seafloor increases sharply. From here onward, the continental slope descends with relative steepness until it reaches the deep ocean floor, which is known as the **abyssal plain**. Sediments cover all of these ocean floors. Upon the continental shelf and continental slope are thick layers of sediment that are derived primarily from the erosion of the rocks of the continents. These sediments spill down over the edge of the continental slope and onto the adjacent margin of the abyssal plain to form a wedge known as the **continental rise**. At an average depth of 4 km (2.5 mi), the abyssal plain covers 94% of the area of the oceans and 64% of the surface of the world, and is therefore the most extensive of all environments. In the Indian and Pacific oceans, the abyssal plain is also fringed by a system of trenches, reaching depths of up to 11,000 m (36,000 ft), where old ocean floor disappears back into the depths of the earth (see p. 154).

Because the physical conditions change as one moves downward, away from the light and warmth of the surface, a number of different depth zones can be distinguished in the waters of the sea, while the shape of the seafloor similarly defines several different regions there (Fig. 9.2a).

Perhaps the most important physical characteristic of the sea is that conditions do not change uniformly from those at the surface, where the water is warmer and therefore less dense, to the colder denser conditions

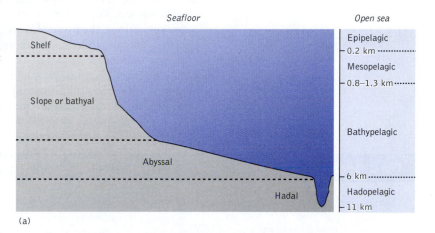

(a)

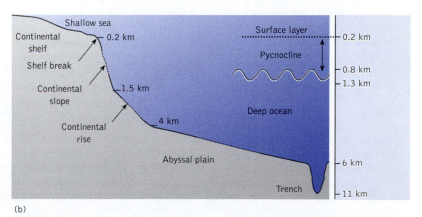

(b)

Fig. 9.2 (a) Diagram of the vertical divisions of the seafloor (left) and open sea (right).

(b) Life zones on the seafloor (left) and in the open sea (right). (Not to scale.)

at greater depths. Instead, there is a comparatively narrow zone of rapid change in density, known as the **pycnocline**. (This is most often caused by a rapid change of temperature, in which case it is also known as a **thermocline**. But it can also be caused by a rapid change in salinity, in which case it is known as a **halocline**. Both can exist in the same area, as in some parts of the western tropical Pacific, where a near-surface halocline lies about 110 m [360 ft] above a thermocline.)

The surface layer of the sea, down to a depth of about 200 m (650 ft), is the most liable to extremes of temperature. This can be as low as $-1.9°C$ (the freezing point of seawater) below the ice cover found in high latitudes, or over 30°C in enclosed low-latitude waters such as the Red Sea. The rapid fall in temperature associated with the pycnocline commences at a depth of between 25 m (80 ft) and 250 m (820 ft) and continues until, at a depth of 800–1,300 m (2,600–4,300 ft), it reaches about 4°C. At greater depths, the temperature decreases slowly until, at most latitudes, it reaches a minimum of 2–3°C at around 3,000 m (9,800 ft). In polar seas, however, it may be as low as $-0.5°C$ in the Norwegian Sea, or $-2.2°C$ in parts of the Southern Ocean. Below 3,000 m, there is only a very slight temperature gradient.

As we shall see, the pycnocline is the most important ecological boundary in the ocean, and it is therefore useful to outline the major features that control its characteristics. The depth at which it begins, as well as its thickness, vary according to latitude, region, and season. Closer to the equator, between about 20°N and 20°S, the pycnocline commences at a shallower depth, and therefore within the lighted zone, shows a greater

The Changing Seas

Guest Author

Box 9.1

By Dr. Martin Angel, National Oceanographic Centre, Southampton, UK

Biogeography is a multidisciplinary science. To understand the present distribution patterns of species and how they may change in response to shifts in climate requires detailed knowledge of how each species' biology and ecology interact with its biological, chemical, and physical environment. The pelagic environment in the open ocean is physically structured in a totally different manner to terrestrial ecosystems, while the modes of variation in time and space of the dominant ecological factors are also quite different. Hence, many of the paradigms of biogeography developed by terrestrial biogeographers either are inappropriate for marine environments or need to be substantially revised.

For example, away from coastal margins there are only two physical boundaries: the surface and the seabed. Oceanic environments are otherwise dominated by gradients rather than boundaries. The dominant gradients relate to depth. Perhaps the most important gradient is caused by light being scattered and absorbed by seawater, so that photosynthetic activity is restricted to the upper 100 m, and the availability of food generally decreases with increasing depth. The deep ocean (> 1,000 m; 3,300 ft), the most voluminous habitat on the planet, is without light.

Another important factor associated with the physiology of the inhabitants is the increase in hydrostatic pressure with depth. Many species are so well adapted to living within specific depth horizons that they are restricted to specific depth environments, such as the well-lit epipelagic realm of the upper 100 m (330 ft), or the twilight zone of the mesopelagic (100–~700 m [330–2,300 ft]), and the permanently dark environments of the bathypelagic and abyssopelagic realms in deep water. Even so, many organisms are able to migrate vertically between these realms daily, seasonally, or during their life cycle. Thus, depth is not necessarily a constraint to the dispersal of species, and indeed by changing depth a species may exploit the differences in the current drifts that occur within the water column. These differences in current flow and direction result from ocean water columns being structured by climatic factors, making ocean environments highly responsive to climatic variability.

The density of seawater is determined by its salinity and temperature. If seawater is cooled and its salt content increased, it becomes denser and sinks to depths with a similar density. Conversely, if it is warmed and diluted with freshwater, it becomes so buoyant that it will not mix with the water that underlies it. The resultant thermocline is often a sharp ecological boundary segregating the epipelagic communities above it from those inhabiting the cooler deep water. Because a strong thermocline prevents vertical mixing, the deeper water cannot supply the nutrients needed for photosynthesis in the surface waters.

In addition to these oceanic gradients, life in the sea is dominated by another feature that is unmatched on land—the worldwide circulation system, or "conveyor belt," that transports water masses and heat around the world (see p. 109). Here there is an interaction between worldwide patterns of climate. On the one hand, these patterns affect the amount of ice that forms in high latitudes and the amount of ocean heating that takes place in low latitudes, thereby controlling the strength of the conveyor belt. On the other hand, the amount of heat that the belt transports in turn affects the pattern of world climate.

A side effect of the conveyor belt is that, where the water is cooled and sinks below the surface, its properties remain unchanged; it forms a water mass with characteristic temperature and salinity properties that can be traced over extensive distances. At any given location in the open ocean, the water column consists of a number of layers of water of different origins whose movements are influenced by density gradients and the effects of the Earth's rotation. Superimposed on these are circulation cells or eddies that are analogous to the cyclones and anticyclones that are such familiar features of our weather in the atmosphere. The local and global distributions of oceanic plankton are strongly influenced by the movements of water masses and the "noise" generated by the eddies.

Why is it important to understand the significance of climate and water movements to plankton distributions? The answer is best illustrated with an example [3]. The long-term monitoring using the Continuous Plankton Recorder has demonstrated that, in the plankton populations in the North Sea and along the west coast of Britain, the formerly dominant copepod *Calanus finmarchicus* has been largely replaced by its relative *Calanus helgolandicus*. *C. finmarchicus* has a remarkable life history. It has a spring/summer generation that is spent in the upper ocean but, as autumn approaches, the summer cohort of immature larvae (copepodites) migrates down to depths of 1,000–2,000 m (3,300–6,600 ft). Here they enter a state of diapause—a sort of hibernation in which their metabolism shuts down and they become physiologically inactive. In the spring they break diapause (the trigger is not known) and migrate back up to the surface, where they mature and spawn. In good years the breeding coincides with the onset of the spring bloom of phytoplankton, when the newly hatched larvae find an abundance of food. In their turn, the larvae are key forage for newly hatched fish larvae, notably those of the cod. Good recruitment to the fish stocks will occur only when the timing of the spawning and hatching of both the copepods and fish eggs coincides. This is because, when the newly hatched fish larvae begin to feed, the amount of food available must exceed a critical density if they are to survive and grow to maturity. In a system subject to climatic variability, any perturbations in the timing of the spawnings will lead to a disastrous mismatch and a failure to recruit. Recent climate shifts have resulted in the poleward shifts of the *C. finmarchicus* populations, and their replacement with *C. helgolandicus*. However, the latter has a very different life cycle and so does not provide the high abundance of larvae at the right season for the fish larvae. There have also been shifts in the deep water movements that carry the overwintering populations of *C. finmarchicus*, so the bulk of its spawning now occurs away from the slope waters that are optimal for the copepods' success. The copepods are no longer so abundant, and substantial changes have taken place in the ecological regime operating in these once highly productive waters.

So the cod's life history, particularly in the northwest Atlantic off the Grand Banks, is inextricably linked with that of the copepod. Break the linkage and the cod can no longer be the successful fish that for centuries played a key role in the socioeconomic health of communities on both sides of the Atlantic. The great fisheries associated with the Grand Banks and around Newfoundland were historically a major source of cod. Overfishing destroyed these fisheries, and off Newfoundland the cod fishery was totally closed in 1992, with serious social impacts. The closure of the fishery was expected to result in the eventual recovery of the cod's stocks—such a recovery of fish stocks occurred in the North Sea after the cessation of fishing during World War II. However, over 20 years later, there are still no signs of a significant recovery. It seems likely that the cause lies in the shifts that are occurring in the biogeographical distribution of the plankton populations—an ominous warning to conservationists, as well as to fishery managers, that they need to maintain vulnerable communities and populations.

Another consequence of the change in the *Calanus* populations appears to be a shift in the distribution of the snake pipefish (*Entelurus aequoreus*) [4]. Prior to 2005, this fish was predominantly an inshore, coastal species, but recently it has developed substantial offshore populations that feed mostly on *Calanus helgolandicus*. These offshore populations are being foraged by breeding seabirds, such as puffins and kittiwakes, with disastrous consequences for chick survival, and sharp declines in the breeding colonies. Furthermore, during the last 20 years, several warm-water fish species have been migrating northward at a rate of 50 km/year. This increases the species richness of the northern fish communities, so that, if this is used as a measure of environmental health, the habitats appear to be healthier. However, this is likely to be an illusion: the climatic shifts are disturbing the ecological balance of both inshore and offshore communities. Thus, while the appearance of increasing numbers of charismatic species such as seahorses is being welcomed, they could also be indicating the occurrence of far-reaching changes that will have unpredictable impacts on marine conservation strategies.

rate of change with depth, and is more resistant to mixing with the adjacent waters. There is also little change in temperature through the year in this near-equatorial band, so that the pycnocline is a permanent feature and lies at a constant depth. Closer to the poles, the depth at which the pycnocline begins lies at a deeper level but also changes according to the time of year, being deeper in winter than in summer. It is only rarely and temporarily in the lighted zone, shows a slower rate of change with depth, and is less resistant to mixing. In near-polar regions, the pycnocline may disappear altogether in winter.

The Basic Biogeography of the Seas

The shape of the ocean basins is the cause of the most basic divide within marine biogeography—between the shallow-sea (or **neritic**) realm and the open-sea realm. The major difference between these two realms is that of scale. The shallow seas occupy a smaller area than the oceans, and each of them is also far smaller than any ocean. But the individual shallow seas also differ more from one another than do the oceans, and each shallow sea contains within itself a greater variety of conditions than any equivalent area of ocean. As we shall see, the patterns of life in the oceanic realm are the results of external factors, such as the Earth's rotation and the heat and light of the sun. The resulting patterns can be seen at the largest scale, for example, at the level of the North Pacific or the South Atlantic Ocean. Although these oceanic patterns also affect the shallow seas, they are less obvious there than local influences such as the nature of the seafloor, the contribution of sediments and freshwater by local rivers and streams, or the pattern of tides. In addition to these physical differences, the various shallow seas are also isolated from one another by wide stretches of ocean. As a result, their biotas are similarly isolated, giving the opportunity for independent evolution and endemicity.

A subsidiary divide within marine biogeography is between the patterns of distribution of **pelagic** organisms, which swim or float within the waters themselves, and those of **benthic** organisms, which live on or in the seafloor. (Pelagic organisms that swim are known as **nektonic**, while those that float are known as **planktonic**.) The distribution of benthic organisms depends on the local characteristics of the seafloor, while that of the pelagic organisms is quite independent of the depth or nature of the seafloor. However, pelagic and benthic species are often dispersed by similar processes and therefore tend to show similar distribution patterns. This is also because both depend on the patterns of productivity in the surface waters, from which organic material falls to the lower levels and, eventually, to the sea bottom.

The Open-Sea Realm

Below the upper few tens of meters of water, which is called the **euphotic zone**, there is not enough sunlight to support photosynthesis. The heating effects of the sunlight, too, are mainly restricted to the upper, wind-mixed layer, termed the **epipelagic zone**, which varies in thickness from a few tens of meters to 200 m (650 ft) (Fig. 9.2a). This warmer layer contains a high concentration of living organisms, whose activity reduces the nutrient content of the water. Below it lies the pycnocline, a zone in which the temperature drops rapidly and where the density of the water therefore also increases rapidly. As a result, the pycnocline is the most important boundary within the ocean waters. However, it is not merely a zone of rapid transition, but also an environment in its own right, with its own characteristic biota that, in this poorly lit zone, is comparable to the shade flora of terrestrial ecology [5].

In the twilight zone below the pycnocline lies the **mesopelagic zone**, which extends to depths of approximately 1,000 m (3,250 ft). Many of the fish that live in this zone during the day move up toward the surface at night, to feed on the richer fauna there. Below the mesopelagic zone lies the **bathypelagic zone**, which extends to a depth of 6,000 m (19,500 ft); it is a zone of total darkness and almost unchanging cool temperatures. The fish of this zone are too far from the surface to migrate there daily. Finally, the **hadopelagic** faunas of the deep trenches are thought to be mainly endemic because many of the trenches are isolated from one another.

The biogeography of the open-sea realm is best approached by first describing the circulation patterns within the oceans, for these have led to differences in their nutrient concentrations, which in turn have affected the patterns of distribution of life there. Much of this has been reviewed by the British oceanographer Martin Angel [6,7].

The Dynamics of the Ocean Basins

The patterns of the waters' movement in the oceans are dominated by stable **gyres**—huge masses of water, filling a large proportion of an ocean basin, which rotate horizontally, with a periodicity that is usually 19–20 years. Their rotation is caused by wind patterns, which themselves result from the uneven distribution of solar energy on the surface of the Earth, as well as from the eastward rotation of the Earth. Heat from the equatorial regions is distributed toward the poles by patterns of wind movement that rotate clockwise in the northern midlatitudes and anticlockwise in the southern midlatitudes (see p. 107). These wind patterns create similar patterns of movement in the waters below, so that warm ocean currents flow away from the equator along the western edges of the oceans, and cold currents flow away from the poles along their eastern edges (see Fig. 9.5). The worldwide pattern of ocean currents is a major element in the transfer of heat around the world. The equatorial waters, in both the Atlantic and the Pacific oceans, have a westward surface current. Because the winds in the Indian Ocean show a seasonal reversal, the "monsoon" (cf. Fig. 3.9, p. 108), the current patterns in the Indian Ocean are similarly variable according to the time of year. The Arctic Ocean is almost totally enclosed by land, so that there is only a little southward dispersal of its waters, mainly to the east and west of Greenland. The Antarctic continent is surrounded by a band of permanently cold, eastwardly flowing water, the Antarctic Circumpolar Current.

In addition to these horizontal movements of the ocean waters, there is also a vertical circulation, driven by differences in water temperature and salinity. As seawater in the polar regions freezes into virtually salt-free ice, the excess salt enriches the water layer immediately below the ice. These waters are therefore unusually saline and dense, causing them to sink downward. In two regions (along the eastern side of Greenland and near the Antarctic Peninsula (see Fig. 3.10, p. 109), this water sinks down to the ocean floor as a coherent body of water known as **bottom water**. Because it is cold, this water is also rich in dissolved oxygen and carbon dioxide, so that it can be recognized by these characteristics as well as by its salinity. As it spreads through the oceans, it therefore brings oxygen to even the deepest waters, far from the oxygen of the surface. This in turn displaces water upward, producing what is known as a **thermohaline circulation**. The time it takes for the ocean waters to complete a cycle of vertical circulation is estimated at 275 years in the Atlantic, 250 years in the Indian Ocean, and 510 years in the Pacific [8].

A similar phenomenon arises from the fact that, near the centers of the great ocean gyres in the North Pacific and North Atlantic, where the weather is normally clear and sunny, the surface loses more water by evaporation than it gains by rainfall. It, too, becomes more saline and dense so that, where it meets neighboring waters to the north and south, it sinks below their lighter, fresher waters along lines known as **convergences** (shown as dashed lines in Fig. 9.5). Convergences are also found where two sets of ocean currents converge, as in the North Pacific Polar Front Convergence and the Southern Subtropical Convergence (shown by dotted areas in Fig. 9.5).

Another cause of vertical water movement are the winds that blow offshore along the western coasts of parts of the Americas, Africa, and Australia. These winds blow the warm surface waters away from the coast, and they are replaced by an upwelling of deeper water. A similar upward movement of water is found at regions known as **divergences** in the Atlantic and Pacific, where there is a shear between currents going in different directions. The Pacific Equatorial Divergence (Fig. 9.5, PEqD) lies to the south of the zone of meeting between the westward equatorial current and the eastward North Pacific Countercurrent (Fig. 9.5, NPEqC). The Antarctic Divergence (the line of small circles in Fig. 9.5) lies between the eastward Antarctic current and the narrow ribbon of westwardly directed Austral Polar Current that lies adjacent to the coastline of Antarctica.

The circulation patterns of the ocean waters, both horizontally and vertically, also create patterns in the concentration of such major nutrients as nitrate, phosphate, and silicate. The patterns of availability of these nutrients in space and time have profound effects on the marine organisms. For example, silicate is vital for the production of the skeleton of the phytoplankton known as diatoms, which are the main source of the rain of organic material that descends into the deeper layers of the ocean. Therefore a close correlation exists between the pattern of regions where waters rise to the surface at upwellings and divergences (Fig. 9.5), the availability of nutrients in the surface waters (Fig. 9.3), and the patterns of productivity in the oceans (Fig. 9.4). Thus, although upwelling areas account for only 0.1% of the ocean surface, these areas provide 50% of the world's fisheries catch. (It must be emphasized that *all* the primary productivity in the oceans is confined to the uppermost few tens of meters as, below this level, the sunlight is absorbed and scattered by the water.) There has long been a debate as to whether the levels of phosphate, or those of nitrate, are the limiting factor in determining

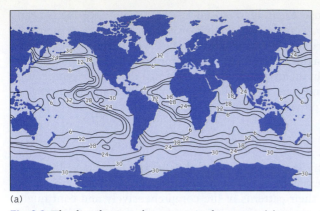

(a)

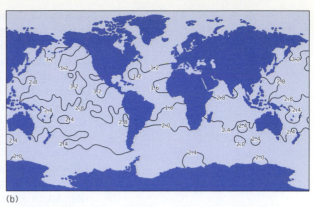

(b)

Fig. 9.3 The distribution of nutrients in the oceans. (a) mean dissolved nitrate at a depth of 150 m; (b) mean dissolved phosphate at a depth of 1,500 m. From Levitus [54] with permission from Elsevier Science.

productivity. One model [9] suggests that each of these has a role, nitrate being more important in the short term of hours or days, while phosphate is more important over the longer term in regulating total ocean productivity.

Another study shows the importance of sea-surface temperature [10]. Satellite measurements of this temperature show that it accounts for nearly 90% of all the geographic variation in the density of the type of plankton known as foraminifera in the Atlantic. Also, because it is closely correlated with temperatures at 50, 100, and 150 m (160, 330, and 490 ft), this is almost as adequate an explanation of the diversity patterns at these depths. The results of this study also show that the diversity of planktonic foraminifera does not simply diminish from the equator to the poles, but instead peaks at middle latitudes in all oceans. This appears to be controlled primarily by the thermal structure of the near-surface levels of the ocean and may be determined by the thickness and depth of the pycnocline. A pycnocline with a gradual temperature change and a deep base may provide more niches per unit of surface area than one with a shallow base and more rapid

Fig. 9.4 Patterns of annual primary production in the oceans. Numbered areas refer to annual productivity of carbon in g/m²/year, as follows: 1, 500–200 g; 2, 200–100 g; 3, 100–60 g; 4, 60–35 g; 5, 35–15 g. After Berger [55].

temperature change. So, in high latitudes, where the pycnocline is nearly absent, there can be little partitioning of niches, and a consequent low diversity of plankton. In intermediate latitudes, where a higher sea-surface temperature and a thick permanent pycnocline prevail, this leads to a maximum diversity. In the tropics, however, although the surface temperature is higher, the pycnocline has a shallow base and a rapid rate of temperature change, so that it provides fewer niches and diversity is lessened.

Patterns of Life in the Ocean Waters

Although the patterns of movement of the surface waters have long been known, only since satellite observations became available has it been possible to monitor their patterns of life comprehensively and continually. The chlorophyll content of the water can now be measured in this way (Plate 4), which allows us to deduce the density of the phytoplankton, the depth of the euphotic zone, and the seasonal cycles in the balance between phytoplankton productivity and loss, which may or may not lead to a seasonal increase in phytoplankton biomass known as a **bloom**. The British oceanographer Alan Longhurst has integrated this biological data with data on the movements of the ocean waters to identify and define what he calls marine biomes and provinces. He recognizes three biogeographical biomes in the oceans (the **polar, westerly winds** and **trade wind biomes**), plus a **coastal biome** that comprises the shallow seas; these biomes are divided into ecological provinces [5]. Figure 9.5 is a simplified, redrawn version of Longhurst's map, showing most of his 33 provinces of the oceanic biomes.

Longhurst's biomes are not really comparable to those on land, however. This is because a terrestrial biome is a *biological* entity, recognized by its plant life, which creates a characteristic environment to which its animal component has become adapted (see p. 103). Except in the deserts and tundra, the physical environment has been profoundly modified by the biological world. In contrast, in the open oceans the physical environment everywhere dominates the life that it contains. Longhurst's biomes differ from one another only in the patterns of their winds and currents and in the ways these vary throughout the year, and therefore offer different annual patterns of opportunity for growth to the plants and animals that are carried in their waters. Another difference between the two systems is that, unlike those on land, the boundaries between Longhurst's marine provinces vary from year to year and from season to season— although the underlying pattern remains stable. So, although Longhurst's usage and system are given here, they should not be taken as suggesting any similarity to the terrestrial systems.

Longhurst's system arises from the pattern of movement of the waters of the world. This results from the pattern of the winds, which is created by the action of the Coriolis force, itself a result of the east-to-west rotation of the world (see p. 107). As a result the winds, and therefore the waters, of the equatorial region move westward, and, as they reach the landmasses to their west, they are deflected toward the poles. If the topography of the world were more uniform, each of the oceans would be occupied by two large gyres, one in the north rotating clockwise and another in the south rotating anticlockwise. The more polar half of each would have westerly winds and ocean currents and would belong to Longhurst's

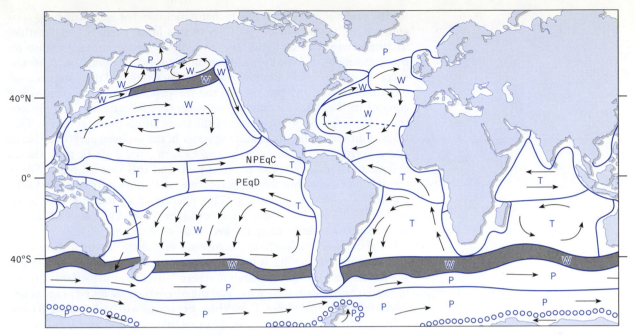

Fig. 9.5 The oceanic biogeographical biomes and provinces, redrawn after Longhurst [5]. Letters in blue (P, T, W) indicate provinces that are placed in the Polar, Trade Winds, or Westerly Winds Biomes, respectively. NPEqC, North Pacific Equatorial Counter-Current Province; PEqD, Pacific Equatorial Divergence. North Pacific Polar Front Convergence and Southern Subtropical Convergence are shown as a dark tint. Antarctic Divergence shown as line of small circles.

westerlies biome, while the more equatorial half would have easterly winds and currents and belong to his trades biome. This pattern does occur in the North Atlantic and North Pacific (Fig. 9.5), but elsewhere the pattern is more complicated. In the South Atlantic, the whole oceanic gyre is placed in the trades biome because it is all under the influence of southeasterly winds; full westerlies only develop south of the latitude of Cape Horn. Both this and the South Pacific gyre are still very poorly known, and Longhurst therefore does not divide either of them into northern and southern provinces. The winds and currents in the Indian Ocean, which Longhurst places in his trades biome, are variable because of the annual monsoon reversal; the winds blow from the northeast from November to March or April, and from the southwest from May to September.

The absence, occurrence, and timing of **planktonic blooms** are an important part of Longhurst's system. These blooms are best understood by imagining an ocean with stable conditions in which the euphotic layer lies at a constant depth, which is also the depth of the pycnocline. Within this euphotic layer, the light and heat of the sun allow continuous growth of the phytoplankton, but their total biomass is kept to a constant, low level by grazing by the **zooplankton** (the tiny animals in the plankton) and by the supply of nutrients, which the phytoplankton normally use up as soon as they are available.

This is, in fact, more or less the pattern in most of Longhurst's trades biome. This mainly comprises the areas where winds and currents are always from the east and where there is also enough year-round sunlight for the warm euphotic zone to be permanent and of approximately constant depth. Because of the stability of this layer, there is little vertical mixing between the surface waters, in which the phytoplankton exhausts its nutrient supply, and the deeper, more nutrient-rich waters. Consequently, there is little seasonal variation in algal productivity, and therefore no algal bloom. However, the province of the trades biome that lies across the equator in the Atlantic provides an exception because in this province there is an increase in the power of the trade winds in the summer. This leads to an accumulation of surface water in the western part of the basin and a corresponding decrease in the eastern part. This in turn allows deeper, nutrient-rich water to rise upward in this eastern region. As a result, it is now within the euphotic zone, so that its nutrients can fuel a summer algal bloom (Plate 4c). (This phenomenon does not take place within the Pacific because it is too broad for there to be enough time for such a seasonal wind change to have such an effect.)

Longhurst's westerlies biome comprises provinces in which there is an algal bloom in the spring. It includes provinces with three rather different seasonal situations: those in high northern latitudes in the Atlantic, those in similar latitudes in the Pacific, and the rest, which are found in lower latitudes in both hemispheres. These three situations will be described in turn.

In the eastern North Atlantic, phytoplankton productivity is limited by both light and nutrients. The bloom therefore only takes place in the spring, when the amount of sunlight increases and the pycnocline rises toward the surface. The rate of increase of the phytoplankton is extremely rapid and is unpredictable. As a result, the population of grazing zooplankton has not been able to evolve an annual pattern of increase tied to that of the phytoplankton; their increase therefore lags behind it rather than controlling it. Instead, the increase in the phytoplankton continues until, after a few days, it has exhausted all the available nutrients. The larger phytoplankton such as diatoms then die or lose their buoyancy, and their remains fall downward through the water column to the abyssal levels. Their place in the upper waters is taken by tiny single-celled animals known as flagellates and by blue-green algae known as cyanobacteria (the **picoplankton**). These are responsible for 80 to 85% of the subsequent primary production. However, because they are too tiny to sink or to be filtered out of the water by zooplankton, they do not form the basis for a classical food web leading to larger animals, such as is found in the Pacific (see below). This area often has a second bloom in the autumn. This may be because mixing of the waters by storms brings new supplies of nutrients to the surface waters or because many of the zooplanktonic grazers have descended to greater depths to overwinter. This autumn bloom is short-lived, for the failing light intensity soon limits photosynthetic activity, and it declines rapidly when the nutrients in the upper level have been used up. Productivity remains at a very low level throughout the winter.

In contrast with this situation, in the high latitudes of the North Pacific the springtime bloom (Plate 4d) does not use up all the available nitrate. This may be because its timing is more predictable, so that the zooplanktonic grazers that have overwintered at depth, such as the copepod *Neocalanus*, rise to the surface at the appropriate time, ready to graze on the phytoplankton and control their numbers. However, it is also possible that the availability of iron is a limiting factor. Whatever may be the cause, the diatom bloom in the upper waters does not "crash." It is therefore available for a longer period of time, into the summer (Plate 4a), as the basis for a food chain upward via zooplankton, fishes, and larger crustaceans, and thence to seabirds, seals, and whales.

The other provinces of Longhurst's westerlies biome (which together cover over half of the oceans) lie at lower latitudes, so that light is not a limiting factor. Phytoplankton productivity therefore increases during the winter as the progressive deepening of the mixed layer recharges the surface layers with nutrients, and declines in early summer as these nutrients are used up.

In the polar biome it is light, rather than nutrients, that limits algal growth. This growth therefore rapidly increases in the spring (Plate 4d) as the sunlight increases in duration and strength and the ice (if any) melts. The peak of this bloom is near to midsummer (Plate 4a), after which it declines because of grazing by zooplankton. However, a second peak takes place in September, when the copepod grazers descend to greater depths to overwinter. This secondary peak is less developed in the Antarctic, where these copepods overwinter closer to the surface, among the phytoplankton.

There are some areas where, although the level of nitrates in the surface waters remains high, this does not lead to increased productivity. These regions, known as the high-nutrient/low-chlorophyll regions, comprise the north subpolar Pacific, the eastern tropical Pacific, and the Southern Ocean around Antarctica. It has been suggested that this is because productivity is continually limited by grazing zooplankton, or that it is due to a lack of some other nutrient, perhaps iron. Certainly, the addition of iron to samples of seawater from the Southern Ocean in the Antarctic caused a quadrupling of the level of productivity [11].

Faunas and Barriers in the Oceans

A major problem in documenting marine biogeography is the difficulty in recognizing the two essential components of faunal subdivision—the barriers between the faunal regions and the differences between the faunas themselves. As we have just seen, it is possible to distinguish major water masses by the patterns of movement of the ocean waters and by the differing annual histories of the biota they contain. The patterns of temperature, nutrient availability, and phytoplanktonic production seen in Longhurst's provinces are of fundamental importance to the marine organisms that inhabit them. One would therefore expect each population to be physiologically and behaviorally adapted to the conditions in its own water mass. Such a set of adaptations could only have evolved if each population is a separate species, physically and genetically distinct from others and therefore separated from them by obvious barriers. In the terrestrial biotic regions, these barriers are the great mountains, deserts, or oceans. It is therefore very surprising to find that the oceanic biomes and provinces appear to be separated by no more than the interface between masses of water moving in different patterns or directions. Surely, one would have thought, these "barriers" must be so permeable as to permit a great deal of dispersal between one biogeographical unit and another. Perhaps there are no important taxonomic differences between their biotas, and the differing annual histories that we see are due merely to physical factors that impose these regimes upon biota that basically differ little from one another taxonomically. In the case of terrestrial biota, the biogeographer would seek to confirm or deny such an hypothesis by comparing the taxonomic composition of the different biotas, starting with their morphology.

At first sight, such an endeavor seems to provide little evidence of major taxonomic differences between the marine biomes and provinces—certainly nothing as dramatic as the totally different mammal families of the continents. There are many statements in the literature as to the wide geographical ranges of many marine taxa and the accompanying lack of diversification into separate species. This may be true in some, perhaps many, cases: very many open-sea species do appear to be widespread. For example, the little nektonic, epipelagic arrow-worm *Pterosagitta draco* is found in tropical and subtropical waters of all the oceans, and this is also true of polychaetes (bristle-worms) and of the pelagic crustaceans known as euphausids. (These form the "krill" on which baleen whales, and many other marine animals, feed and are by weight the most abundant animal on the planet.) Similarly, arrow-worms that live in the deeper mesopelagic and bathypelagic levels range from the subarctic to the subantarctic because there is little latitudinal change in temperature at these greater depths [12]. This pattern is also found in fishes living there.

However, we should be very cautious in generalizing from these cases. For example, there are not only separate subspecies of the arrow-worm *Sagitta serratodentata* in the Atlantic and in the Pacific, but also a separate species, *S. pacifica*, in the southern Pacific. Furthermore, molecular evidence suggests that the uniform morphological characteristics of some of the widespread species conceal considerable biochemical and physiological diversity, so that in reality they have probably diverged into several geographically defined species. For example, Gibbs [13] comments that the mesopelagic fish *Nominostomias*, which had been thought to contain only eight species, is now known to have over 100. Similarly, there is now molecular evidence suggesting that, though many widespread marine species appear uniform throughout their range, they have in reality diverged into several geographically distinct species. For example, work on the sea urchin *Diadema* [14], whose species range throughout the world, has shown that their morphology is an extremely unreliable guide to their species structure. It has even been stated that the morphological differences are so slight that specimens cannot usually be identified without knowing where they were collected! (This is not a situation that brings joy to the heart of a marine biogeographer!) We are still, almost literally, scratching at the surface in our knowledge and understanding of marine biogeography, for much of our new information merely shows the superficiality of our previous beliefs. It is tempting to see this as merely yet another example of how much less we know about life in the sea than about life on land. But the recent discovery, using molecular genetic information, that what had been thought to be only 9 species of earthworm in fact belonged to 14 different species suggests another explanation—that perhaps we have never properly understood the taxonomy of invertebrates at the species level and that the new molecular techniques are merely revealing our ignorance.

The fact that the apparent lack of taxonomic differentiation of marine species is probably illusory is also suggested by the stable behavior of some species of fish. For example, the populations of the Pacific herring *Clupea harengus pallasi* along the Pacific coast mainly spawn in one or other of two locations: San Francisco Bay or Tomales Bay, 50 km further north. The juveniles remain in their home bay for about a year and then swim out to the open ocean, where they stay for two years before returning to the bay where they themselves spawn for the first time. It would appear to be impossible to be sure whether they have remained "faithful" to the bay in which they hatched, but studies of their parasites have shown that this is so [15]. Not only do they show differences in the parasites that they contain, but these differences in turn also suggest that the two populations of fish feed in different locations. This is because the intermediate hosts of the parasites of the Tomales Bay population are whales, which lie offshore, while those of the San Francisco Bay population are seals, marine birds, and sharks, which are found closer inshore. So, again, the apparent lack of barriers in the oceans may merely be the result of our lack of understanding.

We are aware of one major barrier in the world's oceans, however: the eastern part of the present-day Pacific, beyond the Hawaiian Islands, which is almost island-free. The consequent lack of shallow-water environments makes this East Pacific Barrier a formidable obstacle to the dispersal of shallow-water organism, as we shall see later in this chapter. It seems likely that the geological hotspot that causes the Hawaiian Islands (see p. 232) was in its present position even in the Mesozoic, so the then far-wider eastern part of the Panthalassa Ocean must have presented an even greater barrier to the dispersal of organisms between its western and eastern shores.

In the long term, it will be interesting to find the extent to which changes in the patterns of connection of the oceans and in the currents that run between and around them (see Fig. 5.8, p. 163) can be detected in faunal differences. Palaeontologists have long recognized the existence of a Tethyan marine fauna, including both vertebrates (fishes and marine reptiles) and invertebrates, that occupied the comparatively narrow seaway between Laurasia and Gondwana (cf. Fig. 10.2, p. 210). Similarly, the American oceanographer Brian White [16] believes that one can find parallels between the sequence of appearance of different water masses in the Pacific Ocean, due to plate tectonic changes, and the relationships between the different species of some fishes. It is likely that further work will expand on this, to give the faunas of Longhurst's different provinces the same sort of historical and taxonomic basis that has long been known for terrestrial zoogeographical regions. For example, the limits of distribution of many Pacific epipelagic organisms coincide with one another and seem to be related to the patterns of water masses [17]. At an even larger scale of comparison, the marine fauna of the North Atlantic is much less diverse and has fewer endemic forms than that of the North Pacific. The reason for this may be that the Atlantic is a smaller ocean then the Pacific and therefore provides a smaller area within which evolutionary novelties may appear and disperse to other parts of the ocean. The Australian marine biologists Dave Bellwood and Peter Wainwright have already produced a very valuable review of the relationship between these plate tectonic events and the evolution of reef fish faunas [18]. The question of barriers between shallow-water faunas is discussed further in Faunal Breaks within the Shelf Faunas (p. 288).

The Ocean Floor

Patterns of Life on the Ocean Floor

Apart from the intertidal zone between the high water mark and the low water mark, marine biologists recognize four different life zones (Fig. 9.2b) for organisms that live on the sea bottom. (It must be emphasized again that changes in conditions and in faunas are always gradual, so that the boundary between two zones is merely a level of more rapid change rather than a level of abrupt change.) The **shelf zone** comprises the continental shelf below the low water mark. The other levels lie at successively greater depths: the **bathyal zone** (also known as the slope or **archibenthal** zone), the **abyssal zone**, and the **hadal** zone. The hadal zone is easy to define, for it comprises the environment of the ocean trenches, at a depth of over 6 km. However, the depth of the boundaries between the shelf zone, the bathyal zone, and the abyssal zone vary according to season, conditions, and latitude.

The upper level of the bathyal zone is normally at the edge of the continental shelf, at about 200 m (650 ft). Its lower level, and therefore the level of transition to the adjacent abyssal zone, varies considerably. Where the surface waters are cold, as in the Arctic region, the bathyal/abyssal transition is similarly at a shallow level, about 400 m (1,300 ft), and the bathyal fauna itself extends to within 12 m (40 ft) of the surface. At lower latitudes, closer to the equator, where the water temperatures are in general higher, the bathyal/abyssal transition is at a deeper level, usually at about the 900 m (3,000 ft) base of the pycnocline. At this level, the water temperature has dropped to 4°C, while below it drops to 1–2.5°C. The Swedish zoologist Sven Ekman [19], who was one of the founders of marine zoogeography, called this change in the depth of the bathyal/abyssal transition the principle of **equatorial submergence**. The pattern of this variation strongly suggests that the depth at which the transition takes place is dependent on temperature.

The seafloor at all these levels is covered by muds and oozes—organic sediments derived from the remains of organisms living in the waters above. As well as fragments of the bodies of larger organisms, most of these remains are of the zooplankton and (especially) phytoplankton, the latter being known as **phytodetritus**, which sticks together to form large flakes known as "marine snow." At the times of plankton blooms in the waters above, this falls like a blizzard to the seafloor, where it is consumed by the animals that live on or in the sediments there. The intensity of this recycling is shown by the estimate that the entire upper 10 cm (4 in) of organic sediments in the Santa Catalina Basin off California is eaten and excreted by worms every 70 years.

Turning now to the nature of the faunas in these zones, one can distinguish two quite different environments. At one extreme is the fauna living on the continental shelf. Here, conditions vary in both time and space, but temperatures are higher and there is light, providing the energy basis for a rich ecosystem. At the other extreme is the fauna of the abyssal zone, adapted to lightless, cold waters in which the only source of energy is the rain of phytodetritus from above. In each of these zones there live organisms that are specifically adapted to that environment. The bathyal zone is therefore an intermediate zone within which there is a gradual change with depth, from a mainly shelf-like fauna to a mainly abyss-like fauna. The precise pattern of faunal replacement with depth varies from place to place, according to the physiological tolerances and limitations of the individual species, and according to their interactions with local competitors and predators.

The patterns of change with depth in the faunas of the bathyal and abyssal benthos have been intensively studied off the Atlantic coast of North America. As the continental shelf is the point at which the physical changes in the environment take place most rapidly, it is not surprising to find that this is the region where the rate of faunal change is most rapid; it then continues at a slower rate with increasing depth.

Another aspect of these faunas is their degree of diversity. This, again, has been most intensively studied in the western North Atlantic, where the faunas show a very clear pattern [20]. The best data so far on the faunal diversity of the continental slope comes from the work of Frederick Grassle and Nancy Maciolek [1], who took box-cores totaling 8.69 m² (233 ft²) along 176 km (110 mi) of the slope off New Jersey. Not only did these contain nearly 800 species, but also each new core contained additional species. Towing a dredge trawl along the continental slope at 1,400 m (4,600 ft) for an hour produced over 25,000 specimens belonging to 365 species, but the same operation carried out on the abyssal plain at 4,800 m (16,000 ft) produced only 3,700 specimens belonging to 196 species, showing a not-unexpected drop in faunal diversity as one moves into the ocean depths. In many groups, the faunal diversity is low on the continental shelf, high at a mid-bathyal depth (2,000–3,000 m; 6,600–9,800 ft), and low again on the abyssal plain.

The ultimate cause of these changes is likely the gradual change in food and nutrient availability as one moves away from the high-productivity surface waters. The problem has been to identify how this change imposes itself on the community structure of the seabed. Various mechanisms have been suggested, such as changes in the intensity of predation, or competition, arising from differing rates of population growth. An interesting discovery is that a clear correlation exists between faunal diversity and the characteristics of the sediments of the seafloor, especially the diversity of particle size [21]. Because there is no primary production in the lightless deep sea, its economy is based on the organic particles that settle on the bottom, and the fauna that lives on or in the seafloor is dominated by detritivores. Therefore, it may well be that a greater range in particle size provides a greater range of niches for the bathyal benthic fauna.

Whatever the precise pattern (or range of patterns) of relative diversity in the deep-sea benthos, the total diversity on the abyssal plain is very great—even though, because of the low level of food supply, the density of the deep-sea biomass is very low [1, 22]. The ecology of these faunas is so poorly understood that it is difficult to be sure of the reasons for this unexpected faunal richness. It is possible that the lack of barriers to dispersal on the abyssal plain makes it easy for many organisms, which have evolved at different locations on its enormous area, to become widely distributed. The coexistence of the resulting large number of species might be facilitated by another phenomenon. The abyssal benthic environment is often disturbed by the settling of aggregations of plankton or of the remains of larger organisms. It has been suggested that this is so frequent that the community rarely has an opportunity to come to a final balance in which some species become locally extinct due to competition from rival species.

The amount of information on the composition of the fauna of the deep-sea benthos is still very limited and geographically unbalanced. Most of it comes from the western Atlantic, and none is available from central oceanic regions. It is also nearly all derived from soft-bottom communities, little being known of hard-bottom communities. It is therefore impossible to identify different biogeographical areas on the deep seafloor of a single ocean, or even to draw up lists of faunal differences between one ocean and another.

There is even less systematic information on the fauna of the hadal zone, which lies in the great submarine trenches, at depths of 6,000–11,000 m (20,000–36,000 ft). As might be expected from its great depth, its fauna is extremely sparse, even more impoverished than that of the abyssal plain. Yet it does also seem to be different—about 68% of the species, 10% of the genera, and one family are endemic to the hadal zone. Most of the endemic species have a vertical depth range of less than 1,500–2,000 m (5,000–6,600 ft), so there is steady faunal change with increasing depth. These faunas are confined to isolated patches along the deep-sea trenches (cf. Fig. 5.3, p. 156). This has permitted considerable independent evolution of endemic species within each trench fauna, and it is quite possible that their patterns of biogeographical relationship may be similar to that of the hydrothermal vents of the midoceanic rises. Unfortunately, it is still too early for marine biologists investigating these patterns to be able to make firm statements as to what patterns exist and in what variety, or as to how these patterns may vary according to geographical location or seafloor topography.

Similarly, there is, still not enough information on possible faunal differences between the faunas of the floors of the different ocean basins as a whole, although it seems extremely likely that these exist. Earlier studies, mainly by Russian marine biologists, have suggested that only 15% of the deep-sea benthic species occur in more than one ocean, and only 4% in all of them. It is also possible that the pattern of midocean ridges, where hot material rises from the depths of the planet to form new seafloor as the tectonic plates separate (see Fig. 5.3), may similarly act as more minor barriers to faunal movement, and therefore delimit subsidiary faunal areas within the oceans. But the presence, nature, and scale of such possible differences have yet to be established. In continental biogeography, larger-scale patterns of faunal change or difference are often found to have been caused by historical events of evolutionary innovation or extinction, followed by dispersal or vicariance arising from plate tectonic events. Smaller-scale differences are more likely to result from ecological factors. It will be interesting to see the extent to which marine biogeographical research reveals similar patterns in the deep sea. Even though the same processes are likely to be operative in the two environments, the very different scale, and probably very different rates, in the oceans may nevertheless lead to significantly different results.

The Biogeography of Hydrothermal Vent Faunas

In 1977, marine biologists discovered a dramatically different deep-sea environment containing a fauna that shows a fascinating biogeographical pattern. This lies at the midoceanic ridges, which are at an average depth of 2.5–3.5 km (1.5–2 mi). Although the ridges themselves are many hundreds of kilometers wide, they are split by a rift valley only about a kilometer wide, where hot lava is emerging. In some widely scattered areas known as **hydrothermal vents**, each of which covers only a few hundred square meters, the cold seawater penetrates fissures in the surrounding rocks. The water temperature there may reach 400°C (only the enormous pressures at this depth prevent it from turning to steam), and it reacts chemically with the rocks, so that it becomes rich in metals and sulphur. Where this superheated water emerges and is cooled by the surrounding waters of the ocean, these minerals precipitate out of the fluid. Some of them form solid "chimneys," which can be many meters high, while others remain as distinct particles in the rising plume of water, which therefore looks like smoke emerging from the chimney.

Accompanying this extraordinary environment is a unique fauna, whose food web is not based on plants that have trapped the sun's energy, but instead on chemosynthetic bacteria that extract energy from the chemicals dissolved in the hot fluids. Some of these bacteria are consumed by grazing or filtering organisms, while others live symbiotically, rather as photosynthetic algae live in corals (see below, p. 294). They form the base of a fauna consisting mainly of worms, arthropods, and molluscs. The fauna is of low diversity. In the Juan de Fuca section, west of Seattle, there were only 55 species, and 90% of the total number of organisms was contributed by only five species—two gastropod molluscs and three polychaete worms [23]. Such a fauna is typical of other highly disturbed habitats, such as colonizing areas after volcanic eruptions or forest fires.

The vent communities have been found at many different locations in the oceanic ridge system (see p. 154), but their composition varies, providing puzzling and interesting biogeographic problems. One of these problems is the method by which their organisms disperse from vent to vent. It is theoretically possible that their larvae disperse across the abyssal plain from one midoceanic ridge to another, but the highly unusual nature of the vent environment makes that seem unlikely. It is therefore no surprise to find that the degree of similarity between one vent fauna and another is not dependent on the direct distance between them across the abyssal plain. Instead, it depends on the distance between them following a longer path along the pattern of midoceanic ridges and faults [24].

The vent communities of the Eastern Pacific Rise, west of Central and South America, are dominated by giant tube-worms (*Riftia*) up to 2.5 m (8 ft) tall and include also some shrimps. The same community is found 3,200 km (2,000 mi) away in the Juan de Fuca region west of Canada, but there they belong to different genera or species. In fact, 80% of the Juan de Fuca species are endemic, though related to those further south. It is possible that this faunal similarity and relationship developed because these two parts of the oceanic ridge system were once continuous with one another. They were only separated from one another about 30 million years ago, when the westward drift of the North American continent obliterated part of the system, an event that seems to have provoked vicariant evolution in the two now-separated regions of the ridge.

Further away, the vent communities of the North Atlantic ridge are quite different. They lack the tube-worms, and they are dominated by shrimps that belong to the same family as those found in the eastern Pacific. Any connection between these two oceans is most likely to have been east-west via the gap between North and South America before this was closed by the growth of the Panama Isthmus, which became complete about 3 million years ago. But this poses quite different problems: there has never been any midocean ridge system in this area, so that any communication between the two oceans must have been by long-distance dispersal of larvae rather than across shorter gaps between vents along a ridge complex. The only alternative would have been southward and eastward along the ridge system that runs around Antarctica.

The vent fauna of the small ridge near the Mariana Islands in the northwestern Pacific is even more isolated, being 8,000 km (5,000 mi) away from any other ridge system. It is not surprising that it differs from the other two known vent faunas, being dominated by large snails, but 50% of these belong to species known from the other vent faunas. Again, dispersal between the Marianas fauna and the others must either have been long distance, across regions that lack ridges, or long ago along ridge systems that once ran across the southern Pacific until they were progressively smothered by the northward movement of Australasia, which began about 43 million years ago.

The Shallow-Sea Realm

Some of the differences between the shallow-sea (or neritic) realm and the open-sea (or pelagic) realm have already been considered, but some others need to be noted. Even if one subdivides the open ocean into areas corresponding to the movements of the surface waters, each of these is immensely greater than any of the individual units of the shallow-sea realm, most of which are long and narrow, constrained between the coast and the continental edge. Furthermore, each of these shelf seas is heavily influenced by the characteristics of the adjacent land, such as the nature of the coast and the presence of rivers, which may contribute both freshwater and a varying input of sediment. Because of the relative shallowness of the sea, the nature of the seabed also influences conditions through the whole of the overlying water mass, as far as the surface itself; there is no such interaction in the open sea. The shelf seas are therefore also much more heterogeneous than the open ocean.

Two phenomena result from these factors. First, because each shelf fauna may be different from others, and therefore contains its own endemic, locally adapted species, the total number of shelf species is far greater than the total number of pelagic species, which occupy an environment that is far less varied globally. For example, over 970 species of the neritic mysid crustaceans are known, compared with only 86 species of their pelagic relatives, the euphausid crustaceans. Second, there is a fairly sharp distinction between the ranges of the two types of organism, as few of the pelagic species venture into the neritic environment (Fig. 9.6). Also, though the shallow seas account for only 8% of the total area of the oceans, they hold the majority of its diversity of life.

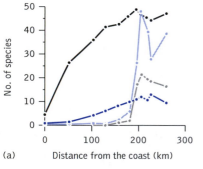

(a)   Distance from the coast (km)

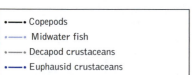

- Copepods
- Midwater fish
- Decapod crustaceans
- Euphausid crustaceans

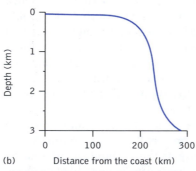

(b)   Distance from the coast (km)

Fig. 9.6 The number of species of four different pelagic groups off Florida, showing how the numbers decrease as one moves from the open ocean and toward the coast, and often peak at, or close to, the shelf-break. After Angel [7].

Although the environment and biogeography of the shelf sea faunas themselves are quite different from those of the open sea, they can nevertheless be assembled into similar biogeographical units (cf. Figs. 9.5 and 9.9). There are two reasons for this. One is that the waters of the open sea usually themselves traverse the shelf areas—and even if they do not, they inevitably influence the temperature of the adjacent shallow seas. The other reason is that any faunal connection between individual shelf faunas can only be via the open ocean waters, which can carry planktonic larvae from place to place. Thus, the shelf areas that are linked together as, for example, parts of the Northern Hemisphere warm-temperate region, share a particular temperature regime because they are connected by an ocean surface current of that temperature range.

Although the primary role of larvae may be to ensure that the bottom-dwelling adult is distributed locally, where the presence of the parents suggests that the environment is favorable, they may also play a role in more widespread dispersal. This can only be successful if the location that the larva reaches is similarly environmentally favorable, and also does not contain a more competent rival or predator. However, the potential efficacy of larval dispersal is shown by the fact that species of benthic invertebrate along the western coasts of the Atlantic are more widely distributed if they have planktonic larvae than if they do not (Fig. 9.7). Of

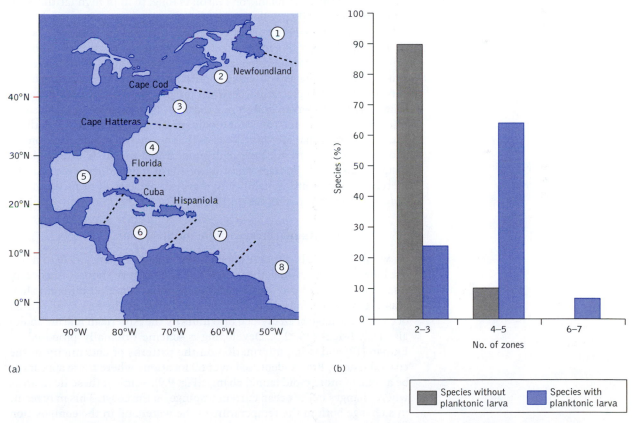

(a)   (b)

Species without planktonic larva   Species with planktonic larva

Fig. 9.7 (a) Biogeographic zones down the western coasts of the Atlantic. (b) Number of these zones that are occupied by invertebrate benthic species, with or without planktonic larvae. After Scheltema [56].

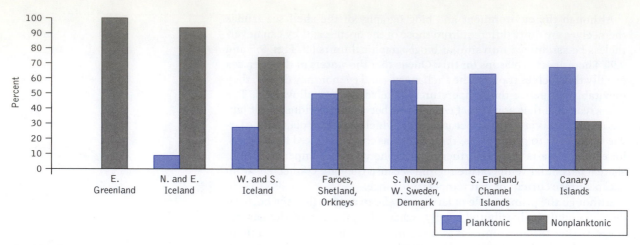

Fig. 9.8 The percentage of gastropod species that either have, or do not have, planktonic larvae, at different latitudes. After Thorson [57].

course, long-lived larvae will need to feed during the days of dispersal, so it is not surprising to find that such larvae are more common in low latitudes, where the phytoplankton season is long, than in high latitudes, in which it is shorter (Fig. 9.8).

However, it may be incorrect to assume that the possession of larvae that can potentially live for a longer time indicates that their function is primarily long-distance dispersal [25]. Research using a combination of statistical modeling and observed population genetic structure suggests a surprisingly poor correlation between larval type and the prevalence of long-distance dispersal, and shows that the average scale of dispersal is often very variable for the same taxon and location. In addition, those taxa with larvae that are short-lived or that normally disperse over only a short distance seem to have a disproportionately high rate of long-distance dispersal. Perhaps one role of these long-lived larvae may be merely to give greater reliability to dispersal over small or medium distances, rather than to provide the potential for long-distance dispersal.

Faunal Breaks within the Shelf Faunas

As we shall see, it is difficult for shelf organisms or their larvae to make the long crossing between the eastern and the western shores of an ocean. As a result, most of the similarities between the faunas of different shelf seas are between faunas at different latitudes on the same side of an ocean. These have been divided into units by the American marine zoologist Jack Briggs [26, 27]. Developing a scheme originally proposed by Ekman [19] and using information on the patterns of endemicity in the coastal faunas, Briggs identified over 60 locations where there appears to be a zone of more rapid faunal change (Fig. 9.9). Some of these lie in areas where changes in the ocean current impinge on the coast. This may result in a change both in the temperature of the water and in the composition of the fauna that arrives within the current. These locations are therefore at the boundaries between different temperature regimes. The remainder

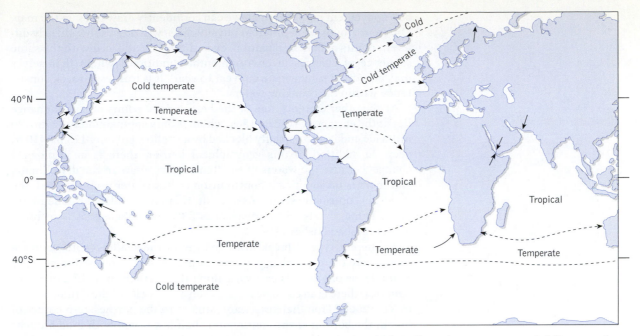

Fig. 9.9 The locations of major faunal changes in shallow-sea communities, shown by small arrows, and their relationships to the boundaries between areas with different surface-water temperatures. After van der Spoel & Heyman [58].

lie within the influence of a single ocean-current temperature regime, and Briggs uses them to divide each regime into regions, some of which are subdivided into provinces. Excluding those that comprise only islands or island groups, he defines a total of 23 regions and 28 provinces.

As would be expected, the clarity of these locations of faunal change varies according to the rapidity of the environmental change that causes them. For example, the sea-temperature change that results from an alteration in the origin of the ocean currents is inevitably diffused over a considerable extent of coastline; the faunal change is therefore similarly gradual. A change of this kind takes place just north of Santa Barbara in Southern California, where Point Conception projects westward and guides the cold, southerly directed California Current away from the coast. It is therefore considered to be the point of transition between the cold-temperate and the warm-temperate regions of the coastal faunas. However, it may be significant that a recent phylogeographic study [28] has failed to find any grouping of genetic population breaks here, even though it appears to be one of the more geographically plausible.

In some other areas, the boundary points between adjacent faunas are even less definite. For example, parts of the west coast of southern Africa are bathed by tropical waters from the north, while others receive cooler waters from the south. As a result, there is a change in the temperature of the surface water from Cape Santa Marta at 14°S to Cape Frio at 18°30′S— a distance of some 480 km (300 mi). Studies of the faunas of jellyfish, polychaete worms, molluscs, crustaceans, and shore fishes indicated a variety of points of faunal change, so that Briggs felt unable to decide on any one firm location. However, we cannot expect that there will always be a convenient point at which we can confidently draw lines on the map. Different groups, with different physiologies, will often show slightly different points of faunal transition, even if they are all faced with the same sequence of changes in environmental conditions (which themselves may also differ somewhat from year to year). It is fruitless to seek consistency within the changing waters.

At the other extreme, one of the best defined points of faunal change lies at the entrance to the Red Sea, which is only approximately 32 km (20 mi) wide and is also partially blocked by a shallow sill only 125 m (410 ft) deep. In addition to this geographical barrier, there is an ecological change; because the waters of the Red Sea evaporate in the dry climate, and there is no significant contribution of freshwater from the land, the Red Sea is unusually saline. As a result, there is considerable endemicity in its fauna: corals 25%, crustaceans 33%, cephalopods 50%, echinoderms 15%, and fishes 17%.

A different type of break in the nature of the shallow-water fauna is seen where there is a change in the nature of the bottom sediments. For example, as one travels eastward along the northern coastline of South America, there is an ecological transition at the delta of the Orinoco River in Venezuela. From here eastward, almost to the northeastern corner of Brazil, the bottom of the continental shelf is covered with mud brought down by the great tropical rivers, whose freshwaters also greatly reduce the salinity of the waters of the shallow sea. The coral reefs so characteristic of the Caribbean region to the west are absent, together with their associated fish fauna, which is replaced by such groups as the sea catfishes and croakers.

The Coastal Faunas of Islands

Although most coastal faunas lie along the edges of the continents or around islands on the continental shelves, others are to be found around the edges of isolated oceanic islands. Most of these lie around volcanic island-arcs or chains that have resulted from the actions of oceanic trenches or hotspots (see p. 169), and most of them lie within the Pacific Ocean. As one might expect, these isolated faunas show a high degree of endemism. Springer [29] studied the patterns of distribution of 179 species of shore fish, belonging to 111 families, from these isolated Pacific islands. He calculated that 20% of them are endemic to the region and that most of these are endemic to a single island. He noted that most of the cartilaginous elasmobranch (shark-like) fishes disappear from the faunal lists as one moves out into the deep Pacific Ocean from the east. This may be because their buoyancy system, based on the presence of a large, oily liver, is less effective than the swim-bladder of bony fish. Perhaps more surprisingly, he found that the number of taxa rapidly diminishes as one moves into the deep ocean from the west, suggesting that these shore fishes find these open wastes of water almost as formidable an obstacle as do land animals and plants. It will be interesting to see whether studies of the shore fishes of the Hawaiian chain show a pattern of relationship similar to that of the insects and plants of the Islands (see p. 235), in which the faunas of the younger islands are derived mostly from those of the older ones.

Transoceanic Links and Barriers between Shelf Faunas

Although currents are potentially capable of carrying the larvae of shelf organisms from one side of an ocean to the other, many larvae live for so short a time that there are comparatively few examples of east-west linkages between their shelf faunas. The eastern Pacific Ocean, because it is so wide (5,400 km; 3,350 mi) and because it contains few islands to provide intermediate havens for shallow-water organisms, is a very effective barrier to many such organisms. Ekman therefore named it the **East Pacific Barrier**. Of the shore fishes that are found either in the Hawaiian Islands or between Mexico and Peru at the eastern end of the Pacific Ocean warm-temperate region, only 6% are found in both. Similarly, Ekman [19] showed that only 2% of the 240 species and 14% of the 11 or 12 genera of echinoderms found in the Indo-West Pacific area had been successful in reaching the west coast of the Americas. (The greater success of the genera is because a new taxon first appears as a new species. It is only later that, if enough time has gone by, or enough new, related species have appeared, the level of novelty is great enough for us to recognize a new genus. Therefore, genera are older than individual species and will have had a greater length of time in which to cross the barrier.)

As always, there are exceptions to this generalization. For example, it has recently been shown [30] that populations of the sea urchin *Echinothrix diadema* in the eastern Pacific islands of Clipperton Atoll and Isla del Coco are genetically so similar to those of Hawaii that there must have been recent and massive gene flow between these locations. This is despite the fact that it normally takes 100–155 days for water to be carried from one area to another in the northern Equatorial Counter-Current—longer than the duration of the life of the echinoderm larva (50–90 days). However, it may be that larvae were able to make the journey in years when the El Niño regime (pp. 404–406) was strong, when the journey time would have been reduced to 50–81 days. Similar relationships between the populations of crabs and starfish in these regions have also been recorded.

The great antiquity of the East Pacific Barrier has been shown by Richard Grigg and Richard Hey of the University of Hawaii [31], who studied the zoogeographical affinities of fossil and living genera of coral. They found that those of the East Pacific are more closely related to those of the West Atlantic than to those of the West Pacific, even for corals living as long ago as the Cretaceous Period. The fact that the Barrier appears to have been effective in inhibiting dispersal across the Pacific in the Mesozoic is not surprising, for the Americas then lay much further east. Thus, the gap between their western coastlines and the island arcs of the western Pacific would have been correspondingly wider.

Shelf faunas also provide evidence on the progressive widening of the Atlantic Ocean. Calculation of the degree of similarity of the shelf faunas on either side of the North Atlantic from the Early Jurassic onward, using the coefficient of faunal similarity (see the caption to Fig. 11.7, p. 337), shows a steady decrease in similarity as they are gradually separated by the widening ocean [32].

The completion of the Panama Isthmus barrier between the western Atlantic and the eastern Pacific about 3 million years ago provides a good example of vicariant evolution in the shelf faunas on either side of the new land connection. Only about a dozen of the approximately 1,000 shore fishes of the region still appear to be identical on either side of the barrier, but marine invertebrates seem to have been slower to evolve into new species: 2.3% of the echinoderm species, 6.5% of the porcellanid crab species, and 10.8% of the sponge species of the region are found on both the eastern and the western shores of the Isthmus [28].

The deep waters of the Atlantic form a similar **Mid-Atlantic Barrier** to the dispersal of shelf organisms between the African tropics and the South American tropics. However, as the Atlantic is narrower than the Pacific, this barrier is less effective; thus in most groups there is a greater proportion of species that are found on both sides of the ocean. For example, in the shore fishes, there are about 900 species on the western shelf and about 434 species on the eastern; of this total, about 120 species (9%) are common to both faunas [26]. Most of these dispersals appear to have been from South America to Africa, perhaps because the greater richness of the South American shelf fauna, in terms of numbers of both species and individuals, makes it more likely that they will succeed in dispersing. Although the surface South Equatorial Current flows westward, these migrants may have used the deeper Equatorial Under-Current, which runs in the opposite direction.

An example of the appearance of a new link between shelf faunas is the one that took place between the Arctic and the North Pacific about 3.5 million years ago, after submergence of the Bering Strait. This led to an exchange of cold-water species, in which the majority (125 species) dispersed from the Pacific to the Arctic Basin (in this case, in the direction of the current flow), and only 16 species dispersed in the reverse direction [33]. (But because of the shallowness of the sea across the Bering Strait, it remained an obstacle to the dispersal of deeper-living plankton.)

The completion in 1869 of the Suez Canal between the Mediterranean Sea and the Red Sea provided a far more recent example of a link between two marine faunas. It has been named the **Lessepsian exchange**, after the French entrepreneur Ferdinand de Lesseps, who was responsible for its construction. The exchange has been very unbalanced for, although 50 species of fish, 40 species of mollusc, and 20 species of crustacean have colonized the Mediterranean from the Red Sea, few if any species have made the reverse journey, which poses an interesting problem as to the cause of this difference [34, 35]. Though both seas are small and almost landlocked, there is an important difference between their faunas. That of the Mediterranean is derived from the rather limited fauna of the cold Atlantic Ocean, whereas that of the Red Sea is derived from the rich, tropical fauna of the Indian Ocean and is therefore better suited to colonizing the warm, shallow waters of the eastern Mediterranean. Furthermore, the species living at the northern end of the Red Sea already inhabit a shallow, sandy or muddy, and unusually saline environment. They are therefore better suited to surviving in the shallow hypersaline lakes of the Suez Canal through which they have to pass in order to reach the Mediterranean.

Latitudinal Patterns in the Shelf Faunas

Marine organisms show a general latitudinal trend of decreasing diversity as one moves away from the tropics. However, much of that is due to the distribution of coral reefs, which have a faunal diversity unparalleled elsewhere in the shallow seas. Their distribution is centered on the tropics and therefore heavily distorts the underlying pattern. In addition, Crame [36] has shown that Antarctic and sub-Antarctic shelf faunas are far richer (and more ancient) than had previously been thought, further emphasizing that this pattern, and its significance, must now be reconsidered. (For a general discussion of latitudinal patterns of diversity, see p. 124.)

Many marine organisms also provide examples of the phenomenon known as **bipolar distribution** (or **antitropical** or **amphitropical distribution**). This term describes a situation in which related organisms are to be found in temperate or polar environments in both the Northern and the Southern Hemispheres, but not in the intervening equatorial region. Whatever may be the reason for terrestrial examples of this pattern, its occurrence in marine faunas has prompted suggestions of specifically marine mechanisms. Charles Darwin suggested that the cooling of the equatorial waters during the Ice Ages had allowed these genera to pass through waters that are now once again too warm for them to inhabit. Brian White [37] has argued that it was instead the warmer temperatures earlier in the Cenozoic that made it impossible for these genera to live in the equatorial waters, so that they now show a relict distribution on either side of this zone. This perspective is supported by Gordon Howes in his useful analysis of the biogeography of the gadoid fishes [38]. Yet another explanation is that of Jack Briggs [39], who links it with his theory that the Indo-West Pacific region is a center of evolutionary origin; he suggests that bipolar genera have become extinct in the equatorial regions because of competition from genera that have newly evolved there. Perhaps the simplest suggestion is that the organisms living in cool waters on either side of the equator have been able to disperse beneath it via cooler waters at greater depth (cf. the principle of equatorial submergence, see p. 282) [40]. An example of this was the catching of a notothenioid fish off Greenland [41]. Fishes of this family are normally confined to cold waters in the Southern Hemisphere. This lone example must have traveled at least 10,000 km (6,200 mi) submerged at depths of 500–1,500 m (1,600–5,000 ft) to remain within its normally preferred temperature range. Some species of plankton show precisely this pattern of permanent distribution. For example, the chaetognath *Eukrohnia hamata* is only found in near-surface waters at latitudes greater than 60° whereas, between those latitudes, where the surface waters are warmer, it is only found at depths of about 1,000 m (3,300 ft).

In his review of this topic, Crame [42] pointed out that most theories had tacitly assumed two premises: first, that these patterns had originated within the last 5 million years and were possibly related to the climatic changes of the Ice Ages, and second, that the two now-separate taxa had achieved this pattern as a result of dispersal rather than of vicariance. Concentrating on the distribution of marine molluscs, which have a good fossil record extending over 245 million years, Crame identified three main periods of bipolar distributions. The first was in the Jurassic-Cretaceous and seems to have been caused by vicariance resulting from the breakup of Pangaea. The second was in the Oligocene-Miocene and may have been caused by vicariance resulting from the cooling temperatures at that time, which allowed temperate taxa to spread across the equator, only to become extinct there when temperatures rose again. The third period was during the Plio-Pleistocene Ice Ages, which, together with the closing of the Panama

Isthmus, caused increased cooling and upwelling along the equatorial divergences in both the Pacific and the Atlantic, allowing dispersal of temperate forms from one hemisphere to the other. Our now steadily increasing knowledge of past patterns of climatic change on both land and sea is likely to lead to further understanding of the problem of bipolar distributions.

Coral Reefs

For many reasons, corals provide fascinating and unique aspects of marine biogeography. One of the most complex and diverse environments on Earth, they are clearly definable in nature, with limits of distribution that are simply explained by fundamental aspects of their biology. They therefore provide a good example of the extent to which marine patterns can be explained when the taxonomic and environmental aspects are simpler than elsewhere in the sea. On the other hand, interpretation of their patterns of diversity raises fundamental problems. Finally, because they are easily recognizable in the fossil record, historical biogeography can contribute more to our understanding of patterns of coral distribution than to most other aspects of marine biogeography. Two books, one by the Australian marine biologist Charlie Veron [43] and the other edited by the American Charles Birkeland [44], have dealt with many aspects of the biology and history of corals (although Veron's theories [45] of coral evolution are controversial).

Coral reefs provide a complex, three-dimensional environment that is home for an immense diversity of marine organisms [46], including 25% of the diversity of life in the oceans, and comprise the greatest diversity of species of vertebrate per square meter known on Earth. To date, 35,000–60,000 different species of reef-dwelling organisms have been described, and this is probably only a fraction of the total number. Between 1950 and 1994, the number of species of fish, molluscs, echinoderms, and corals known from the Cocos (Keeling) Islands in the Indian Ocean tripled. Though they make up only 1% of the area of the oceans, they include 25% of its species, including at least 4,000 species of fish, almost one-third of the total number of species of marine fishes.

The types of coral that form reefs are known as **hermatypic corals**, and they are a spectacular example of animal–plant symbiosis. The animals are colonial organisms called **hydrozoans**, in which the individuals, or polyps, resemble tiny sea anemones and feed on zooplankton. The reefs result from the activities of these polyps in secreting a basal calcium carbonate skeleton, to which the individual polyp is attached, and which is continuous with that of its neighbors. The plants are a type of algae known as **zooxanthellae**, which live within the tissues of the hydrozoans.

The biology of the corals limits their distribution to particular circumstances of nutrient levels, temperature and light. Corals are found in areas where the nutrient levels are so low that there is too little primary productivity from free-living algae or phytoplankton to provide the basis for a diverse ecosystem. Corals can flourish in this environment because their zooxanthellate algae are living within the hydrozoans, where nutrient levels are high. Of the other two factors, temperature is more important than light, as is shown by the fact that some corals can grow in deeper water as long as the temperature level is adequate. Coral reefs are only found where there is a minimum sea-surface temperature of at least 18°C, sustained over long periods of time, with a maximum of 30–34°C (Fig. 9.10). As a result, relatively diverse coral reef assemblages are found up to about 30° of north and south latitude, with extremes in Japan at 35°N, Bermuda at 32°N, and Lord Howe Island at 32°S, but the majority is found between the latitudinal zones where the temperature never drops below 20°C. Hermatypic corals also cannot

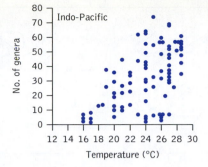

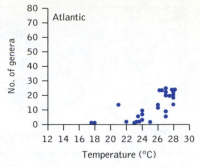

Fig. 9.10 The number of genera of coral at different mean annual sea-surface temperatures in the Indo-Pacific and Atlantic oceans. From Rosen [59].

flourish where there is significant sedimentation, for this impedes the light that is vital for their photosynthetic algae.

Many reef organisms have planktonic larvae, and it has been generally assumed that these would be readily dispersed over distances of hundreds of kilometers, given reasonable longevity and current speed. However, work by a group of American marine ecologists [47] suggests that we may need to reconsider these assumptions. Using molecular techniques, they examined variations in the genetic structure of populations of a shrimp, *Haptosquilla*, taken from 11 reef systems in the East Indies, where the Java Sea and Flores Sea lie between the northern islands of Borneo and Sulawesi and the southern string of Sumatra, Java, and smaller islands to the east (cf. Fig. 11.9, p. 348) The planktonic larvae of the shrimp live for 4–6 weeks and, since oceanic currents in the region are strong, it had been supposed that they would easily have been able to disperse for at least 600 km (370 mi). Surprisingly, however, these studies showed a sharp genetic break between the shrimp populations to the south of the Java-Flores seas and those that lay to the north.

Another aspect of the biogeography of corals has been discussed by the American marine biologist Gustav Paulay [48]: the comparative poverty of the eastern ends of the oceans, compared with the western. There are several different reasons for this contrast. One major influence is the pattern of ocean currents, for the upwellings of cool, nutrient-rich waters that take place along the eastern margins of the oceans (see Figs. 9.3, 9.4) inhibit coral growth in those regions. In addition, most of the warm, equatorial ocean currents are directed westward; when they reach the edge of the continent, they diverge both northward and southward, bringing warmer waters to higher latitudes. Another contributory factor is that the continental shelves, on which most coral reefs lie, are much narrower along the western margins of Africa and the Americas than along the eastern margin of Asia and in the Caribbean region. However, not all corals grow on the continental shelves; some grow around the margins of volcanic islands. Since these are most common in the western parts of the Pacific Ocean, this provides another increment to the greater reef area of that region. As a result of all of these factors, 85% of the area of coral reef lies in the Indo-Pacific Ocean and only 15% in the Atlantic Ocean. Similarly, the reefs of the eastern Pacific are only a few meters (or yards) thick, while those of the western Pacific are up to more than a kilometer (0.6 mi) in thickness.

A different factor causes the gap in the distribution of coral reefs along the tropical northern coastline of South America. The westerly winds of the equatorial Atlantic bring heavy rains to the low-lying river basins of

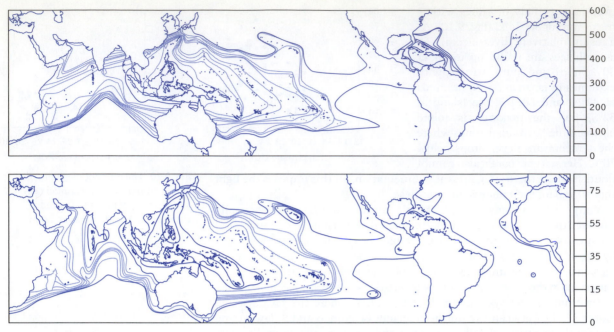

Fig. 9.11 The similar patterns of tropical marine species richness found in coral reef fishes (above) and cowrie molluscs (below). From Bellwood & Meyer [60].

tropical South America, which drain back to the sea via the great rivers. The dilution of the seawater by this freshwater, and the immense discharge of sediment onto the seabed, make this coastal area inimical to the growth of corals. The impact of this phenomenon has been increased by the deforestation of the Amazon Basin, which has resulted in the waters of the great rivers bearing even more sediment and also having an increased concentration of nutriments, while fires in Central America have had a similar effect on corals in the neighboring part of the Caribbean.

The diversity of most coral reef organisms decreases both longitudinally and latitudinally as one moves away from the Indo-Australian Archipelago, or IAA (the group of islands stretching from Sumatra to Papua/New Guinea and north to the Philippines), which is therefore one of the world's major hotspots of biodiversity (Figs. 9.11, 9.12, 9.14). (One locality in the East Indies has been reported to have over 1,000 fish species—more than occur in the entire tropical Atlantic.) As in the case of terrestrial gradients of diversity (see Chapter 4), earlier explanations of this phenomenon focused on possible ecological and evolutionary factors, of which there are three main models: center of origin, center of overlap, and center of accumulation.

The center of origin model suggests that speciation rates are unusually high in the IAA but that the various species have different abilities to disperse outward from it, so causing the gradient in diversity as one moves outward from its center. Various factors have been invoked as explanations of this high speciation rate, including a high, consistent rate of solar energy input and an abundant area of reef that therefore provides many habitats within which speciation can take place. Changes in sea level,

(a)

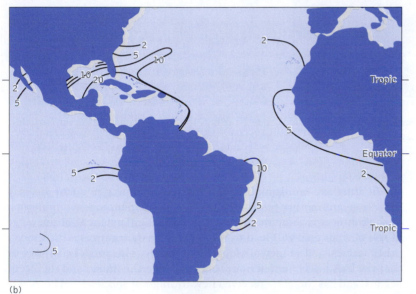

(b)

Fig. 9.12 Contours of generic diversity in corals, combining the distribution ranges of all the genera. (a) The Indian Ocean and West Pacific regions. (b) The East Pacific and Atlantic Oceans. After Veron [43].

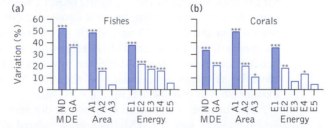

Fig. 9.13 Proportion of variation in reef fish and coral species richness at 67 sites, explained by the mid-domain effect (MDE) and eight environmental variables that relate to energy or habitat area. ND, distance of site from the mid-domain relative to domain size; GA, great arc distance of site from the mid-domain. A1, reef area; A2, area of substratum 0–30 m deep; A3, area of substratum 30–200 m deep. E1, mean annual sea-surface temperature; E2, solar irradiance range; E3, sea-surface temperature range; E4, mean solar irradiance; E5, productivity. *** $P < 0.001$; ** $P < 0.01$; * $P < 0.05$. Dark bars indicate variables selected for additional analyses. From Bellwood et al. [50].

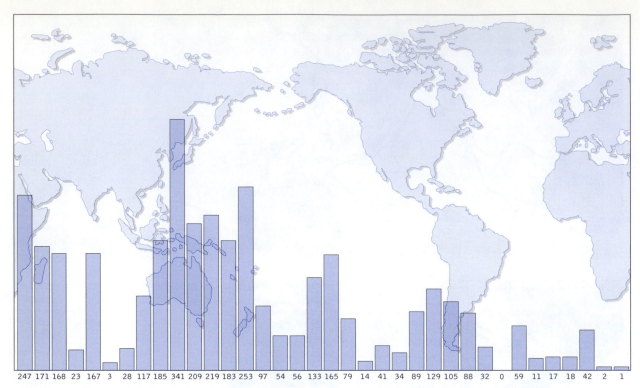

247 171 168 23 167 3 28 117 185 341 209 219 183 253 97 54 56 133 165 79 14 41 34 89 129 105 88 32 0 59 11 17 18 42 2 1

Fig. 9.14 Longitudinal gradients in fish species richness. The columns represent the total number of fish species (from a sample of 799 species) that occur in each 10°-wide band of longitude. Note how the diversity increases in the latitudes that include the West Indies and Caribbean, where there are many coral reefs; cf. Fig. 12.7c. After McAllister et al. [61].

subdividing and reuniting the reefs, thereby enabling frequent vicariant evolution, are another possible factor. This latter phenomenon is also part of the center of overlap model: the subsequent enlargement of the ranges of the new species will lead to the overlap of their ranges with those of other species, and so increase local biodiversity. This model also suggests that the IAA has benefited from being between the Indian and the Pacific oceans, and has therefore received new immigrant species from both.

The center of accumulation model suggests two other possible reasons for the high number of species in the IAA, in addition to its higher speciation rates. The first is that, because equatorial currents flow from east to west across the Pacific Ocean, species that may have arisen on its many islands will tend to extend their ranges toward the IAA [49]. Second, because of its large area and high energy input, species in the IAA may have larger population sizes and therefore be less likely to become extinct, so that it is also a center of survival.

As in the case of terrestrial gradients of diversity, the mid-domain effect (or MDE) (see Chap. 4, p. 133) has been suggested as a possible factor in the gradients shown in the IAA. The Australian marine biologist Dave Bellwood and his colleagues [50] have recently analyzed the proportions of the variation in reef fish and coral species diversity that are explained by different aspects of the MDE and of two environmental effects (reef area and solar energy input). As shown in Fig. 9.13, they found that in the case of both corals and fishes the most important of these aspects were distance from the center of the domain (ND), the area of reef (A1), and mean sea-surface temperature (SST, E1). A model combining the roles of these three factors in explaining species richness was then calculated. In fish, ND was the most important single factor, explaining 51% of the variation, followed by reef area and then SST (explaining, respectively, an additional 16% and 2%, to make a total of 69%). In corals, reef area was the most important single factor, explaining 51% of the variation, followed by distance from the center of the domain, and then SST (explaining, respectively, an additional 7% and 3%, to make a total of 61%). The authors comment that the similarity of the extent to which this analysis explains the patterns in two groups that differ markedly in their mobility, source of energy, and manner of reproduction suggests that the results are likely to apply to many other tropical marine organisms. This suggestion has recently been supported by other Australian workers [51].

It would be a truism to note that a large part of our analyses of the patterns and distributions of reef faunas inevitably relies on an adequate understanding of their genetics and taxonomy. However, research carried out by the American marine biologists Christopher Meyer, Jonathan Geller, and Gustav Paulay [52] demonstrates the complete unreliability of our "knowledge" of these crucial questions. They studied the genetics and distribution of the gastropod mollusc *Astralium rhodosteum* over 40 islands or locations over the 11,000-km (6,800-mi) distance from Thailand to Eastern Polynesia. They found that this apparent species is in reality a species complex that, despite the fact that its larva is nonfeeding and short-lived, and so apparently having only limited powers of dispersal, has developed endemic clades on every Pacific archipelago they studied. It comprises two major clades and at least 30 geographically isolated minor clades. One of these minor clades extends for over 750 km (465 mi), but some others are separated from one another by less then 100 km (60 mi). Though its powers of distribution allow the mollusc to colonize every island, at the same time the distribution of each clade is sufficiently limited for each one to have been able to evolve into a separate evolutionary unit, in strikingly allopatric fashion. In addition, colonization of an island by one clade seems to create barriers to its subsequent colonization by another. These clades are also almost indistinguishable from one another morphologically, having virtually identical shells—so the true biogeography of this "species" could not have been evaluated without genetic studies.

Meyer, Geller, and Paulay comment that the data from this study do not support the idea that any of the current theories of the origins of the diversity of the IAA provides the unique key to its biogeography. Instead, this diversity is the product of multiple processes in space and time, including such ecological factors as the differences between oceanic and shelf environments. (Of course, it also has to be recognized that a fundamental reason for the great diversity of reef life in this region is its geological history. Volcanic activity and sea-level changes have led to fragmentation of the larger islands each time sea levels rose, providing many opportunities for the appearance of new endemic taxa by vicariance.) It therefore seems unlikely that any of the theories currently proposed provides the whole answer. Rather than trying to decide between any of them, it might be better to wait until a much greater quantity and variety of reliable data, based on molecular genetic studies, has been gathered, and then try to establish under what circumstances each process has been important in producing the patterns that we see today.

Finally, we must remember that fundamental aspects of coral biogeography must have changed over geological time. The coral faunas of the eastern Pacific, western Atlantic, and eastern Atlantic, together with their interrelationships, have been greatly affected by the continental movements that closed the old Tethys Sea during the Cenozoic (see above, p. 281). In the Late Oligocene, about 25 million years ago, Africa had not yet met Europe (cf. p. 342) and South America was still separate from North America. Tropical waters therefore extended from the western Atlantic through the Tethys Sea between Africa and Europe and across the Indian Ocean to the Pacific (though the East Pacific Barrier would have prevented tropical marine faunas from extending all the way across the Pacific). As a result, the coral fauna of the proto-Caribbean gap between the Americas had 25 genera in common with that of the Mediterranean seaway; four genera were endemic to the proto-Caribbean, and nine to the Mediterranean. But the closure of the Mediterranean seaway (known as the Terminal Tethyan Event) 18–12 million years ago, followed by the completion of the Panama Isthmus about 3 million years ago, caused major changes in the environment of this whole stretch of tropical waters, as well as separating the fauna of the eastern Pacific from that of the Caribbean. Many genera became extinct in the western Atlantic, but most of these are still to be found in the IAA region. New taxa have also evolved in each of the now-separate areas, further reducing the similarity of their coral faunas. The corals of the IAA region and of the western Atlantic now differ greatly; only 8 of their combined total of 107 genera are found in both, and they share none of their species of zooxanthellae. Similarly, only one-third of the genera and none of the species of the eastern Pacific are also found in the Caribbean region on the other side of the Panama Isthmus. The coral fauna of the Eastern Atlantic also suffered badly from extinction after the closure of the Mediterranean seaway, losing some 85% of its coral genera, but long-distance dispersal has led to some exchange of taxa with the Caribbean region.

Recent research has also shown that plate tectonic activity may have been responsible for moving the biological hotspot of the IAA for thousands of miles [53]. A combination of fossil and molecular evidence strongly suggests that 42–39 mya in the Late Eocene, the highest diversity of reefs was found in the narrowing Tethys Sea between Africa and Europe (cf. Fig. 10.8c, p. 342), with its hotspot in southwestern Europe but extending eastward along the coastline of Arabia. The hotspot is exactly where the approach of Africa to Europe had led to an increase in shallow-water habitats. By 23–16 mya in the Early Miocene (cf. Fig. 10.8e), the diversity in the western Tethys had diminished as the closer approach of Africa to Europe had now eliminated most of the shallow seas, and the center of the Tethyan hotspot had moved to Arabia, still in the earlier stage of approach between Africa and Asia. By this time, the northward movement of Australia had led to its collision with Pacific island-arcs and the southeast margin of Asia, providing shallow-water habitats for a new focus of diversity, and at the same time placing the region in the tropical environment that encourages diversity. The fusion of Arabia to Asia led to the elimination of the old center of diversity there. However, the great variety of shallow-water habitats in the IAA, with its complex of oceanic and island-arc islands and microcontinental fragments, has led to the hotspot that we see there now. Nevertheless, its fauna still shows its relationship to the earlier, Miocene, fauna of the region: the vast majority of those taxa for which a fossil record is available first appeared long ago, at times ranging from the Eocene to the Pliocene [53].

Even over the last few million years, the coral reef faunas have been affected by changes in connections between the oceans. For example, although most of the coastal marine groups of the eastern Pacific are highly diverse, are endemic to that region at the species level, and are closely related to the biota on the eastern side of the Panama Isthmus, the coral reef biota shows a very different pattern. It is much less diverse than that of the IAA, but is closely related to it—all of the east Pacific coral genera and over 90% of the species are shared with the IAA. This relationship pattern in the corals is probably the result of the loss of many coral taxa in the eastern Pacific after the Panama Isthmus closed, followed by dispersal of some of the coral taxa from the IAA to the eastern Pacific.

Summary

1 The composition, structure, and ecology of marine biota are still poorly understood, and their environment differs greatly from that of terrestrial organisms. Nevertheless, it is becoming clear that marine biogeography is basically similar to that of the land, although the boundaries between the units are more gradual and are not fixed in their locations. On the other hand, marine organisms do not seem to show the same degree of correlation between morphology and species definition, which makes the study of their biogeography far more difficult than that of terrestrial species.

2 The major subdivision is between the open-sea realm and the shallow-sea realm.

3 The open-sea realm is divided vertically into zones based on light, temperature, and nutrient availability. The surface waters are divided into biomes and provinces, which are related to the rotating patterns of ocean circulation and the patterns of productivity.

4 The units in the shallow-sea realm are much smaller than those in the open-sea realm because they are dependent on the characteristics of the local land, rivers, sediments, seabed, tides, and ocean currents.

5 Coral reefs provide the most diverse environments within the seas and the clearest examples of gradients of marine diversity.

The remaining chapters in the book are concerned with the analysis and history of the biogeography of the continents. This commences with the patterns found some 200 million years ago, continues through the succeeding millions of years of change in continental patterns, faunas, floras, and climates, and ends with an attempt to understand the likely results of the rapid changes that are taking place on our planet today.

Further Reading

Angel MV, ed. Reports of the Scientific Committee on Oceanic Research. 1, The ecology of the deep-sea floor. 2, Pelagic biogeography. *Prog Oceanogr* 1994; 34: 79–284.

Gage JD, Taylor PA. *Deep Sea Biology*. Cambridge: Cambridge University Press, 1991.

Nybakken JW. *Marine Biology. An Ecological Approach*, 4th ed. Menlo Park, CA: Addison Wesley Longman, 1997.

Ormond RFG, Gage JD, Angel MV, eds. *Marine Biodiversity. Patterns and Processes*. Cambridge: Cambridge University Press, 1997.

References

1 Grassle JF, Maciolek NJ. Deep-sea species richness: regional and local diversity estimates from quantitative bottom samples. *Am Naturalist* 1992; 139: 313–341.

2 Grant WS, Leslie RW. Inter-ocean dispersal is an important mechanism in the zoogeography of hakes (Pisces: *Merluccius* spp.). *J Biogeogr* 2001; 28: 699–721.

3 Beaugrand F, Reid PC, Ibañez F, Lindley JA, Edwards M. Reorganisation of North Atlantic marine copepod diversity and climate. *Science* 2002; 296: 1692–1694.

4 Harris M., et al. A major increase in snake pipefish (*Entelurus aequoreus*) in northern European Seas since 2003: potential implications for seabird breeding. *Marine Biology* 2007; 151: 973–983.

5 Longhurst A. *Ecological Geography of the Sea*. London: Academic Press, 2nd ed., 2006.

6 Angel MV. Spatial distribution of marine organisms: patterns and processes. In: Edwards PJR, May NR, Webb NR, eds. *Large Scale Ecology and Conservation Biology*, pp. 59–109. British Ecological Society Symposium no. 35. Oxford: Blackwell Science, 1994.

7 Angel MV. Pelagic biodiversity. In: Ormond RFG, Gage JD, Angel MV, eds. *Marine Biodiversity Patterns and Processes*, pp. 35–68. Cambridge, Cambridge University Press, 1997.

8 Stuiver M, Quay PD, Ostlund HG. Abyssal water carbon-14 distribution and ages of the world's oceans. *Science* 1983; 139: 572–576.

9 Tyrrell T. The relative influences of nitrogen and phosphorus on oceanic primary production. *Nature* 1999; 400: 525–531.

10 Rutherford S, D'Hondt S, Prell W. Environmental controls on the geographic distribution of zooplankton diversity. *Nature* 1999; 400: 749–753.

11 Martin JH, Fitzwater SE, Gordon RM. Iron deficiency limits phytoplankton growth in Antarctic waters. *Global Biogeochem. Cycles* 1990; 4, 5–12.

12 Pierrot-Bults AC. Biological diversity in oceanic macrozooplankton: more than merely counting species. In: Ormond RFG, Gage JD, Angel MV, eds. *Marine Biodiversity Patterns and Processes*, pp. 69–93. Cambridge: Cambridge University Press, 1997.

13 Gibbs RH. The stomioid fish genus *Eustomias* and the oceanic species concept. In: *Pelagic Biogeography*. UNESCO Technical Papers in Marine Science no. 49, 1985: 98–103.

14 Lessios HA, Kessing BD, Pearse JS. Population structure and speciation in tropical seas: global phylogeny of the sea urchin *Diadema*. *Evolution* 2001; 55: 955–975.

15 Moser M, Hsieh J. Biological tags for stock separation in Pacific Herring *Clupea harengus pallasi* in California. *Journal of Parasitology* 1992; 78, 54–60.

16 White BN. Vicariance biogeography of the open-ocean Pacific. *Prog Oceanogr* 1994; 34: 257–282.

17 McGowan JA. The biogeography of pelagic ecosystems. In: *Pelagic Biogeography*. UNESCO Technical Papers in Marine Science no. 49, 1985: 191–200.

18 Bellwood DR, Wainwright PC. History and biogeography of fishes on coral reefs. In: PF Sale, ed. *Coral Reef Fishes—Dynamics and Diversity in a Complex Ecosystem*, pp. 3–15. San Diego: Academic Press, 2002.

19 Ekman S. *Zoogeography of the Sea*. London: Sidgwick & Jackson, 1958.

20 Rex MA, Etter RJ, Stuart CT. Large-scale patterns of species diversity in the deep-sea benthos. In: Ormond RFG, Gage JD, Angel MV, eds. *Marine Biodiversity Patterns and Processes*, pp. 94–121. Cambridge: Cambridge University Press, 1997.

21 Etter RJ, Grassle JF. Patterns of species diversity in the deep sea as a function of sediment particle size diversity. *Nature* 1992; 360: 576–578.

22 Grassle JF. Deep sea benthic diversity. *Bioscience* 1991; 41: 464–469.

23 Tsurumi M. Diversity at hydrothermal vents. *Global Ecology and Biogeogr* 2003; 12: 181–190.

24 Tunnicliffe V, Fowler MR. Influence of sea-floor spreading on the global hydrothermal vent fauna. *Nature* 1996; 379: 531–533.

25 Kinlan BP, Gaines SD, Lester SE. Propagule dispersal and the scales of marine community processes. *Diversity and Distributions* 2005; 11: 139–148.

26 Briggs JC. *Marine Zoogeography*. New York: McGraw-Hill, 1974.

27 Briggs JC. *Global Biogeography*. Amsterdam: Elsevier, 1995.

28 Dawson MN. Phylogeography in coastal marine animals: a solution from California? *J Biogeogr* 2001; 28: 723–736.

29 Springer VG. Pacific plate biogeography, with special reference to shore fishes. *Smithsonian Contributions to Zoology* 1982; 367: 1–182.

30 Lessios HA, Kessing BD, Robertson DR. Massive gene flow across the world's most potent marine barrier. *Proc Royal Soc Lond B* 1998; 265: 583–588.

31 Grigg R, Hey R. Paleoceanography of the tropical Eastern Pacific Ocean. *Science* 1992; 255: 172–178.

32 Fallaw WC. Trans-North Atlantic similarity among Mesozoic and Cenozoic invertebrates correlated with widening of the ocean basin. *Geology* 1979; 7: 398–400.

33 Vermeij GJ. Anatomy of an invasion: the trans-Arctic exchange. *Paleobiology* 1991; 17: 281–307.

34 Edwards AJ. Zoogeography of Red Sea fishes. In: Williams AS, Head SM, eds. *Key Environments: Red Sea*, pp. 279–286. Oxford; Pergamon Press. 1987.

35 Golani D. The sandy shore of the Red Sea—launching pad for the Lessepsian (Suez Canal) migrant fish colonizers of the eastern Mediterranean. *J Biogeogr* 1993; 20: 579–585.

36 Crame JA. An evolutionary framework for the polar regions. *J Biogeogr* 1997; 24: 1–9.

37 White BN. The isthmian link, antitropicality and American biogeography: distributional history of the Atherinopsidae (Pisces; Atherinidae). *Syst Zool* 1986; 35: 176–194.

38 Howes GJ. Biogeography of gadoid fishes. *J Biogeogr* 1991; 18: 595–622.

39 Briggs JC. Antitropical distribution and evolution in the Indo-West Pacific Ocean. *Syst Zool* 1987; 36: 237–247.

40 Boltovskoy D. The sedimentary record of pelagic biogeography. *Prog Oceanogr* 1994; 34: 135–160.

41 Møller PR, Nielsen JG, Fossen I. Patagonian toothfish found off Greenland. *Nature, Lond* 2003; 421: 599.

42 Crame JA. Bipolar molluscs and their evolutionary implications. *J Biogeogr* 1993; 20:145–161.

43 Veron JEN. *Corals in Space and Time*. Sydney: University of New South Wales Press, 1995.

44 Birkeland C., ed. *Life and Death of Coral Reefs*. New York: Chapman & Hall, 1997.

45 Paulay G. Circulating theories of coral biogeography. *J Biogeogr* 1996; 23: 279–282.

46 Kohn AJ. Why are coral reef communities so diverse? In: Ormond RFG, Gage JD, Angel MV, eds. *Marine Biodiversity Patterns and Processes*, pp. 201–215. Cambridge: Cambridge University Press, 1997.

47 Barber PH, Palumbi SR, Erdmann MV, Moosa MK. A marine Wallace's line? *Nature, Lond* 2000; 406: 692–693.

48 Paulay G. Diversity and distribution of reef organisms. In: Birkeland C, ed. *Life and Death of Coral Reefs*, pp. 298–353. New York: Chapman & Hall, 1997.

49 Jokiel P, Martinelli FJ. The vortex model of coral reef biogeography. *J Biogeogr* 1992; 19: 449–458.

50 Bellwood DR, Hughes TP, Connolly SR, Tanner J. Environmental and geometric constraints on Indo-Pacific coral reef diversity. *Ecology Letters* 2005; 8: 643–651.

51 Mellin C, Bradshaw CJA, Meekan MG, Caley MJ. Environmental and spatial predictors of species richness and abundance in coral reef fishes. *Global Ecol. Biogeogr.* 2010; 19: 212–222.

52 Meyer CP, Geller JB, Paulay G. Fine scale endemism on coral reefs: archipelagic differentiation in turbinid gastropods. *Evolution* 2005; 59: 113–125.

53 Renema W. et al. Hopping hotspots: global shifts in marine biodiversity. *Science* 2008; 321: 654–657.

54 Levitus S, Conkright ME, Reid JL, Najjar RG, Mantyla A. Distribution of nitrate, phosphate and silicate in the world oceans. *Prog Oceanogr* 1993; 31: 245–274.

55 Berger WH. Global maps of ocean productivity. In: Berger WH, Smetacek VS, Wefer G (eds.) *Productivity of the Ocean: Past and Present*, pp. 429–455. London: Wiley, 1989.

56 Scheltema RS. Planktonic and non-planktonic development among prosobranch gastropods and its relationships to the geographic ranges of species. In: Rylands JS, Tyler PA, eds. *Reproduction, Genetics and Distribution of Marine Organisms*. 23rd European Marine Biology Symposium, Fredensburg, 1989. Denmark: Olsen & Olsen, 1989.

57 Thorson G. Reproductive and larval ecology of marine bottom invertebrates. *Biol Rev* 1950; 25: 1–45.

58 van der Spoel S, Heyman RP. *A Comparative Atlas of Zooplankton: Biological Patterns in the Oceans*. Utrecht: Bunge, 1983.

59 Rosen BR. Reef coral biogeography and climate through the late Cainozoic: just islands in the sun or a critical pattern of islands? *Special Issue, Geological Journal* 1984; 11: 201–262.

60 Bellwood DR, Meyer CP. Searching for heat in a marine hotspot. *J Biogeogr* 2009; 36: 569–576.

61 McAllister DE, Schueler FW, Roberts CM, Hawkins JP. Mapping and GIS analysis of the global distribution of coral reef fishes on an equal-area grid. In: Miller R, ed. *Advances in Mapping the Diversity of Nature*, pp. 155–175, London: Chapman & Hall, 1994.

Living in the Past

Chapter 10

The past is a foreign country; they do things differently there.
(*The Go-between*, by L. P. Hartley)

This chapter explains how the very different geographies, climates, faunas, and floras of our planet gradually changed, over the last 400 mya, into those that we see today. At first there was a pattern of separate continents, which later united into a single Pangaea landmass, followed by renewed fragmentation and some collisions. The early history of mammals and flowering plants is described, together with the changing climates and floras from the Mid-Cretaceous up to the dramatic cooling at the end of the Eocene.

It is interesting and instructive to attempt to explain the patterns of distribution of life by using the ecological approach that has been explained in Chapters 2–4. But even the most superficial analysis of the distributions of the different groups of animals and plants shows that they are quite differently distributed among the continents, and biogeographers also want to understand how this has come about. This historical approach to continental biogeography is the subject of the next two chapters.

Although very few groups have precisely the same pattern of geographical distribution, there are some zones that mark the limits of distribution of many groups. This is because these zones are barrier regions, where conditions are so inhospitable to most organisms that few of them can live there. For terrestrial animals, any stretch of sea or ocean proves to be a barrier of this kind—except for flying animals, whose distribution is for this reason obviously wider than that of solely terrestrial forms. Extremes of temperature, such as exist in deserts or in high mountains, constitute similar (though less effective) barriers to the spread of plants and animals.

These three types of barrier—oceans, mountain chains, and large deserts—therefore provide the major discontinuities in the patterns of the spread of organisms around the world. Oceans completely surround Australia. They also virtually isolate South America and North America from each other and completely separate them from other continents. Seas, and the extensive deserts of North Africa and the Middle East, effectively isolate Africa from Eurasia. India and Southeast Asia are similarly isolated from the rest of Asia by the vast, high Tibetan Plateau, of which the Himalayas are the southern fringe, together with the Asian deserts that lie to the north. Each of these land areas, together with any nearby islands to which its fauna or flora has been able to spread, is therefore comparatively isolated. It is not surprising to find that today's patterns of distribution of both the faunas (faunal provinces or zoogeographical regions) and the floras (floral regions) largely reflect this pattern of geographical barriers.

Before the detailed composition of these faunal provinces and floral regions can be understood fully, it is first necessary to follow the ways in which today's patterns of geography, climate, and distribution of life came into existence. From what has been discussed in earlier chapters, it is clear that the differences between the faunas and floras of different areas might be due to a number of factors. First, any new group of organisms will appear first in one particular area. If it competes with another, previously established group in that area, the expansion in the range of distribution of the new group may be accompanied by contraction in that of the old. However, once it has spread to the limits of its province or region, whether or not it is able to spread into the next will depend initially on whether it is able to surmount the geographical barrier between them, or to adapt to the different climatic conditions to be found there. (Even if it is able to cross to the next province or region, however, it may be unable to establish itself because of the presence there of another group that is better adapted to that particular environment.) Of course, changes in the climate or geographical pattern could lead to changes in the patterns of distribution of life. For example, gradual climatic changes, affecting the whole world, could cause the gradual northward or southward migrations of floras and faunas because these extended into newly favorable areas and died out in areas where the climate was no longer hospitable. Similarly, the possibilities of migration between different areas could change if vital links between them became broken by the appearance of new barriers, or if new links appeared.

Early Land Life on the Moving Continents

Our understanding of the movements of the continents and of the timing of the different episodes of continental fragmentation or union is now fairly detailed, and the distribution of fossil organisms correlates very well with the varying patterns of land. The earliest time at which there is enough evidence to discern patterns of life is the Early Devonian, 380 million years ago (Fig. 10.1), by which time separate floras and fish faunas can be distinguished in the northerly placed Siberian continent, the equatorially placed Euramerican continent, and the eastern region of Gondwana [1,2]. Early amphibians are found in near-equatorial positions in both Euramerica and Australia [3,4], where the climate would have encouraged a rich growth both of plants and of the invertebrates that the earliest terrestrial vertebrates would have fed upon. The fossil record suggests that all the early amphibian and reptile groups evolved in the continent called Euramerica, where they were largely confined to a narrow warm, humid equatorial zone, bordered by dry subtropical belts, until that continent collided with others in the Late Carboniferous [5].

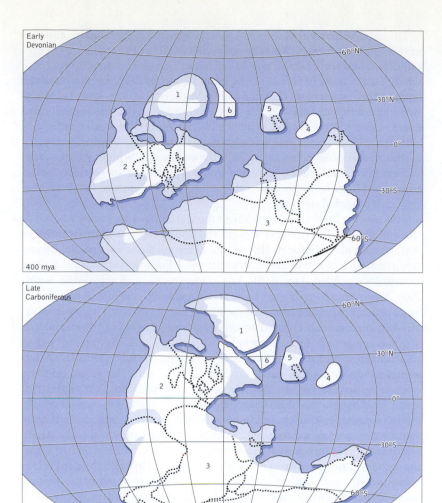

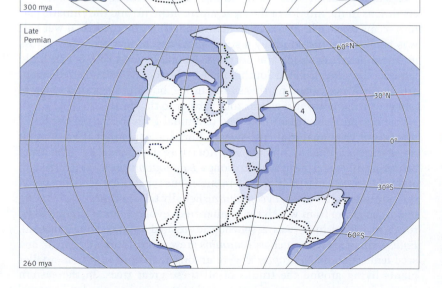

Fig. 10.1 World geography at three different stages in the past. Tripel-Winkel projection. Dark tint indicates ocean; outlines of epicontinental seas (light shading) after Smith et al. [40]. Dotted lines indicate modern coastlines. Continental positions after Metcalfe [5]. 1, Siberia; 2, Euramerica; 3, Gondwana; 4, southern China; 5, northern China; 6, Kazakhstan.

The great expansion of the land plants began in the Devonian, and this may well itself have had an effect on the world's climate. Today, vegetated surfaces decrease the **albedo** (or reflectivity) of an area by 10–15%, while plants recycle much of the rainfall (up to 50% in the Amazon Basin). Theory suggests that the photosynthetic activities of the plants would have reduced the carbon dioxide content of the atmosphere, and there is also both botanical evidence [6] and geological evidence [7] that this took place. This would have led to an "ice-house" effect (the reverse of the greenhouse effect of higher temperatures, which results from *increased* carbon dioxide). This enormous buildup of the world's vegetation might have been the cause of the next significant event, which was the onset of global cooling.

This cooling began in the middle of the Carboniferous Period and led to the appearance of ice sheets around the South Pole, similar to those of Antarctica today. Through this period of time, the whole of Gondwana was rotating clockwise and moving toward Euramerica, with which it eventually collided (Fig. 10.2). As Gondwana moved across the South Pole, the glaciated area moved across its surface. Although the whole South Polar area must have been icy cold, the proportion that was actually glaciated probably varied according to the position of the Pole itself. When this was near to the edge of the supercontinent, the adjacent ocean would have provided enough moisture to create the heavy snowfalls that would have formed extensive continental ice sheets. But when the Pole lay further inland, away from the ocean, it is possible that the inland areas, though cold, would not have been heavily glaciated.

This polar glaciation caused the latitudinal ranges of the Carboniferous floras to be compressed toward the equator. In the equatorial region there was a great swampy tropical rainforest, rather like that of the Amazon Basin today. This lay across Euramerica and was fed by rains from the warm westward equatorial ocean current that would have washed the eastern shore of that continent. Surprisingly, the distribution of different types of rock laid down at this time shows no sign of the presence of monsoonal, dry conditions in the interior of the great supercontinent, and the climatic pattern seems to have been essentially latitudinal [8]. The equatorial wet belt was bordered to the south by a subtropical desert that stretched across northern South America and northern Africa; beyond this lay the cold lands around the glaciated South Pole. Another desert belt covered northern North America and northeastern Europe, but Siberia (still a separate island continent) lay in a wetter zone further to the north.

The absence of dormant buds and of annual growth rings in the fossil remains of the equatorial coal-swamp flora indicates that it grew in an unvarying, seasonless climate. The flora was dominated by great trees belonging to several quite distinct groups. *Lepidodendron*, 40 m (130 ft) tall, and *Sigillaria*, 30 m (100 ft) tall, were enormous types of lycopod (related to the tiny living club-moss, *Lycopodium*). Equally tall *Cordaites* was a member of the group from which the conifer trees evolved, and *Calamites*, up to 15 m (50 ft) high, was a sphenopsid, related to the living horsetail, *Equisetum*. Tree ferns such as *Psaronius* grew up to 12 m (40 ft) high, and seed ferns such as *Neuropteris* were among the most common smaller plants living around the trunks of all these great trees. In the eastern United States, and in parts of Britain and central Europe, the land covered

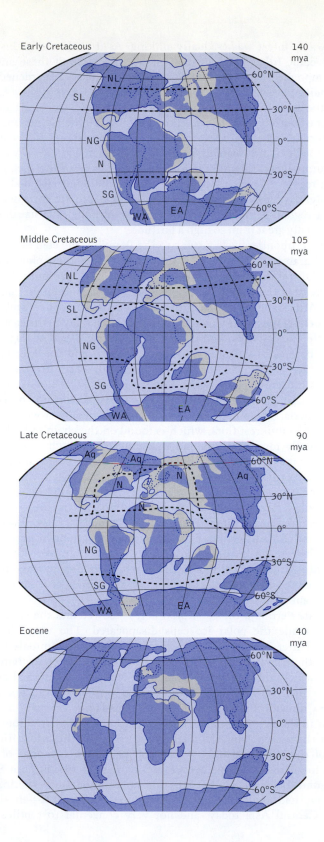

Fig. 10.2 World geography at four different stages in the past. Tripel-Winkel projection; outlines of epicontinental seas (light shading) after Smith et al. [40]. Dotted lines indicate modern coastlines. Continental positions after Cambridge Palaeomap Services [41]. Floral provinces after Vakhrameev [42]. Aq, *Aquilapollenites*; EA, East Antarctica; N, Normapolles; NG, Northern Gondwanan; NL, Northern Laurasian; SG, Southern Gondwanan; SL, Southern Laurasian; WA, West Antarctica.

by this swamp forest was gradually sinking. As it sank, the basins that formed became filled with the accumulated remains of these ancient trees. Compressed by the overlying sediments, dried and hardened, the plant remains have become the coal deposits of these regions.

Far to the south of the equatorial coal-swamp flora, the lands around the growing South Polar Ice Sheets bore a different flora, lacking many of the northern trees and with fewer ferns and seed ferns. It is known as the *Glossopteris* flora, after a genus of woody shrub that is found only in that region. Surprisingly, this flora is found as close as 5° to the then South Pole, in a region that must have been not only cold but also had a very short winter day length. The very marked seasonal growth rings of plants from these regions are therefore not surprising, although the size of these is unexpectedly great from so high a latitude.

After the Carboniferous gave way to the Permian, the coal swamps of southern Euramerica disappeared and deserts lay in their place by the Late Permian. This was partly because these regions had moved northward, away from the equator, and partly because the mountain ranges of northern Africa and eastern North America had extended and risen higher, blocking the moist winds from the ocean that lay to the east.

Four different Early Permian floras can be distinguished [9] (Fig. 10.1). In the north, the still-separate continents of Siberia and Kazakhstan bore the temperate "Angara" flora, with *Cordaites*-like conifers, and also herbaceous horsetail plants, ferns, and seed ferns. The flora is richest toward the eastern coast and becomes less diverse toward the colder north of Siberia. The second flora is the rich, varied, ever-wet tropical "Cathaysian" rainforest flora, which is found on the chain of land areas that eventually formed China and Indochina and which at that time extended out into the Pacific. This flora was made up of sphenopsid and *Lepidodendron* trees, *Gigantopteris* lianas, and many types of seed fern; conifers and *Cordaites* were rare. The old coal swamps of southern Euramerica had disappeared, and the rest of Euramerica bore a seasonal tropical "Euramerican" flora. Finally, the Gondwana flora, descended from the earlier Late Carboniferous *Glossopteris* flora, occupied the whole of that great supercontinent. It was almost certainly divided up into subsidiary floras according to the varying climates of the different parts of that great region, but as yet too little is known of it for us to be able to recognize these.

In the Late Permian, the equatorial belt narrowed and the subtropics expanded, while the polar ice caps disappeared. The world became warmer and drier, causing changes in both the plants and animals as they adapted to these new environmental conditions, new groups evolving and spreading widely through the world, while many of the original groups became extinct. As one might expect, land vertebrates did not reach Siberia or China until after those landmasses had joined the world supercontinent in the Late Permian. Then, as in the Late Carboniferous, evidence for the presence of arid areas in the center of the midlatitudinal regions of the supercontinent is lacking. In fact, rich faunas of Late Permian fossil reptiles have been found in regions of southern South America and Africa that, according to climatic modeling experiments, would have had annual temperature changes of 40–50°C—similar to those of Central Asia today, and not very congenial to reptiles [10].

However, many of the rivers drained internally into great lakes, whose waters would have had a cooling, stabilizing effect on the climate.

One World—for a While

The coalescence of the different continental fragments to form Pangaea led to climatic changes. In general, the world became steadily warmer and drier during the Triassic. The disappearance of the oceans that had previously separated the continents, and the formation of the enormous supercontinent of Gondwana, had left vast tracts of land now far from the oceans and the moist winds that originated there. Furthermore, the new, lofty mountain ranges that still marked the regions where Euramerica, Siberia, and China had collided provided physical and climatic barriers to the dispersal of their floras and faunas.

In the floras, we can identify a general evolutionary change in which older types of tree, such as those belonging to the lycopods and sphenopsids and *Calamites* disappeared. They were replaced either by the radiation of existing types of tree, such as caytonias and ginkgos, or by new groups such as cycads and bennettitaleans. A new type of fern, the osmundas, also appeared. This floral change was complete by the end of the Triassic Period. However, there were also local variations; for example, in the Gondwana flora, the Early Triassic *Glossopteris* was replaced by the seed fern *Dicroidium*.

In the Jurassic and Early Cretaceous, floras seem to have gradually become more similar to one another, approaching the modern pattern in which there are gradual latitudinal changes governed by climate, the patterns of dominance of different groups changing from lower latitudes to higher latitudes. So, two floral provinces can be distinguished in the northern landmass Laurasia during this period of time [11] (Fig. 10.2). The Northern Laurasian (or Siberian-Canadian) province had a moderate, warm, seasonal climate with lower floral diversity and more deciduous plants such as conifers, while the Southern Laurasian (or Indo-European) subtropical province does not show these features. The southern boundary between these two floral provinces shifted northward between the Late Jurassic and the Early Cretaceous. This seems to have been caused by a climatic change, the southern part of Asia becoming increasingly hot and arid; this is sometimes quoted as evidence for a general increase in global warmth. However, there is no evidence for such an increase elsewhere, and it seems instead to reflect a more local phenomenon. It may therefore instead be related to changes in the temperatures of the ocean currents adjacent to southern Asia, and therefore to the rainfall of this area.

These Asian Mesozoic floras extended to high (about 70°) northern and southern latitudes, to areas that, though clearly warm, must have had seasonal very brief periods of daylight [12]. On the other hand, oxygen-isotope measurements of the composition of the fossil shells of marine Cretaceous plankton (see p. 371) show that the intermediate to deep waters of these oceans were 1.5°C warmer than those of today. On land, the presence and spread of plants, dinosaurs, and early mammals through high-latitude routes such as the Bering region and Antarctica support these observations.

To examine the biogeographical history of the vertebrate land animals (amphibians and reptiles) of Pangaea, one must return to the Permian and Triassic. By the middle of the Permian, these animals appear to have been able to disperse through regions of different climate, for Pangaea soon came to contain a fairly uniform fauna, with little sign of distinct faunal regions [13]. Great changes took place in this worldwide fauna during the Triassic [14]. The bulk of the Permian faunas had been made up of mammal-like reptiles and other older types of reptile, but these disappeared during the Early and Middle Triassic. They were at first replaced by a radiation of early reptiles known as archosaurs. However, these in turn were soon replaced (in the Late Triassic) by their own descendants, the dinosaurs, which came to dominate the world throughout the Jurassic and Cretaceous.

Those dinosaur groups that appeared in the Triassic, Jurassic, and perhaps the earliest Cretaceous were able to spread throughout Pangaea; examples are the carnivorous ceratosauroids and allosauroids and the great quadrupedal herbivorous titanosaurs. However, South America gradually broke away from North America in the early Cretaceous, severing the link between the Northern and Southern Hemisphere continents. Pangaea then gradually became divided into separate continents. As a result, the extent to which the groups that evolved later were able to spread differs according to the time at which each appeared [15, 16]. Those that had evolved earlier, before the subdivisions of the land, were able to spread more widely than those that appeared later.

The different landmasses of the Northern Hemisphere became interconnected in varying patterns throughout the last 180 mya (cf. Fig. 5.7, p. 160). At the beginning of this period of time, in the Middle Jurassic, all these areas, from western North America through Eurasia to eastern Siberia, were connected to form a single landmass, Laurasia (Fig. 5.7A; Plate 3b), though the two ends of this chain were separated by the equivalent of the Bering Strait. This allowed the Triassic and Early Jurassic dinosaurs to range throughout the whole northern area that, in this ice-free world, probably had a fairly uniform climate and flora; they were also able to spread into Gondwana. As a result, most of the dinosaur groups that evolved in the Jurassic and Early Cretaceous (ostrich dinosaurs, dome-headed dinosaurs, dromaeosaurs, primitive duck-billed dinosaurs, and larger carnivorous tyrannosauroids) were able to disperse throughout the undivided Northern Hemisphere. There, they were able to replace older groups such as the ceratosauroids, allosauroids, and titanosaurs. But because Gondwana was now separate, they were unable to reach that southern landmass, and the older groups were therefore able to survive there.

By the beginning of the Late Cretaceous, 90–80 mya, the shallow Mid-Continental Seaway in North America and Turgai Sea in Eurasia had subdivided each of these continents into western and eastern sections, but Siberia and Alaska had become connected to one another. The result was the establishment of two continents, Euramerica and Asiamerica (Fig. 5.7B; Plate 3c). The distribution of dinosaur groups that evolved in the Late Cretaceous reflects this new geography. For example, the ancestor of the rhinoceros-like herbivorous ceratopsian dinosaurs evolved in Asia in the Late Cretaceous, when that area was isolated (Fig. 10.2). The group later spread to North America when Asia became linked to it, and more advanced ceratopsians evolved there toward the end of the Cretaceous. However, these did not spread back to Asia, which suggests that a marine barrier had by then separated the two continents.

Our knowledge of the dinosaurs of the Southern Hemisphere is more limited than that of the Northern Hemisphere (though much is currently being discovered both there and in China). We know that sauropods and more advanced duck-billed dinosaurs were present in the Late Cretaceous of South America. It is possible that their ancestors had reached South America earlier in the Cretaceous, from Eurasia via Africa before Africa split from South America, as has been suggested for early placental mammals. Though it seems likely that a chain of volcanic islands lay between North and South America during the Cretaceous (cf. Fig. 11.13), it seems unlikely that animals the size of dinosaurs could have dispersed via this route, though the chain might have formed an adequate filter route for the dispersal of placental mammals from North America to South America.

Whatever may have been the precise nature of that link, the environment of the whole region between the two Americas must have been transformed and devastated by the results of the great meteor that struck just off the coast of northern Yucatan in northern Mexico 64.5 mya. The meteor was at least 10 km (6 mi) in diameter and caused a crater 200–300 km (125–190 mi) in diameter. The impact caused a fireball many hundreds of kilometers across, and burning hot winds must have swept around the world. This, together with the red-hot ejecta from the crater, would have started forest fires in many areas, and the thickness of the resulting soot layer that is still visible in the geological record suggests that 90% of the world's forests may have burned. The impact also threw 1,000–4,000 km^3 (400–1600 mi^3) of heated limestone rock into the atmosphere. The combination of this and the smoke from the forest fires would have caused an initial darkening of the world's skies and a temperature drop of about 10°C for about six months. This would have been followed by a greenhouse effect temperature rise of 3–10°C, caused by the carbon dioxide and sulphur dioxide released by the heated carbonate rock, which would have lasted several tens of years. All of this would have profound and complicated effects on the world's weather [17].

The most obvious biological result of these changes was the total extinction of the dinosaurs, but even the marine biota was profoundly affected, with the extinction of the great marine reptiles (plesiosaurs and ichthyosaurs), the ammonites and many groups of marine plankton. The effects on the world's flora are described in a later section of this chapter. These changes in the world's biota were recognized long ago, and led to the establishment of the geological boundary between the Cretaceous and the Tertiary (or Cenozoic). It is therefore known as the Cretaceous/Tertiary Boundary Event.

The Biogeography of the Earliest Mammals

The biogeographical history of the early mammals poses some interesting problems [18]. As noted in Box 10.1, the earliest records of both marsupials and placentals are from the Early Cretaceous of Asia. In Eurasia, the placentals were dominant throughout the Cretaceous. In

How to Evaluate the Fossil Record: The Early History of the Marsupial and Placental Mammals

The first, most primitive, mammals appeared in the Triassic Period, not long after the first dinosaurs, but they almost certainly laid eggs, as do the living monotremes (the platypus and spiny anteater of Australia). The modern mammals are divided into two major groups, the marsupials and the placentals. In the marsupial type of mammal the young leave the uterus at a very early stage and then complete their development in the mother's pouch, whereas in the placental mammals the whole period of embryonic development takes place in the uterus. Although the placentals spread to all parts of the world except Australia, the marsupials only succeeded in surviving in the Southern Hemisphere. Unfortunately, our understanding of the distribution patterns of these two groups is hampered by our incomplete knowledge of the timing of some crucial plate tectonic events and of the composition of some early faunas.

The history of the two groups can only be followed from their fossil record, so it is necessary at this point to emphasize what can and cannot be deduced from the presence of a particular fossil at a particular time. To take, for example, the history of the placentals, we can see in the fossil record that their diversity gradually diminishes as we pass back in time through the early Cenozoic and into the Late Cretaceous (Fig. 10.3). This suggests that their latest common ancestor (P in Fig. 10.4) lived in the Late Cretaceous, a little later than the latest common ancestor of the modern marsupials (M in Fig. 10.4). The common ancestor of any group, plus all of its descendants, are known as the "crown group." Thus the mainly Cenozoic radiations of the modern marsupials and of the placentals are their crown groups. Members of those groups that lived earlier in time are known as its "stem group." Members of the stem group of the placentals, for example, would have had at least some placental characteristics, but we cannot be sure which, for it is unlikely that all of them evolved

simultaneously. Instead, these features would have evolved as individual characters, or character-complexes, over millions of years.

We can be confident that any feature that we find in all the living placental groups was also present in their common ancestor; otherwise we should have to put forward the unlikely hypothesis that these features had evolved independently in each lineage. The fossil record also allows us to ascertain how characteristics of the skeleton and teeth of the placentals gradually evolved during their earlier history. But this record does not give us information on other features, such as those of their physiology or reproduction. So, though we can be sure that the common ancestor of all the placentals had the placental method of reproduction, we cannot tell how and when this evolved from the marsupial method. We therefore cannot be sure that any fossil mammal that is not part of the crown group of placentals had the placental method of reproduction; it might still have had the marsupial method.

The evolutionary histories of the stem groups of the marsupials and placentals converge back in time to the common ancestor of all the modern mammals (A in Fig. 10.4), which we can identify from such features as its dentition, but which must also have had the marsupial method of development. This animal must have lived over 125 million years ago, for Early Cretaceous rocks of that age contain the earliest known marsupial (*Sinodelphys*) and the earliest known placental (*Eomaia*) [19]. Though other fossil mammals are known, which lived during the Jurassic or Cretaceous, they were probably egg-layers, like their modern Australian descendants, the platypus and spiny anteater. Molecular work (see below) suggests that the ancestors of these two modern mammals diverged from the marsupial and placental lineage 166 mya, in the Early Jurassic.

North America, however, the marsupials radiated in the Late Cretaceous; placentals were absent from that continent for most of that period of time and did not attain appreciable diversity there until the last 10 million years of the Cretaceous. By the Eocene, 55 million years ago, marsupials had arrived in both Europe and Africa, but later became extinct in both the Northern Hemisphere and Africa. Neither marsupials nor placentals were able to reach India or Madagascar in the Cretaceous, for those two areas of land, still joined together, had separated from the rest of Gondwana in the Jurassic.

There are huge gaps in our understanding of the biogeographical history of mammals in South America, Antarctica, and Australia. (Both West and East Antarctica had warm climates at this time, with forests extending

close to the South Pole.) Both marsupials and placentals are known
from the Late Cretaceous of South
America, and are also found in West
Antarctica (in the Antarctic Peninsula,
which stretches up to South America),
which was still connected to it at
that time (see also Fig. 10.2). But we
have no fossil record of the mammals of the much larger continent of
East Antarctica, nor of those of the
Early Tertiary of Australia. As a
result, in many cases we do not
know when the ancestors of the later
groups that we recognize evolved or
diversified, nor do we know their
patterns of dispersal. Recent genetic
work [20] suggests that there were
two dispersals of South American
marsupials across Antarctica to
Australia, the peramelids (bandi-
coots) being related to the South
American *Caenolestes*, while the remainder of the Australian forms (now
known as the Eometatheria) are related to the little South American
Dromiciops. (Alternatively, perhaps *Dromiciops* was the result of a dispersal in the opposite direction). The marsupials of the two continents
were later widely separated by the extinction of mammals in Antarctica,
caused by its glaciation. When we eventually find diverse Early Tertiary
marsupials in Antarctica, it will be interesting to see whether the continent contains two different faunas (e.g., South American forms in
Western Antarctica, and Australian forms in Eastern Antarctica), separated by a geographic barrier that the early marsupials had to cross and
that we at present cannot identify. Such a barrier might also have been
responsible for the absence of placentals from Australia: there is no trace
of them alongside the varied marsupials that we know from there from
the Oligocene onward. (Because it is normally the placentals, rather then
the marsupials, that have succeeded whenever the two groups have been
in competition in other parts of the world, most biogeographers have
tended to assume that placentals are absent from Australia in the Tertiary
because they never reached that continent.)

Figure 10.5 illustrates an important current problem in interpreting
the fossil record of the mammals that, as we shall see, has implications
for the interpretation of their biogeography. Quite consistently, the
molecular work suggests that the different groups diverged from one
another at a very much earlier date than the dates of their earliest fossil
members would suggest. If this is really the case, then the early history
of all these groups has left no trace in the fossil record. It is unlikely, of
course, that the earliest *known* fossil of any lineage is in fact the truly
first one, so there are doubtless many earlier fossils awaiting discovery.
Nevertheless, the continuing process of exploration and discovery of
fossils (particularly in China and the Southern Hemisphere) has now

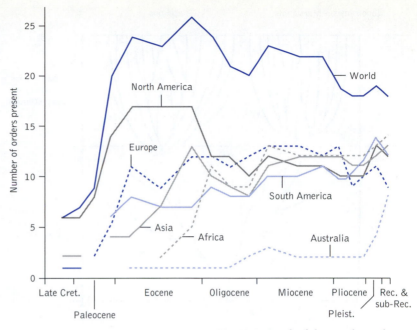

Fig. 10.3 Graph of the numbers of
mammalian orders through time for
the world as a whole and for each
continent. Pleist: Pleistocene; Rec:
Recent. After Lillegraven [43].

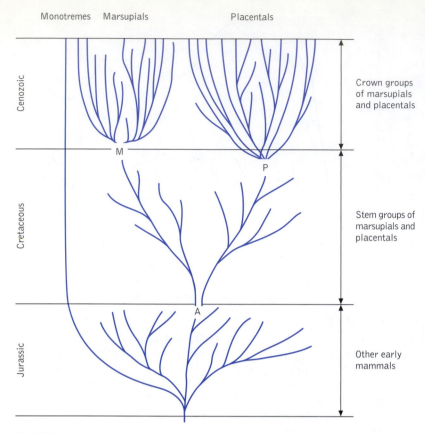

Monotremes Marsupials Placentals

Cenozoic

Cretaceous

Jurassic

M

P

A

Crown groups
of marsupials
and placentals

Stem groups of
marsupials and
placentals

Other early
mammals

Fig. 10.4 The history of the mammals through time. See text for explanation.

made this a very unlikely explanation. It is also inconceivable that so many lineages, involving so many quite different types of mammal, with quite different ways of life, would leave no record of their presence over such long periods of time and in every continent of the world. Furthermore, it is far more plausible that the meteor strike at the Cretaceous/Tertiary boundary (or K/T boundary), with the widespread extinctions and environmental changes that resulted, provided the ecological opportunity for the radiation of other groups, than that these events have left little trace in the evolutionary history of these groups.

A similar discrepancy between the fossil record and the divergence times suggested by the molecular work is also found in work on fossil plants and birds, suggesting that any cause is a general one, rather than linked to the nature of a particular group. A possible solution is the suggestion that work based on the rate of change of molecular clocks does not take into account the possibility that these may speed up very greatly during a period of rapid adaptive radiation. Such a period is likely to take place when the members of the crown group start to radiate from one another, for rapid rates of change in the organisms may be reflected by increased rates of molecular change. Any underestimates of the rate will also result in larger and larger total errors as the timescale increases.

Another possible explanation for the widespread discrepancies between the molecular and the fossil datings of the times of divergence arises from the concept of the stem group versus the crown group. It is quite possible, and even likely, that the earliest members of any lineage that later became successful and diverse, such as the different orders of mammal, had few of the specializations that later led to their success. We must also remember that not all of these specializations will have been identifiable in the skeletal remains that are their only representation in the fossil record. So these early members may have been difficult to identify and rare until ecological changes such as the end-Cretaceous meteor strike provided them with opportunities for rapid capitalization of their latent advantages, which only then would become apparent in the fossil record.

The discrepancy between the apparent dates of divergence of the different groups of mammal based on fossil data and those suggested by molecular

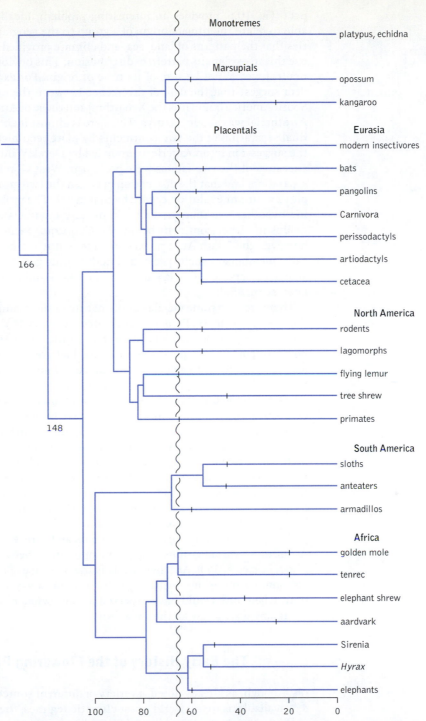

Fig. 10.5 The timescale of mammalian evolution and its relationship to biogeography. The cladistic diagram shows the relationships between the different groups. They are joined together at the times suggested by molecular studies; the short vertical bar on each lineage indicates the age of the oldest known fossil of that group. NB: The timescale is only linear back to 100 my. The vertical wavy line at 65 my indicates the date of the K/T boundary. [Based on information and figures in refs. 18, 21, 44, 45.]

data (Fig. 10.5) provides an interesting problem. Ideally, one should be able to relate the time of origin of a group to the geographical opportunities that the patterns of land, sea, and climate provided at that time, and use this to explain its pattern of distribution. This obviously becomes difficult if we cannot be sure of its time of origin. For example, molecular data suggest that the date of divergence between the early placentals of South America (known as the Xenarthra) and those of Africa (known as the Afrotheria) was about 100 mya. This date is almost identical to that of the final separation of the two continents by plate tectonics. This has led to the suggestion by an American group led by Derek Wildman [21] that the ancestors of the two faunas were present in West Gondwana before that separation, and that the distinction between the two groups was an example of vicariant evolution caused by that event. The problem in this inviting hypothesis is that we do not know any members of either of these groups in either continent before the Cenozoic, 65 mya. By that time, however, the South Atlantic was so wide that it seems unlikely that the early members of such specialized herbivorous groups would have been able to disperse across it (though both monkeys and rodents did so a little later; see p. 358!).

There are also some puzzles in the nature of the mammal faunas of the different continents. Perhaps the greatest mystery is the restricted ecological diversity of the early placental families of Africa. Those of the northern continental mass seem to have had great ecological potential, for they radiated into a very diverse range of families that fed on leaves, fruit, seeds, and nuts, and invertebrates, as well as evolving into carnivores and hoofed herbivores (and bats and whales). Hardly any of these diverse ecological opportunities seem to have been taken by the early placentals of Africa, as can be seen by merely noting those that were occupied by the families of African origin shown in Fig. 10.5. These vacant niches in Africa were only filled later, by the arrival of other families from Europe about 50 mya (see p. 342). Similarly, in South America they were filled by a varied radiation of two groups. Some of them were derived from over 20 families of hoofed herbivorous placental whose ancestors had arrived there from North America, while others, which were marsupials, evolved from the arboreal opossums and carnivorous borhyaenids. All of these varied groups, except the opossums, became extinct in the Pliocene (see p. 360). Another puzzle is the presence of a 50-million-year-old South American anteater in Germany, far away in time and space from where one would have expected it. (But science would cease to be intriguing if we knew all the answers!)

The Early History of the Flowering Plants

Although, as already noted, a variety of different sources of evidence are available as to the world's past climatic regimes, the fossilized leaves of flowering plants provide an additional source after they became common and diverse, in the middle Cretaceous. In areas of high mean annual temperature and rainfall, the leaves have "entire" margins, not subdivided into lobes or teeth; they are large and leathery, and are often heart-shaped, with tapering, pointed tips, and a joint at the base of the leaf. These features are less common in floras from areas of low mean annual temperature and rainfall. Further information can be gained from the climate preferred by the living representatives of the families of plant found in each flora [22]. However, it would be unwise to assume that living species could never have existed beyond the range of environmental conditions within which they are found today. They may in the past have lived in environments from which they later became excluded by competitive interaction with other, newly evolved, types. In any case, many of the Cretaceous and Early Eocene flowering plants belonged to groups that later became extinct or that were only transitional to modern genera, so that the climatic implications of their presence are unclear. Finally, the fossil data are often biased toward near-coastal areas of deposition, whose climates are usually milder and less seasonal than those of more inland areas.

As noted earlier (p. 103), botanists analyzing the floras of the world can distinguish biomes (such as forest, woodland, prairie/savanna, etc.) that are alike in their climate and soil, each of which bears a community of plants that are similar in their structure and appearance. Any given biome is characteristic of its own particular preferred climate and soil, and their distribution through time can therefore be a useful indication of the distribution of climates. Most can be recognized in each of the continents, though the precise families of plant that are found in them may be quite different, so that the constitution of each biome may vary from place to place. For example, the five major rainforest biomes of today (in South America, Africa, Madagascar, Southeast Asia, and New Guinea) are quite distinct from one another in their detailed taxonomic composition. The fossil record shows not only that different combinations of ecologically compatible species can exist, but also that there were some biomes quite unlike any found today, such as the Eocene flora of the London Clay (see below, p. 323) or the Oligocene semideciduous tropical dry forest of southeastern North America (p. 325).

The climatic preferences of plants are used in the botanists' recognition of three categories: **megatherms** and **mesotherms** prefer, respectively, mean annual temperatures of above 20°C and between 20°C and 13°C, while **microtherms** prefer mean annual temperatures of below 13°C. The luxuriant vegetation of the megathermal tropical rainforest can be clearly recognized in the fossil record, and the history of this biome and of its typical families has been clearly explained in the book by the British palaeobotanist Robert Morley [23]. This information and his maps (Fig. 10.8) provide a great deal of information on the effects of continental movement and climatic change on the changing floras, as described below.

Sparse remains of angiosperms are found in the late Jurassic and very early Cretaceous. The first may have been early successional herbs or shrubs in areas recently disturbed by erosion and deposition, only later diversifying ecologically to occupy streamside and aquatic habitats, the forest understory, and early successional thickets [24]. It is only later in the Early Cretaceous, 125 mya (Fig. 10.6), that the fossil record shows a diversity of angiosperm families, found in the tropical regions of South America and Africa, which were still joined together at that time [25–27]. Even at the end of the Cretaceous, although angiosperms formed 60–80% of the low-latitude floras, they comprised only 30–50% of those in high latitudes. The rise of the angiosperms was paralleled by a corresponding

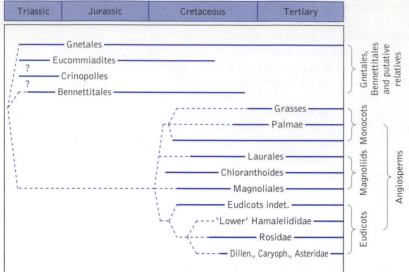

Fig. 10.6 Simplified phylogenetic tree showing the distribution in time of the earliest identifiable groups of flowering plants. After Crane, Friis, & Pedersen [46].

reduction in the numbers and variety of mosses, club mosses, horsetails, ferns, and cycads, but there was less change in the overall diversity of conifers (see Fig. 10.7). Understanding the dispersal of the flowering plants presents fewer problems than understanding that of mammals, for it commenced earlier, before the breakup of the supercontinents had progressed very far.

From the Mid-Cretaceous onward, the Earth's climate changed in two ways: it became cooler, changing from a generally warm world to one with polar ice sheets, and it became more seasonal. The most comprehensively studied floras of the Cretaceous and Cenozoic are those of the North American continent, described by the American palaeobotanist Alan Graham (see Further Reading). The Cretaceous cooling is clearly shown in a series of floras, ranging over 30 my, from about 70°N in Alaska [28]. The earliest contains the remains of a forest dominated by ferns and by gymnosperms such as cycads, ginkgos, and conifers. The nearest living relatives of this flora are found in forests at moderate heights in warm-temperate areas at about 25–30°N. By the time of the last of these middle Cretaceous Alaskan floras, the flora had changed in two ways. First, the angiosperms had by this time diversified to such an extent that they dominated the flora. Second, this forest was similar to that found today at a latitude of 35–40°N—much further north than the living relatives of the earlier flora. These differences suggest that the climate of northern Alaska was already becoming cooler in the middle Cretaceous. By this time, too, latitudinal differences in floras have become apparent, higher latitudes bearing larger-leaved deciduous vegetation, suggesting moist mesothermal forest, while lower latitudes have thicker, smaller-leaved evergreen vegetation, suggesting subhumid megathermal forest. By the end of the Cretaceous, other Alaskan floras show that the mean temperature had dropped by about 5°C and that the diversity of the flowering plants had dropped very greatly [29].

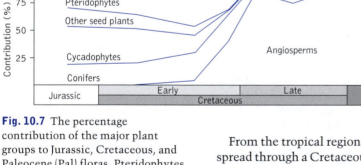

Fig. 10.7 The percentage contribution of the major plant groups to Jurassic, Cretaceous, and Paleocene (Pal) floras. Pteridophytes include ferns and lycopods; cycadophytes include cycads, bennettitaleans, and pinnate-leaved seed ferns. After Crane & Lidgard [26]. Copyright 1989 by the AAAS.

From the tropical regions of South America and Africa, the angiosperms spread through a Cretaceous world in which broadly similar floras existed across wide latitudes in both hemispheres (Fig. 10.8a) [30]. In the Northern Hemisphere there were three such angiosperm floras. At high latitudes lay a broad-leaved deciduous forest, whose conifers included the endemic family Pinaceae. The middle latitudes of western North America bore a broad-leaved evergreen mesothermal forest with ferns and

320   Chapter 10

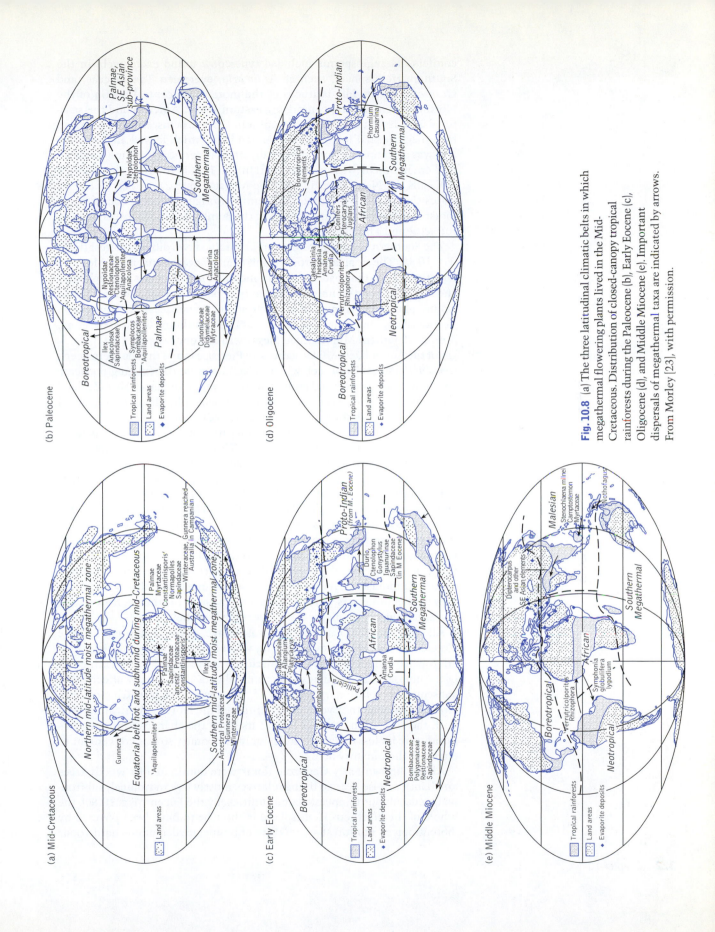

Fig. 10.8 (a) The three latitudinal climatic belts in which megathermal flowering plants lived in the Mid-Cretaceous. Distribution of closed-canopy tropical rainforests during the Paleocene (b), Early Eocene (c), Oligocene (d), and Middle Miocene (e). Important dispersals of megathermal taxa are indicated by arrows. From Morley [23], with permission.

conifers. Its angiosperms included types now found exclusively in the Southern Hemisphere, such as *Araucaria*, evergreen Taxodiaceae, and *Gunnera*. The middle latitudes of the more south-central region of the Northern Hemisphere, comprising eastern North America and the scattered islands that then made up southern Europe, is instead characterized by a megathermal forest flora: this had large, thick-leaved angiosperms and was accompanied by a greater diversity of ferns, suggesting a subhumid environment with year-round rainfall. This is the flora that Morley [30] calls the Northern Mid-Latitude Megathermal Zone (Fig. 10.8a); he states that this was an important centre of origin and radiation of many moist megathermal groups, including the Bombacaceae, Menispermaceae, Rutaceae and Theaceae.

The Southern Hemisphere similarly contained two different forest floras (Fig. 10.8a). The tropical Northern Gondwana megathermal flora, Morley's Equatorial Belt, that had previously included many cycads and ephedras and few ferns, now contained a diversity of pollen of types found today in palms, plus early Gunneraceae, Myrtaceae, Annonaceae, Leguminosae, Restionaceae, and Dipterocarpaceae. The Southern Gondwana flora, Morley's Southern Mid-Latitude Megathermal Zone, was the center of origin of the Aquifoliaceae, Olacaceae, and Proteaceae and the southern beech tree, *Nothofagus* (Fagaceae), which lived alongside the existing podocarp and araucarian conifers and ferns. This flora, very like that of New Zealand today, is not known from southern Africa, which by then was separated from the rest of Gondwana by a wide stretch of ocean.

The three belts of Late Cretaceous megathermal forest were separated by subtropical zones of high pressure. Though there would have been some dispersal between them, evolution and modification within each were mainly independent of events in the others. The result was the appearance of the Northern (or "Boreotropical") and Southern Megathermal Provinces and the equatorial Palmae Province (Fig. 10.8b). They were probably mainly single-story forests, and they were the environment within which the great quadrupedal herbivorous sauropod dinosaurs evolved and lived, together with the huge carnivorous tyrannosauroids that preyed upon them. It was not until the Early Paleocene that tropical closed multi-canopy rainforests developed independently in each of them, while the now extinct dinosaurs were replaced by a diversity of arboreal fruit-eating mammals that could disperse the larger seeds of their trees.

All the world's ecosystems must have been disrupted by the Cretaceous/Tertiary boundary event, when the meteorite strike in northern Mexico caused worldwide climatic changes that led to the extinction of the dinosaurs and of many other animal taxa. This greatly affected the subsequent evolutionary history of the mammals (see Chapter 11). However, though the event also annihilated the North American forests, with the extinction of up to 75% of its flora, the changes to the world's floras were far less extensive and largely temporary, so it is convenient to continue their history here.

All the known Early Cenozoic floras were dominated by woody trees and lianas, whose nearest living relatives mostly live in tropical, subtropical, or deciduous temperate communities. By the Eocene (Fig. 10.8c), the whole of the Northern Hemisphere is thought to have been covered by 'boreotropical' rainforest—a mixture of temperate deciduous hardwoods and evergreen megathermal tropical elements, containing many new plants that are common today, all living in a warm-temperate climate. Forests extended close to the poles in this warm, ice-free world [31]. It is therefore not surprising to find that taxa that are today restricted to particular latitudes were, in the Eocene, found much further from the equator, or existed over a much wider band of latitudes. For example, the diverse fauna of Ellesmere Island in the Canadian Arctic (81°N) included mammals, snakes, lizards, and tortoises [32, 33]. The forests of such high northern latitudes were unlike any modern forests, for they included elements of more southern temperate broad-leaved deciduous forest, as well as characteristically northern needle-leaved conifers such as pine, larch, and spruce. Tropical or subtropical floras existed in northern North America and Eurasia, extending to latitudes of 45–50°N. Most of the Early Cenozoic flowering plants were therefore megatherms or mesotherms.

The two Megathermal Provinces expanded toward their respective poles during the Late Paleocene/Early Eocene, when the world became much warmer—the **Paleocene-Eocene Thermal Maximum** (see Fig. 5.9 a, p. 323), which was caused by a sudden increase in the CO_2 content of the atmosphere, perhaps due to a release of methane from the Atlantic Ocean. This climatic change was accompanied by major changes in both floras and mammalian faunas. In the Northern Megathermal Province this climatic change, plus the appearance of the de Geer route between North America and Europe, (see p. 363) facilitated the exchange of many of the megatherms and mammals between the two continents. However, after the ending of the land bridges between the two continents in the Early Eocene, these two floras and faunas diverged. That of Europe is well known from the fossils of the London Clay, then at a latitude of about 45°N; these fossils contain the seeds and fruits of 350 species of plant, belonging to over 150 genera [34]. The flora includes magnolias, vines, dogwood, laurel, bay, and cinnamon, as well as the palm trees *Nipa* and *Sabal* and the conifer *Sequoia*; there are no growth rings in this flora. Though its closest analogue today would be a subtropical rainforest, its composition is not identical to that of any single modern flora. Some of these plants are now found in the tropics, especially in Southeast Asia, while others live in temperate conditions like those of east-central China today. The accompanying fauna, which includes crocodiles and turtles, is similar to that found today in the tropics. At this time, the dominance of the Palmae in the equatorial province became reduced, but the nature of its vegetation is not yet fully understood. This flora also is unlike anything that we see today. In contrast, the climate of North America through this time became much drier, and its closed-canopy rainforests were replaced by open woodland with savanna patches. The Paleocene-Eocene Thermal Maximum would also have facilitated the exchange of plants and mammals between South America and Australia across Antarctica.

By the Mid-Late Eocene, the floras of tropical South America and West Africa were becoming progressively more different as the climate cooled and the two continents were moving further apart, their floras becoming recognizable as the beginnings of today's South American (or "Neotropical") and African floras. Many of the taxa that these two floras had shared in the

Late Cretaceous became rarer, but the remainder make up what has been called a common "amphi-Atlantic" element. It was also at this time that the Indian flora, with its dipterocarps, dispersed into Southeast Asia, after the two continents collided.

The World Cools—The Terminal Eocene Event

So far in our story of the biogeography of the past, the influence of plate tectonics on the changes has been mainly the direct and obvious one resulting from continental movement—though the spread of epicontinental seas caused by increases in seafloor spreading activity (see pp. 161–162) also had some effect. However, the next change in the biogeographic environment of the world was rather different. Because it took place at the boundary between the Eocene and the Oligocene, 36–37 mya, it is known as the **Terminal Eocene Event** (see Prothero, in Further Reading) (cf. Fig. 5.10a). (The timing of this event at the geological boundary is not a coincidence, as it was the recognition of the differences between the fossil faunas and floras before and after the event that led geologists to give different names to the two periods of time.) This was also the time when both South America and Australia were breaking away from Antarctica. By changing the patterns of circulation of the ocean currents (see p. 349), this may have led to the beginning of the glaciation of Antarctica, in turn causing the sudden and sharp cooling of the world's climate. However, though the separation of Australia from Antarctica did take place at this time, that of South America occurred much later, and it has been suggested that the cause of the Antarctic glaciation was instead a drop in the CO_2 content of the atmosphere [35]. This had been declining during the early Cenozoic from a high level, two to five times its present level, which it reached in the later Cenozoic. This resultant cooling would have gradually reached a point at which snow fell on the mountain chains of East Antarctica, which would have reflected back the sun's heat, this feedback effect leading to a rapid growth of the ice sheets.

Whatever may have been its cause, the climatic change at the end of the Eocene involved both a decrease in the mean annual temperature, and an increase in the mean annual temperature range (i.e., a decrease in equability). For example, within 1 or 2 million years the mean annual temperature of the Pacific Northwest of North America dropped from about 22°C to about 12°C, while the mean annual temperature range increased from about 7°C to nearly 24°C, and annual rainfall almost halved. In this short space of time, the band of broad-leaved evergreen forests shifted southward, from 40–60°N to 20–35°N. They were replaced by temperate broad-leaved deciduous forests and woodlands (Fig. 10.8d). It is not surprising to find that the floral diversity became much lower, for many families must have become extinct, in these areas at least. The degree of cooling at this time was lower in Europe than in North America, and there is less evidence of drying, largely because most of the European floras of that time are known from areas not far from the coasts. One result of this increasing seasonality of the climate was that the old closed-canopy boreotropical forests of North America were replaced by thinner mixed-mesophytic forest with temperate deciduous elements. This provided an evolutionary opportunity for the appearance of new types of nonwoody flowering plant, and therefore of herb-dominated flowering plant communities. As a result, the plant biomes started to become more like those of today, both in their systematics and in their structural composition. For example, the warm-temperate deciduous forests of the lowlands of North America, and the coniferous forests of its highlands were similar to those of today. But there were also exceptions: for example, the tropical rainforest of southeastern North America was replaced by a semideciduous tropical dry forest unlike anything that can be seen in the world today.

Another opportunity arose with the spread of arid environments. In the Late Eocene, the warm seas had easily evaporated to provide moist, rain-bearing winds. As a result, the regional zones of arid and semiarid climate in the subtropics, where today the high-pressure areas of descending dry air produce deserts (see p. 107), had not yet appeared, although there were deserts in the continental interiors. In the later Eocene, the gradually cooling, drier climates had led to the first appearance of such biomes as chaparral–woodland–savanna in North America. After the Terminal Eocene Event, the colder seas provided less rain, and there was a great expansion of arid-land vegetation [36]. In Eurasia, this climatic change is accompanied by a marked change in the mammal faunas, to smaller types such as rodents and rabbits, and the old forest types become extinct [37].

Not surprisingly, many megathermal plants became extinct at this time, as colder climates expanded from the poles toward the equator. Because of sea barriers, the megathermal elements in North America and Europe could not follow the warm climates southward. As a result, most of the megatherms of the Northern Megathermal Province disappeared and were replaced by microthermal vegetation. Though some of the megatherms of Asia were able to disperse southward into southern China, the more southern regions of India and Southeast Asia were affected by changing patterns of atmospheric circulation caused by the rise of the Himalayas; their rainforests were mainly replaced by more seasonal, monsoonal vegetation. Because Australia had drifted northward into warmer latitudes, a few of its megatherms survived along its northeast coast. In contrast, most of the megatherms of southern South America and South Africa disappeared as the climates of these regions cooled. However, the floras of South America and Australia still contained, as they do today, descendants of the old Late Cretaceous Southern Mid-Latitude Megathermal Zone, with such families as the proteas, myrtles, *Nothofagus* and the "southern conifers" such as *Araucaria*, *Podocarpus*, and *Dacrydium*. These forests also covered at least the periphery of Antarctica, but the inland flora of that continent is unknown.

The climatic deterioration that began in the Early Oligocene not only steadily reduced the area occupied by megathermal plants and altered the biota of the two northern continents, but also reduced the variety of plants that could disperse through the Bering link (although the North Pacific was bordered by continuous broad-leaved deciduous forests until at least the Middle Miocene). It led to the expansion of the Northern Hemisphere gymnosperm family Pinaceae—pine, fir, spruce, and larch. It was at one time thought that the climatic cooling caused a wholesale southward movement, through the whole of the Northern Hemisphere, of an "Arcto-Tertiary flora." This was thought to have evolved in the Arctic during the Cretaceous and to have survived until today, little changed, in southeastern North America and east-central Asia. However, this concept has not been supported by later knowledge of the floral history of the Alaskan region. Instead, the Northern Hemisphere angiosperm floras appear to have adapted to the Late Cenozoic climatic change in three ways: by the adaptation of some genera to changed, cooler climates; by the restriction of the range of some genera and their replacement by other already existing genera that preferred a cooler climate; and by the evolution of new genera that preferred those cooler climates. (As already noted, we cannot assume that the structure and composition of biomes of the past are similar to those of today.)

Much of the Late Cenozoic microthermal vegetation of the Northern Hemisphere appears to have evolved there from ancestors within the same area. Since there was little exchange of plants between North America and Eurasia during this time, these two floras steadily diverged. It has also been suggested that they shared a common Madro-Tertiary dry flora, exchanged via a low- to midlatitude dry corridor. Though later work on the times of divergence of members of the flora in the two continents, and on the distribution of fossil relatives, do not support the concept, there is as yet no convincing alternative explanation [38].

The American botanist William Weber [39] has provided evidence that the mountainous areas of Asia and North America share many taxa of flowering plant, lichens, and mosses, and believes that they were part of a single "Oroboreal" (or "Northern mountain") flora. He believes that the degrees of difference and endemism between their floras today are due only to later extinction and reduction in the ranges of the taxa.

When climates improved again, in the Mid-Miocene, the rainforests of India/Southeast Asia expanded again, but now included the relics of the northern megathermal rainforests as well as the dipterocarps, which had arrived on the Indian plate (Fig. 10.8e). This flora was the beginning of the modern Indo-Pacific floral region (see p. 338). When, in the Pliocene, the surviving North American megatherms were able to disperse into Central and South America over the new Panama land bridge, the survivors of the old Northern Megathermal Province were now to be found only there and in Southeast Asia, forming an "amphi-Pacific" element in these two floras. When climates became drier, in the late Cenozoic, much of the Indian rainforest was replaced by deciduous forests, grasslands, savannas, and savanna woodlands.

We have now followed the biogeographical history of the terrestrial flora, and some of its fauna, from their beginnings in the Devonian to after the great Cretaceous/Tertiary extinction event 65 mya, and the climatic and floral history into the Cenozoic. In the next chapter, we turn to consider the Cenozoic biogeographic history of the mammals, and the changes in flowering plant biogeography later in the Cenozoic.

1 The patterns of distribution of animals and plants are controlled mainly by the patterns of geography—by the positions of oceans, shallow epicontinental seas, mountains, and deserts. Because these were different in the past, mainly because of the movement of continents due to plate tectonics, the patterns of distribution of life in the past were different also.

2 The problems involved in trying to estimate the date of appearance of the ancestors of any living group are explained, and the discrepancies between estimates derived from the fossil record and those derived from molecular clocks are discussed.

3 The contrasting patterns of success of the marsupial and the placental mammals in the different continents seem to have developed because placentals did not colonize the South America/Antarctica/Australia chain of continents until after the marsupials of those continents had radiated into a great variety of families and these continents had started to separate from one another.

4 Climate changes have also had a great effect on these patterns. In the Late Cretaceous and Early Cenozoic, the world's climate was warmer than it is today. Many organisms were then able to spread via high-latitude routes that are now closed both by climatic changes and by the separation of continents. The turning point came with the great cooling at the end of the Eocene, after which warmth-loving plants and their ecosystems either became more restricted in their distribution or were replaced by those adapted to cooler climates.

5 It is now possible to follow in considerable detail the way in which the changing patterns of geography and climate led to changes in the floras of the different continents and to their diversification. Increasing aridity led to the replacement of forests by woodland and grasslands. These in their turn caused changes in the nature of the mammal faunas that lived in them.

Summary

Further Reading

Culver SJ, Rawson PS, eds. *Biotic Response to Global Change: The Last 145 Million Years.* London: Natural History Museum, 2000.

Graham A. *Late Cretaceous and Cenozoic History of North American Vegetation.* New York: Oxford University Press, 1999.

Morley RJ. *Origin and Evolution of Tropical Rain Forests.* New York: Wiley, 2000.

Prothero DR. *The Eocene-Oligocene Transition. Paradise Lost.* New York: Columbia University Press, 1994.

References

1 Cox CB. Vertebrate palaeodistributional patterns and continental drift. *J Biogeogr* 1974; 1: 75–94.

2 Edwards D. Constraints on Silurian and Early Devonian phytogeographic analysis based on megafossils. In: McKerrow WS, Scotese CR, eds. *Palaeozoic Palaeogeography and Biogeography*, pp. 233–242. Geological Society Memoir no. 12, 1990.

3 Young GC. Devonian vertebrate distribution patterns and cladistic analysis of palaeogeographic hypotheses. In: McKerrow WS, Scotese CR, eds. *Palaeozoic Palaeogeography and Biogeography*, pp. 243–255. Geological Society Memoir no. 12, 1990.

4 Milner AR. Biogeography of Palaeozoic tetrapods. In: Long JA, ed. *Palaeozoic Vertebrate Biostratigraphy and Biogeograph*, pp. 324–353. London: Belhaven, 1993.

5 Metcalfe I. Palaeozoic and Mesozoic geological evolution of the SE Asian region. In: Hall R, Holloway JD, eds. *Biogeography and Geological Evolution of SE Asia*. pp. 25–41. Leiden: Backhuys, 1998.

6 Berman DS, Sumida SS, Lombard RE. Biogeography of primitive vertebrates. In: Sumida SS, Martin KLM, eds. *Amniote Origins: Completing the Translation to Land*, pp. 85–139. London: Academic Press, 1997.

7 Chaloner WG, McElwain J. The fossil plant record and global climatic change. *Rev Palaeobotany Palynol* 1997; 95: 73–82.

8 Mora CL, Driese SG, Colarusso LA. Middle to Late Paleozoic atmospheric CO_2 levels from soil carbonate and organic matter. *Science* 1996; 271: 1105–1107.

9 Ziegler AM. Phytogeographic patterns and continental configurations during the Permian Period. In: McKerrow WS, Scotese CR, eds. *Palaeozoic Palaeogeography and Biogeography*, pp. 367–379. Geological Society Memoir no. 12, 1990.

10 Crowley TJ, Hyde WT, Short DA. Seasonal cycle variations on the supercontinent of Pangaea. *Geology* 1989; 17: 457–460.

11 Batten DJ. Palynology, climate and the development of Late Cretaceous floral provinces in the Northern Hemisphere: a review. In: Brenchley P, ed. *Fossils and Climate*, pp. 127–164. Geological Journal, Special Issue no. 11. London: Wiley, 1984.

12 Chaloner WG, Lacey WS. The distribution of Late Palaeozoic floras. *Spec Papers Palaeontol* 1973; 12: 271–289.

13 Cox CB. Triassic tetrapods. In: Hallam A, ed. *Atlas of Palaeobiogeography*, pp. 213–223. Amsterdam: Elsevier, 1973.

14 Cox CB. Changes in terrestrial vertebrate faunas during the Mesozoic. In: Harland WB, ed. *The Fossil Record*, pp. 71–89. London; Geological Society, 1967.

15 Upchurch P, Hunn CA, Norman DB. An analysis of dinosaurian biogeography: evidence for the existence of vicariance and dispersal patterns caused by geological events. *Proc Roy Soc* B 2002; 269: 613–621.

16 Serrano PC. The evolution of dinosaurs. *Science* 1999; 284: 2137–2147.

17 Crowley TJ, North GR. Palaeoclimatology. *Oxford Monogr Geol Geophysics* 1991, 1–330.

18 Cifelli RL, Davis BM. Marsupial origins. *Science* 2003; 302: 1899–2000.

19 Luo Z-Y, Ji Q, Wible JR, Yuan C-X. An Early Cretaceous tribosphenic mammal and metatherian evolution. *Science* 2003; 302: 1934–1940.

20 Asher RJ, Horovitz I, Sanchez-Villagra MR. First combined cladistic analysis of marsupial mammal interrelationships. *Molecular Phylogenetics and Evolution* 2004;33: 240–250.

21 Wildman DE et al. Genomics, biogeography, and the diversification of placental mammals. *Proc Nat Acad Sci USA* 2007; 104: 14395–14400.

22 Collinson ME. Cenozoic evolution of modern plant communities and vegetation In: Culver SJ, Rawson PS, eds. *Biotic Response to Global Change: The Last 145 Million Years*. London: Natural History Museum, 2000.

23 Morley RJ. *Origin and Evolution of Tropical Rain Forests*. New York: Wiley, 2000.

24 Crane PR. Vegetational consequences of the angiosperm diversification. In: Friis EM, Chaloner WG, Crane PR, eds. *The Origins of Angiosperms and Their Biological Consequences*, pp. 107–144. Cambridge: Cambridge University Press, 1987.

25 Crane PR, Lidgard S. Angiosperm diversification and paleolatitudinal gradients in Cretaceous floristic diversity. *Science* 1989; 246: 675–678.

26 Crane PR, Lidgard S. Angiosperm radiation and patterns of Cretaceous palynological diversity. In: Taylor PD, Larwood GP, eds. *Major Evolutionary Radiations*, pp. 377–407. Systematics Association Special Volume no. 42, 1990.

27 Lidgard S, Crane PR. Angiosperm diversification and Cretaceous floristic trends: a comparison of palynofloras and leaf macrofloras. *Paleobiology* 1990; 16: 77–93.

28 Smiley CJ. Cretaceous floras from Kuk River area, Alaska; stratigraphic and climatic interpretations. *Bull Geol Soc Am* 1966; 77: 1–14.

29 Parrish JT, Spicer RA. Late Cretaceous vegetation: a near-polar temperature curve. *Geology* 1988; 16: 22–25.

30 Morley RJ. Interplate dispersal paths for megathermal angiosperms. *Perspectives in Plant Ecology, Evolution and Systematics*. 2003; 6: 5–20.

31 Tallis JH. *Plant Community History: Long-Term Changes in Plant Distribution and Diversity*. London: Chapman & Hall, 1991.

32 McKenna MC. Eocene paleolatitude, climate, and mammals of Ellesmere Island. *Palaeogeogr Palaeoclimatol Palaeoecol* 1980; 30: 349–362.

33 Estes R, Hutchinson JH. Eocene lower vertebrates from Ellesmere Island, Canadian Arctic Archipelago. *Palaeogeogr Palaeoclimatol Palaeoecol* 1980; 30: 325–347.

34 Collinson ME. *Fossil Plants of the London Clay*. London: Palaeontological Association, 1983.

35 De Couto RM, Pollard D. Rapid Cenozoic glaciation of Antarctica induced by declining atmospheric CO_2. *Nature* 2003; 421: 245–249.

36 Wolfe JA. Some aspects of plant geography of the Northern Hemisphere during the Late Cretaceous and Tertiary. *Ann Missouri Bot Garden* 1975; 62: 264–279.

37 Meng J, McKenna MC. Faunal turnovers of Paleogene mammals from the Mongolian Plateau. *Nature* 1998; 394: 364–367.

38 Graham A. *Late Cretaceous and Cenozoic History of North American Vegetation*. New York: Oxford University Press, 1999.

39 Weber WA. The Middle Asian element in the southern Rocky Mountain flora of the western United States: a critical biogeographical review. *J Biogeogr* 2003; 30: 649–685.

40 Smith AG, Smith DG, Funnell BM. *Atlas of Mesozoic and Cenozoic Coastlines*. Cambridge: Cambridge University Press, 1994.

41 Cambridge Paleomap Services. ATLAS version 3.3. Cambridge: Cambridge Paleomap Services, 1993.

42 Vakhrameev VA. *Jurassic and Cretaceous Floras and Climates of the Earth*. Cambridge: Cambridge University Press, 1994.

43 Lillegraven JA. Ordinal and familial diversity of Cenozoic mammals. *Taxon* 1972; 21: 261–274.

44 Springer MS, Murphy WJ, Eizirik E, O'Brien SJ. Molecular evidence for major placental clades. In Rose KD, Archibald JD, eds. *The Rise of the Placental Mammals*. 2005. Baltimore, MD: Johns Hopkins University Press.

45 Bininda-Edmonds ORP et al. The delayed rise of present-day mammals. *Nature* 2007; 446: 507–512.

46 Crane PR, Friis EM, Pedersen KR. The origin and early diversification of angiosperms. *Nature* 1995; 374: 27–34.

The Geography
of Life Today

*I*n the last chapter, we saw how earlier forms of life were distributed over the very different geographies that then patterned the face of our planet, and how the major groups that we see today (the flowering plants and mammals) came into existence and occupied the world. In this chapter, we follow the histories of these groups as they diversified and spread through the still-changing pattern of continents and climates to occupy the regions we recognize today.

The currently accepted systems of biogeographical regions have their roots in the 19th century, when our growing knowledge of the world allowed biologists to realize that its surface could be divided into separate areas that differed in their endemic animals and plants. Naturally enough, these divisions were based on the distribution of dominant, easily visible groups. Thus, Candolle in 1820, followed by Engler in 1879, used the patterns of distribution of flowering plants as the basis for a system of floral regions, while Sclater in 1858, working on birds, and Wallace in 1860–1876, working on mammals, defined the system of zoogeographical regions. Apart from some modifications to the botanical scheme by Good [1] and Takhtajan [2], and quite minor changes to the zoogeographical scheme by Darlington [3], these 19th-century interpretations survived almost unchanged until the end of the 20th century.

As we now know, the nature of the biota of mammals and of flowering plants that eventually developed on each continent was the result of an interaction between several factors: their early history of origin and diversification, the gradual fragmentation of the continents (especially in the Southern Hemisphere), and the climatic changes that took place during the Cenozoic. In trying to understand the geographical origins of particular elements in the biota of these faunal and floral regions, it is important to realize that they may in the past have dispersed via now-broken land connections that also had climates far more congenial than those found there today. In particular, the warmer climates of the Early Cenozoic allowed animals and plants to spread and disperse via high-latitude routes that are now too cold to allow this. These routes were the Bering land bridge between Alaska and Siberia, and the land connections between eastern North America and Europe via Greenland (see p. 363). During the period of greatest warmth in the Cenozoic, the Late Paleocene/Early Eocene, the average temperature of the coldest month in these southern Arctic regions was 10–12°C, and frost was absent or rare [4]. These are probably the routes taken by Old World tropical genera that are found today also in the tropical flora of Mexico and the Panama Isthmus [5]. Although climates had cooled somewhat by the Miocene, in the Late Cenozoic, it was still warm enough for some mammals that we now view as typically African (such as lions, hyenas, giraffids, and macaques) to be found also in southern Europe and there live alongside typically Eurasian genera [6].

This last point also indicates that in the past there would normally have been a gradual transitional zone between neighboring regions. The clear distinctions that can be made today between the different zoogeographical regions or the different floral kingdoms, with sharp lines drawn between them on the world maps, are the results of coincidences of current geography and recent climatic changes. As we shall see, the Ice Ages of the last couple of million years progressively reduced the northward extent of the range of the more warmth-loving flowering plants and mammals. As they were pushed southward, however, they found their way blocked in North America by the narrow filter of the Panama Isthmus, and blocked in Eurasia by the combination of the Mediterranean Sea, the deserts of North Africa, the Middle East and southern Asia, and the Himalayan Mountain Range. As a result, many of these families became extinct north of those barriers. When the climate recovered during the interglacial periods, those same barriers prevented them from returning northward. As a result, instead of there being broad zones of gradual transition between neighboring faunal or floral regions, they are separated by areas within which climate or geography forms a barrier between very different biotas.

The 19th-century naturalists based their schemes of biogeography on the mammals and the flowering plants because these two groups are the visually dominant elements in the faunas and floras of the world, easily recognizable and identifiable. Their biogeographic patterns differ from one another not because they represent the patterns of all animals and of all plants, for they do not, but because they represent the two extremes in the biogeographic patterns of terrestrial organisms. Terrestrial mammals are very poor at dispersing across ocean barriers and onto oceanic islands, and their biogeographical regions are therefore delimited by the edges of the continental shelves. In contrast, many flowering plant families have seeds that are especially adapted for dispersal by the wind and, in particular, those of Southeast Asia have dispersed widely eastward through New Guinea and across almost all of the islands of the Pacific Ocean. As a result, the Indo-Pacific floral kingdom includes that vast area of ocean.

Few other groups of animals or plants have so far been biogeographically analyzed in similar detail, though a start has been made on some groups, such as dung beetles [7] and birds. One might expect that those organisms with good powers of dispersal, such as mosses, ferns, and flying insects, will show patterns more similar to those of the flowering plants, while those with more limited powers will show patterns more like those of mammals. However, the mammalian patterns are particularly well understood because of their excellent and decipherable fossil history. This allows us to identify, with considerable confidence, where the groups originated, when and where they diversified, and when and by what route they dispersed to new areas. No other major group is as yet known in this degree of detail, and the historical biogeography of mammals may well always be the standard against which those of other groups are evaluated, as far as continental patterns are concerned.

We can now turn to outline the global patterns of biogeographical regions of the mammals and flowering plants, and to examine how the early history of these groups created those patterns. This is followed by a detailed account of what took place in each region.

Mammals: The Final Patterns

As we have already seen (p. 315), both marsupial and placental mammals had spread through Asiamerica into South America in the Late Cretaceous and earliest Cenozoic. Details of their times and directions of dispersal between the continents and the movements of the continents themselves are shown in Fig. 11.1, which also shows the times of opening or closing of seas on the continents, or of oceans between them. Compare this figure with Fig. 5.7, which shows Gondwana's pattern of progressive breakup in the form of a dated cladogram.

Figure 11.2 shows the pattern of zoogeographical regions that is recognized here. This follows Wallace's 1860–1876 system, except that the boundaries of the Oriental and Australian regions in the East Indies follow the edges of the Asian and Australian continental shelves, rather than the two regions meeting along a line within the East Indies pattern of islands (see Box 11.1, p. 347). However, some of the names given to the regions by Sclater and Wallace (Palaearctic, Nearctic, Neotropical) in the 19th century reflected the fashion in those days of coining Classical polysyllables for most new scientific concepts and objects. With only trivial differences of boundary, these regions are effectively Eurasia, North America, and South America, and these familiar names are used here.

Because both South America and Africa were largely isolated from other continents in the Late Cretaceous and Early Cenozoic, each developed a characteristic mammalian fauna. In the Late Cenozoic, India and Southeast Asia had a fauna similar to that of Africa. But their different climatic histories, and the appearance of desert and mountains separating the two areas, led to the divergence of their mammalian faunas. As a result, a separate Oriental zoogeographic region is recognized. The Australian region, with its unique marsupials, forms another distinct fauna. The two landmasses of the Northern Hemisphere have mammalian faunas that differ somewhat from one another, although both are similar in having been greatly impoverished by the climatic change of the Pleistocene Ice Ages.

The distribution pattern of terrestrial mammals in the Late Cenozoic Miocene/Pliocene Epochs is shown in Table 11.1. The final pattern found today is slightly different from this because elephants became extinct in the Northern Hemisphere during the Pleistocene, and because edentates and marsupials dispersed to North America via the Panama land bridge. The final total of orders for each region in Table 11.1 takes account of these changes. The last line of Table 11.1 also shows the total number of terrestrial families of mammal in each region. These figures therefore exclude whales, sirenians (dugongs and manatees), pinnipeds (seals, etc.), and bats, as well as humans and the mammals they took with them in their travels (such as the dingo and rabbit in Australia).

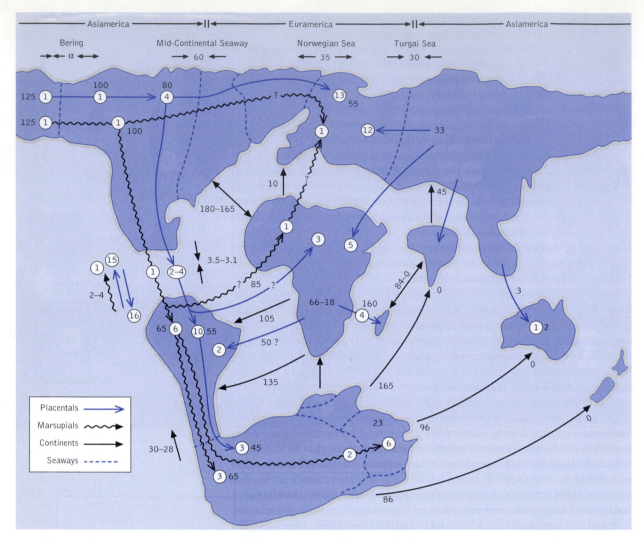

Fig. 11.1 Diagram to illustrate the main events in the historical biogeography of mammalian dispersal. Arrows indicate the directions of dispersal of mammalian families or of continents. Dotted lines indicate the positions of the Bering, Mid-Continental, Labrador, Greenland, and Turgai seas, with arrows indicating whether the sea opened or closed at the time indicated by the adjacent figure (in millions of years ago). (The alpha symbol at the Bering Sea indicates that it opened and closed a number of times.) The position of the Labrador Sea, west of Greenland, is shown as a dotted line, but no further information is given as it is irrelevant to mammalian dispersal. Figures in circles indicate the number of mammal families in the area in question, the adjoining figure indicating the relevant time in millions of years ago. No attempt has been made to give the total numbers of families in the different continents at any period later than 55 million years ago.

The individual families within the orders of mammal show considerable variations in their success at dispersal. A few have been extremely successful. Nine families have dispersed to all the regions except the Australian: soricids (shrews), sciurids (squirrels, chipmunks, marmots), cricetids (hamsters, lemmings, voles, field mice), leporids (hares and rabbits), cervids (deer), ursids (bears), canids (dogs), felids (cats), and mustelids (weasels, badgers, skunks, etc.). In addition, the bovids (cattle, sheep, impala, eland, etc.) have dispersed to all the regions except South America and Australia, and the murids (typical rats and mice) have dispersed everywhere except North and South America. This group of 11 families can conveniently be called the "wanderers." Their inclusion in any analysis of the patterns of distribution of the living families of terrestrial mammal tends to blur the underlying patterns of relationship of these zoogeographic regions. These "wanderers" have therefore been excluded from Fig. 11.3, which shows the distribution of the remaining 79 families. (The Oriental region is shown twice, so that the single family shared with South America (the relict distribution of camelids) can be shown.)

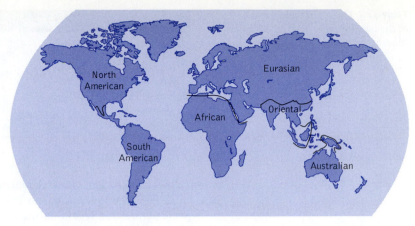

Fig. 11.2 Biogeographical regions of a group that does not disperse readily—the mammals.

As can be seen in Fig. 11.3, the majority of all the families of terrestrial mammal (51 out of 90, i.e., 57%) are endemic to one region or another. The degree of endemicity of the mammals in each of the different regions is calculated in Table 11.2. (Rodents, the most successful of all the orders of mammal, contribute 19 of these endemic families: 2 North American, 1 Eurasian, 10 South American, and 6 African.) It is clear from Fig. 11.3 and Table 11.2 that the degree of distinctiveness of the six zoogeographic regions of today, if judged by the endemicity of their mammals, varies greatly. These figures are the result of three main factors: isolation, climate, and ecological diversity.

The results of the long isolation of the Australian and South American regions are obvious from the distribution of the families in the Miocene, when the other four regions were all interconnected (Fig. 11.4). Comparison of Figs. 11.3 and 11.4 also shows how many South American families became extinct after the Pliocene connection with North America (see

Table 11.1 Distribution of terrestrial mammals during the Late Cenozoic (Miocene–Pliocene).

	Africa	Oriental	Eurasia	North America	South America	Australia
Rodents	×	×	×	×	×	×
Insectivores, carnivores, lagomorphs	×	×	×	×	×	
Perissodactyls, artiodactyls, elephants	×	×	×	×	×	
Primates	×	×	×		×	
Pangolins	×	×				
Conies, elephant-shrews, aardvarks	×					
Edentates					×	
Marsupials					×	×
Monotremes						×
Total number of orders today*	12	9	7	8	9	3
Total number of terrestrial families today*	44	31	29	23	32	11

*The final totals of orders and families also take account of the Quaternary extinctions and dispersals (see text).

Table 11.2 Degree of endemicity of the families of terrestrial mammal. Number of endemic families × 100 ÷ total number of families.

Region	Endemicity
Australia	10 × 100 ÷ 11 = 91%
South America	15 × 100 ÷ 32 = 47%
Africa	16 × 100 ÷ 44 = 36%
Holarctic	7 × 100 ÷ 37 = 19%
North America	3 × 100 ÷ 23 = 13%
Oriental	4 × 100 ÷ 31 = 13%
Eurasia	1 × 100 ÷ 30 = 3%

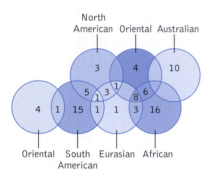

Fig. 11.3 Venn diagram showing the interrelationships of the families of terrestrial mammal of their six zoogeographic regions today, excluding the 11 "wandering" families.

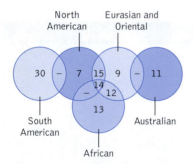

Fig. 11.4 Venn diagram showing the interrelationships of the families of terrestrial mammal of the continents at the end of the Miocene, excluding the 11 "wandering" families.

below, p. 359). The North American region was connected to Eurasia via the high-latitude Bering region, and many mammal groups were able to disperse across this region during those warmer times. Nevertheless, these groups did not include the tropical and subtropical groups that then ranged throughout Africa and southern Eurasia—including parts of Eurasia well north of the present limits of the Oriental region, which was therefore not recognizable as a separate zoogeographic region at that time. The Pleistocene glaciations of the Northern Hemisphere caused the extinction of many of the mammal faunas of both North America and Eurasia. These two regions therefore contain few mammal families, and these families are also still found mostly in the adjoining southern regions. As a result, North America and Eurasia contain few endemic mammal families. Tropical and subtropical Old World mammal families are therefore now found only in the Oriental and African regions. Without the three families found only in Madagascar, the degree of endemicity of the mammals of the African region would be slightly lower (13 × 100 ÷ 41 = 32%), but it would still be significantly higher than that of the Oriental region. Interestingly this difference is also found in birds: 13 families are endemic to the African region, but one is endemic to the Oriental region. These differences are probably mainly the result of the greater latitudinal spread of Africa, which extends from the equator northward through the Sahara Desert to approximately 30°N, and southward to 35°S, and therefore includes a wider range of environments than India, which only extends between 14°N and 35°N. But the greater area of Africa may also have played a part in this, providing more space for the evolution of novel groups.

It is also valuable to study the interaction between plate tectonics, evolution, and biogeography in a little more detail, to see how and when the different elements of the modern patterns of distribution were established. One of the best documented groups is the superfamily Equoidea, whose modern representatives include the horses, zebras, and asses (Fig. 11.5). Their early evolution was on the Euramerican plate and in particular in North America after that continent became separate. The genus *Hipparion* spread across the Bering region to Eurasia in the Late Miocene and thence to Africa when that continent became connected to Eurasia (cf. Fig. 11.8). The genus *Equus* appeared in the Pliocene and gave rise to three distinct lineages: the asses, the zebrids, and, from the zebrids, the true horses. Four different species of zebrid are still found in Africa, and the quagga, related to the true horse, became extinct there during the 19th century. Both *Hippidion* and true horses reached South America after the

Panama Isthmus formed, but, surprisingly, all the equids of the New World became extinct during the Pleistocene.

Eisenberg [8] has made an interesting analysis of the extent to which the different ecological niches have been filled in the different zoogeographical regions. Because East Africa has an extensive area of grasslands and savanna, 21% of the mammal genera of Africa are browsers or grazers, compared with only 9% of those in South America. Similarly, because great areas of South America are covered by rainforests, 22% of the mammal genera of northern South America are fruit-eating or omnivorous, compared with only 11% in southern Africa. For the same reason, northern South America also has a high proportion of mammals (bats) that feed on insects, either in the air or on leaves. Surprisingly, only two genera of Southeast Asian bat feed on insects living on leaves, compared with nine in South America.

Comparison of data of this kind provokes interesting questions. Australia and northern South America contain similar proportions of arboreal genera, although Australia today has a much smaller area of forests. Eisenberg suggests that this may be an inheritance from the past, when Australia was more heavily forested, so that many marsupials became arboreal. Similarly, he speculates on the possible reasons for the fact that there is a comparatively small number of fruit-eaters/omnivores in Australia. This may be because there are fewer fruit trees in Australia, or because parrot-type birds there have been more successful in that niche than have the mammals. It is also possible that the Australian fruit trees, which are less taxonomically diverse than those of South America and live in a more seasonal climate, may be more seasonal than those of South America, so that fruit is not a reliable, year-round source of food. This may also be the reason there are many more bats in northern South America.

In a thought-provoking article, the ecologists Cris Cristoffer and Carlos Peres [9] compare the composition of the forests and its herbivores in the Old World with those of the tropics of South America. They point out that in the

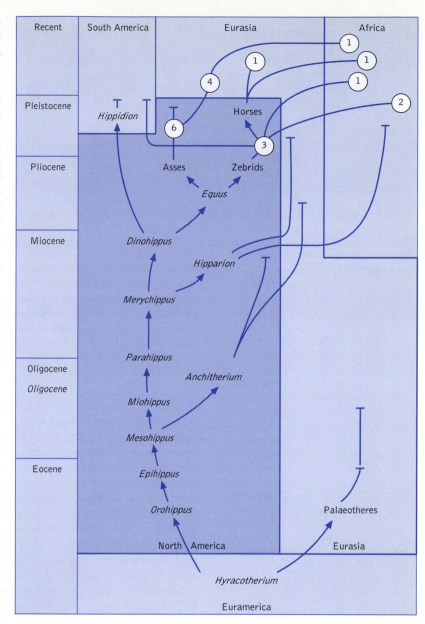

Fig. 11.5 The biogeographic history of horses (Equidae and Palaeotheriidae), showing how most of their diversification took place in North America (tinted in blue), followed by later dispersal to other continents. The circled figures show the number of genera of horse, ass, or zebra that existed in each continent at particular times. Extinct lineages are shown with a terminal cross-bar. Modified after MacFadden [54].

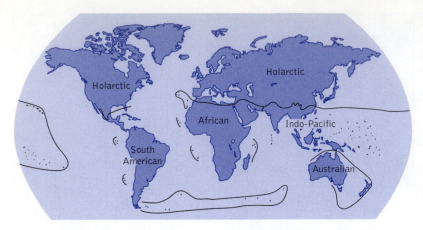

Fig. 11.6 Biogeographical regions of a group that disperses readily—the flowering plants.

Old World forests there are fewer lianas and a higher proportion of large and robust trees. The size of these trees allows them to withstand the foraging of the large herbivores, such as elephants and rhinos, that consume or damage smaller trees, breaking off branches or pushing the tree over. They note that the Old World also has a larger area of savanna, and they speculate that the savanna herbivores could have provided a source for the evolution of the large herbivores of the Old World forests. These have a much greater diversity and biomass of large terrestrial herbivores than those of South America, which instead has a greater number of larger arboreal herbivores and a greater number of smaller herbivores, both vertebrate and invertebrate.

Another interesting aspect of comparative mammalian biogeography is that Australia has very few large carnivores. Flannery [10] points out that there may be several reasons for this. Much of that continent is unproductive desert or semidesert, and the vegetation in general is of low productivity because of Australia's impoverished soils (see p. 347). In addition, Australia's annual rainfall is also extremely variable, because it is highly affected by the El Niño events in the Pacific Ocean (see p. 404). All this leads to wide fluctuations in the populations of Australian herbivores. As a result, the carnivores that prey on them and that are always less numerous than their prey must be very vulnerable to extinction. Flannery notes that Australia has an unusually high number of reptilian predators, such as pythons and varanid lizards, and had giant members of these groups in the Pleistocene Period, as well as a large land crocodile. He suggests that the reptiles, being cold-blooded and therefore not needing to consume as much food as their warm-blooded mammalian competitors, may have been better able to survive periods of starvation and so to retain a higher, safer level of population. Milewski and Diamond [11] have suggested that the lack of micronutrients containing iodine, cobalt, and selenium in Australia's impoverished soils may similarly have affected the evolution of large herbivores in the continent, for these may be of limited size, fecundity, and intelligence compared with those of other continents.

A Consistent Pattern of Floral Regions

Figure 11.6 shows the pattern of floral regions that is recognized in this book. In the main, this pattern follows the system laid down by the Russian botanist Armen Takhtajan [2], who defined 6 kingdoms, 12 subkingdoms, and 37 floristic regions. However, it differs from Takhtajan's system in several important respects, which are argued in greater detail elsewhere [12].

First, for reasons that are explained on p. 344, the flora of the tiny Cape region of South Africa is not recognized here as a distinct floral kingdom. Second, Takhtajan continued the practice, which originated with the German botanist Engler in 1879, of recognizing an "Antarctic" plant kingdom that comprised southernmost South America, Tasmania, part of New Zealand, and a number of islands scattered in the southernmost Atlantic, Indian, and Pacific oceans (cf. Fig. 1.1). However, the floral similarities between these areas are merely the result of their having some survivors of the old South Gondwana flora (p. 322 and Fig. 10.8) as subsidiary elements in their floras. The forests of this flora consisted of mainly podocarp gymnosperms and mainly evergreen angiosperms (the southern beech tree, *Nothofagus*, being particularly characteristic), plus various herbaceous and shrubby angiosperms. This flora seems to have survived, reasonably intact, in southernmost South America, in western Tasmania, and perhaps also in New Zealand (see also p. 351). Elements of it are also found in scattered locations further north, such as in New Caledonia and the mountains of New Guinea. As we shall see, the flora of the Australian region is derived mainly from the original "Antarctic" flora, although it has been much altered in adapting to the steadily increasing aridity of that continent. Thus the elements of the old Antarctic flora now have merely a relict distribution around its descendant, the highly modified Australian flora. It is anomalous to recognize this climatic and relict distribution as a part of the system of modern flowering plant kingdoms, all of which otherwise correspond to large, compact, and easily defined areas of the globe. Accordingly, the Antarctic floral kingdom is not recognized here.

Third, the flora of Africa south of the Sahara, and that of India and Southeast Asia, are recognized as separate floral kingdoms, instead of being united as a single floral kingdom. This reflects the fact that the floras of these two regions are as different from one another as, for example, the African flora is different from the flora of South America. Although there is, at the family level, much similarity between the floras of the three areas (cf. Fig. 11.7), the African flora has been greatly changed by the Cenozoic uplifting, and consequent comparative aridity, of much of southern and eastern Africa; this now lies at a height of over 1,000 m (3,300 ft)

Mammal families

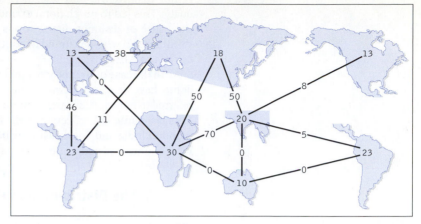

Flowering plant families

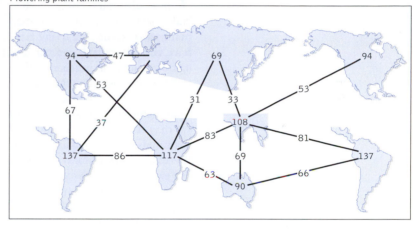

Fig. 11.7 A comparison of the faunal and floral similarities between the different regions. Figures within each region show the number of families found there. (Families found only in Madagascar have been omitted.) Figures linking the regions show the coefficients of biotic similarity, 100C/N, of the biota of these regions: C is the number of families common to the two regions being compared, while N is the number of families in whichever of the two regions has the smaller fauna. Above, mammal families (excluding the "wandering" families). Below, flowering plant families (excluding those with a worldwide distribution); data taken from Heywood [55].

rather than at the 400–500 m (1,300–1,600 ft) level it occupied previously. This led to much floral extinction and impoverishment, and may be the cause of disjunct distributions such as that of the primitive family Winteraceae, which is found in South America, Australia, and Madagascar, but not in Africa.

As in the case of the zoogeographical regions, the names of the plant kingdoms have been modernized here. Accordingly, these three floras are referred to here as the South American, African, and Indo-Pacific kingdoms. This last name is a reflection of the fact that many flowering plant families found in India and Southeast Asia have dispersed widely eastward through New Guinea and across almost all of the islands of the Pacific Ocean.

The Distribution of Flowering Plants Today

The distribution patterns of the families of flowering plant are much more difficult to interpret than those of the families of mammal, for several reasons. First, there are many more flowering plants than there are mammals—some 300 living families, 13,000 genera, and 250,000 species have been described, compared with only 100 families, 1,050 genera, and 4,400 species of living mammal.

Another problem is that the distribution patterns of the flowering plant families are very diverse. This may be partly because they are much better than mammals at dispersal across ocean barriers, since successful dispersal may require only a single airborne seed instead of a breeding pair of mammals or a pregnant female. Most families of flowering plant are therefore much more widely distributed than most families of mammal. For example, almost everywhere in the world, four flowering plant families are among the six most numerous: the Asteraceae (daisies, sunflowers, etc.), Poaceae (grasses), Fabaceae (peas, clover, vetches, etc.), and Cyperaceae (sedges). Only 56 (18%) of the families of flowering plant are endemic to one particular region (and most of these are relatively unimportant families restricted to only a few genera), while 51 (57%) of the families of terrestrial mammal are similarly endemic. The more widespread nature of angiosperm distribution is also shown by the fact that, of the 302 families of flowering plant (excluding the four marine families), 86 are found worldwide and 28 others are found in each region except the Holarctic (or Boreal) and Antarctic regions. The differences between the floras of the various tropical regions are therefore less clearcut than the differences between their mammals.

An even more fundamental problem lies in the comparative uncertainty of flowering plant systematics. Because the fossil record of plants is less complete and less well understood than that of mammals, assessments of the relationships of groups above the level of the genus have in the past been based on subjective judgments of the relevance of morphological features of the flowers, leaves, stem, pattern of branching, and so on. Different workers have come to quite different conclusions as to the contents of different groups and as to their relationships—and each different grouping may, of course, lead to a different pattern of biogeography for the group. The new methods that use molecular information to evaluate relationships have already suggested the breakup of some traditional family groupings [13]. Some of these had previously included taxa that were widely separated in the Northern and Southern Hemispheres, providing an apparent problem of disjunct distributions. The new methods suggest that the disjunct groups are not, after all, closely related, and so the problem disappeared.

Mammalian versus Flowering Plant Geography: Comparisons and Contrasts

As might be expected, the systems of faunal and floral regions are very similar to one another (compare Figs. 11.2 and 11.6). This is because the continental movements and climatic changes of the Cenozoic have produced patterns of continents and climates that have had similar effects on the distributions of the two groups. However, there are differences in the degree of closeness of relationship between the mammals of the different regions and between the flowering plants of the different regions. As has already been noted, the relationships between the zoogeographic regions are seen more clearly after the deletion of the 11 families of "wandering" mammal. Similarly, those of the flowering plant regions are also shown more clearly if the families with a worldwide distribution are first excluded. If the resulting patterns are now compared (Fig. 11.7), some very interesting differences can be seen.

The first point is that there is a much greater similarity between the flowering plant floras of South America, Africa, the Oriental region and Australia than between the mammalian faunas of these regions. Only in the case of the African/Oriental comparison are the plant and animal figures at a similar level. Three factors seem to have caused these differences.

The most important factor is probably the fact that flowering plants are much better than mammals at dispersing across ocean barriers. The extent of their spread across the Pacific (about 180 genera of flowering plant have reached the most isolated island group, Hawaii) shows that they can cross even quite wide stretches of ocean, especially where intermediate island stepping-stones were available. Mammals are much less competent at such dispersal.

The results of these differences in dispersal ability were increased by the second factor—the fact that the families of flowering plants evolved and dispersed earlier than the families of mammals. New palaeobotanical techniques have made it possible to retrieve and identify complete and partial flowers from Middle Cretaceous sediments [14]. These show that several different extant families had appeared by the Middle Cretaceous, about 120 million years ago, and at least a dozen by 95 to 75 million years ago. The angiosperms therefore commenced their dispersal across the world much earlier than the mammals and so had a much greater chance of reaching the different continents before they had drifted too far apart. In contrast, the diversification and dispersal of modern mammals only began in the earliest Cenozoic, 66 to 55 million years ago, by which time the continents had drifted further apart and were more difficult to reach.

However, those mammals that did succeed as colonists were able, in the isolation of each continent, to diverge into a number of unique, endemic groups that show little similarity to those in other continents—edentates, New World monkeys, and caviomorph rodents in South America; elephants, elephant-shrews, conies, and aardvarks in Africa; marsupials in Australia.

The third factor is that there has been more extinction and replacement during the history of mammals than during that of flowering plants. For example, in addition to the approximately 100 living families of mammal, over 300 other families evolved and became extinct during the Cenozoic—some 70% of the families of mammal died out completely. Some of these were previously widespread families, which were replaced in the now-separate continents by new, endemic families. In other cases, the family became extinct only in some areas, so that it now had a disjunct distribution, as in the camel–llama group (see p. 361). Another example of the influence of extinction is seen if one compares the similarities between the mammal faunas of North and South America before and after the Pleistocene extinctions (see pp. 359–361). All these phenomena reduced the levels of similarity between the faunal regions. In contrast, the flowering plants have experienced much less extinction; in fact, there is as yet no record of extinction for any major group of angiosperm. As a result of all these factors, plant taxa are much longer-lived than are the mammalian taxa. For example, the distribution of the southern beech tree, *Nothofagus*, shows that it evolved in the Late Cretaceous, at least 70 million years ago, while the average longevity of mammalian genera has been estimated at only 8 million years.

For all these reasons, it is not surprising that the floras of the different continents show more similarities to one another than do their mammalian faunas. However, there is one exception: the almost identical levels of similarity for the two groups when the African and Oriental regions are compared. The floral similarity here is not surprising, for it is at the same general level as the similarities between the other tropical regions—South America vs. the Ethiopian region, and South America versus the Indomalesian region. It is therefore the faunal similarity between the African and Oriental regions that is unexpectedly high. This is probably because of the faunal exchange that took place between Africa and Eurasia after the two continents became connected in the Miocene (see p. 342), and before deserts spread through the Middle East.

The greater similarity between the floras of the tropical regions explains yet another difference. This is the fact that there is more similarity between the floras of North America and Eurasia, which are linked in a single Boreal floral region, than there is between the mammal faunas of these two regions. As we shall see, both regions lost nearly all of their subtropical biota during the Ice Ages. After the last Ice Age had ended, both animals and plants started to spread northward to recolonize the newly warmed lands of the Northern Hemisphere. However, there was already much more similarity between the plants of South America and Africa than between their mammals. Their dispersal northward therefore produced a corresponding similarity between the floras of North America and Europe. No such similarity resulted from the northward spread of the very different mammals of South America and of Africa.

Finally, the high similarity between the floras of North America and Africa is probably because a large number of South American tropical families that are also found in Africa have spread into the almost subtropical southeastern region of North America. The Cenozoic historical biogeography of each of the main regions, as well as the large islands of New Zealand and Madagascar, will now be considered in turn.

The Old World Tropics: Africa, India, and Southeast Asia

Two original major landmasses contributed to this area, and both were originally part of Gondwana (see Figs. 10.1 and 10.2). India collided with southern Asia in the Early Eocene, 52–45 million years ago, its continuing northward movement throwing up the Himalayan Mountains and raising the Tibetan Plateau. Africa became united with the northern continents during the Mid-Miocene, 16–10 mya. But Africa had never lain far from southern Eurasia, and it was probably from here that tropical flowering plants dispersed to the tropics of southern Asia. Many elements of the tropical biota doubtless became widespread throughout the region and, before the Late Cenozoic cooling of the Northern Hemisphere, would also have ranged northward into higher latitudes of Eurasia. However, that cooling, together with the spread of seas and deserts in the Middle East, led to a new division of the Old World tropics into a western division, made up of Africa alone, and an eastern Oriental section made up of India and Southeast Asia.

It is not yet possible to trace in full the separate contributions of Africa, India, and Southeast Asia to the final Old World tropical biota. As always, the fossil record of mammals is easier to interpret and can then be used as a guide for the probable biogeographic histories of the angiosperm floras. However, because they have better powers of dispersal, a greater variety of flowering plants than of mammals were able to disperse from Eurasia into Africa.

Africa

The best-known and most distinctive of the Old World tropical mammal faunas is that of Africa. As explained in Chapter 10, molecular studies of the interrelationships of the different orders of placentals suggest that the common ancestor of a group of African placentals entered the continent in the Late Cretaceous. This group is made up of the elephants, hyracoids (conies), aquatic sirenians (sea cows), elephant shrews, aardvarks, Cape golden moles, insectivorous tenrecs, and the extinct embrithopods; they have been placed together in a superorder Afrotheria that is endemic to Africa [15].

At the end of the Cretaceous, the equatorial belt of Africa was covered by the megathermal rainforest of the Palmae Province (see Fig. 10.7), but substantial extinction took place as a result of the climatic deterioration caused by the Cretaceous/Tertiary meteorite strike which killed off the dinosaurs. Though this flora later diversified, there was a further reduction in diversity at the time of the Terminal Eocene Cooling Event, followed by another period of diversification. The Late Miocene cooling and drying led to expansion of the African savannas, exacerbated by the fact

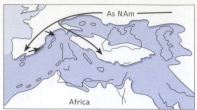

Late Oligocene/Early Miocene

(a)

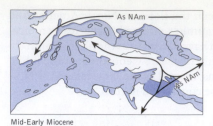

Mid-Early Miocene

(b)

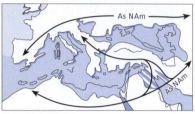

Start of Late Miocene

(c)

Near end of Late Miocene

(d)

Fig. 11.8 Reconstructions of the Mediterranean area at different times during the Cenozoic. Light tint, sea; dark tint, evaporitic deposits laid down as the seas dried up. Arrows show directions of mammal dispersals; As, Asia; NAm, North America. Some present-day geographical outlines have been added to aid recognition and location. After Steininger et al. [56], with permission.

that Africa was uplifted from the Mid-Tertiary, so that today only 15% of its area is low enough to be within the altitudinal range of rainforest. Widespread extinctions therefore took place within the African rainforest.

In the early Cenozoic, the shallow seas that separated Africa from Eurasia (Fig. 11.8) were not the only barrier to biotic exchange between the two continents, for northern Africa lay in the northern midlatitude arid belt. Nevertheless, some placentals managed to enter Africa from the north. An invasion by early primates and creodonts (early carnivorous mammals) may have occurred near the Paleocene–Eocene boundary 56 mya [16], for such an event is needed to explain the more diverse fauna found in the Late Eocene to Early Oligocene of northern Africa. Apart from members of the Afrotheria, these faunas include creodonts, rodents, the cloven-hoofed artiodactyls, and the earliest members of the anthropoid primate line (that later evolved into apes and human beings). Though they are today represented only by the rabbit-like conies, in the Oligocene (and perhaps earlier) the hyracoids were the dominant small and medium-sized terrestrial herbivores of the continent.

After this early arrival of some placental families, the placental fauna of Africa did not receive any further additions until much later in the Tertiary, about 19 mya in the Miocene, when a firm land connection with Eurasia took place in what is now the Middle East (Fig. 11.8b). (This long period of isolation is presumably due to the arid belt mentioned earlier.) The closure of the seaway between the two continents interrupted the circumequatorial world oceanic circulation. This in turn may have been the cause of the climatic deterioration that took place in central Europe at that time. Elephants, creodonts, and primates, and cricetid rodents passed across the new land bridge from Africa to Asia, while carnivorans, suids (pigs), and bovids (cattle, antelope, etc.) entered Africa, causing the extinction of many of the earlier types of African mammal.

Although the seaway reopened later in the Miocene, so that the climate through central Europe became warmer and moister, the seaway was not enough of a barrier to prevent the dispersal to Eurasia of new types of primate, elephant, and suid that had evolved in Africa. The seaway was finally broken near the beginning of the Late Miocene, 12 million years ago, by rising mountains in Arabia, Turkey, and the Middle East. It was at this time that the early horse *Hipparion*, which had evolved in North America, appears in both Eurasia and Africa (see Fig. 11.5), while rhinoceros, hyenas, and sabre-toothed cats dispersed from Africa to Eurasia (Fig. 11.8c). A final, dramatic event was the closure of the western connection between the Mediterranean and the Atlantic near the end of the Late Miocene, 6 million years ago. It was caused by a worldwide fall in sea levels (caused by an increase in the polar ice caps), as well as by the rise of mountains in both Spain and northwest Africa. This particularly affected the western part of the Mediterranean Sea, as far east as Italy, as there are no major rivers to provide water for this region—unlike the eastern part, which received inflows from western Asia via the Black Sea, and from Africa via River Nile. As a result, over the next 2 million years the western Mediterranean from time to time dried up completely, leaving an immense plain 3,000 m (10,000 ft) below sea level, covered with a thick deposit of rock salt (Fig. 11.8d).

The results of the Miocene interchange of mammals between Africa and tropical Eurasia can still be seen today, for a number of groups are found exclusively in these two areas. In nearly every case, however, they contain different genera. For example, the African rhinoceros, elephant, and porcupine all belong to genera different from those found in the Oriental region. Similarly, the lemurs of Madagascar and the chimpanzees and gorillas of Africa are not found in the Oriental region, where these groups are represented by the lorises and by the orangutan and gibbon. The scaly anteater (*Manis*) is an exception, for the same genus is found in both areas. However, some of the groups that are found only in these two areas today once had wider distributions. For example, elephants originated in Africa in the Eocene, migrated into Eurasia in the Early Miocene, and migrated back later in the Miocene [17]. The ancestors of the mammoths and modern Asian elephant probably evolved in Africa 4 to 5 million years ago and migrated back into Eurasia 2 million years later. Although they spread through much of the world, the mammoths became extinct everywhere during the Ice Ages.

Differences between the mammal faunas of the tropics of Africa and Eurasia might have developed in any case, because of the distance between the two areas. But the two faunas also became more isolated from one another by the development of the Red Sea in the Pliocene and by the extension of deserts in the Middle East. These differences became more acute in the Late Miocene because the fauna of East Africa had to adapt to the cooling and consequent increased dryness of that region, as a result of which it became covered by woodland and bushland with a ground cover of herbs and grass. This led to a great radiation of browsing and, especially, grazing mammals, especially the bovids (cattle, sheep, antelope, gnu, impala, etc.); there are 76 species of bovid in Africa, 29 of which belong to only 10 genera, compared with only 37 species in Asia. Together with the giraffes, wart-hogs, and zebras, they form huge herds in East Africa and are commonly thought of as the "typical" fauna of Africa. But in reality these are latecomers to the African scene; their ancestors are not known in Africa until the Middle Miocene. Our own genus, *Homo*, also seems to have evolved in this environment (see p. 414) and human activities were probably responsible for the characteristic flora of the seasonally arid, overgrazed and burned grasslands of Africa [18]. Together with the general drying of the world climate at that time, the drying of East Africa reduced the eastward extent of the African rainforest, which now became restricted to West Africa and the Congo Basin.

This decrease in its area is probably one of the reasons why the African tropical flora is much less diverse than that of South America and Southeast Asia. The Pleistocene glaciations of the Northern Hemisphere were reflected by dry, cool periods in Africa, which led to fluctuating subdivision and reunion of the rainforests.

The drier climates that resulted from the changes in North Atlantic ocean currents caused by the closure of the Panama Isthmus 3 mya led to the development of the modern Mediterranean climate with its hot, dry summers and cool, wet winters, and the expansion of sclerophyll evergreen woodlands in the region. This drier European environment received many Asian mammals that inhabited steppe and savanna conditions, while the African hippopotamuses spread to its river systems, and African rhinoceroses lived on its plains. The fauna of northern Africa was therefore similar to that of the rest of Africa during most of the Tertiary. The Sahara Desert, whose 8 million km^2 (3 million sq mi) area is greater than Australia and dominates the map of Africa today, only began to appear in the Pliocene, but eventually changed the climate of North Africa, whose fauna has lost nearly all of its African character.

At the margins of Africa, two areas contain biota that merit special consideration: the flora of the Cape region of southern Africa, and the fauna of Madagascar.

The Cape Flora

Until recently, it had been customary for plant biogeographers to recognize the flora of the Cape region of southern Africa as a separate floral Kingdom, thus placing it at the same level of importance as the floras of each of the major continents of the world, or of the whole of the temperate Northern Hemisphere [12]. However, the Cape region is one of several regions that have a Mediterranean-type climate, the others being California, coastal Chile, southwestern Australia and the Mediterranean Basin itself [19]. The Cape flora is therefore merely one of several floras that have resulted from similar ecological/evolutionary histories, leading to high levels of endemicity, and is now regarded as a plant province or region, rather than a kingdom. These five regions occupy less than 5% of the Earth's surface, yet contain around 48,250 species of flowering plant (almost 20% of the world total), as well as exceptionally high numbers of rare and locally endemic plants. Before the beginning of global cooling and drying in the Pliocene, all five regions were covered by subtropical forest, but now have a mixture of floras that include some relicts of the former forest, plus sclerophyllous shrublands and woodlands, with drought- and fire-adapted lineages predominating.

Madagascar

With its area of 587,000 km^2 (226,500 sq mi), Madagascar is second only to New Guinea in size. Its plate tectonic history is complex. It became separate from Africa in the Jurassic c. 160 mya, after a narrow ocean gradually extended clockwise around Africa (see Fig. 10.2), but remained attached to India as that continent separated from the rest of Gondwana about

130 mya and moved northward. Madagascar reached its present position relative to Africa by 121 mya and separated from India and became an island about 88 mya, while India continued its journey toward Asia. This long period of isolation explains the high degree of endemicity in the biota of Madagascar: about 96% of its 4,200 species of trees and shrubs are endemic, as are 9,700 species of plant and 770 species of vertebrate. The eastern part of the island still contains a large area of rainforest, which is species-rich, probably because it did not suffer as badly from prolonged drying during the glacial periods as did that of Africa. The history and biota of Madagascar has recently been comprehensively described [20].

Although the Cretaceous biota of Madagascar must have been similar to that of India, we have no fossil evidence of that, and recent molecular evidence suggests that the flowering plants of Madagascar are more likely to have arrived by dispersal from Africa [21]. The dispersal characteristics of the Madagascar flora suggest that most of them entered by wind dispersal, for which the Mozambique Channel, only 380 km (240 mi) wide at its narrowest, cannot have been a great barrier.

Since it was never narrower than today, the Mozambique Channel has always been wide enough to provide a formidable barrier for terrestrial African mammals. Madagascar's native terrestrial mammal fauna is made up of 101 species, nearly all of which belong to only four groups: the Insectivora (tenrecs), Primates (lemurs), Carnivora (fossas), and Rodentia. In each of these groups, the Madagascan species all belong to a single lineage. (The only other terrestrial mammals that reached the islands naturally are a pygmy hippopotamus that became extinct during the Pleistocene, and a river hog.) It now seems possible that the Madagascan rodents may have reached the island from Asia, not Africa, and went on to invade the African mainland [22]. The remaining three groups of Madagascan mammals represent only a small part of the African mammal fauna, and molecular techniques suggest that the dates at which these different groups of mammal diverged from their non-Madagascan ancestors, after arrival in the island, varied from 60 to 18 my ago [23]. Both of these facts suggest that they arrived by sweepstakes dispersal (see p. 42) across the Mozambique Channel. Judith Masters of the Natal Museum in South Africa and her co-workers [24] have argued strongly that the strong southerly-directed ocean currents in the Channel would have made such eastwards dispersals impossible. However, Africa and Madagascar lay much further south earlier in the Cenozoic, and recent work on palaeo-oceanographic modelling [25] suggests that the ocean current systems would then have been very different. From the beginning of the Cenozoic until c.20 my ago, strong surface currents flowed from northeast Mozambique and Tanzania eastwards towards Madagascar, favoring sweepstakes dispersal. The present patterns, which instead favor dispersals from Madagascar to Africa, provide a ready explanation for the discovery, based on molecular work, that chameleons appear to have evolved in Madagascar and spread from there to Africa and other oceanic islands [26]. The coastal lizards *Cryptoblepharus* of Madagascar seem to have dispersed there from Indonesia or Australia [27] using the South Equatorial Current.

The presence of two lineages of the freshwater cichlid fishes in Madagascar provides another complication. Molecular studies suggest that these are most closely related to the cichlids of India rather than to those of Africa [28].

However, the time of divergence between the cichlids of the two areas, 29–25 million years ago, is far too late for them to be the result of vicariant evolution.

Because its immigrants came from only a few groups of mammals and birds, the successful colonists were able to occupy niches that would not normally be available to them—a phenomenon that is often found in such island biotas (see p. 228). Some of the now extinct types of lemur included the tree sloths and ground sloths of South America, as well as the koalas of Australia and gorillas, while the Carnivora, which belong to the mongoose family, included something very like a lion. The birds include the massive extinct herbivorous flightless elephant bird, as well as radiations of endemic passerine birds—the shrike—like vangas and the warblers [20].

It is obvious from all the above that molecular work is continually providing both new solutions and new problems in interpreting the biogeography of Madagascar. It will be fascinating to see how these studies progress in the future.

India and Southeast Asia

Because it separated from Gondwana in the Jurassic, before the flowering plants, marsupials, or placental mammals had evolved, the early biota of India would have been like that of the rest of Gondwana at that time. Thus, it would have had dinosaurs, as well as early types of Mesozoic mammal, of a more primitive evolutionary grade than the marsupials and placentals (as these primitive types were widespread in the Mesozoic world). In the Late Cretaceous, still-united India/Madagascar were part of the Palmae Province (cf. Fig. 10.7). Until the two land areas split apart 88 mya and for a while thereafter, India must have received flowering plants from Africa via that island. Eventually, India would have received progressively more flowering plants from Asia, as it approached and eventually collided with that continent in the Middle Eocene: there is clear evidence of the typically Indian or Gondwanan types of pollen in Southeast Asia after the collision [29]. These new angiosperms replaced much of the former flora of Southeast Asia, and those that survived are still present in the flora that is found in both regions today. This forms the Indo-Malesian floral kingdom and has also extended eastward far into the islands of the Pacific Ocean. After the Mid-Tertiary global cooling, the rainforests of Southeast Asia and China also received many frost-sensitive plants from the midlatitudes of Asia. A similar process led to the appearance of northern frost-sensitive plants in northern South America after the completion of the Panama Isthmus. The resulting similarity between elements in the forests of South America and China, sometimes referred to as "amphi-Pacific" elements, are thus the result of the southward migration of these plants at either end of the Northern Hemisphere, rather than being evidence for some "Pacific Ocean baseline," as suggested by panbiogeographers (cf. p. 19).

Unfortunately, the mammal fauna of India in the Cretaceous and Paleocene, before its collision with Asia, is unknown, but it must later have been quickly colonized by the placental mammals of that continent. The later history of exchange of mammals between India/Eurasia and Africa has been described above. About 3 million years ago, a similar increase in aridity, related to the uplift of the Himalayan Mountains to new heights, led

The series of islands between the mainlands of Asia and Australia contains a transition between the Asian flowering plants and placental mammals, and the Australian flowering plants and marsupial mammals. To plant biogeographers, the whole area is the Malaysian province of the Indo-Pacific floral kingdom, which extends eastward to include New Guinea and most of the islands of the Pacific Ocean. Zoologists, on the other hand, found that New Guinea contained marsupials, but very few placentals, and a predominantly Australian bird fauna. In the 19th century, the zoogeographer Alfred Russel Wallace had suggested a line of faunal demarcation, later called Wallace's Line, which separated the predominantly Asian bird fauna from the more eastern, predominantly Australian bird fauna. This line, which runs close to the Asian continental shelf, has in the past therefore been recognized as the boundary between the Oriental and the Australian zoogeographic regions. However, the area between the Asian and the Australian continental shelves contains relatively few mammals of any kind or origin. Later zoogeographers proposed six different variants of Wallace's Line [30]. The debate on where to draw a line diverted attention from the real interest of the area, which is the extent to which animals or plants have been able to enter or cross this pattern of islands from either direction. It is therefore best to draw the boundary of the Oriental and Australian faunal regions at the continental shelves, as is done in the other faunal regions, and to exclude the intervening area, which some earlier biogeographers have named Wallacea (Fig. 11.9).

The biogeographical problems of Wallacea are not as straightforward as a simple inspection of the modern map might suggest. One example is provided by the island of Sulawesi, which has a varied mammal fauna and complex geological origin [30]. The western part of the island was part of Borneo until it drifted away in the Eocene; it may have been a major source of Asian plants for the islands to the east, and for New Guinea [29]. In contrast, the eastern parts of Sulawesi were originally fragments of Australasia and only joined the rest of the island in the Early Miocene. The Makasar Strait between Borneo and Sulawesi is 104 km (65 mi) wide today. Sulawesi has a varied mammal fauna, including bats, rats, shrews, tarsiers, monkeys, porcupines, squirrels, civets, pigs, deer, and a fossil pygmy elephant. Many of these are endemic and so were not introduced by humans; they probably crossed the Makasar Strait during the Pleistocene, when lower sea levels would have reduced its width to only 40 km (25 mi). Although some types of mammal found in Sulawesi are found in islands further to the east, most were probably taken by humans; only the bats and rats appear to have made these additional travels unaided. Sulawesi is also the only island in Wallacea where there is a natural overlap between Asian and Australian mammals, for it contains two species of the marsupial phalangers. (Phalangers are the only marsupials that have dispersed into Wallacea, being found also on several other islands.) In general, the biotas of Wallacea are less diverse than those to the east or west, because of the difficulties of colonization across a series of ocean straits, and because of the inherent vulnerabilities of island biotas (see Chapter 8).

to the northern parts of the Indian subcontinent becoming much drier and to an increase in the number of grazing mammals such as horses, antelope, and camels, and of elephants.

Between Southeast Asia and the great island continent of Australia lies a complex pattern of islands that has been colonized from each of these two areas (see Box 11.1).

Australia

The history, characteristics, and biogeographical affinities of the biota of Australia are the most unusual and interesting of any in the world, and their explanation requires a very rewarding understanding of the interplay between continental movement, climatic change, and biotic dispersal [29, 32, 33]. The continent's only mountains lie in a narrow belt along its eastern edge, and they are mainly old (around 200 my). As a result, there have been no new mountains whose erosion might have provided new sediments and minerals to the vast flat expanses of the rest of the continent. Their soils therefore have about half the levels of nitrates

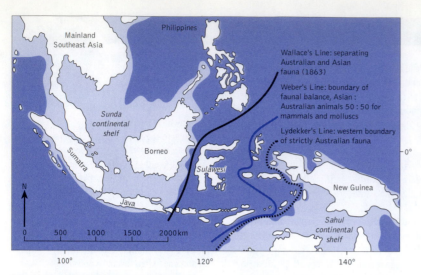

Fig. 11.9 Map of the East Indies. The continental shelves are shown in light blue, and the deeper ocean in dark blue, shaded. Three of the "Lines" of faunal division are shown and explained. "Wallacea" is the area that lies between Wallace's Line and Lydekker's Line. After Moss & Wilson [31].

In the map:
Mainland Southeast Asia
Philippines
Wallace's Line: separating Australian and Asian fauna (1863)
Weber's Line: boundary of faunal balance, Asian : Australian animals 50 : 50 for mammals and molluscs
Lydekker's Line: western boundary of strictly Australian fauna
Sunda continental shelf
Sumatra
Borneo
Sulawesi
Java
New Guinea
Sahul continental shelf
N
0 500 1000 1500 2000km
100° 120° 140°
0°

and phosphates of equivalent soils elsewhere. This situation has been exacerbated by the continent's plate tectonic history—a massive northward shift across the climatic belts, from an original far-south position with high rainfall, to its present position where low rainfall has turned its interior into a vast semi-arid zone.

By the Middle and Late Cretaceous, Gondwana had lost both India and Africa and retained only a narrow connection with South America, so that Antarctica/Australia was the most substantial remnant of the old supercontinent. Its biota included marsupial mammals, araucarian and podocarp conifers, ginkgoes, bennettitaleans, cycads, and some of the earliest flowering plants (including early members of the *Nothofagus* and *Protea* families). They were living in far-south latitudes which, though ice-free, had the almost sunless winters and almost continuously light summers seen in the Antarctic today. The Cretaceous and early Tertiary flora is known almost exclusively from southern Australia, where there is evidence of low-diversity, moist temperate rainforests dominated by gymnosperms, unlike any vegetation known today; *Nothofagus* later became more common. Comparison with later times, when there is evidence from other Australian environments, suggests that there were dense evergreen rainforests in more northern regions. In central Australia there were floras like that of today's sclerophyll and monsoonal tropics biomes, dominated by proteas rather than *Nothofagus*, and with a greater variety of angiosperms.

Australia and Antarctica started to split apart about 97 million years ago, but at first they separated quite slowly. They therefore remained parts of a single weather system, so that some of the warmth that Australia received was circulated by wind systems southward to its Antarctic neighbor. As a result, and also because at this time atmospheric CO_2 levels and global temperatures were quite high, there was no Antarctic ice cap. There was an extensive warm sea inlet between the two continents, so the climate of this region must have been warm with high rainfall. By 50 mya, there is evidence of major meso-megathermal gymnosperm-dominated rainforest there, similar to today's aseasonal-wet biome, with *Nothofagus*, and coastal mangroves. The old cool-temperate *Nothofagus* forests were now mainly confined to the southern part of Australia, and *Ginkgo* became extinct at about this time.

This, then, was the genial environment of Australia in the early Cenozoic, when its marsupials were undergoing their great radiation, whose diversity paralleled that of the absent placentals. It was also the time when the passerine birds started their diversification in the continent, prior to dispersing to Asia in the Eocene, where their descendants the oscines (songbirds) started their worldwide radiation that today comprises nearly half of all living species of bird (see p. 198).

Australia only started to move rapidly northward 46 million years ago, in the Eocene. By the Early Oligocene, around 35 million years ago, it had separated sufficiently far from Antarctica for the deepwater Antarctic Circumpolar Current of cooler water and associated westerly winds to become established. This was probably a major factor in the marked worldwide cooling that took place at this time. Now isolated from the warmth of Australia, Antarctica cooled and its ice sheets started to form. There would also have been less evaporation from the now cooler seas around Australia, reducing the rainfall on that continent, with a consequent increase in its arid, desert areas.

Today, a major characteristic of Australia is its "sclerophyll" flora, 45% of whose genera are endemic to the continent. **Sclerophyll** (or **scleromorph**) plants grow slowly, readily stop growing, and have a small total leaf area, made up of small, broad, evergreen, leathery leaves. It was long thought that this flora had evolved in response to the drying of the continent that took place as it continued to move northward into the 30°S high-pressure zone of low rainfall. However, plants with these characteristics appeared in the Australian record far earlier than this, 60–55 mya, and seem to have evolved from the rainforest flora, for all the larger families with sclerophyll types are also found in the rainforests. It is now clear that the sclerophyll habit started as an adaptation to Australia's low soil nutrients, and only later turned out to be adaptive also to increasing aridity. The most spectacularly successful sclerophyll forms are the gum-tree genus *Eucalyptus* (Myrtaceae), which includes over 500 species, and the family Proteaceae.

Australia became increasingly arid from the Late Oligocene and into the Miocene. This aridity was accompanied by the reduced dominance of plants characteristic of the aseasonal-wet biome (*Nothofagus* and conifers) and the increasing dominance of *Eucalyptus*. Over this same period of time, burning became an important aspect of Australian ecology. All of these changes led to the appearance of open grasslands and savanna in central Australia containing radiations of the eucalypts, casuarinas, pea-flowered legumes and *Acacia* (Fabaceae). By the end of the Pliocene, the *Eucalyptus* forests had changed to become dry, rather than wet, sclerophyll forests, and much of the continent had been taken over by the huge Eremean biome with its open, arid shrublands, woodlands, and grasslands, into which such herbs as brassicas and chenopods had arrived from elsewhere (Fig. 11.10).

This plate tectonic and climatic history has therefore combined to make Australia the driest of all continents; two-thirds of it have an annual rainfall of less than 500 mm, and one-third has less than 250 mm.

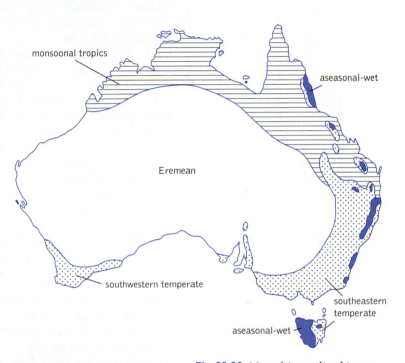

Fig. 11.10 Map of Australian biomes. From Crisp, Cooke, & Steane [33]. Southeastern temperate: sclerophyll eucalypt forest, woodlands and heath, seasonally dry. Southwestern temperate: sclerophyll eucalypt forest, woodlands and heath, Mediterranean climate.

Monsoonal tropics: savannah, mostly sclerophylll eucalypt and acacia, seasonally dry. Eremean: arid shrubland, low woodland and grassland.

Aseasonal wet: year-round high rainfall, tropical to temperate or subalpine, closed-canopy rainforest on volcanic soils, to heath on poor soils.

This also explains why, although both the rainforest and the sclerophyll floras once covered much greater areas of Australia, they are today found only in a relict distribution, in scattered peripheral areas.

The mammals of Australia had to adapt to these climatic and vegetational changes. Unfortunately, their fossil record in Australia does not begin until the Mid-Miocene. The herbivores of this fauna were mostly browsers, for the modern Eremean grassland ecosystem with grazers (especially the diverse kangaroos and wallabies) did not appear until the Pliocene. Uniquely, Australia's mammal fauna is made up of marsupials alone (plus the few egg-laying monotremes), and they have radiated into a great variety of forms, occupying the niches that placentals have filled everywhere else in the world. The marsupial equivalents of rats, mice, squirrels, jerboas, moles, badgers, anteaters, rabbits, cats, wolves, and bears all look very like their placental counterparts—only the kangaroo looks quite unlike its placental equivalent, the horse.

The sclerophyll vegetation of Australia is low in nutrients and high in toxic biochemicals to deter the herbivore. The effects of this in depressing the population density of herbivores are shown by the brush opossum, *Trichosurus vulpecula*, whose density in the very different vegetation of New Zealand is five to six times greater than its density in its native Australia [34]. The effects of the low nutrients on the Australian carnivores and on the size of its herbivores have already been noted (see p. 336).

Although Australia had parted from its original Gondwana relationships, its northward movement into the Pacific eventually brought its northern edge close to a great oceanic trench (see p. 155), where old ocean-crust material was sinking downward. This lighter material therefore came to underlie the northern edge of the Australian continent, causing it to rise from the Middle Miocene onward, and form the mountains of New Guinea, which is the largest island in the world (nearly 800,000 km^2, 300,000 mi^2). These mountains provided a high, cool environment that today is the wettest area on Earth, although it is close to the driest continent. This mountainous region was colonized by the Australian rainforest flora, including *Nothofagus*, but the surrounding lowlands of New Guinea were colonized by a mixture of Asian and Australian plants [28]. Asian placental mammals, too, started to spread eastward but, apart from humans and the aerial bats, only the rats spread naturally as far as Australia, where their 50 species now form 50% of the diversity of the Australian land mammal fauna. Human beings probably arrived 60,000 to 40,000 years ago and brought the domestic dog (the ancestor of the dingo) about 3,500 years ago.

New Zealand

New Zealand is unique. Its total area, 270,000 km^2 (104,000 sq mi), makes it far larger than any other Pacific oceanic island, and it is therefore not so vulnerable to the fluctuations of population and diversity that characterize them (see Chapter 8). It is also unlike other Pacific oceanic islands in being a fragment of old Gondwana, rather than being entirely volcanic, and has suffered from a series of changes in area. When it first separated from Gondwana in the Late Cretaceous, about

82 million years ago, it was part of a minicontinent known as Zealandia [35, 36], almost half the size of Australia, and stretching as far as New Caledonia to the northwest and Chatham Island to the southeast (Fig. 11.11). This small landmass drifted northeastward to its present position, separated from Australia by the Tasman Sea, 1,400 km (870 mi) wide. Most of it sank below sea level 30–25 mya, during the Oligocene, and it is possible that all of it was submerged at this time. The geological evidence for any part of New Zealand having been dry land at the end of the Oligocene, 26 mya, is very weak, and at present the biogeographical evidence on this seems more clear-cut. However that may be, much or all of the land that we see today emerged only after the end of the Oligocene, when the Alpine Fault that runs the length of the islands was initiated by tectonic plate boundary collision. More recently, the New Zealand biota was greatly affected by the Ice Age climatic cooling, which caused extensive glaciation there and must have led to many extinctions.

New Zealand's isolated position has made the interpretation of its flora and fauna an interesting battlefield. On one side are the panbiogeography proponents of vicariance, for whom patterns of scattered, disjunct distribution are the result of the subsequent fragmentation of an originally single homeland, so that the members of the present biota of New Zealand are merely the descendants of that original biota. On the other side are the proponents of dispersal, for whom these patterns instead resulted from the organisms crossing the intervening barriers after they formed. (See pp. 18–20 for a discussion of this controversy.)

Recent molecular studies that provide the dates of divergence of the New Zealand lineages from their closest external relatives have shown convincingly that New Zealand's organisms (especially the plants) reached there in comparatively recent times, and therefore by dispersal [36]. The biological evidence for the persistence of dry land in New Zealand since before the Oligocene comes down to a relatively few animals (and no plants). Probably the strongest evidence is the presence of the two flightless ratite birds, the moa and kiwi, which could neither have flown nor swum to New Zealand. Both the unusual tuatara lizard *Sphenodon* (whose ancestors were in the island in the Miocene) and the primitive frog *Leiopelma* are evolutionary relics of once widespread groups. This may also be true of the only fossil mammal known from New Zealand; found in Miocene rocks, it is probably a relict of the early egg-laying mammals (see Fig. 10.4, p. 316), as are the monotremes of Australia.

With regard to the plants, the origin of New Zealand's vascular plant flora has been discussed by Mike Pole of the University of Tasmania [38]. The nature of the earliest flora, from the Late Cretaceous, reflects New Zealand's cold, high-latitude position at that time, while the excellent record from the Miocene (by which date the islands were already extremely isolated) is of floras characteristic of rainforest or sclerophyll forest/woodland. This Miocene flora must have been the result of dispersal from Australia, which also then had a similar climate and flora but, like the earlier flora, it is quite unlike the flora of New Zealand today. Its Miocene flora

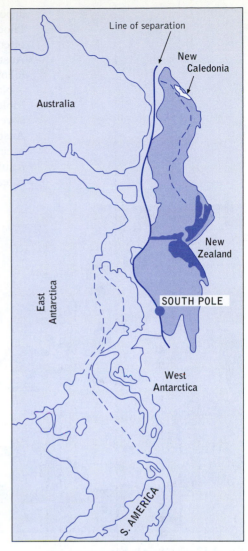

Fig. 11.11 The original position of Zealandia (shaded) close to the South Pole, before it broke away from Gondwana. From McDowell [36].

appears to have become largely extinct following the climatic cooling at the Pliocene/Pleistocene boundary. Its modern flora was likely the result of often multiple, long-distance dispersal from the present-day flora of Australia, aided by the strong west winds that blow around Antarctica, followed by major radiations of the successful immigrants within Australia.

Not all of New Zealand's immigrants, however, arrived from the west. Molecular phylogenetic results suggest that a few plants have dispersed in a westerly direction, against the prevailing currents of wind and water, to Australia and New Guinea. These may be the result of dispersal by oceanic birds, which fly for long distances. An even more surprising result suggests that the little creeping flowering plant *Tetrachondra* spread to New Zealand from South America. However, this may not have been directly across the thousands of miles of ocean that now separate the two areas of land; fossil wood of *Nothofagus* has been found in Pliocene deposits (5–2 million years old) in Antarctica, so this continent could have provided an intermediate stepping-stone.

As a result of all these factors, the New Zealand flora is highly endemic at the species level (86% of its less than 2,500 species are endemic), but with no endemic families, perhaps because as yet there has not been sufficient time for their evolution to have taken place. The flora also shows especially high endemicity in such plants as ferns and orchids and the tree *Metrosideros*, all of which are characteristic of floras of distant islands, as is the fact that some of the herbs have evolved into endemic trees (see Chapter 8).

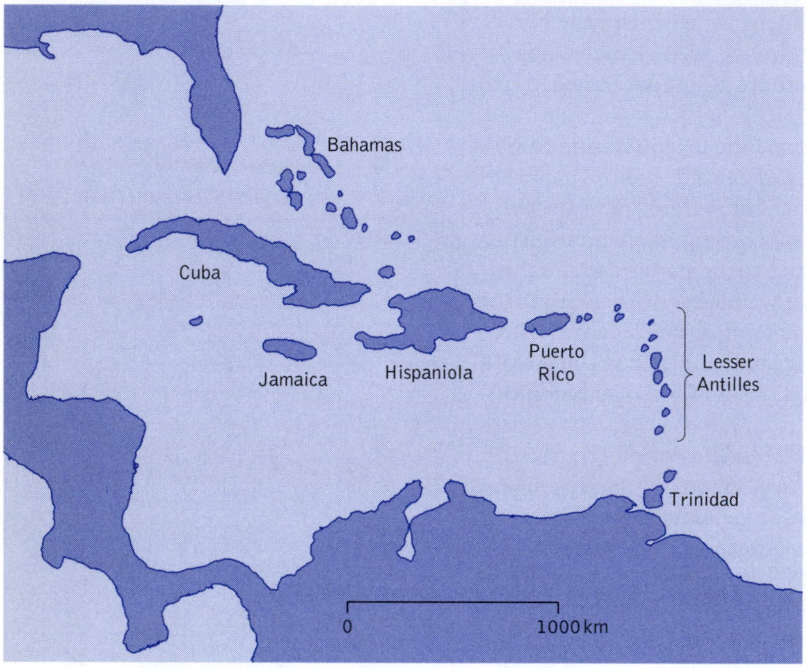

Fig. 11.12 The West Indies.

The West Indies

Though superficially similar in their geographical position, poised between two continents (Fig. 11.12), the islands of the Caribbean do not present the same fascinating problems as the islands of Wallacea because they do not lie between continents with totally different faunas and floras. Instead, they have posed a different, interesting problem: are their faunas and floras primarily the result of vicariance, due to evolution on islands that arose by the breakup of a larger "proto-Antillean" landmass, or are they the result of dispersal from the neighboring continents? This question, and the geological history of the region, have been reviewed by the American biogeographer Blair Hedges [39; see also Guest Author Box 11.2].

By Dr. S. Blair Hedges, Department of Biology, Pennsylvania State University

The islands of the Caribbean have provided a classic test of the two major mechanisms of historical biogeography: vicariance and dispersal. Formed in the Mid-Cretaceous (~100 mya), they have had a long and complex geologic history that included an early connection between North and South America (proto-Antilles) and a catastrophic asteroid impact (~66 mya). During the Cenozoic, some large islands (Greater Antilles) broke apart and fused, a stable carbonate platform (Bahamas Bank) kept up with sea-level changes, and a chain of volcanic islands (Lesser Antilles) migrated slowly from west to east. Soon after the theory of plate tectonics became accepted, it was recognized that the current biota of the islands may be the fragmented (vicariant) remnant of a once continuous proto-Antillean biota. For the last three decades, a debate has ensued over the importance of vicariance versus dispersal in the origin of the biota.

The answer has not come easily, nor has it been agreed upon by everyone. Nonetheless, most research suggests that the entire living biota of the Caribbean islands arrived by dispersal and not through the geologic breakup of an ancient landmass. Initially, it was thought that the key information to answer the question would come from the phylogenetic relationships of organisms. In part, this thinking emerged from the popularity—in the 1980s—of the field of cladistics (and vicariance biogeography) which emphasizes relationships of organisms over most other types of data. Undoubtedly, relationships are important, but the problem with this line of thinking is that the branching order of species might match the geologic breakup of land areas, but the timing could be very different. So it was soon realized that data on the times of divergence of organisms from their closest relatives on the mainland were critical data. The fossil record in this region is poor, but molecular clocks provided those data.

Molecular clocks (see Chapter 7, p. 215) need calibration against some external events, such as well-dated fossils, or geological events, such as the time of emergence of an island above sea level—for this is the earliest time at which it could be occupied by terrestrial organisms. For the Caribbean islands, it turned out that relationships were not important in answering the basic question of vicariance versus dispersal. This is because nearly all of the times of divergence measured by molecular clocks, for many different groups of terrestrial vertebrates, have been too young to have resulted from a Late Cretaceous vicariance event. Instead, the times were scattered throughout the Cenozoic, almost randomly, and in accord with a mechanism that relies on chance events such as dispersal. However, relationships were useful in determining the source area of dispersal. For most terrestrial vertebrates that cannot fly or otherwise disperse over water on their own powers, their closest relatives are in South America, while a majority of the birds, bats, and freshwater fishes in the Caribbean islands appear to have come from North and Central America.

Other diverse evidence, too, supports a dispersal origin for the Caribbean biota. Most important among them is the taxonomic composition of the endemic groups. There are some enormous adaptive radiations, often with species filling niches different from those in the same genus on the mainland. For example, some of the smallest and largest species of major groups (e.g., cycads, swallowtail butterflies, frogs, lizards, and snakes) occur on islands in the Caribbean. Yet at the same time, many major groups are absent, such as salamanders and caecilian amphibians, marsupials, rabbits, armadillos, and carnivorous placental mammals. The fossil record, including that of the 15–20 my old Dominican amber, which includes the remains of insects, frogs, lizards, and small mammals, shows a similar taxonomic composition. This is best interpreted as a strong filter effect, whereby a few colonists survive long-distance dispersal and then radiate into a diversity of unoccupied ecological niches. This same evidence also argues against an origin for the biota by way of a Mid-Cenozoic (~34 mya) land bridge from South America, which has also been suggested. Such a land bridge would not have acted as a strong filter and would have allowed many other groups to enter the archipelago that we do not in fact find there.

Ocean currents and geographic proximity best explain the source areas of the island colonists identified in molecular phylogenies. Water flows almost unidirectionally from east and southeast to west and northwest in the Caribbean—and this was true even prior to the uplift of the Isthmus of Panama. As a result, flotsam ejected carried down from the rivers in northern and northeastern South America will end up in the Caribbean, if it continues to float. For example, even though Cuba is much closer to North America than to South America, it is much easier for a lizard to arrive in Cuba by floating on vegetation from South America; this is reflected in the composition of the Cuban lizard fauna. But for organisms that can fly or swim, the geographically closer areas are the more likely sources, and the common air current direction in the Caribbean—northeast to southwest—might even assist dispersers flying from North America.

Two lineages of island vertebrates that show old (Cretaceous) times of divergence from their closest relatives on the mainland, using molecular clocks, have been debated as possible examples of proto-Antillean vicariance. These are the giant shrews (solenodons) of Cuba and Hispaniola and the night lizards (Xantusiidae) of Cuba. While an ancient origin cannot be ruled out, both groups are biogeographic relics, for their mainland fossil record demonstrates a wider distribution in the past. This raises the possibility—not normally considered for other groups—that they diverged more recently from close relatives on the mainland that are now extinct and hence inaccessible to molecular clocks. Some geologists also are uncertain about whether there was any continuously emergent land in the Caribbean before the late Eocene (~37 mya), which would have been necessary for maintaining such lineages. Moreover, it is not clear how these organisms might have survived the end-Cretaceous asteroid impact, which occurred a short distance away. The origin of these two groups will likely continue to be debated.

Now we know that flotsam was critical for the origin of the Caribbean terrestrial biota, but surprisingly little is known about this mode of dispersal across ocean waters. How abundant are floating islands? How long do they stay afloat, and how far do they travel? What organisms do they typically carry? There are many anecdotal accounts of floating islands but almost no scientific studies. Analysis of satellite imagery, GPS tracking, and taxonomic surveys of floating islands might answer some of these questions. Whatever the details, we can be certain that long-distance dispersal by flotsam did occur and that fragile animals—such as small frogs—successfully colonized Caribbean islands millions of years ago after riding the ocean waves for weeks on a jumble of logs.

What is now the Caribbean region began as merely a gap between North America and South America (Fig. 11.13). In this gap, there was opposition between the expansion resulting from the mid-Atlantic spreading ridge to the east, and that resulting from the East Pacific spreading ridge to the west. Part of this expansion was taken up by an eastern trench, where old seafloor disappeared into the depths of the earth, and this would have been accompanied by volcanic activity and the appearance of a chain of volcanic islands. As the Americas moved westward, this whole system was left behind in a more eastern position, and today forms the Lesser Antillean island chain (see Fig. 11.12) and trench system. Here the Atlantic plate seafloor is still being consumed, and this marks the eastern boundary of the new, small Caribbean tectonic plate. (The Scotia Sea

Late Cretaceous 150 mya

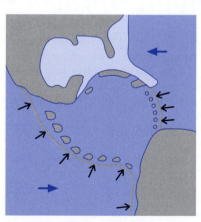

Eocene 50 mya

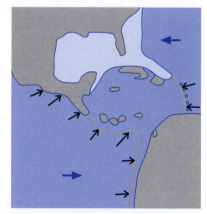

Early Miocene 20 mya

Fig. 11.13 The formation of the West Indies. Dark gray tint indicates dry land, light blue tint indicates shallow water, dark blue tint indicates deep water. Filled arrows indicate plate motion: AP, Atlantic Plate, CP Caribbean Plate, PP Pacific Plate. Smaller arrows indicate where ocean crust is being consumed, leading to the appearance of volcanic islands. Modifed, after Huggett [57], with permission of Routledge Press.

today provides a model of this situation, for it is formed from a small portion of the South Pacific plate that extended between the Antarctic Peninsula and South America as the latter moved westward.) As this took place, a new system of trench and volcanic islands formed to the west, where seafloor produced by the East Pacific spreading center was consumed. This marks the western boundary of the Caribbean plate, and its volcanic islands gradually coalesced to form the Panama Isthmus, linking North and South America (Fig. 11.14). There was also, in the Early Cenozoic, a northward component of movement as South America approached North America, resulting in the appearance of the Greater Antilles (the larger islands of Cuba, Jamaica, Hispaniola, and Puerto Rico) along the northern margin of the Caribbean plate.

It is unfortunately very difficult to be sure which islands, or parts of islands, were above sea level at particular times. This is because this would have been the result of the complex interaction of the timing and intensity of volcanic activity, the processes of uplift versus erosion, and changes in sea level (at the beginning of the Oligocene, 32 million years ago, the sea level dropped by 160 m [520 ft]!). The history of the Caribbean islands is therefore much more poorly understood than that of the Hawaiian Islands (cf. p. 232).

Although Hispaniola appears to have been formed by the fusion of two smaller islands, the remaining Caribbean islands remained separate units that individually appeared or disappeared. This makes it likely that the organisms found there must have arrived by overwater dispersal followed by independent evolution, rather than by vicariance. As Hedges points out, the nature of the faunas of the Caribbean islands strongly supports this interpretation. The ecological diversity of the islands is largely based on the radiation of a comparatively small number of higher taxa, some of which contain a very large number of species. This suggests that a small number of immigrants, finding themselves in environments lacking their normal competitors, were able to diversify opportunistically into the vacant ecological niches. Today we recognize over 1,300 native species of freshwater or terrestrial vertebrate in the islands, of which 75% are endemic (see Table 11.3).

The results of these radiations are impressive. The eleutherodactylid family of frogs has radiated into 161 species, all of which are endemic to the islands. The anguid, iguanid, gekkonid, and teiid families of lizard have produced 338 species, all but three of which are endemic. The boid, dipsadid, leptotyphlopid, tropidophiid, and typhlopid families of snake have given rise to 129 species, of which all but two are endemic. However, it must be remembered that these species are distributed over the 29 islands of the Lesser Antilles, plus the Bahamas and the four larger islands of the Greater Antilles, and that most species are restricted to one island, and often to a small area within that island. There are also interesting examples of parallel evolution, most clearly seen in the lizards of the genus *Anolis* that live on trees on the Greater Antilles. Ecologists have distinguished at least six morphological types of this lizard, differing in characteristics of body size, limb proportions, and so on, each of which occupies a different part of the tree (crown; twigs; upper, middle or lower trunk; on the ground or on bushes). Cuba and Hispaniola contain all six types, Puerto Rico contains five, and Jamaica

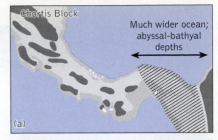

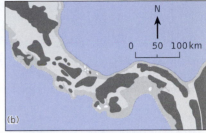

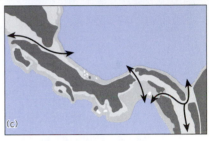

Fig. 11.14 The formation of the Panama Isthmus. Dark gray tint indicates dry land, shallow marine sediments indicated by dots, and deep ocean sediments by horizontal parallel lines. From Coates & Abando [58], with permission of The University of Chicago Press.
(a) Middle Miocene, 16 to 15 million years ago.
(b) Late Miocene, 7 to 6 million years ago.
(c) Late Pliocene, approximately 3 million years ago.

Table 11.3 The taxonomic diversity of native West Indian land vertebrates. From Hedges [39]

Group	Orders	Families[a]	Genera Total	Genera Endemic	Genera % Endemic	Species Total	Species Endemic	Species %Endemic
Freshwater fishes	6	9	14	6	43	74	71	96
Amphibians	1	4	6	1	17	173	171	99
Reptiles	3	19	50	9	18	499	478	96
Birds	15	49	204	38	19	425	150	35
Mammals:								
Bats	1	7	32	8	25	58	29	50
Other[b]	4	9	36	33	92	90	90	100
TOTALS	30	97	342	95	28	1319	989	75

[a] Includes one endemic family of birds (Todidae) and four of mammals (Capromyidae, Heptaxodontidae, Nesophontidae, and Solenodontidae).
[b] Edentates, insectivores, primates, and rodents.

four, but each type is represented by one or more independently evolved species on each island, so that there is a total of 128 species of *Anolis* on the four islands.

The West Indies contain 672 species of amphibian and reptile, which belong to about 75 different evolutionary lineages. Of these, 49 are related to South American taxa, and only four to North American taxa (the origins of the remaining 22 lineages are Africa, Central America, or uncertain). This pattern is probably the result of dispersal within a Caribbean area in which the patterns of both sea currents and wind currents (sometimes of hurricane strength) run from east to west. It is also significant that the great rivers of northern South America, such as the Orinoco, open to the east of the Caribbean; others opened to the south of the area earlier in the Cenozoic, and all may have acted as sources for flotsam that may have borne amphibians and reptiles. In contrast, most of the 425 lineages of birds of the Caribbean region are related to those of North America, probably because the islands are closer to that continent, and also were used as over-wintering refuges. Finally, the Caribbean islands were also colonized by members of the exclusively South American mammals, the caviomorph rodents (four lineages), ground sloths (two lineages), and New World monkeys (one lineage), which evolved into endemic genera and species.

Hedges points out that the geological and ecological picture suggesting that these land vertebrates arrived in the West Indian islands by dispersal, rather than by vicariance, is supported by the times of divergence of these lineages. Molecular methods studying the amino-acid sequence divergence of the protein serum albumin (see Chapter 7), which has a remarkably constant rate of evolution, and also DNA-sequence "clocks," suggest that these lineages diverged at a variety of times during the mid-Tertiary (Fig. 11.15). The geological evidence on the palaeogeography of the Caribbean indicates that its pattern of large and small islands during that time was similar to that of today, though not identical. This evidence therefore does not support suggestions that there had been immigration to the region over a land bridge that later broke up into independent islands. Such a pattern would instead have resulted in a clustering of divergence times around a few dates corresponding to the times of breakup.

In addition to these biogeographic relationships between the West Indies and other parts of the New World, two reptiles (the amphisbaenian *Cadea* and the gecko *Tarentola*) appear to have arrived in the West Indies from the Mediterranean region about 40 million years ago.

The West Indies islands thus provide a fascinating picture in which geological, faunal and molecular studies seem to be converging on a convincing story of dispersal into and through a region of complex history. It is also an interesting contrast to the East Indies or Wallacea, where the history of the neighboring continents has been quite different.

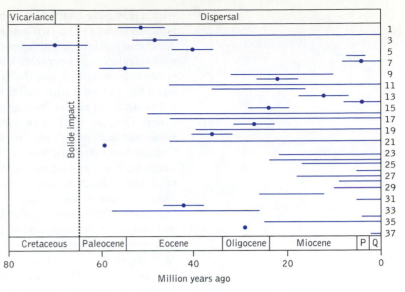

Fig. 11.15 Times of origin of 37 independent lineages of endemic West Indian amphibians and reptiles. From Hedges [39].

South America

The history of the biota of South America has been dominated by the effects of the great earth-engine of plate tectonics. Some of these were the result of the alternate establishing and breaking of links between South America and North America. As noted earlier in the description of the origin of the Caribbean islands, there had been islands in that region ever since the Americas started to move westward in the Early Cretaceous. This may have been the route by which some North American dinosaurs dispersed southward (see p. 313) in the Mid-Cretaceous, accompanied by some early types of marsupial and placental mammal. But this connection has always been tenuous, sometimes permitting passage and sometimes becoming broken. This led to a cycle of immigration, isolation, and evolution, and final new immigration and extinction. The other effects were the results of the westward drift of the continent, which led to mountain-building and consequent climatic changes.

The Early Cenozoic

The mammal fauna of South America today is characterized by a few marsupials (opossums) and a diversity of edentate placental mammals known as the Xenarthra (sloths, armadillos, and South American anteaters) that are hardly known elsewhere. Otherwise, it does not seem particularly unlike other mammal faunas, for it includes members of most of the placental orders (see Table 11.1). However, before the extinctions and immigrations, South America had a unique mammal fauna of strange marsupials and of ungulate placentals. Later, in the earliest Cenozoic, 10 different groups of mammal and a giant flightless bird had migrated northward into North America. Their passage across the Caribbean plate may have been aided by low sea levels at that time, which may have changed the chain of volcanic islands into a temporary isthmus between the two continents. By the early Cenozoic, the South American placentals had diversified into 22 families of unusual herbivorous ungulate, the edentates had evolved into the anteaters, sloths, and armoured armadillos, and the marsupials had evolved into the opossums and a group of carnivores called the borhyaenids.

The northern part of the continent in which this fauna lived was, as we saw in Chapter 10, largely covered by megathermal rainforests of the Equatorial Belt or Palmae Province (cf. Fig. 10.8). The Terminal Eocene Cooling Event does not seem to have led to any major change to drier environments here, as happened in the Far East, so that the present Amazon rainforest is the direct descendant of this early megathermal forest. In the Paleocene, there was a narrow band of similar forest in the Southern Megathermal Province. This expanded southward during the Eocene thermal maximum, but disappeared during the Terminal Eocene Cooling Event. It was replaced by microthermal forest containing the southern beech tree *Nothofagus* and other characteristic Southern Hemisphere elements, such as members of the Proteaceae, Restionaceae, and Gunneraceae, and conifers such as *Araucaria*, *Podocarpus*, and *Dacrydium* (all of which survived through the Cenozoic both in southern South America and in Australia).

Two other endemic South America mammal groups are the New World monkeys and the caviomorph rodents (which include the guinea pig and the capybara), whose closest relatives live in Africa. The earliest fossils of these two groups are from the Early Oligocene, 37 my ago, a time when South America had already drifted a considerable distance away from Africa. However, recent molecular work [40] on the date of divergence of these groups from their African relatives suggests that they arrived much earlier, c.50 my ago, when the ocean gap between the two continents was significantly smaller. As today, the ocean currents in the equatorial South Atlantic would have been westwards (cf. Fig. 5.8, p. 163), aiding such a transoceanic rafting.

Ever since it started to separate from Africa in the Middle Cretaceous, South America had been moving westward toward an oceanic trench that lay in the eastern part of the South Pacific. Along this trench, old eastern Pacific seafloor was being drawn back into the earth (see Fig. 5.2, p. 155). When South America eventually reached the trench, this tectonic movement caused volcanic and earthquake activity that led to the appearance of the Andean mountain chain. Today, nearly 8,000 km (5,000 mi) long, this is the longest continental mountain chain in the world and rises to heights of up to nearly 7,000 m (23,000 ft). Mountain-formation began in the south about 40 mya, the southern Andes reaching a height of less than 1,000 m (3,300 ft) in the Early Cenozoic, and gradually spread northward, the most northeastern Andes being the last to appear, about 11 mya. The rise of these mountains caused a reversal in the direction of flow of northwestern rivers such as the Amazon and the Orinoco, and led to vicariant evolution in the biota of that region.

The rain shadows caused by the Miocene uplift of the Andes led to the replacement of much of the former forests by woodlands and grasslands, and the associated radiation of the endemic South American placental herbivores. This is also the time at which cool-temperate plants first appear in northern South America, having dispersed from the north. Grasslands as a major element in the flora first appeared in South America about 30 mya (earlier than in North America, where this did not happen until 25–20 mya). The Early Oligocene fauna of South America is the oldest in the world that is dominated by grazing herbivores.

The Late Cenozoic/Pleistocene

In addition to their westward movement, North and South America also moved roughly 300 km (180 mi) closer together from the Middle Eocene to the Middle Miocene. From the Middle Miocene onward, the water level along the westward margin of the Caribbean plate became steadily shallower. Volcanic islands formed and enlarged to create an increasingly complete link between the two continents (Fig. 11.14). The first biogeographical evidence of this link came in the Late Miocene and Early Pliocene, when two families from each continent crossed this island chain to reach the other continent. The Panama Isthmus became a complete land bridge for the first time about 3 million years ago, in the Middle Pliocene.

The Pliocene witnessed another geological event that had a major impact on the biota of South America. This was the great Pliocene uplift of the Andes, which doubled their height from 2,000 to 4,000 m (6,500 to 13,000 ft). As a result, the extensive intermingling of the faunas of North and South America that took place after the final completion of the Panama land bridge also took place at a time of profound ecological change in the South American continent. The fascinating interplay of geology and biogeography has been analyzed in rewarding detail, particularly by the American palaeontologists Larry Marshall and David Webb [41–43], who have called it the Great American Interchange.

Before the Interchange, each continent had 26 families of land mammal, and about 16 families from each dispersed to the other continent. Of the North American mammals, 29 genera dispersed southward, mainly in the Late Pliocene/Early Pleistocene, about 2.5 million years ago, plus a few about 1 million years later. At the same time, nine genera of South American mammal dispersed northward; about 1.5 mya, they were followed northward by opossums, porcupines, and armadillos (Fig. 11.16).

The fauna that was exchanged between the two continents was one that was adapted to a savanna, open-country environment, suggesting that this was the environment of the connecting Panama Isthmus when the exchange took place. That inference is supported by the fact that several types of birds and xerophilous shrubs that are typical of savannas are now found both north and south of the Isthmus.

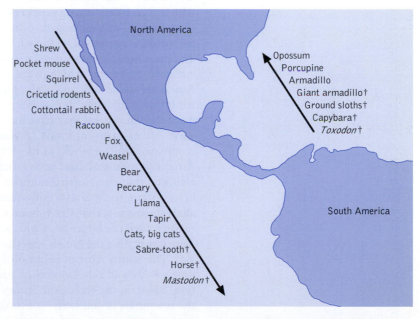

Fig. 11.16 The migrants of the Great American Interchange. A cross after the name indicates that it later became extinct in its new continent.

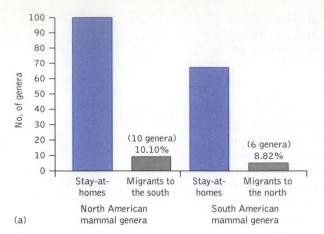

(a)

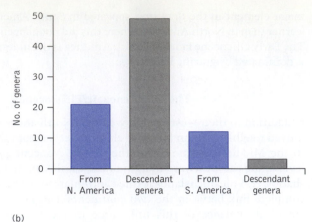

(b)

Fig. 11.17 Left, the numbers of Pliocene genera of North and South American mammal that stayed at home versus the numbers that emigrated to the other continent.

Right, relative success of the immigrants during the Pleistocene. The dark bars show the number of genera that had arrived from the other continent; the light bars show the number of genera that resulted from that immigration. After Marshall et al. [59].

Although similar proportions of the two faunas (9–10% of the genera) emigrated, the North American emigrants were far more successful than those from South America (Fig 11.17). In South America, 85 (50%) of the living land mammal genera are descended from South American immigrants, while the corresponding figure for North America is only 29 (21%). Easily the most successful immigrants were the cricetid rodents, which diversified into 45 genera in South America. But several other families, such as canids, horses, camelids (which evolved into llamas), and peccaries, contributed to the success of North American families there. The most notable extinction in South America was that of all 13 genera of endemic ungulate, which were unable to cope with the new North American carnivores, or with competition from the immigrant North American ungulates. It has been suggested that the North American mammals in general may have had greater success because they were the survivors of many millions of years of competition between mammal groups within the whole of the Northern Hemisphere, while the South American mammals had been protected in their isolated continent. It is also true that the North American ungulates had proportionately larger brains than their South American counterparts. However, statistical analysis suggests that size was an important factor: the types that became extinct were the larger ones and, because the South American mammals were larger than the North American ones, more of them became extinct [44]. The South African palaeontologist Elizabeth Vrba of South Africa, now working at Yale University, has offered a more sophisticated suggestion, based on the ecological history of the region [45]. She points out that the mammals of southern North America contained many more lineages adapted to the open, arid savanna environment that was spreading in the region, as the result of a general cooling of the climate. This alone gave them an advantage over the South American mammals, which were mainly adapted to a closed forest environment—which, as a result of this climatic change, was becoming fragmented and changing to savanna. So the South American mammals were suffering from a dual, climate-induced change in their environment that worked in favor of their North American competitors.

A final phase in the transformation of the South American land mammal fauna came during the Pleistocene. The climatic change that in the Northern Hemisphere caused the Ice Ages and the associated biotic changes (see Chapter 12) also caused a number of extinctions in South America. These included the last of the giant ground sloths, the giant armadillo-like glyptodonts, and the herbivorous South American ungulate *Toxodon* that had made a fleeting appearance in southern North America. In addition, the mastodont elephants and the horses, which had immigrated from North America, died out both there and in South America—although both survived in the Old World.

The tapirs and llamas that had originally dispersed from North America became extinct there but survived in South America. As a result, these two groups now show a disjunct, relict distribution in which their surviving relatives (tapirs and camels) are found in Asia. Perhaps it is not surprising that few of the South American forms that had dispersed to North America when it was warm were able to survive there in the colder environment: only the opossum, armadillo, and porcupine survived.

The final result of the ecological and biogeographical changes is thus a South American mammal fauna that shows little trace of its original inheritance from the Early Cenozoic. The characteristic flowering plant families today include the xerophytic Cactaceae, the Bromeliaceae (the latter include the pineapple and are often xerophytic), the Tropaeolaceae (including the garden nasturtium), and the Caricaceae (pawpaw or papaya tree).

The Rainforests of South America

The adjoining basins of the Amazon and Orinoco rivers (which together are about the size of the continental United States) contain approximately half of today's rainforests, the highest number of genera of palms, the richest of the world's vertebrate biotas, and more species of fish than in the whole of the North Atlantic Ocean. (South America was not as badly affected by the Terminal Eocene cooling as were the other tropical regions, and it has been suggested that this may have been one of the reasons why it was able to retain a richer flora.) The biogeographic patterns to be found in the area have been much debated. This debate began with the suggestion by the American zoologist Jurgen Haffer [46] that the biogeography of many Amazon forest birds showed two features. First, there are a few (about six) areas that each contain clusters of endemic species with rather restricted and similar ranges; these centers of endemism together contain about 150 species of birds, which make up 25% of the forest bird fauna of the Amazon Basin. Second, Haffer also found evidence, between these centers, of zones where the related species of the centers of endemism hybridized. He hypothesized that, during the Quaternary glacial periods of the Northern Hemisphere, the centers of endemicity had been islands of persistently high rainfall (and therefore of rainforest), surrounded by areas of grassland (Fig. 11.18). Many bird species had been able to survive in these refugia, in each of which they had evolved toward becoming separate species (rather like the glacial relicts of the Northern Hemisphere, cf. p. 52). However, they had not yet become

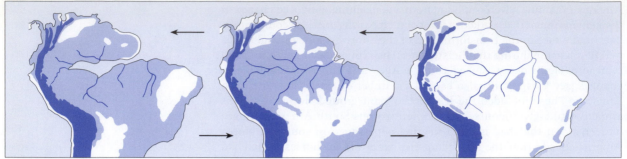

Fig. 11.18 Maps of northern South America showing the suggested extent of the lowland rainforests (light blue) during the different phases of the glacial cycle. Dark blue areas represent land above 1,000 m (3,300 ft), and white represents grassland. The coastlines have been adjusted to reflect the changes in sea levels, but the pattern of the Amazon Basin drainage is left unaltered in all three maps, to facilitate comparison. After Lynch [60].

fully separate species before the climatic improvement had allowed the forest and its fauna to spread again over the intervening grasslands. As a result, when the related forms met again, they were still able to hybridize.

Haffer's theory provided a neat and simple explanation of the situation as it was seen in 1969, but later studies have shown both the weaknesses in his theory and the great variety of factors that have influenced the biogeographic patterns that we see today—and therefore provide an object lesson for many other biogeographic studies. This is most clearly seen in recent work on the parrot genus *Pionopsitta* [47]. The introduction of cladistic and molecular studies shows the errors in earlier classifications (and therefore also of the biogeographic patterns that they suggest). The revised taxonomy reveals the presence of two separate clades, east and west of the northern Andes, and whose Pliocene dates of divergence suggest that they resulted from vicariance caused by that geological event— and so, much earlier than the Quaternary events that Haffer suggests as causes.

Other work, on curassow birds and howler monkeys, shows a similar pattern of relatedness to the Andean orogenies. A similar lack of correspondence of biogeographic pattern and Quaternary events is demonstrated by the work of Leonora Costa [48], who investigated the comparative phylogeography of 11 monophyletic groups of small mammals living in the forests of Amazonia, in the separate forest of the Atlantic coast, and in the intervening area. She found that the times of diversification of 11 different lineages of rodent and opossum varied greatly, but that many of them were far older than the Pleistocene.

Research on quite different topics shows that the climatic history of the Amazon Basin [49] may have been dominated by rhythms in the strength of the low-level atmospheric jet stream, rather than showing the same pattern and timescale as the Ice Ages of the Northern Hemisphere—though the latter may have affected the climates of the Andes themselves. Finally, a variety of geological and palaeoecological data show that there has been biome stability in the Amazon Basin throughout the late Tertiary and Quaternary, with no evidence of increased areas of savanna, and that the climatic changes of the Ice Ages resulted only in population changes within the plant communities, rather than biome replacements [50].

All in all, this varied work highlights the extent to which a previously unsuspected variety of factors may interact in producing biogeographic patterns, and must act as a strong warning that simplistic answers are probably inadequate. Though this may be especially true of the Amazon region, it would be sensible to assume that this lesson also applies to other areas and continents.

The Northern Hemisphere: Holarctic Mammals and Boreal Plants

In contrast to the complexity of the geological and geographical history of the Southern Hemisphere, with large areas of land splitting far apart from one another, moving through significant bands of latitude and climate, and colliding with one another, that of the Northern Hemisphere has been comparatively uniform. Although shallow epicontinental seas and the developing North Atlantic have from time to time subdivided the land areas of North America and Eurasia into different patterns (cf. Fig. 5.7), the two continents have never been far apart, so that dispersal between them has usually been fairly easy. Furthermore, all these areas lay to the east or west of one another, at almost identical latitudes, and their faunas and floras were therefore similar in their nature and adaptations. Consequently, the biogeographic results of plate tectonics in the Northern Hemisphere were far less complex and dramatic than those in the Southern Hemisphere [51]. In addition, their faunas and floras have in the recent past all suffered from the severe climatic effects of the Ice Ages. As a result, they are the two great regions of temperate- and cold-adapted biota. Although the two continents are usually distinguished as being separate "North American" and "Eurasian" zoogeographic regions, they are sometimes considered as a single "Holarctic" or 'Boreal' region.

For most of the last 200 my, until about 30 mya, much of southern Europe was covered by a shallow epicontinental sea whose precise limits varied according to sea levels. The area was therefore rather like the East Indies today—an archipelago of islands of varying size that formed only a partial barrier to the dispersal of animals and very little barrier to the dispersal of plants. Further north, two different routes connected North America/Greenland from Europe until the Eocene [52] (Fig. 11.19). The Thulean route was along a ridge of land that is now submerged, while the de Geer route lay further north. Because of their position in fairly high northern latitudes, these routes acted as a filter to the exchange of mammals between North America and Europe, its intensity varying according to the climate.

Within North America, the Mid-Continental Seaway and Rocky Mountains, with their rain shadow to the east, in turn formed a barrier between western and eastern North America from the Cretaceous until the Early Oligocene, 30 mya, by which time erosion had leveled the early Rocky Mountains to a plain, so that east/west dispersal within North America was once again possible (cf. Fig. 5.7). However, biogeographic relations within the Northern Hemisphere also changed at this time because of the withdrawal of the seas from Eurasia. This allowed some mammal groups to enter Europe from Asia about 30 million years ago, a faunal change known as la Grande Coupure [53]. From then on, the high-latitude

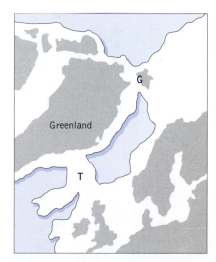

Fig. 11.19 Connections between North America and Europe in the Eocene. G, de Geer route; T, Thulean route (see text). After Tarling [61].

Bering link was the only route between Siberia and Alaska. The disappearance of the Turgai Sea at first also allowed some elements of the European boreotropical rainforest to spread into Eastern Asia. However, it also brought a drier climate to Central Asia. As a result, a drought-tolerant deciduous flora evolved there and later replaced the mixed-mesophytic forests of Europe, while the European boreotropical elements found refuge in the megathermal rainforests of Southeast Asia.

In the Late Oligocene, 25 mya, renewed mountain-building in North America caused a new rain shadow, with cooler and drier climates over the Great Plains, leading to the development of the grassland biome there. This likely caused a great deal of vicariance in the biota on either side of the mountains. The forests with their mesophytic elements became restricted to the western side of the mountains and to the eastern seaboard, and eventually disappeared during the Pleistocene glaciations.

The biogeographic importance of the climate of the Bering region can be seen in its influence on faunal exchange between North America and Eurasia. When the climate became cool, as in the Early Oligocene, few mammals crossed. When it improved again a little later in the Oligocene, a number of Asian mammals dispersed to North America. Some of these had evolved within Eurasia, while others had dispersed to that continent from Africa. The final climatic deterioration in the Bering region began in Miocene times and may have been related to the increase in Antarctic glaciation at that time. From then on, most of the mammals that dispersed were large forms and, even more significantly, types that are tolerant of cooler temperatures; such warmth-loving forms as apes and giraffes could not reach North America. This climatically based exclusion became progressively more restrictive, until in the Pleistocene only such hardy forms as the mammoth, bison, mountain sheep, mountain goat, musk ox, and human beings were able to cross. The final break between Siberia and Alaska took place 13,000–14,000 years ago.

Despite the long history of intermittent connection between the North American and Eurasian regions, each has a few endemic groups, while other groups did reach both regions, but later became extinct in one of them. Prong-horn antelopes, pocket gophers, pocket mice and sewellels (the last three groups are all rodents) are unknown in the Eurasian region, whereas hedgehogs, wild pigs, and murid rodents (typical mice and rats) are absent from the North American region. The domestic pig has been introduced to North America by human beings, as have mice and rats at various times. Horses crossed the Bering connection to Eurasia, but then became extinct in their North American homeland, probably because of competition from the immigrant bison. Horses were therefore unknown to the American Indians until they were introduced by the Spanish conquistadores in the 16th century.

Although the Ice Ages did not begin until the end of the Pliocene, the steadily cooling climates had already exerted a great influence on the floras of the northern continents. For example, there was a considerable change in the European flora during the Pliocene. The modern flora of the region contains over 60% of its Late Pliocene flora, but only 10% of that of the Early Pliocene. The intervening 3 million years had therefore seen a rapid change in the flora of Europe, as it adapted to climatic changes that became greatly exaggerated as the Pleistocene Ice Ages commenced. These Ice Ages stripped North America and Eurasia of virtually all tropical and subtropical animals and plants. This happened so recently that the faunas and floras have as yet had no time to develop any new, characteristic groups. Since they also have no old relict groups, such as the marsupials, it is the poverty and the hardiness of their faunas that distinguish them from those of other regions. Many groups of animals are absent altogether and, of the groups that are present, only the hardier members have been able to survive. Even these become progressively fewer toward the colder, Arctic latitudes. In North America there is, in addition, a similar thinning out of the fauna in the higher, colder zones of the Rocky Mountains; this is a general feature of the fauna and flora of high mountains.

The Eurasian fauna was almost completely isolated from the warmer lands to the south by the Himalayas and by the deserts of North Africa and southern Asia, and has therefore received hardly any infiltrators to add variety. The situation was eventually rather different in the Western Hemisphere. During the Early Cenozoic there had been hardly any exchange of animals between North and South America, presumably because there was still a wide ocean gap between the two continents. After the Panama Isthmus was completed at the end of the Cenozoic (see Fig. 11.14), many North American mammals dispersed to South America. However, North America was successfully colonized by only three types of South American mammal (opossums, armadillos, and anteaters), along with a number of birds such as hummingbirds, mockingbirds, and New World vultures.

For plants, for some reason, the situation was reversed. Instead of surviving in the Panama Isthmus, few of the North American megathermal plants survived the Late Cretaceous climatic cooling. The bulk of the lowland vegetation of Central America is, instead, of South American origin, perhaps because that great tropical region had produced an enormous variety of tropical plants. The mesothermal plants of North America were more successful at colonizing South America, presumably using the cooler mountainous spine of Central America as their route from the Rockies to the Andes.

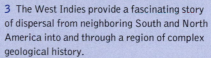

Summary

1 Today, the major elements in the distribution patterns of living mammals and flowering plants are the individual continents, each showing significant differences from the others. However, this is the result of comparatively recent tectonic and climatic events, such as the Ice Ages, which greatly reduced the variety of animals and plants in the northern continents. Barriers of sea, mountain, and desert prevented them from returning northward when the climate later improved. As a result, the tropical and subtropical biota of Africa, India, Southeast Asia, and South America are very different from the impoverished biota of Eurasia and North America.

2 Australia separated early from the rest of Gondwana and was isolated until only about 10 mya, when New Guinea at its northern edge became close to Southeast Asia. Australia has very few native placental mammals but a great variety of marsupials. Because of the steadily increasing aridity of Australia, its original "Antarctic" flora, descended from the Cretaceous Southern Gondwanan cool-temperate flora, was progressively transformed into a sclerophyll flora, which was in turn replaced by grassland and savanna. The Antarctic flora therefore now has a relict distribution, from Patagonia to New Zealand and the mountains of New Guinea.

3 The West Indies provide a fascinating story of dispersal from neighboring South and North America into and through a region of complex geological history.

4 South America too was isolated until only a few million years ago, when the Panama Isthmus was completed. Most of its older fauna of unusual herbivorous placental mammals and carnivorous marsupials then became extinct due to competition from immigrant North American mammals. The rich tropical flora of South America was more successful and colonized the lowlands of Central America.

5 The floras of North America and Eurasia each adapted to the climatic changes of the Late Cenozoic and to the Ice Ages of the Pleistocene, both by evolutionary change and by changes in their patterns of distribution.

This chapter completes the review of the history of the appearance, evolution, and development of the faunas and floras of the different continents until 2 million years ago, which were mainly gradual and long term. In contrast, their subsequent history was rapidly and fundamentally changed by the sudden climatic somersaults of the Ice Ages, which are described in the next chapter.

Further Reading

Culver SJ, Rawson PS, eds. *Biotic Response to Global Change: The Last 145 Million Years.* London: Natural History Museum, 2000.

Goldblatt P, ed. *Biological Relationships between Africa and South America.* New Haven, CT: Yale University Press, 1993.

Hall R, Holloway JD, eds. *Biogeography and Geological Evolution of SE Asia.* Leiden: Backhuys, 1998.

Pennington RT, Cronk QCB, Richardson JA, eds. Plant phylogeny and the origin of major biomes; a discussion meeting. *Phil Trans R Soc* B 2004; 359: 1453–1656.

References

1 Good R. *The Geography of the Flowering Plants*, 4th ed. London: Longman, 1974.

2 Takhtajan A. *Floristic Regions of the World.* Berkeley: University of California Press, 1986.

3 Darlington P. *Zoogeography: The Geographical Distribution of Animals.* New York: Wiley, 1957.

4 Tiffney BH. An estimate of the Early Tertiary paleoclimate of the southern Arctic. In: Boulter MC, Fisher HC, eds. *Cenozoic Plants and Climates of the Arctic*, pp. 267–295. NATO Advanced Science Institutes Series I, 27, 1994.

5 Graham A. Development of affinities between Mexican/Central American and northern South American lowland and lower montane vegetation during the Tertiary. In: Churchill SP, Balslev H, Forero E, Luteyn JL, eds. *Biodiversity and Conservation of Neotropical Montane Forests*, pp. 11–22. New York: New York Botanical Garden, 1995.

6 Pickford M, Morales J. Biostratigraphy and palaeobiogeography of East Africa and the Iberian peninsula. *Palaeogeogr Palaeoclimatol Palaeoecol* 1994; 112: 297–322.

7 Davis ALV, Scholtz CH, Philips TK. Historical biogeography of scarabaeine dung beetles. *J Biogeogr* 2002; 29: 1217–1256.

8 Eisenberg JF. *The Mammalian Radiations. An Analysis of Trends in Evolution, Adaptation and Behavior.* Chicago: University of Chicago Press, 1981.

9 Cristoffer C, Peres CA. Elephants versus butterflies: the ecological role of large herbivores in the evolutionary history of two tropical worlds. *J Biogeogr* 2003; 30: 1357–1380.

10 Flannery T. The mystery of the Meganesian meat-eaters. *Aust Nat Hist* 1991; 23: 722–729.

11 Milewski AV, Diamond RE. Why are very large herbivores absent from Australia? A new theory of micronutrients. *J Biogeogr* 2000; 27: 957–978.

12 Cox CB. The biogeographic regions reconsidered. *J Biogeogr* 2001; 28: 511–523.

13 Soltis DE, Soltis PS, Nickrent DL, et al. Angiosperm phylogeny inferred from 185 ribosomal DNA sequences. *Ann Missouri Bot Garden* 1997; 84: 1–49.

14 Crane PR, Herendeen PS. Cretaceous floras containing angiosperm flowers and fruits from eastern North America. *Palaeoecol Palaeoclimatol Palaeoecol* 1996; 90: 319–337.

15 Springer MS et al. Endemic African mammals shake the phylogenetic tree. *Nature, Lond* 1997; 388: 61–64.

16 Gheerbrant E. On the early biogeographical history of the African placentals. *Hist Biol* 1990; 4: 107–116.

17 Kalb JE. Fossil elephantids, Awash paleolake basins, and the Afar triple junction, Ethiopia. *Palaeogeogr Palaeoclimatol Palaeoecol* 1995; 114: 357–368.

18 Retallak GJ. Middle Miocene fossil plants from Fort Ternan (Kenya) and evolution of African grasslands. *Paleobiology* 1992; 18: 383–400.

19 Cowling R, Rundel PW, Lamont BB, Arroyo MK, Arianoutsou M. Plant diversity in Mediterranean-climate regions. *Trends Ecol Evol* 1996; 11: 362–366.

20 Goldman SM, Benstead JP, eds. *The Natural History of Madagascar*, pp. 1130–1134. Chicago: University of Chicago Press, 2003.

21 Renner SS. Multiple Miocene Melastomataceae dispersal between Madagascar, Africa and India. *Phil Trans R Soc* B 2004; 359: 1485–1494.

22 Jansa SA, Goodman SM, Tucker PK. Molecular phylogeny and biogeography of the native rodents of Madagascar (Muridae: Nesomyinae); a test of the single-origin hypothesis. *Cladistics* 1999; 15: 253–270.

23 Poux C, et al. Asynchronous colonization of Madagascar by the four endemic clades of primates, tenrecs, carnivores, and rodents as inferred from nuclear genes. *Syst. Biol.* 2005; 54: 719–730.

24 Stankiewicz J, Thiart C, Masters JC, de Wit MJ. Did lemurs have sweepstake tickets? An exploration of Simpson's model for the colonization of Madagascar by mammals. *J Biogeogr* 2006; 33: 221–235.

25 Ali JR, Huber M. Mammalian biodiversity on Madagascar controlled by ocean currents. *Nature* 2010; 463: 653–656.

26 Raxworthy CJ, Forstner MRJ, Nussbaum RJ. Chameleon radiation by oceanic dispersal. *Nature, Lond* 2002; 415: 784–787.

27 Rocha S, Carretero MA, Vences M, Glaw F, Harris DJ. Deciphering patterns of transoceanic dispersal: the evolutionary origin and biogeography of coastal lizards (*Cryptoblepharus*) in the Western Indian Ocean region. *J Biogeogr* 2006; 33: 13–22.

28 Vences M et al. Reconciling fossils and molecules: Cenozoic divergence of cichlid fishes and the biogeography of Madagascar. *J Biogeogr* 2001: 28; 1091–1099.

29 Morley RJ. *Origin and Evolution of Tropical Rainforests*. New York: Wiley, 1999.

30 Simpson GG. Too many lines; the limits of the Oriental and Australian zoogeographic regions. *Proc Am Phil Soc* 1977; 121: 107–120.

31 Moss SJ, Wilson MEJ. Biogeographic implications of the Tertiary palaeogeographic evolution of Sulawesi and Borneo. In: Hall R, Holloway JD, eds. *Biogeography and Geological Evolution of SE Asia*, pp. 133–163. Leiden: Backhuys, 1998.

32 Hill RS. Origins of the southeastern Australian vegetation. *Phil Trans Roy Soc Lond* 2004; B359: 1537–1549.

33 Crisp M, Cook L, Steane D. Radiation of the Australian flora: what can comparisons of molecular phylogenies across multiple taxa tell us about the evolution of diversity in present-day communities? *Phil Trans Roy Soc Lond* 2004; B359: 1551–1571.

34 Tyndale-Biscoe CH. Ecology of small marsupials. In: Stoddart DM, ed. *Ecology of Small Mammals*, pp. 342–379. London: Chapman & Hall, 1979.

35 Trewick SA, Paterson AM, Campbell HJ. Hello New Zealand. *J Biogeogr* 2007; 34: 1–6.

36 McDowell RM. Process and pattern in the biogeography of New Zealand—a global microcosm? *J Biogeogr* 2008; 35: 197–212.

37 Winkworth RC, Wagstaff SJ, Glenny D, Lockhart PJ. Plant dispersal N.E.W.S. from New Zealand. *Trends in Ecology and Evolution* 2002; 17: 514–520.

38 Pole MS. The New Zealand flora—entirely long-distance dispersal? *J Biogeogr* 1994; 21: 625–655.

39 Hedges SB, Paleogeography of the Antilles and origin of West Indian terrestrial vertebrates. *Ann Miss Bot Garden* 2006; 93: 231–244.

40 Rowe DL, Dunn KA, Adkins RM, Honeycutt RL. Molecular clocks keep dispersal hypotheses afloat: evidence for transAtlantic rafting by rodents. *J Biogeogr* 2010; 37: 305–324.

41 Marshall LG. The Great American Interchange—an invasion-induced crisis for South American mammals. In: Nitecki MH, ed. *Third Spring Systematic Symposium: Crises in Ecological and Evolutionary Time*, pp. 133–229. New York: Academic Press, 1981.

42 Webb SD, Marshall LG. Historical biogeography of Recent South American land mammals. In MA Mares, HG Genoways, eds. *Mammalian Biogeography in South America*, pp. 39–52. Spec Publ Series 6, Pymatuning Lab. of Ecology, University of Pittsburgh, 1982.

43 Webb SD. Late Cenozoic mammal dispersals between the Americas. In: Stehli FG, Webb SD, eds. *The Great American Biotic Interchange*, pp. 357–386. New York: Plenum, 1985.

44 Lessa EP, Fariña RA. Reassessment of extinction patterns among the Late Pleistocene mammals of South America. *Palaeontology* 1996; 39: 651–659.

45 Vrba ES. Mammals as a key to evolutionary theory. *Jl Mammalology* 1992; 73: 1–28.

46 Haffer J. Speciation of Amazonian forest birds. *Science* 1969; 165: 131–137.

47 Ribas CC, Gaban-Lima R, Miyaki CY, Cracraft J. Historical biogeography and diversification within the Neotropical parrot genus *Pionopsitta* (Aves: Psittacidae). *J Biogeogr* 2005; 32: 1409–1427.

48 Costa LP. The historical bridge between the Amazon and the Atlantic forest of Brazil: a study of molecular phylogeography with small mammals. *J Biogeogr* 2003; 30: 71–86.

49 Bush MB. Of orogeny, precipitation, precession and parrots. *J Biogeogr* 2005; 32: 1301–1302.

50 Colinvaux PA, Irion G, Räsänen ME, Bush MB, Nunes de Mello JAS. A paradigm to be discarded: geological and paleoecological data falsify the Haffer and Prance refuge hypothesis of Amazonian speciation. *Amazoniana* 2001; 16: 609–646.

51 Sanmartin I, Enghoff H, Ronquist F. Patterns of animal dispersal, vicariance and diversification in the Holarctic. *Biol J Linn Soc Lond* 2001; 73: 345–390.

52 McKenna MC. Cenozoic paleogeography of North Atlantic land bridges. In: Bott MHP et al., eds. *Structure and Development of the Greenland-Scotland Ridge*, pp. 351–399. New York: Plenum, 1983.

53 Prothero DR. *The Eocene-Oligocene Transition: Paradise Lost.* New York: Columbia University Press, 1994.

54 MacFadden BJ. *Horses.* Cambridge: Cambridge University Press, 1992.

55 Heywood VG, ed. *Flowering Plants of the World.* Oxford: Oxford University Press, 1978.

56 Steininger FF, Rabeder G, Rogl F. Land mammal distribution in the Mediterranean Neogene: a consequence of geokinematic and climatic events. In: Stanley DJ, Wezel FC, eds. *Geological Evolution of the Mediterranean Basin*, pp. 559–571. New York: Springer, 1985.

57 Huggett RJ. *Fundamentals of Biogeography*. London: Routledge, 1998.

58 Coates AG, Obando J. The geologic evolution of the Central American Isthmus. In: Jackson JBC, Budd AF, Coates AG, eds. *Evolution and Environment in Tropical America*, pp. 21–56. Chicago: University of Chicago Press, 1996.

59 Marshall LG, Webb SD, Sepkoski JJ, Raup DM. Mammalian evolution and the Great American Interchange. *Science* 1982; 215: 1351–1357.

60 Lynch JD. Refugia. In: Myers AA, Giller PS, eds. *Analytical Biogeography*, pp. 311–342. London: Chapman & Hall, 1988.

61 Tarling DH. Land bridges and plate tectonics. *Mémoir Spéciale Géobios* 1982; 6: 361–374.

Ice and Change

The preceding chapters have demonstrated that an understanding of the biogeography of the modern world requires a knowledge of past events. Most of the changes considered so far relate to the distant past and the processes of changing continental arrangements and the evolution of the major groups of living organisms. But the current distribution of plants and animals has been greatly affected by relatively recent events in the Earth's history, especially those of the last two million years when extensive ice has periodically covered many parts of the Earth's surface. The world's major ice caps were already in existence around 42 million years ago, having begun forming in Antarctica during the Eocene [1]. Global temperatures dropped very rapidly at the end of the Eocene [2], and this appears to correspond with the first formation of an ice cap on Greenland in the Northern Hemisphere [3].

Many landform features in the temperate areas of the world show that major, geologically rapid changes in climate have taken place since the Pliocene. The general cooling of world climate that started early in the Tertiary continued into the Quaternary; the boundary between the two is placed at about 2 million years ago, but difficulties in definition as well as in dating techniques and geological correlation leave this date open to some doubt. The definition of the boundary comes from Italian marine sediments, where the appearance of fossils of cold-water organisms (certain foraminifera and molluscs) suggests a fairly sudden cooling of the climate, which has been dated at 1.8 million years. Similar evidence of cooling has been found in sediments from the Netherlands, and this is believed to mark the end of the final stage of the Pliocene (locally termed the Reuverian) and the first stage of the Pleistocene (the Pretiglian) [4]. Evidence from sediment cores in the North Atlantic indicates that debris was being carried into deep water by ice rafts as long ago as 2.4 million years, and many geologists feel that this would be a more appropriate date for the Pliocene/Pleistocene boundary. This is now the most widely accepted date for the boundary. But evidence from Norway suggests that Scandinavian glaciers extended down to sea level as long as 5.5 million years ago [5]; the point in time when we mark the opening of the Quaternary is therefore inevitably disputed.

At various stages during the Pleistocene, ice covered Canada and parts of the United States, northern Europe, and Asia. In addition, independent centers of **glaciation** were formed in lower-latitude mountains, such as the Alps, Himalayas, Andes, the East African mountains, and in New Zealand. At the height of its development, about 80% of glacial ice lay in the Northern Hemisphere. A number of present-day geological features show the effects of such glaciations. One of the most conspicuous of these is the glacial drift deposit, boulder clay, or **till** covering large areas and sometimes extending to great depths. This is usually a clay-rich material containing quantities of rounded and scarred boulders and pebbles, and geologists consider it to be the detritus deposited during the melting and retreat of a glacier. The most important feature of this till, and the one by which it may be distinguished from other geological deposits, is that its constituents are completely mixed; the finest clay and small pebbles are found together with large boulders. Often the rocks found in such deposits originated many hundreds of miles away and were carried there by the slow-moving glaciers. Fossils are rare, but occasional sandy pockets have been found that contain mollusc shells of an Arctic type. Some enclosed bands of peat or freshwater sediments within these tills provide evidence of the warmer intervals. They often show that there were phases of locally increased plant productivity, and they may contain fossils indicative of warmer climates.

Many of the valleys of hilly, glaciated areas have distinctive, smoothly rounded profiles because they were scoured into that shape by the abrasive pressure of the moving ice. In places, the ice movement has left deep scratches on the rocks over which it has passed, and tributary valleys may end abruptly, high up a main valley side, because the ice has removed the lower ends of the tributary valleys. Such landscape features provide the geomorphologist with evidence of past glaciation. Understanding the significance of the shapes of hills and valleys depended on a geological principle put forward by the Scottish geologist James Hutton (1726–1797), who expounded the idea of **uniformitarianism**. In essence, this states that present-day conditions can be used as a key for understanding past processes and there is no need to interpret geology in the light of supposed global catastrophes, such as the Biblical Flood of Noah. Using uniformitarianism as his basic premise, the Swiss geologist Louis Agassiz (1807–1873) saw how glaciers could affect landforms and proposed that areas now clear of ice had once been covered by great depths of ice in an Ice Age. The validity of this proposal soon became evident following further research in Europe and North America.

Immediately outside the areas of past glaciation were regions of treeless tundra that experienced **periglacial** conditions (Fig. 12.1). The present-day equivalents of these areas are very cold, and their soils are constantly disturbed by the action of frost. When water freezes in the soil, it expands, raising the surface of the ground into a series of domes and ridges. Stones within the soil lose heat rapidly when the temperature falls, and the water freezing around them has the effect of forcing them to the surface, where they often become arranged in **stone stripes** and **polygons**. Similar patterns are produced by **ice wedges** that form in ground subjected to very low temperatures. Some of these patterns, which are so evident in present-day areas of periglacial climate, can be found in parts of the world that are now much warmer. For example, they have been discovered in

Fig. 12.1 Contorted birch tree on the tundra in northern Lapland, marking the northern limit of tree growth. Lying beyond the permanent ice, these regions experience periglacial conditions.

eastern parts of the United States and in western Europe as a result of air photographic survey. Such "fossil" periglacial features show that, as the glaciers expanded, so the periglacial zones were pushed before them toward the equator.

Climatic Wiggles

The Pleistocene Epoch has not been one long cold spell. Careful examination of tills and the orientation of stones embedded within them soon showed that several advances of ice had taken place during the Pleistocene, often moving in different directions. Occasional layers of organic material were sometimes discovered trapped between tills and other deposits, and these have provided fossil evidence of warm periods alternating with the cold. Where terrestrial sequences of deposits are reasonably complete and undisturbed, as in the eastern part of England (East Anglia) and parts of the Netherlands, it has been possible to construct schemes to describe these alternations of warm and cold episodes, to name them, and to determine their relationships in time. But in many parts of the world this has not proved at all easy, and the correlation of events, as reflected in terrestrial deposits, between different areas has often been speculative and unsatisfactory, mainly because of the difficulty experienced in obtaining secure dates for the deposits. At one time, for example, geologists considered that there were four episodes of ice advance in Europe, defined mainly by sequences of tills in the Alps. The four glaciations were named Günz, Mindel, Riss, and Würm. This is now regarded as a gross oversimplification of the true situation, as there have been far more climatic fluctuations in the Pleistocene than this simple model suggests.

Because of the difficulties experienced in climatic reconstruction using land-based evidence, attention has turned to the seas, where marine sediments provide a more complete and uninterrupted sequence. The retrieval of long, deep ocean cores of sediment has provided an opportunity to follow the rise and fall of various members of plankton communities in the past, particularly those, like the **foraminifera**, which, though tiny, have robust outer cases that survive the long process of sedimentation to the ocean floor and there accumulate as fossil assemblages. Some members of the foraminifera, like some species of *Globigerina* and *Globorotalia*, are sensitive to ocean temperature, so their relative abundance in the fossil record provides evidence of past climates.

An even more powerful tool for reconstructing long-term climatic changes has been the use of **oxygen isotopes** retained in the fossil material of the sediments. "Normal" oxygen (^{16}O) is far more abundant than the heavier form of oxygen (^{18}O). For example, the heavy form comprises only about 0.2% of the oxygen incorporated into the structure of water (H_2O). Water evaporates from the sea, but those molecules containing ^{18}O condense from a vapor form rather more readily than their lighter counterparts, so this heavy form tends to return rapidly to the oceans. Water containing ^{16}O, on the other hand, remains in the atmosphere as vapor for longer and is more likely to fall eventually over the ice caps and to become

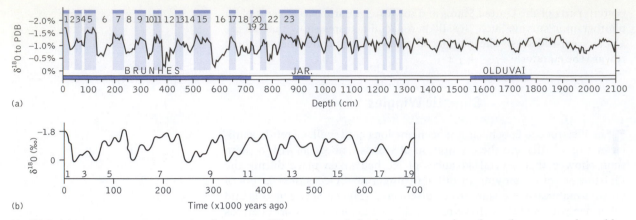

Fig. 12.2 Oxygen isotope curve covering the past 3 million years. The timescale reads from right to left. Peaks represent periods of relative warmth, and troughs correspond to cold episodes [9].

incorporated into these as ice. Under cold conditions, the volume of global ice increases, and this (since it is formed largely from precipitation) tends to lock up more of the ^{16}O, leaving the oceans richer in ^{18}O. So the ratio of ^{18}O:^{16}O left in the oceans increases during periods of cold. This ratio is then reflected in the skeletons of foraminifera and other planktonic organisms and is deposited in the ocean beds. Analysis of the oxygen isotope ratios in ocean sediments thus provides a long and continuous record of changing water temperatures going back millions of years [6]. It has even been possible to use these methods for the analysis of oxygen isotope ratios in inland areas, as in the gradual deposition of calcite in the Devil's Hole fault in Nevada [7].

As a result of such oxygen isotope studies of a series of cores in the Caribbean and Atlantic oceans, Cesare Emiliani of the University of Miami, Florida, was the first researcher to be able to construct **palaeotemperature** curves for the ocean surface waters [8]. A summary curve for the past 3 million years is shown in Fig. 12.2, and it is quite obvious from this diagram that the climatic changes in this latter part of the Pleistocene have been numerous and complex [9]. Translating oxygen isotope ratios into mean annual temperature fluctuations requires a number of assumptions about global temperature and ice volumes, but it is generally agreed that the difference in mean annual temperature between a glacial maximum (the troughs in this diagram) and a period of interglacial warmth (the peaks) is about 8–10°C.

Examination of the changing isotope ratios in Fig. 12.2, which have been assembled from numerous studies of deep-sea sediments and ice cores, shows that the fluctuations in temperature become stronger as time proceeds (note that the time sequence runs from right to left). Gentle wanderings around a slowly falling mean value gradually become more pronounced as greater extremes of temperature are experienced. Sharper changes also become more apparent, with the temperature changing both radically and rapidly. One can also see that the course of change, though showing a general wave-like form, is not simple or regular, but has many minor wiggle patterns imposed on it. This latter point implies that we should not expect the terrestrial record in currently temperate lands such as North America or central Europe to show a simple alternation of temperate conditions with arctic ones, but a much more varied pattern in which cold and warmth alternate in varying degrees and intermediate conditions are often found.

Interglacials and Interstadials

The fluctuations in global temperature represented in the ocean cores are reflected in the terrestrial geological sequence by glacial and interglacial deposits. A warm episode (usually represented by an organic, peaty deposit) sandwiched between two glacial events (often represented by tills) and that achieved sufficient warmth for a long enough duration for temperate vegetation to establish itself, is termed an **interglacial**. The sequence of events demonstrated in the fossil material of such an interglacial shows a progressive change from high arctic conditions (virtually no life) through subarctic (tundra vegetation) to boreal (birch and pine forest) to temperate (deciduous forest) and then back through boreal to arctic conditions once more. If the warm event is of only short duration, or if the temperatures attained are not sufficiently high, then the vegetation changes may only reach a boreal stage of development. In this case it is termed an **interstadial**. We are currently living in the most recent interglacial (termed by geologists the **Holocene**).

Interglacials are often times of increased biological productivity (except in the more arid parts of the world), and they are often represented in temperate geological sequences as bands of organic material (Fig. 12.3). This material usually contains the fossil remains of the plants, animals, and microbes that existed at or near the site during its formation, and it is this evidence, often stratified in a time sequence, that allows us to reconstruct past conditions and habitats. One of the most valuable sources of fossil evidence for this purpose have been the **pollen grains** from plants that are preserved in the sediment and that reflect the vegetation of the area at that time. Pollen grains are sculptured in such a way that they are often recognizable with a considerable degree of precision. They are also produced in large numbers (especially those of wind-pollinated plants) and are widely dispersed. Finally, they are preserved very effectively in waterlogged sediments, such as peat and lake deposits. Therefore the analysis of pollen grain assemblages, termed **palynology**, can provide much information about vegetation and hence about climate and other environmental factors [10].

Cores of sediment from lakes, peat bogs, or even from ice, are extracted intact, and then samples are taken at different depths. Pollen and spores are preserved in these samples and can be concentrated by removing or dissolving the matrix material. Eventually, these microfossils are dense enough to be counted with the aid of a light microscope, and the proportions of different types can be determined.

Fossil pollen data are usually presented in the form of a pollen diagram, and Fig. 12.4 shows a pollen diagram belonging to the last (so-called Ipswichian) interglacial in Britain. The vertical axis is the depth of deposit, which is directly related to the age of the samples, so the diagram

Fig. 12.3 Organic sediments from an interglacial (dark band), resting on gravel of a former raised beach and covered by deposits laid down under periglacial conditions. A cliff exposure at West Angle, Pembrokeshire, Wales [11].

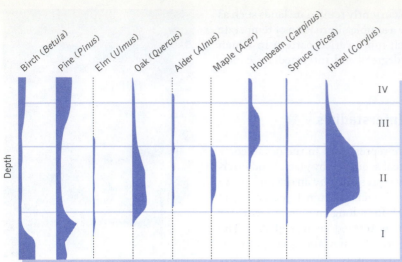

Fig. 12.4 Pollen diagram from sediments of the last (Ipswichian) interglacial in eastern Britain. Only tree taxa are shown. The depth axis is related to age, with oldest sediments at the base.

should be read from the bottom up. The sequence begins with boreal trees, birch, and pine, which are then replaced by deciduous trees such as elm, oak, alder, maple, and hazel. Later in the sequence these trees go into decline, to be replaced by hornbeam, and then pine and birch once more. The details of such a sequence will obviously vary from one locality to another. Figure 12.5 shows the pollen diagram from the same interglacial further east in Europe, in Poland, where it is called the Eemian interglacial. It is conventional to name interglacials and other geological episodes with local names that can later be used in correlation between different areas. In the Polish Eemian interglacial lime, spruce and fir play a more important role than in equivalent western sites. The vegetation sequence thus varies considerably between regions and also between different interglacials, but all such sequences show a consistent pattern, for they all pass through a predictable series of developmental stages. Often these are shown on such diagrams by dividing them into four zones (usually labelled in Roman numerals, I–IV): pretemperate, early temperate, late-temperate, and post-temperate, respectively. Pollen diagrams from our present interglacial suggest that we are well advanced into the late-temperate stage.

The last interglacial is the easiest to identify stratigraphically because it occurred in the very recent past. Dating it precisely is difficult because the radiocarbon method becomes less accurate as one proceeds back in time, and dating methods relying on argon isotopes are dependent on the presence of volcanic material [12]. Nevertheless, it is believed that the last interglacial began about 130,000 years ago (although a recently studied and well-dated site from Nevada suggests commencement at 147,000 years ago, and this has caused considerable controversy among geologists and palaeoclimatologists [13]). It ended some 11,500 years ago with the beginning of the last major glaciation. This means that we can identify the terrestrial record of the last interglacial with the final dip in the oxygen isotope curve seen in Fig. 12.2. Earlier interglacials are much more problematic; they are usually given local names when they are first

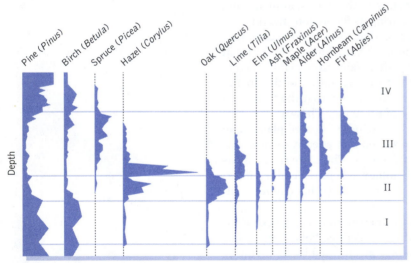

Fig. 12.5 Pollen diagram from sediments of the last (Eemian; equivalent to the British Ipswichian) interglacial in Poland. Note the greater importance of lime, spruce, and fir than is the case in the British Ipswichian.

described, and attempts are later made to correlate them, often on the basis of the fossils they contain. Some possible correlations are shown in Fig. 12.6, but any correlation of terrestrial deposits in the absence of firm dating can only be tentative, especially in the case of older glacials and interglacials.

Biological Changes in the Pleistocene

With the expansion of the ice sheets in high latitudes, the global pattern of vegetation was considerably disturbed. Many areas now occupied by temperate deciduous forests were either glaciated or bore tundra vegetation. For example, most of the north European plain probably had no deciduous oak forest during the glacial advances. The situation in Europe was made more complex by the additional centers of glaciation in the Alps and the Pyrenees. These would have resulted in the zone compression and often the ultimate local extinction of species of warmth-demanding plants and animals during the glacial peaks. Figure 12.7 shows the broad vegetation types that occupied Europe during interglacial and glacial times. During the interglacials, tundra species became restricted in distribution due to their inability to cope with high summer temperatures and their failure to compete with more robust, productive species. High-altitude sites and disturbed areas sometimes served as **refugia** within which groups of such species may have survived in isolated localities. Similarly, during glacials particularly favorable sites that were sheltered, south-facing, or oceanic and relatively frost-free may have acted as refugia for warmth-demanding species through the time of cold. In Europe, for example, deciduous forest species are thought to have survived glacial episodes in what is now Spain and Portugal [14], Italy, Greece, the Danube Delta, Turkey, and around the Black and Caspian seas [15].

Animal species with good dispersal mechanisms were able to shift their populations to cope with the new conditions more easily than plants. Plants were dependent on the movement of their fruits and seeds to spread into new areas, and meanwhile old populations died out as conditions became less favorable. The fact that each species faced its own peculiar problems, both in terms of climatic requirements and dispersal capacity, means that species must have become shuffled into new assemblies (or "communities," see Chapter 4) during times of change. Figure 12.7 expresses vegetation as life forms (with examples), so the picture is one of major biome shifts under different climatic regimes, but

Temp.	North America	Alps	Northwest Europe	Britain	Approx. date
C	Wisconsin	Würm	Weichsel	Devensian	←10 000 ←70 000
W	Sangamon	Riss/Würm	Eemian	Ipswichian	
C	Illinoian	Riss	?Saale	?Wolstonian	←250 000
W	Yasmouth	Mindel/Riss	?Holstein	?Hoxnian	←500 000
C	Kansan	Mindel	?Elster	?Anglian	
W	Aftonian	Günz/Mindel	?Cromerian complex	?Cromerian	
C	Nebraskan	Günz	?Menapian		←1×10⁶
W		Donau/Günz	?Waalian		
C		Donau	?Eburonian		
W		Biber/Donau	?Tiglian		
C		Biber	?Pretiglian		←2.4×10⁶
W			Reuverian		

Fig. 12.6 Conventional correlations assumed between local glacial and interglacial events in the Pleistocene. Local complexities render such correlations tentative, particularly in the earlier stages. C, cold; W, warm.

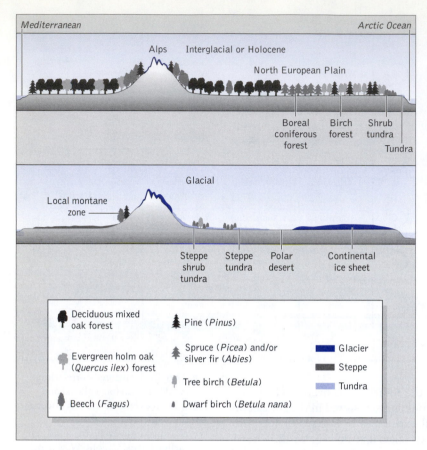

Mediterranean   Arctic Ocean

Alps   Interglacial or Holocene

North European Plain

Boreal   Birch   Shrub
coniferous   forest   tundra
forest
Tundra

Glacial

Local montane
zone

Steppe   Steppe   Polar   Continental
shrub   tundra   desert   ice sheet
tundra

Deciduous mixed   Pine (*Pinus*)
oak forest

                         Spruce (*Picea*) and/or   Glacier
Evergreen holm oak   silver fir (*Abies*)
(*Quercus ilex*) forest   Steppe
                         Tree birch (*Betula*)   Tundra

Beech (*Fagus*)   Dwarf birch (*Betula nana*)

Fig. 12.7 A simplified diagrammatic representation of the vegetation belts of Europe during glacial and interglacial time. The profiles run south/north from the Mediterranean Basin to the Arctic Ocean. During glacial times the temperate forest became confined to refuge areas mainly south of the Alps.

the constitution of these biomes changed according to the patterns of movement of different tree species. The same is true of the animal components of the biomes. An extensive study [16] of the ranges of North American mammals during Late Quaternary times has shown that individual species shifted at different times, in different directions, and at different rates during the climatic swings. Only during the last few thousand years have the familiar modern groupings of plants and animals assembled themselves, so the idea that species assemblages moved as intact "communities" must be abandoned in the light of these findings.

Extinctions occurred during this process of successive climatic changes and distributional shifts. In Europe many of the warmth-preferring genera and species so abundant in the Tertiary Epoch were lost to the flora. The hemlock (*Tsuga*) and the tulip tree (*Liriodendron*) were lost in this way, but both survived in North America, where the generally north-south orientation of the major mountain chains (the Rockies and the Appalachians) allowed the southward migration of sensitive species during the glacials and their survival in what is now Central America. The east-west orientation of the Alps and Pyrenees in Europe permitted no such easy escape to the south. The wing nut tree (*Pterocarya*) was also extinguished in Europe, but it has survived in Asia, in the Caucasus, and in Iran. Studies relating to the histories of many tree species during the latter part of the Pleistocene have greatly assisted in explaining the present-day distribution of trees and the composition of forests [17].

The Last Glacial

The most recent glacial stage lasted from approximately 115,000 years ago until about 10,000 years ago, so our current experience of a warm Earth is rather unusual as far as recent geological history is concerned. However, even a glacial is not a time of uniform cold, and there have been numerous warmer interruptions to the prevailing cold climate. Many interstadials are recorded within the last glacial. Figure 12.8 shows a detailed oxygen isotope curve of an ice core taken from the Greenland Ice Cap [18]. The instability of the temperature during the glacial is evident, and it is clear that there were many short-lived warm episodes right the way through the glacial. The cycles of alternating warmth and cold during the last glacial have been termed **Dansgaard-Oeschger Cycles**, after the names of the scientists who first described them. Each cold event lasts about 1,000 to 2,000 years, separated by intervals of approximately 7,000 years, and the cold phase is usually accompanied by massive discharges of icebergs into the North Atlantic Ocean, first discovered by Hartmut Heinrich in 1988 and now called **Heinrich Events**. These icebergs carried debris, such as sand and limestone fragments, out into the Atlantic where they were eventually melted and released their loads of detritus, which joined the marine sediments and left a record of the frequency of Heinrich Events. The sediments also record a lowered abundance of plankton fossils, such as foraminifera, during Heinrich Events, suggesting reduced oceanic productivity. In addition, there is evidence of major changes in ocean salinity at the time of their occurrence, and this has led some oceanographers to propose that the entire circulation pattern of the Atlantic Ocean changed during the Dansgaard-Oeschger Cycles [19], an idea that is discussed in more detail later in this chapter.

The plant and animal life on land also responded to the climatic fluctuations during the last glacial. Continuous sediment records that contain fossil pollen and allow us to trace terrestrial vegetation back through the last glacial into the preceding interglacial are relatively uncommon. Many possible sites were actually scoured by glaciers and so have lost their record, but there are some deep lake sites, often associated with old volcanic craters, where a complete record is available. One such profile that takes the vegetation record back about 125,000 years is shown in Fig. 12.9. This is a pollen diagram from Carp Lake, situated on the eastern side of the Cascade Mountains in the Pacific Northwest of America [20]. It lies just outside the limit of the great ice sheet (the Laurentide Ice Sheet) that covered much of the northern part of North America during the final Wisconsin glacial. It currently lies at the boundary of two biomes, the sagebrush (*Artemisia*) steppe and the montane pine forest (*Pinus ponderosa*), and its closeness to this boundary (called an **ecotone**) means that the vegetation will have been particularly sensitive to climatic swings in the past. The basal part of the diagram (Zone 11) shows the open pine and oak (*Quercus*) forest of the last interglacial, with some indications that conditions were then warmer and drier than at present. There follows, for the bulk of the diagram, alternating peaks of *Artemisia* and pine, showing the constant alteration of the vegetation over this sensitive ecotone as the climate warmed and cooled. At one stage (Zone 7, about 85,000–74,000 years ago), there was an episode of such warmth that an open mixed forest including oak established itself. It is likely that this interlude was warmer and wetter during the

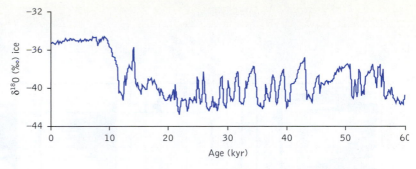

Fig. 12.8 Oxygen isotope curve from an ice core extracted from the Greenland Ice Cap. (Less negative values of ^{18}O in ice indicate higher temperatures.) The sequence covers a period of 60,000 years, approximately the latter half of the last glacial event. The instability of the temperature, as reflected by the isotope curve, is evident, the cold period being frequently interrupted by short episodes of higher temperature. From Dansgaard et al. [19].

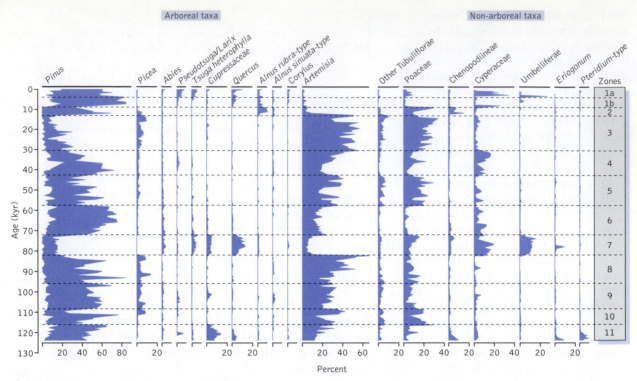

Arboreal taxa

Non-arboreal taxa

Fig. 12.9 Pollen diagram from Carp Lake, a crater lake in the eastern Cascade Mountains of the northwestern United States. The sediments cover approximately 125,000 years and display the erratic response of vegetation to climatic instability during the last (Wisconsin) glacial. From Whitlock & Bartlein [20].

summers than at present. Cooler conditions then returned until the glacial ended and the current vegetation established itself in the last 10,000 years (Zone 1).

This remarkable record demonstrates that the temperature fluctuations recorded in the ice-core profile of oxygen isotopes (see Fig. 12.8) were reflected in the response of vegetation and that the instability of the glacial climate caused constant adjustments in the ranges of species and in the boundaries of biomes.

Even in the tropics, such as northern Australia, vegetation changed markedly in response to climatic fluctuations over the past 120,000 years, as discussed in Concept Box 12.1 and Fig. 12.10.

Tropical aridity seems to have been widespread when the higher latitudes were experiencing their glacial episodes, as is confirmed by evidence from tropical Africa, India, and South America. Much of the humid tropical forest of the Zaire Basin was probably replaced by dry grassland and savanna during the glacial, although some forest may well have survived in riverine and lakeside situations. The temperature records from tropical locations, such as East Africa, follow closely the trends observed in other parts of the globe. Pollen analyses from lake sediments along the Uganda/Zaire border [23] have shown that rainforest

In the tropics, the effect of ice was not experienced directly except on the very high mountains, but the climate was generally colder, and changes in vegetation reflected in pollen diagrams from tropical regions suggest that there were important variations in precipitation. In Queensland, Australia, for example, Peter Kershaw [21] has analyzed the sediments of a volcanic crater lake from a rainforest area, and the results are shown in Fig. 12.10. Dating at this site is difficult, but the total span of the diagram is thought to cover about 120,000 years, extending back into the closing stages of the last interglacial, a similar time coverage to that of the Carp Lake diagram from the Pacific Northwest of North America (Fig. 12.9). During the last interglacial (basal zone E3), this Australian site was occupied by many of the rainforest tree genera that currently occupy the area, but at the time when the final glacial began in high latitudes, the rainforest was replaced by forest of simpler structure in which the palm-like trees of the genus *Cordyline* played an important part. There was then a brief reversion (dated at about 86,000–79,000 years ago—very close to the warm spell in the North American diagram) when rainforest became reestablished. But then

much more arid conditions set in, and the forest became increasingly dominated by gymnosperm trees, such as the monkey puzzle (*Araucaria*). Between about 26,000 and 10,000 years ago, the climate became very dry, and the former forests took on a sclerophyllous form, being dominated by *Casuarina*, a tree currently associated with hot dry conditions. At the end of the "glacial," however, there was a very rapid change in vegetation, with the invasion of rainforest trees once again. The period of maximum aridity at this site, 26,000 to 10,000 years ago, encompasses the period of maximum extent of the Northern Hemisphere glaciers at about 22,000 years ago.

In the Nullabor Plain, south-central Australia, studies on fossil vertebrates show that a high proportion of these organisms became extinct during the Pleistocene [22], but climate on its own is unlikely to account for this loss in biodiversity. An increasing incidence of bushfires is a more likely immediate cause of the losses, and the impact of humans (who arrived in Australia around 40,000 years ago) cannot be excluded as a contributory factor.

was replaced by dry scrub and grassland to the east of the Western Rift Valley during the height of the last glacial episode. Any refugial fragments must have been situated further west, in the lowlands of the Congo Basin.

Figure 12.11 shows a proposed reconstruction of the approximate areas of rainforest that existed at the time of maximum glacial extent in the high latitudes (about 22,000 years ago). From this it can be seen that rainforest was much fragmented as a result of the glacial drought in tropical latitudes [24]. Many areas currently occupied by rainforest were greatly modified at this time. This is an important point to bear in mind when considering the high species diversity of the rainforests (see Chapter 2). Most rainforests have not enjoyed a long and uninterrupted history but have been disrupted by global climatic changes, especially cold and drought. Even some of the regions (like the Uganda/Zaire borders), which are now hotspots of biodiversity, bore a very different vegetation during the last glacial. In the Amazon Basin of South America, however, there is still considerable debate about the vegetation of the last glacial maximum. At one time it was believed that savanna woodlands occupied much of the region, but current opinion [25] favors dry seasonal forests with an admixture of montane, cold-tolerant tree species. During the glacials the temperate forests were forced to occupy new areas at lower latitudes, but the tropical rainforests had nowhere to which they could retreat, so they became fragmented and dismembered as their composition was drastically altered (see Chapter 11). Some ecologists

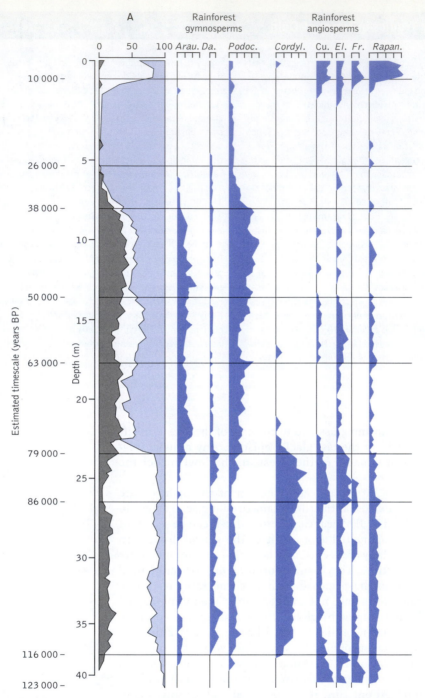

Fig. 12.10 (Right and facing page.) Pollen diagram from Lynch's Crater, Queensland, Australia. The column on the left shows the percentage of rainforest gymnosperms (gray); rainforest angiosperms (white); sclerophyll taxa (blue). The frequencies of pollen of all taxa are shown as percentages of the dry land plant pollen total; each division represents 10% of the pollen sum. Abbreviations for taxa: Arau., *Araucaria*; Da., *Dacrydium*; Podoc., *Podocarpus*; Cordyl., *Cordyline*; Cu., *Cunoniaceae*; El., *Elaeocarpus*; Fr., *Freycinettia*; Rapan., *Rapanea*; Casuar., *Casuarina*; Euc., *Eucalyptus*. After Kershaw [21].

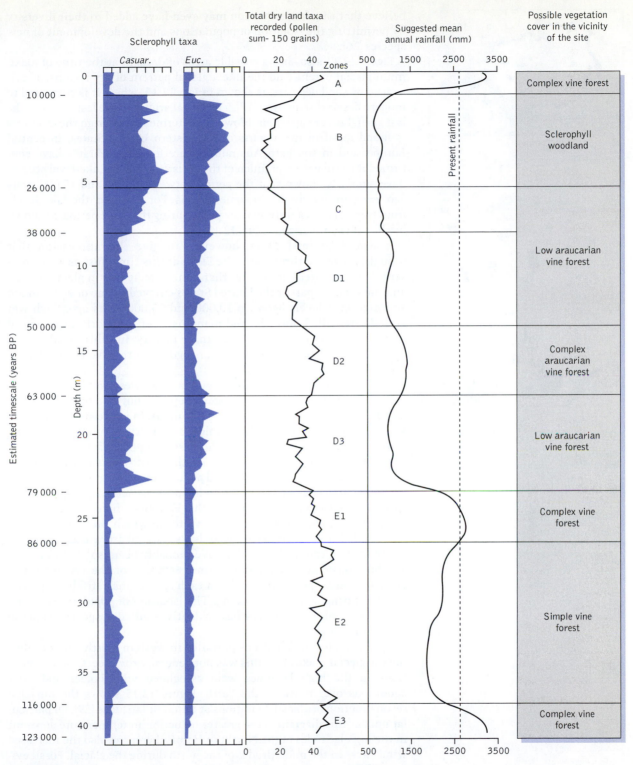

Sclerophyll taxa

Casuar.   *Euc.*

Total dry land taxa
recorded (pollen
sum- 150 grains)

Suggested mean
annual rainfall (mm)

Possible vegetation
cover in the vicinity
of the site

Zones

Present rainfall

Estimated timescale (years BP)

Depth (m)

Years BP	Depth	Zone	Vegetation
0	0	A	Complex vine forest
10 000		B	Sclerophyll woodland
26 000	5	C	
38 000			Low araucarian vine forest
	10	D1	
50 000	15	D2	Complex araucarian vine forest
63 000	20	D3	Low araucarian vine forest
79 000	25	E1	Complex vine forest
86 000	30	E2	Simple vine forest
35			
116 000	40	E3	Complex vine forest
123 000			

Fig. 12.10 *(cont'd)*

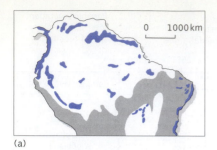

(a)

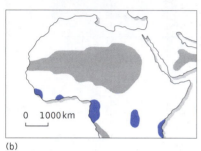

(b)

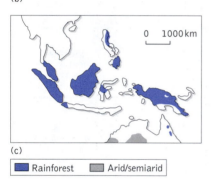

(c)

■ Rainforest	■ Arid/semiarid

Fig. 12.11 Possible distribution of regions of humid rainforest and arid/semiarid areas 22,000 years ago, when the glaciation of the higher latitudes was at its maximum. From Tallis [24].

believe that this fragmentation may even have added to their diversity by permitting the isolation of populations and the development of new species [26].

Figure 12.11 also shows possible areas of drought at the time of maximum glacial extent, and the likely global pattern of dry regions, as evidenced by sand deposits, is shown in Fig. 12.12, where it is compared to modern desert distribution [27]. The global aridity of the last stages of the last glacial is very apparent. Blown sand, termed **loess**, from these deserts is found fossil in many parts of the eastern United States, in central Europe, and in southern Australia, where fossil vertebrates have confirmed the arid nature of much of the Pleistocene [22]. Lake-level studies have similarly shown that the period between 20,000 and 15,000 years ago was particularly dry in some regions. For example, the lake levels from tropical Africa were at a low point during this time period, as shown by the data summarized in Fig. 12.13.

It would be misleading, however, to give the impression that drought prevailed throughout the Earth at this time. Just as at the present day some areas are wetter than others, so it was in glacial times, and the western part of the United States enjoyed a time of wet climate and high lake levels between 10,000 and 25,000 years ago. Such wet periods are called **pluvials**, and many parts of the world experienced pluvials at various times in Quaternary history. In subtropical Africa, for example, 55,000 and 90,000 years ago seem to have been times of pluvial climate.

Evidence for the existence of pluvial lakes in western North America is provided not only by the geological deposits but also by the present-day distribution of certain freshwater animals. In western Nevada there are many large lake basins that are now nearly dry, but in the remaining waterholes there live species of the desert pupfish (*Cyprinodon*) (Fig. 12.14). Over 20 populations of the pupfish are known in an area of about 3,000 square miles [28]. The isolated populations have gradually evolved into what are considered four different species, each adapted to its own specific environment, rather like the Hawaiian honey-creepers (see Chapter 8). In many respects, the wet sites in which these fish live can be regarded as evolutionary "islands" separated from each other by unfavorable terrain. The species have probably been isolated from one another since the last pluvial at the beginning of the present interglacial, whereas populations within each species may still be in partial contact during periods of flooding. The isolation of populations during the Pleistocene and Holocene has evidently resulted in speciation in the case of these fish.

Although pluvial conditions prevailed in western North America during the glacial maximum, this was not generally the case. Glacial conditions in the high latitudes were associated with colder and drier conditions over much of the Earth. Figure 12.15 shows the modeled reconstruction of glacial conditions at different latitudes [29]. Of particular note in this diagram is the greater reduction in temperature apparent in the high latitudes of the Northern Hemisphere, and also the increased wind stress in the midlatitudes of the north during the glacial. Fossil evidence from various parts of the world dating from the last glacial confirm

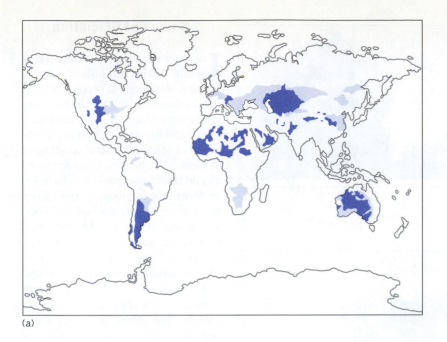

(a)

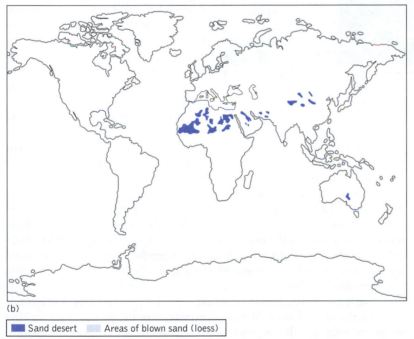

(b)

■ Sand desert   ░ Areas of blown sand (loess)

Fig. 12.12 (a) Proposed distribution of sand deserts at the height of the last glacial compared with (b) their modern distribution. Mobile sand (loess) was a feature of many areas during the glacial episode. From Wells [27].

that the glacial stages were times of severe disruption to the entire biosphere. This fact, coupled with the likelihood that we are still locked into the oscillating climatic system that has operated over the past 2 million years, makes it imperative that we understand the mechanisms that have generated the glacial/ interglacial cycle.

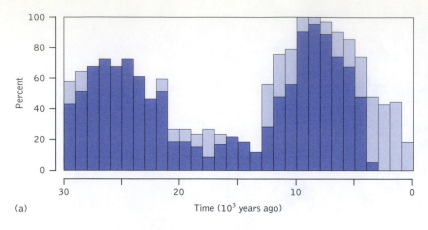

(a)

Time (10³ years ago)

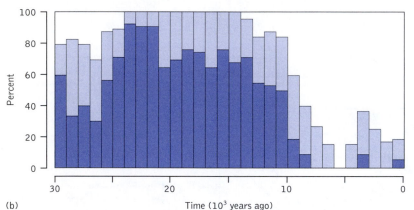

(b)

Time (10³ years ago)

Fig. 12.13 Lake levels over the past 30,000 years in (a) tropical Africa and (b) the western United States. The bars represent the proportion of lakes studied with high (dark shading), intermediate (light shading), or low (white) lake levels. From Tallis [24].

Causes of Glaciation

Ice Ages are relatively rare events in the Earth's 4.6 billion years of history. Although the polar regions receive less energy from the sun than the equatorial regions, they have been supplied with warmth by a free circulation of ocean currents through most of the world's history. Only occasionally do landmasses pass over the poles, or form obstructions to the movement of waters into the high latitudes, which can result in the formation of polar ice caps. Table 12.1 (p. 386) shows the approximate timing of glacial episodes during Earth history.

The table shows that ice ages tend to occur approximately every 150 million years, although there was not one during the Jurassic Period, 150 million years ago, probably because there was no major continental mass over the South Pole at that time. During the most recent ice age, the movement of Antarctica into its position over the South Pole led to the development of a Southern Polar Ice Cap perhaps as long ago as 42 million years. Falling global temperature then led to the development of the Greenland Ice Cap around 34 million years ago. Rearrangements of the Northern Hemisphere landmasses have subsequently resulted in the isolation of the Arctic Ocean so that it received relatively little influence from warm-water currents, leading to its becoming frozen some 3 to 5 million years ago. The presence of two massive polar ice caps increased the albedo, or reflectivity, of the Earth, for whereas the Earth as a whole reflects about 40% of the energy that falls on it, the ice caps reflect about 80%. The formation of two such ice caps therefore significantly reduced the amount of energy retained by the Earth. The scene was set for the development of an Ice Age.

It is still necessary, however, to explain why the recent Ice Age has not been a period of uniform cold, but has consisted of a sequence of alternating warm and cold episodes. A proposal to explain this pattern was put forward in the 1930s by the Yugoslav physicist Milutin Milankovich and has become widely accepted by climatologists. Milankovich constructed a model based on the fact that the Earth's orbit around the sun is elliptical and that the shape of the ellipse changes in space in a regular fashion, from more circular to strongly elliptical. When the orbit is fairly circular there will be a more regular input of energy to the Earth through the year, whereas when it is strongly elliptical the contrast between winter and summer energy supply will be much more pronounced. It takes about 96,000 years to complete a cycle of this change in orbital shape, as shown in Fig. 12.16.

A second source of variation is produced by the tilt of the Earth's axis relative to the sun, which again affects the impact of seasonal changes, with a cycle duration of about 42,000 years. The third consideration is a wobble of the Earth's axis around its basic tilt angle, which shows a cycle of about 21,000 years. All of these cycles affect the strength of solar intensity that is received by the Earth, and the pattern of climatic change should, according to Milankovich, be a predictable consequence of summing the effects of these three **Milankovich Cycles**.

Geophysicists working on the chemistry of ocean-floor sediments have expended much effort in seeking evidence for cyclic periodicity in past ocean temperatures and in checking the cycles found against those proposed by Milankovich. In 1976 Jim Hays, John Imbrie, and Nick Shackleton [30] were able to confirm that all three levels of Milankovich Cycles could be detected in the marine sediments. The Milankovich pattern has now been found in sediments dating back 8 million years, but one important change has been detected in their effects. Whereas 8 million years ago the effects of the 96,000-year cycle were weak, in the last 2 million years it is this cycle that has been very strong and that has dominated the glacial/interglacial sequence. Additional factors must have amplified this particular cycle in recent times.

One possible explanation for the current exaggeration of the effects of the 96,000-year cycle is that ice masses themselves are responsible. The ice grows slowly and decays relatively quickly, which can itself modify the global climate. Computer models have been constructed that take account of this effect, and they produce a better fit to observed data than the Milankovich cycles on their own.

Detailed testing of the Milankovitch theory is, of course, dependent on well-dated records, and at present the dating of climatic fluctuations in the Pleistocene is crude. The difference in the dating of the beginning of the last interglacial, for example, varies by over 17,000 years, depending on the materials used. Until dating is more firmly established, full confirmation of the correlation between orbital cycles and climate cannot be achieved.

Solar forcing, in which variations in astronomical conditions determine global climate, as reflected in the Milankovitch Cycles, appears to be a major element underlying the observed pattern of glacials and interglacials in the past 2.4 million years. There are complicating factors, however, many

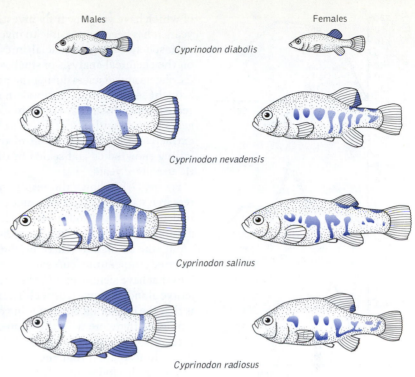

Males

Females

Cyprinodon diabolis

Cyprinodon nevadensis

Cyprinodon salinus

Cyprinodon radiosus

Fig. 12.14 The four species of desert pupfish that have evolved in the streams and thermal springs of the Death Valley region. Males are bright iridescent blue; females are greenish. (Their different markings are shown by the blue tint.) After Brown [28].

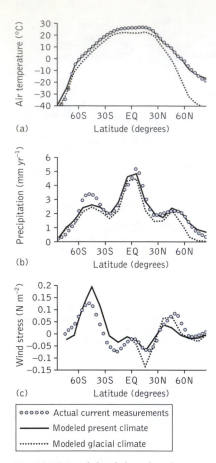

(a) Air temperature (°C) — Latitude (degrees)

(b) Precipitation (mm yr^{-1}) — Latitude (degrees)

(c) Wind stress (N m^{-2}) — Latitude (degrees)

∘∘∘∘∘∘ Actual current measurements
——— Modeled present climate
············ Modeled glacial climate

Fig. 12.15 Models of glacial and current climatic conditions (annual means) at different latitudes compared with actual modern measurements. Note that the colder, drier, windier conditions during the glacial are especially pronounced in the Northern Hemisphere. From Ganopolski et al. [29].

Table 12.1 The occurrence of ice ages in the history of the Earth.

of which have yet to be fully investigated and explained. Over the past 25 years it has become possible to investigate the composition of the Earth's atmosphere during the glacial/interglacial cycles. The technique depends on the chemical analysis of small bubbles of air trapped within the ice of ice sheets and glaciers during the past and retained in a 'fossil' form until brought to the surface by modern coring. The extracted ice is carefully sealed from contamination by modern air and is crushed to force out the gases trapped within it at different levels. It is assumed that the gas thus extracted is a true reflection of the contemporaneous air enclosed by falling snowflakes and sealed in the ice over the course of hundreds of thousands of years.

Figure 12.17 shows the results of an analysis of ice from the Vostok research station in eastern Antarctica [31]. The ice core is 3,300 m (10,800 ft) in depth (top scale) and covers the last 400,000 years of the Earth's history (bottom scale). Superimposed on these scales is the projected solar energy input, the **insolation** (curve e), the oxygen isotope record (curve d), the inferred temperature (curve b), and two of the gases found within the bubbles that have significance in climatic fluctuations, namely, carbon dioxide (curve a) and methane (curve c). The first point to notice about this diagram is that the insolation curve ties in very acceptably with the oxygen isotope curve and therefore with the temperature curve. Major peaks in temperature, of which there are five including the current, correspond well with peaks in insolation. But the two atmospheric gases, carbon dioxide and methane, also follow closely the inferred temperature curve, both rising in concentration during the interglacials and falling during the glacials. Significantly, both of these compounds are **greenhouse gases**; both have a high capacity for energy absorbance in the infrared region of the spectrum and so are able to act as thermal insulators for the Earth.

We must try to explain the correspondence of these gases with global temperature change and also assess their influence on such changes, since they are capable of altering the energy exchange system of the planet. One could explain the rise in these gases during times of warmth by considering the additional respiratory activity of organisms, particularly microbes, during the interglacials. Respiration of all organisms produces carbon dioxide, and microbial decay is stimulated by warmth, especially in the cooler regions of the Earth where glacial conditions may have held back decomposition. The weathering of rocks, such as limestone, in soils would also increase during times of warmth, resulting in additional carbon dioxide production. Methane is produced in a variety of ways, by the activities of termites, by generation within the rumens of large herbivores, and by

Era	Geological Period	Approximate Age (Mya)
Cenozoic 10,000 years	Pliocene/Pleistocene	10
Palaeozoic	Carboniferous/Permian	300
	Ordovician	450
Proterozoic (Precambrian)		600
		750
		900
		2,300

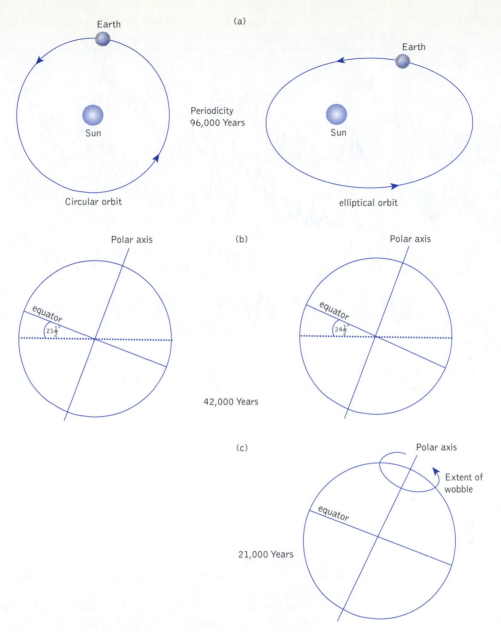

Fig. 12.16 The three variations in orbital and axial conditions that affect the solar radiation arriving at the surface of the Earth, as proposed by Milankovitch. (a) The eccentric nature of the Earth's orbit around the sun, which has a periodicity of about 96,000 years. (b) The variation of the tilt of the Earth on its axis, having a periodicity of about 42,000 years. (c) The wobble of the Earth on its axis, which affects the season of tilt and has a periodicity of about 21,000 years.

the incomplete decomposition of organic matter in wetlands. But on the reverse side, warmer conditions could result in additional plant productivity and biomass accumulation, which would lead to more accumulation of organic matter at the expense of atmospheric carbon dioxide. Changing levels of carbon dioxide in the atmosphere may even have resulted in

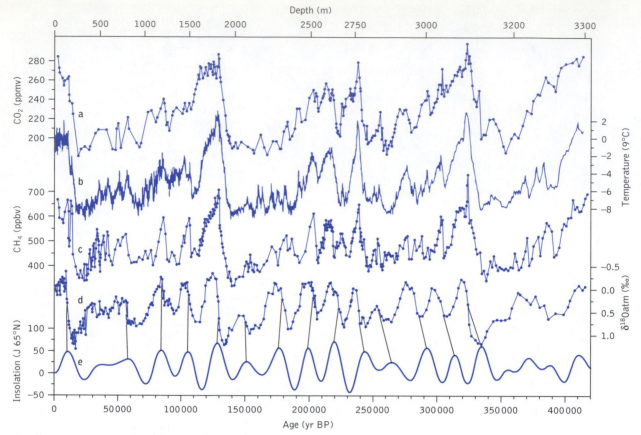

Fig. 12.17 Analyses of gas bubbles in an ice core taken from the Antarctic Ice Cap at the Vostok site. The gases extracted are considered fossil samples of the ambient atmosphere of the past. The concentrations of carbon dioxide (curve a) and methane (curve c) are shown, together with a reconstruction of temperature change (curve b) derived from oxygen isotope studies (curve d). At the bottom of the diagram (curve e) is a proposed curve for insolation based on Milankovitch cycle predictions. From Petit et al. [31].

structural changes in plants (see Concept Box 12.2). The role of the oceans is also likely to be important here. At present, it is estimated that about a third of the carbon dioxide entering the atmosphere each year ends up in the oceans, largely because many of the microscopic planktonic organisms (especially the coccolithophorids and the foraminifera) construct shells of calcium carbonate that sink to the ocean floors after their death. The carbonate they use is ultimately derived from carbon dioxide from the atmosphere dissolving in the surface waters and then being taken up by the plankton. The question is, was this oceanic sink for carbon greater during the interglacials, resulting in the extraction of carbon from the atmosphere? If oceanic circulation changes during a glacial, resulting in upwelling and the conveyance of deep, nutrient-rich waters to the surface, then planktonic growth would be stimulated and a biological pump would be brought into action, taking carbon out of the atmosphere and depositing it in the ocean sediments [32]. It is also possible that changes in deep-sea currents could have flushed carbon-rich waters from the abyssal depths at the commencement of interglacials, thus increasing greenhouse gases in the atmosphere and enhancing global warming. Evidence from the North Pacific seems to favor this argument [33].

Whatever the precise mechanism at work, the outcome is a close correspondence between atmospheric carbon dioxide (and methane) and atmospheric temperature. The raised levels of these gases during the interglacials act as a positive feedback, enhancing the higher temperatures because of their thermal, greenhouse properties in the atmosphere.

Wallace S. Broecker of Columbia University has emphasized the importance of the circulation of the Earth's ocean currents, coupled with those of the atmosphere, to explain the rapidity with which some of the climatic shifts in the Pleistocene have come about [36]. His theory does not contradict the ideas of Milankovitch, but supplements them. Much of the warmth of the North Atlantic is brought into the region by highly saline waters heading northward at intermediate depths of about 800 m (2,600 ft). This current effectively redistributes the tropical warmth into the high latitudes, as seen in Chapter 3 (Fig. 3.10). Broecker estimates that this input of energy into the North Atlantic is equivalent to 30% of the annual solar energy input to the area. Fossil evidence, however, indicates that this oceanic conveyor belt became switched off during the glacial episodes, and the reduction in the heat transfer to high latitudes led to the development of the extended ice sheets. If this were so, it would explain the severe and relatively sudden drop in temperature in the high latitudes of the Northern Hemisphere seen in Fig. 12.17. Smaller changes in the thermohaline circulation patterns of the North Atlantic could account for the climatic oscillations experienced during the last glaciation, the Dansgaard-Oeschger cycles [37].

One further suggestion is that massive volcanic eruptions may precede and initiate the process of glaciation [38]. There is certainly some evidence for a correlation between glacial advances and periods of volcanic activity during the last 42,000 years in New Zealand, Japan, and South America. Volcanic eruptions produce large quantities of dust, which are thrown high into the atmosphere. This has the effect of reducing the amount of solar energy arriving at the Earth's surface, and dust particles also serve as nuclei on which condensation of water droplets occurs, thus increasing precipitation. Both of these consequences would favor glacial development. Attempts to correlate volcanic ash content with evidence of climatic changes in ocean sediments, however, have not met with much success. There has, however, been an increase in the general frequency of volcanic activity during the last 2 million years, which correlates with a time of overall global cold conditions.

The Current Interglacial: A False Start

Following the maximum extent of the most recent glaciation at about 22,000 years ago, the climate began to warm, as shown in the oxygen isotope record in Fig. 12.8, and the great ice sheets started to recede. Every indication pointed to the beginning of a new interglacial. But the increasing warmth was quite suddenly interrupted by a return to extremely cold conditions, forming a cycle reminiscent of the Dansgaard-Oeschger Cycles of the earlier part of the last glacial.

The instability of climate around 14,000 to 10,000 years ago was first noted by geologists in Denmark, who found that the sediments of lakes dating from the transition between the glacial and the current interglacial exhibited some unusual features. The inorganic clays, typical of sediments formed in lakes surrounded by arctic tundra vegetation, where soils are easily eroded and the vegetation has a low organic productivity, were replaced by increasingly organic sediments as the development of local vegetation stabilized the mineral soils, and the aquatic productivity of the lakes led to an increasing organic content in the sediments. But this process, reflecting the increased warmth of the climate, was evidently interrupted, for the sediments then reverted to heavy clay, often with angular fragments of rock denoting a return to severe periglacial or locally even glacial climatic conditions. Above this layer, the organic sediments reappeared, which is interpreted as a sign of increased plant productivity associated with climatic warming. This time the increased temperature led to the development of our present interglacial. The evidence for the climatic interruption proved to be a consistent feature of sediments of this age throughout northwestern Europe.

The cold episode that caused the deposition of these sediments was severe enough to lead to the regrowth of many glaciers. Geomorphologists have shown that the glaciers of Scandinavia and Scotland extended considerably during this episode, while small glaciers began to form on north-facing slopes in more southerly mountains, from Wales to the Pyrenees. The event was called the **Younger Dryas**, because the fossil leaves of the arctic plant themountain avens (*Dryas octopetala*) were found in abundance in the clay layers during the original Danish studies (Fig. 12.18). Radiocarbon dating of the Younger Dryas from a range of sites indicates that it took place between 12,700 and 11,500 solar (calendar) years ago [39]. The precise dating of this event is in some doubt because radiocarbon dating lacks precision during this stage in Earth's history; the calibration curve for converting radiocarbon years into solar years, based on tree growth rings, has a flat plateau at this time.

Although stratigraphic and fossil evidence for the Younger Dryas cold phase was abundant in northwestern Europe, it was more difficult to discern in the sediments of the European Alps, especially on the southern side. On the western side of the Atlantic its influence, though apparent on the eastern seaboard of North America, is difficult to detect when one moves inland. There is some evidence for cooling at this time in the northern

Fig. 12.18 Mountain avens (*Dryas octopetala*), an arctic–alpine plant, remains of which were found in late-glacial sediments from Denmark and which has lent its name to the cold event called the Younger Dryas stadial.

Pacific, as far south as California [40], and even in southern Chile [41], but again the episode is less strongly recorded than in the North Atlantic region. Although evidence for climatic cooling at this time has been collected from around the world, the evidence suggests that this event was centered on the North Atlantic. In addition, oceanic information from fossils in the sediments shows that a cold front of polar water did indeed extend as far south as Spain and Portugal during the Younger Dryas stadial.

One possible explanation of how this climatic reversion, centered in the North Atlantic, may have come about was offered by Claes Rooth of the University of Miami and was later supported by the data of Wallace Broecker [42] and his co-workers. They suggested that the change resulted from an alteration in the pattern of discharge of meltwater from the great Laurentide glacier of North America. As this ice mass, covering the whole of eastern Canada, melted, its waters initially flowed south to the Gulf of Mexico. During the Younger Dryas stadial, however, it is proposed that the meltwater was rerouted via the St. Lawrence into the North Atlantic (Fig. 12.19). Not only would this bring large volumes of cold water into the North Atlantic, but the freshwater would also dilute the saline waters on the oceanic conveyor belt (see Chapter 4) and could disrupt the global movement of this conveyor. Such modification would explain why the Younger Dryas was most strongly felt in the regions around the North Atlantic, where the warm influence of the conveyor has its greatest impact. But if meltwater flow and salinity changes were responsible for switching off the conveyor, we would expect the onset of the Younger Dryas to be accompanied by a rapid rise in sea level. The work of R. G. Fairbanks [43] suggests that ice sheets were not melting rapidly at the time of the onset of the Younger Dryas, so the meltwater hypothesis is not supported. Subsequent work on the growth of deep-sea corals in the North Atlantic also confirms these findings [44], and the Broecker model looks less likely to be the complete explanation for these climatic changes. The cold episode certainly took place, however, and the record of marine sediments confirms that oceanic circulation changed [45], but the relationship between climate change, oceanic circulation, terrestrial hydrology, and the composition of the atmosphere still needs to be clarified [46].

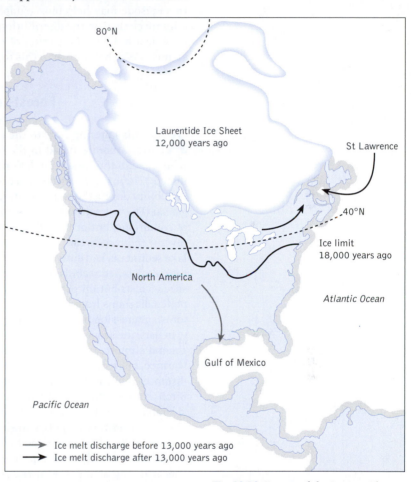

Fig. 12.19 Extent of the Laurentide Ice Sheet in the final stages of the last Ice Age. At its greatest extent the ice sheet terminated south of the Great Lakes, so that 18,000 years ago its meltwater outflow is likely to have been south into the Gulf of Mexico. Once it had retreated north of the St. Lawrence River, about 12,000 years ago, the meltwater became diverted into the North Atlantic.

Perhaps the most interesting and important aspect of the Younger Dryas is that it demonstrates how quickly the Earth's climate can shift from interglacial to glacial mode and back again. The brief warm spell before the Younger Dryas, called the **Allerød interstadial**, was sufficiently hot to bring beetles with Mediterranean affinities as far north in western Europe as the middle of England. Yet within a few centuries glacial conditions once more prevailed in the area. Perhaps even more remarkable is the rapidity of change at the end of the Younger Dryas. Evidence suggests that within a period of 50 years the mean annual temperature rose by 7°C [39], and some have suggested an even more rapid transition. Close study of this episode may help biogeographers to understand the impact of rapid climate change on the distribution patterns of living organisms, and may provide a basis for projection of the possible impact of current climate changes. We shall return to this topic in Chapter 14.

Forests on the Move

After this faltering start to the current interglacial, the generally raised temperature (recorded in the oxygen isotope ratios of the ocean sediments and in the ice accumulations of the ice caps) was consistently maintained. The oxygen isotope curves indicate that there was less climatic instability during the warmer interglacial than had been the case during the glacial. The vegetation and animal life of the Earth had to adjust to a new set of conditions, but at least the increased warmth was fairly consistent for several thousand years. The pollen grains preserved in the accumulating lake sediments laid down since the end of the Younger Dryas (the Holocene) provide detailed records of the arrival and expansion of tree populations as species distribution patterns changed and forests were reconstituted. The pollen diagrams from Minnesota shown in Fig 12.20 [47] illustrate a typical forest progression from the Northern Hemisphere temperate region.

In part, the sequence of trees (*Picea, Betula, Quercus, Ulmus*) reflects the general climatic tolerances of these tree genera responding as the climate warmed, but it is also a function of their relative speed of migration and the distances they had to cover to invade land exposed by the retreating glaciers. Birch (*Betula* spp.), for example, has light, airborne fruits that can travel considerable distances. It is also able to produce fruits when only a few years old, which permits a rapid expansion of its range. Add to this the fact that it is reasonably cold-tolerant and may have survived quite close to the glacial limits, and it is only to be expected that it should be one of the first trees to appear in any abundance in the postglacial pollen record. One must also remember that birch produces large quantities of pollen, so it could be over-represented in the pollen record. Many factors must be considered before a pollen sequence can be interpreted in terms of changing climate. Some elegant techniques have been employed to try to translate pollen densities in sediments into estimates of the population densities. In this way one can trace the population expansion of trees as they invaded new areas [48].

Overall, it is evident from the oxygen isotope data that there was a warming of the climate, reaching its maximum during the early stages of the interglacial, and cooling over the past 5,000 years. The most warmth-demanding trees, however, such as lime (*Tilia* spp.) in western Europe,

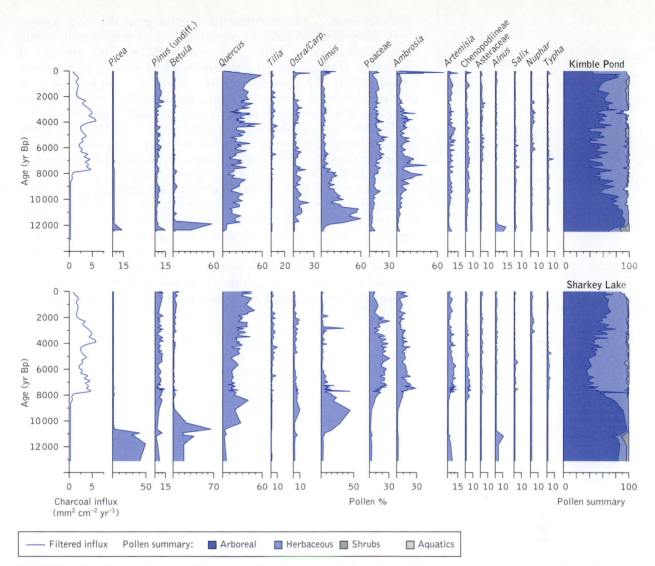

Fig. 12.20 Pollen diagrams from two lake sites in Minnesota. Quantities of charcoal within the sediments are also shown. These sites lie in the transition region between deciduous forest and prairie biomes. From 8,000 years ago trees decline, and grasses and sagebrush elements increase, together with charcoal derived from fire. From Camill et al. [47].

were relatively late arrivals, possibly in part because of their slow rates of spread. Vegetation changes in the later parts of the Holocene are difficult to interpret because the general climatic cooling was associated with varying patterns of precipitation and aridity in different parts of the world. In some areas, such as western Europe, the intensification of human agriculture during the later part of the Holocene was associated with forest clearance, and this obscures any climatic signal in the pollen record. The diagrams from Minnesota show increasing quantities of grass (Poaceae) and ragweed (*Ambrosia*) pollen accompanied by an input of charcoal, derived from fire in the surrounding vegetation, to the lake sediments over the past 8,000 years. These two lake sites lie on the borderline between two major biomes—temperate forest and prairie grassland—so it is possible that the increased fire frequency is associated with climatic conditions, the warmer, drier climate favoring the spread of prairie grasslands at the expense of deciduous forest. Around 4,000 years ago, this trend was reversed. Charcoal levels began to fall, and oak began to increase in its abundance at the expense of the prairie herbs as cooler and more humid conditions took over once more, reducing fire abundance and allowing the development of open oak woodland. Not until very recent times, in the upper layers of sediment at Kimble Pond, do we see an impact on the vegetation that clearly has its origin in human activity. The sudden expansion of ragweed indicates the arrival of settlers from Europe.

Large numbers of pollen diagrams covering the current interglacial are now available from all over the world, and many of them are firmly dated by means of radiocarbon. This has made it possible to study the movement of individual species and genera of plants by constructing pollen maps for particular periods in the past. In this way, one can follow the spread of trees, for example, and observe the routes along which they have dispersed and the way in which they have reassembled themselves into reconstituted forests. Figure 12.21 shows some examples of this type of

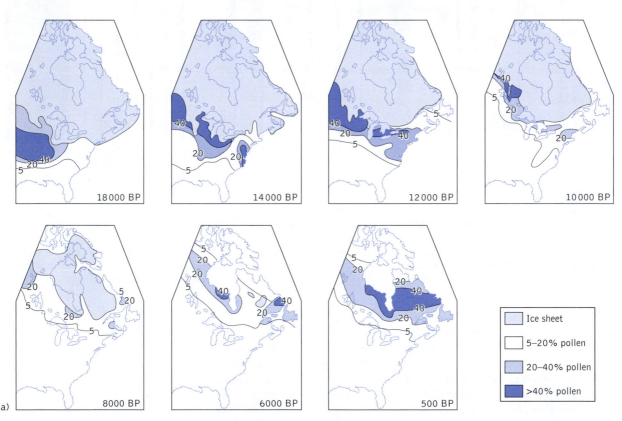

(a)

Fig. 12.21 Spread of selected trees through North America since the last glacial maximum. Dates are in radiocarbon years before present (bp); they have not been corrected to solar years. Contours ("isopolls") join sites of equal pollen representation in lake sediments from the appropriate time period. (a) Spruce (*Picea*). (b) Pine (*Pinus*). (c) Oak (*Quercus*). From Jacobson et al. [49].

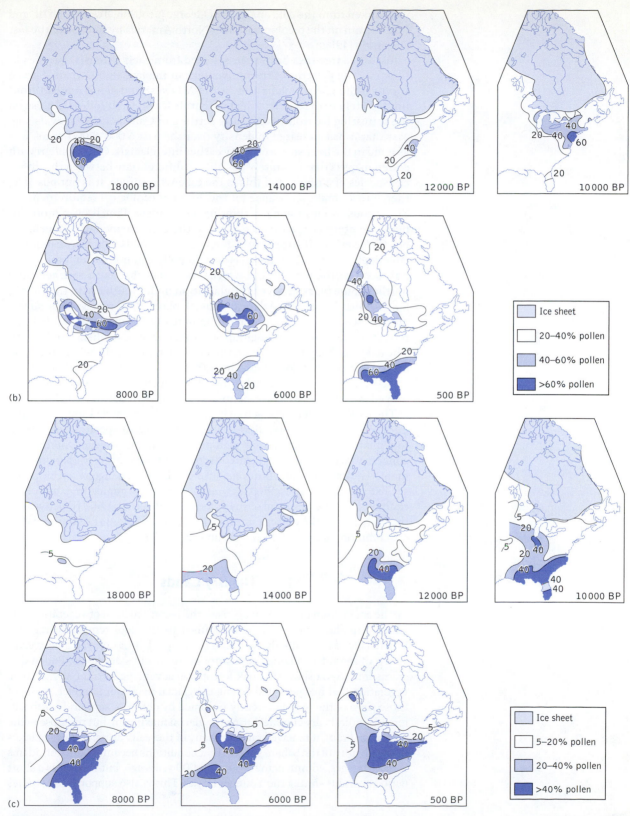

Fig. 12.21 (cont'd)

work taken from the data analyses of George Jacobson, Tom Webb III, and Eric Grimm on the recolonization of North America by trees after the last glaciation [49].

Just three tree taxa have been selected from their extensive studies to illustrate the patterns of tree movement in the last 18,000 years. Spruce survived the glacial maximum in the Midwest and along a broad front immediately to the south of the Laurentide Ice Sheet. It followed close on the retreating ice front, eventually settling in Canada. Pollen levels suggest that it has increased in density over the past 6,000–8,000 years, just as it did in the later stages of many earlier interglacials. One problem with pollen maps is that it is rarely possible to differentiate between the different species of a genus, and this is the case with spruce. It is not possible, therefore, to make allowance for the different ecological requirements of the various species when interpreting pollen maps. Pine is even more difficult to interpret than spruce because there are many native species in North America, all with very different requirements, and they cannot be effectively separated on the basis of their pollen grains. Other fossil material, such as their needles, however, indicates that both southern and northern pine types survived the glaciation in the Southeast of the United States, mainly on the Atlantic coastal plain. The two groups subsequently separated, one group heading north to invade areas left bare by retreating ice, and the other group becoming firmly established in Florida. The oaks had found refuge from the glaciation in Florida, where they had achieved a marked dominance by 8,000 radiocarbon years (about 9,000 solar years) ago. Subsequently, they have declined in Florida and have moved into their current stronghold to the south of the Great Lakes area.

The technique of mapping tree movements has provided some valuable information about the rapidity with which different species can respond to climate change, which may well prove useful knowledge when we are concerned with predicting future responses to our currently changing climate. The rates of spread in response to climate change, even of large-seeded species such as the oaks, are surprisingly rapid. In Europe, for example, the rate of spread of the oaks reached 500 m (1,600 ft) per year in the early part of the current interglacial. Rates of 300 m (1,000 ft) per year are common for many tree and shrub species [50].

The Dry Lands

The application of pollen analysis to the reconstruction of vegetation history in the Holocene is not confined to the cool temperate regions. Where lakes exist and sediments can be sampled, it has proved a very valuable tool in understanding the history of the world's dry lands. Pollen diagrams from Syria show that oak forest was advancing into the dry, steppic vegetation that had persisted during the glacial maximum. In fact, many of those parts of the world currently occupied by desert or semiarid scrub suffered a similarly dry climate during the glacial maximum, but the close of the glaciation brought renewed rains to many of these dry areas. Studies of lakes in the vicinity of the Sahara suggest that conditions became more humid in a series of stages, commencing around 14,000 years ago, but there was a short period of aridity during the Younger Dryas. This is also supported by studies of the discharge rate of the Congo River and isotopic ratios in fossil plant waxes from this time [51]. In northeastern Nigeria, swamp forest vegetation occupied the hollows between what are now the dunes of modern savanna grasslands [52]. The early part of the Holocene thus provided a time of wetness for many currently desert areas, extending from Africa through Arabia to India. Lakes existed in the middle of the very arid Rajasthan Desert of northwest India, and the surge of freshwater down the Nile created stratified waters in the eastern Mediterranean, the low-density freshwater lying over the top of the high-density saltwater. As a result, the lower layers became depleted in oxygen, and black anoxic sediments, called **sapropels**, were deposited [53, 54].

The climatic wetness in the African tropics at this time permitted the northward extension of savanna and rainforest into formerly dry areas, and Fig. 12.22 summarizes data from many pollen diagrams taken from sites in the southern Sahara [55]. The age axis on this diagram is expressed in radiocarbon years: at 10,000 radiocarbon years these are approximately 1,000 years younger than the true solar dates. The expansion of the more humid biomes in the early Holocene, around 9,000 radiocarbon years ago, can be seen on this diagram, followed by their contraction when aridity set in once more about 5,000 years ago.

The Sahara Desert is quite rich in ancient rock paintings, some dating back over 8,000 years. These older ones depict big-game animals now associated with the savanna grasslands, confirming the evidence of the pollen. Between 7,500 and 4,500 years ago, the painters of the pictures were evidently pastoralists who depicted their cattle on rocks in locations where cattle could certainly not graze today. After that time, the pictures of cows were replaced by camels and horses as the arid climate became more severe, and the activities and the domesticated animals of the local peoples altered accordingly.

Many of the great deserts of the world were evidently initiated by climatic changes in the latter half of the Holocene. The involvement of increased human activity during this time, however, has obviously complicated the picture, and some researchers believe that human exploitation of the limited arid land resources, even in prehistoric times, may have contributed to the development of deserts, as in the case of the Rajasthan Desert, where the increasing aridity coincided with the cultural

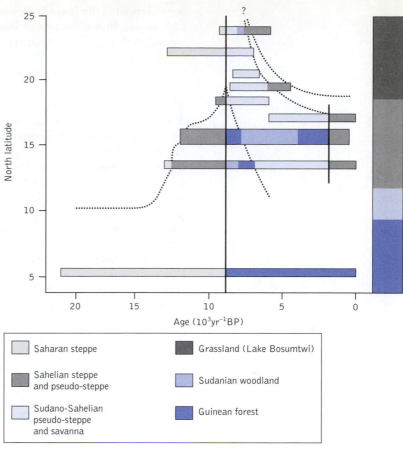

Fig. 12.22 Changes in the vegetation in northern tropical Africa over the past 20,000 years. The present latitudinal zonation of vegetation is shown on the right of the diagram and it can be seen that this zonation was displaced about 5°N during the period 9,000 to 7,000 years ago, in the early stages of the current interglacial. From Lezine [55].

Legend:
- Saharan steppe
- Sahelian steppe and pseudo-steppe
- Sudano-Sahelian pseudo-steppe and savanna
- Grassland (Lake Bosumtwi)
- Sudanian woodland
- Guinean forest

development of the Indus Valley civilization [56]. It is very difficult, however, to determine the role of humankind in the development of desert spread under such circumstances.

Changing Sea Levels

Interglacial warm episodes are times of changing sea levels, and the current warm period is no exception. The melting of glacial ice has released considerable quantities of water into the oceans, and this, combined with the increase in water volume as its temperature rises, has resulted in a **eustatic** rise in sea level relative to the land. This may have amounted to as much as 100 m (330 ft) in places. On the other hand, the loss of ice caps over those landmasses that acted as centers of glaciation relieved the Earth's crust of a weight burden, resulting in an **isostatic** upwarping of the land surface with respect to sea level. The relative importance of these eustatic and isostatic processes varied from one place to another, depending on how great a load of ice an area had borne. In western Europe, the result was a general rise in the level of the southern North Sea and the English Channel with respect to the local land surface. In this way Britain, which was a peninsula of the European mainland during the glaciation, gradually became an island. Prior to this, the rivers Rhine, Thames, Somme, and Seine had all converged to flow west into the Atlantic, between the hilly areas that were later to become Cornwall and Brittany [57]. Evidence from submerged peat beds in the Netherlands suggests a rapid rise in sea level between about 10,000 and 6,000 years ago, which has subsequently slowed down gradually. By this latter date, England's links with continental Europe had been severed. At this time many plants with slow migration rates had still not crossed into Britain and were thus excluded from the British flora. The separation of Ireland from Britain occurred earlier because there are deeper channels between these landmasses, and many species native to Britain have not established themselves as far west as Ireland. As a result, plants such as the lime tree (*Tilia cordata*) and herb paris (*Paris quadrifolia*) are not found growing wild in Ireland. Other plants, however, were more successful and invaded Ireland along the western seaboard of Europe before sea levels fragmented this route (see Concept Box 12.3).

The Legacy of Lusitania

Concept Box 12.3

There is one group of western European plants that has proved of great interest to plant geographers and that succeeded in reaching Ireland before the rising sea level separated that country from the rest of the British Isles, and this is known as the Lusitanian flora. Lusitania was the name for a province of the Roman Empire consisting of Portugal and part of Spain and, as its name suggests, the Lusitanian flora has affinities with that of the Iberian Peninsula. Some of the plants, such as the strawberry tree (*Arbutus unedo*) (see Chapter 3) and giant butterwort (*Pinguicula grandiflora*), are not found growing wild in mainland Britain. Others, such as the Cornish heath (*Erica vagans*) and the pale butterwort (*Pinguicula lusitanica*), are found in southwestern England as well as in Ireland. It therefore seems likely that these plants spread from Spain and Portugal up the Atlantic seaboard of Europe in early postglacial times but were subsequently cut off by the rising sea levels. It is not impossible that some or all of the species may have survived the last glaciation in oceanic southwest Ireland, but direct proof of this is lacking.

Some mammals also failed to make the crossing into Ireland, and it is difficult to explain why they failed when closely related species succeeded. For example, the pygmy shrew (*Sorex minutus*) reached Ireland, yet the common shrew (*S. araneus*) did not (Fig. 12.23). Perhaps this is related to their habitat preferences, for the pygmy shrew is found on moorlands and might have survived better than the common shrew if the conditions of the land bridge were wet, peaty, and acidic. The stoat (*Mustella ermines*) also reached Ireland, but the weasel (*M. nivalis*) did not. Arrival in this case may have been due to sheer chance. A single pregnant female arriving on a raft of floating vegetation could have been sufficient to populate the island. Discoveries in some archaeological sites in Ireland of small mammals, such as the wood mouse (*Apodemus sylvaticus*), raise the possibility that some plants and animals could have been carried over the water by prehistoric humans. The arrival of some large mammals on isolated islands could also be an outcome of human transport, even in preagricultural times. Mesolithic (Middle Stone Age) people in the British Isles, for example, may have been responsible for carrying the red deer (*Cervus elaphus*) to Ireland and to other offshore islands such as Shetland. The red deer or elk was the major prey animal of these people, who did not have truly domesticated animals and plants, and the transport of young animals would have presented few difficulties even in small, primitive boats.

(a)
(b)

Fig. 12.23 Distribution maps of two mammal species in Europe: (a) pygmy shrew (*Sorex minutus*), (b) common shrew (*S. araneus*). The pygmy shrew reached Ireland, but the common shrew failed to do so.

Rising sea levels during the present interglacial were responsible for the severing of land connections in many other parts of the world also. For example, Siberia and Alaska were connected across what are now the Bering Straits, which in places are only 80 km (50 mi) across and 50 m (160 ft) deep. This high-latitude land bridge would have been a suitable dispersal route only for arctic species, but it is believed to have been the route by which humans entered the North American continent (see Chapter 13).

A Time of Warmth

The period of maximum warmth during the present interglacial lasted from early in the interglacial until about 5,500 years ago. At this time, warmth-demanding species extended farther north than they do at present. For example, the hazel (*Corylus avellana*) was found considerably further north in Sweden and Finland than it is today. In North America, fossils of the plains spadefoot toad (*Scaphiopus bombifrons*) have been found 100 km (60 mi) north of its current distribution limits. These examples indicate that conditions have become cooler since that time. The remains of tree stumps, buried beneath peat deposits at high altitude on mountains and far north of the tree line in the Canadian Arctic, also bear witness to more favorable conditions in former times. Things are not always what they seem, however, and one has to remember the possible involvement of humans in the clearance of forests and the modification of habitats. Humans may have played an important part, for example, in the forest clearance that led to the formation of many of the so-called blanket mires of western Europe [58]. By clearing the trees, they created a new set of hydrological conditions, and the saturated soils began to accumulate peat. Continued burning and grazing by prehistoric pastoralists ensured that the forests were not able to reinvade. But the lack of trees does not mean that the climate is no longer suitable for their growth. When grazing animals are removed from these bogs, the woodland is often able to regenerate. In the case of deserts, it is also often difficult to appreciate the extent to which people and their domestic grazers currently limit the distribution of plants and animals.

Climate and fire also interact with one another, as in the prairie region of central North America. Measurements of fire frequency in the past history of the prairies show that the time between fires increases when the climate is cooler, so the composition of vegetation may be determined by climate but in an indirect way [59]. But the use of fire as a management tool by Native Americans also complicates the history and ecological impact of fire [60]. In the Mediterranean region of Europe, the extent of oak woodland was once greater than it is today, but here also the activities of humans have had a strong influence for many thousands of years, and late Holocene climatic influences on the vegetation are therefore difficult to discern.

The spread of forest in temperate areas during the early Holocene, caused by the increasing warmth, created heavily shaded, unsuitable conditions for many of the plants that had previously been widespread at the close of the glacial stage. Some of these, the arctic–alpine species, are also physiologically unsuited to warmer conditions. Many such plants, for example, the mountain avens (*Dryas octopetala*), grow poorly when the summer temperatures are high (above 23°C for Britain and 27°C for Scandinavia). The climatic changes that occurred during the Holocene therefore proved harmful to such species, and many of them became restricted to higher altitudes, especially in the lower latitudes. Other plant species are more tolerant of high temperatures but are incapable of survival under dense shade. Low-latitude, low-altitude habitats that became covered by forests may have become unsuitable for the continued growth of these species, and many of them also became restricted to mountains, where competition from shade-casting trees and shrubs did not occur to the same extent. It is likely, however, that some clearings and open areas were created in the forest by storms, catastrophes, and the impact of large grazing animals. Some ecologists feel that the temperate woodlands were relatively open in structure, forming a kind of wood pasture during the mid-Holocene as a result of the pressures imposed by large grazing animals, but the evidence from pollen analysis and other sources does not support this view [61]. Lowland environments that for some reason bore no forest must have provided some suitable places of refuge, however. Coastal dunes, river cliffs, habitats disturbed by periodic flooding, and steep slopes all supplied sufficiently unstable conditions to hold back forest development and allow the survival of some of these open-habitat organisms, whether herbs, insects, or molluscs.

An outcome of these processes of vegetation change was the development of relict distribution patterns (see Chapter 4). Sometimes the separation of species into scattered populations, even though it has lasted only about 10,000 years, has encouraged genetic divergence, as in the case of the sea plantain (*Plantago maritima*) in Europe. This species has survived in both alpine and coastal habitats, but the different selective pressures of the two environments have resulted in physiological divergence between the two races. The coastal race is able to cope with high salinity but tends to be more frost-sensitive than the montane race. The development of molecular techniques for studying the genetic composition of organisms has led to extensive documentation of populations becoming fragmented by the changes of the Pleistocene and Holocene, resulting in the formation of distinct races. The white-tailed eagle (*Haliaeetus albicilla*), for example, shows a separation into two genetic groups, one from the west of its range, in Europe, and the other from the far east of its range, in Japan and eastern Asia [62]. Environmental changes thus fuel the pace of evolution.

Some species, limited by competitive inadequacies rather than by climatic factors, and favored by physiological or behavioral adaptability, have taken advantage of the disturbed conditions provided by human settlements and agriculture. These plants, which fared so poorly during forested times in the temperate latitudes, have become latter-day weeds, pests, and opportunists. Thus, climatic change and human habitat disturbance have interacted to provide different histories for each species.

Climatic Cooling

Climatic reconstruction and modeling can be based on many sources of evidence. Pollen analysis of lake sediments provides some useful data, but results of this kind, though giving information on past vegetation changes, are not very precise in supplying data about climate. Vegetation tends to respond rather slowly to climate change and is also subject to other factors, especially the influence of human activities. For example, pollen stratigraphic evidence from the Amazon Basin in eastern Brazil has indicated a change from closed forest to open savanna around 5,000 years ago [63]. This could be associated with climatic change, such as drier conditions, and the increased incidence of charcoal in the corresponding sediments could be regarded as supportive of this view. On the other hand, the role of local human populations in burning the forest cannot be discounted, so the evidence is inconclusive regarding climate change. A more precise means of detecting climate change than by using shifts in vegetation is evidently required.

Oxygen isotope curves are useful and have supplied valuable information on temperature variations. In Greenland, for example, the oxygen isotope curve shows warm conditions between about AD 700 and 1200, followed by colder conditions until the late 19th century. This accords well with historical records for these centuries.

One additional technique that is proving increasingly valuable and reliable is the measurement of the rate of peat growth in bogs. Peat accumulates in acid bogs because the rate of decomposition fails to keep pace with the rate of litter deposition at the bog surface. The wetness and acidity of the bog vegetation lead to low microbial activity, so that decomposition is slow. Under wetter conditions, the bog mosses grow rapidly and decomposition is curtailed, so the peat formed under such conditions looks fresh and undecomposed and is often light in color, whereas well-humified peat tends to be dark. The main problem with using this approach as a climate proxy is that of correlation of horizons between different sites. Local factors, such as microclimatic conditions or patterns of surface-water movement and pool formation, could vary between sites. Radiocarbon dating, particularly by the use of atomic mass spectroscopy, which enables very small samples of peat to be dated, has led to a considerable rise in interest in the use of peat growth rates over wide areas as a means of reconstructing climatic change [64, 65]. On the basis of such studies, times of cooler, wetter conditions have been identified from around 5,400 years ago. A time of particularly strong bog growth in the first millennium BC (especially around 800 to 400 BC) has been found through much of western Europe.

Studies of tree rings also provide a means of reconstructing past climates. Wood from the trees of temperate climates, and other types of climate with strong seasonal variation where there is a distinct alternation of summer and winter, shows annual growth rings; the width of a ring corresponds to the amount of growth the tree has achieved in that season. Care is needed in interpreting tree rings from dry climates, however, as increased growth may be related to irregular episodes of water supply. Generally, years in which the climate is suboptimal for the tree species under study produce narrower growth rings. The growth of black spruce (*Picea mariana*) in northern Canada, for example [66], shows a series of increasingly narrow annual rings between AD 1500 and 1650, indicating declining growth conditions. In this case, the rings probably reflect lower temperatures because this period of time marks the commencement of a severe stage in the so-called **Little Ice Age**. This was a lengthy spell of generally low temperature between 1200 and 1850 that is apparent over a very wide area in the Northern Hemisphere.

Other sources of evidence of climate change are provided by lake levels that are sensitive to the balance of water input, discharge, and evaporation. In the northern Great Plains region of North America, drought periods can result in evaporation increase, which in turn leads to a rise in the salinity of the water. Changes in a lake's salinity have many effects on its biota, especially the microscopic planktonic diatoms. Studies of lake salinity changes in North Dakota [67] show a sharp decline in salt content around AD 1150. The climatic change that caused the Little Ice Age in the North Atlantic region resulted in increased precipitation and lower salinity here in the continental region of North America (see Fig. 12.24).

By using a combination of all the available sources of evidence about recent climate change, it has proved possible [68] to construct a curve for the likely mean temperature of the Northern Hemisphere over the past 2,000 years (see Fig. 12.25). There was a peak in warmth around AD 1000 to 1150, often termed the **Medieval Warm Period**, followed by the cold of the Little Ice Age. During the last 200 years, instrumental records have been available for many parts of the world, so recent changes in temperature are well documented.

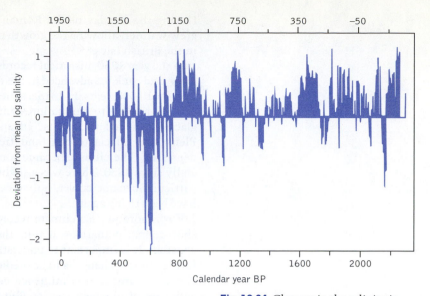

Fig. 12.24 Changes in the salinity in the sediments of Moon Lake, North Dakota (Great Plains), determined from analysis of fossil diatoms. Saline episodes, caused by drought and evaporation, are more frequent prior to AD 1200 than they have been since. From Laird et al. [67].

Recorded History

As soon as human beings appeared on the scene, they often inadvertently began to leave information about climate and its changes. Early records provide clues rather than precise information, such as the ancient rock drawings of hunting scenes discovered in the Sahara, which indicate that its climate was much less arid at the time they were made (early Holocene) than it is today. With the development of writing, more precise information concerning climate was recorded. For example, there are reports of pack ice in the Arctic seas near Iceland in 325 BC, indicating the very low winter temperatures at the time.

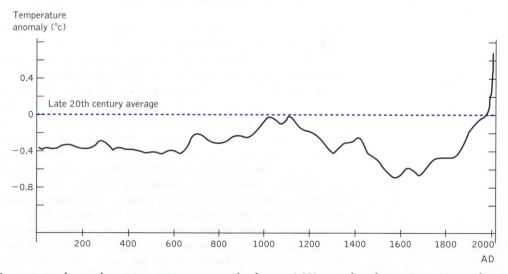

Fig. 12.25 Reconstructed annual average temperature curve for the past 2,000 years, based on various sources of proxy evidence.

During the heyday of the Roman Empire, however, climate steadily grew warmer, allowing the growth of such crops as grapes (*Vitis vinifera*) in the British Isles.

In the year 1250, historical records suggest that alpine glaciers began to grow, and pack ice advanced in the Arctic seas to its most southerly position for 10,000 years. In 1315 a series of poor summers began in northern Europe, leading to crop failures and famine. During the Little Ice Age, glaciers reached their most advanced positions since the end of the Pleistocene glacial epoch, and tree lines in the Alps were severely depressed. The climate became somewhat warmer after 1700 and especially since 1850. There was a slight cooling after 1940, when winters in particular became colder, but since 1970 average temperatures globally have been rising again.

With more precise climate records from different parts of the world, short-term fluctuations within the longer term trends have become apparent. It is important to understand these patterns and their causes if future climate change is to be predicted. As will be discussed in the next chapter, many factors influence current global climate, including the activities of our own species. But there are also relatively short-term shifts in climate that exhibit a pattern of their own. For example, the energy output from the Sun itself varies, as reflected in the number of dark sunspots on its surface. Very few sunspots were present in the 17th century, known by astrophysicists as the Maunder Minimum; calculations suggest a decrease in solar energy output at that time of about 0.4%. This was a time when the Little Ice Age was at its height. There is a general 11-year cycle in sunspot activity that could affect climatic conditions on Earth, but overall change in solar output over the last 150 years can only account for a small proportion of the observed rise in global temperature [69].

Atmosphere and Oceans: Short-Term Climate Change

Many aspects of global climate are influenced by movements of air and water masses, powered by winds and ocean currents. The oceanic conveyor described in Chapter 4 is an example of how ocean currents redistribute the energy received by the Earth from the Sun and greatly modify the climate of different regions. Changes in salinity, as we have seen, can strongly influence this global movement of water, resulting in rapid climate change.

Some changes in the movement of the oceans are periodic and cause cycles of climate change, some of which are local and others global. An example that has received much attention from climatologists of late is the **El Niño Southern Oscillation** (ENSO). The west coast of South America periodically experiences particularly warm dry conditions for several months, often beginning around Christmas time, and these have been called El Niño, meaning literally the male child, inferring the Christ Child. The event occurs every few years—usually every three to seven years—but varies in frequency and intensity. The general pattern of air movement in the region is dominated by the Trade Winds, which blow offshore from the west coast of South America and westward over the

Pacific Ocean toward Southeast Asia (see Fig. 3.9), and under normal conditions these easterly winds are strong. The winds drive the surface waters of the Pacific Ocean westward, resulting in the upwelling of deep, cold water along the coast, so that the **thermocline** (the boundary between warm, less dense water and the deeper, denser water) comes close to the sea surface. On the other side of the Pacific in Indonesia, the constant influx of warm tropical water creates a deep thermocline and even has the effect of elevating the sea level in this region by about 60 cm. The strong easterly winds bring warm, moist air that creates heavy rainfall over the islands of Southeast Asia, as shown in Fig. 12.26.

During El Niño the east winds are weaker, so that the upwelling along the west coast of South America is less pronounced, the thermocline remains deep, and the surface waters may rise in temperature by up to 7°C. The upwelling of deep water brings nutrient-rich waters to the surface and creates a highly productive ecosystem, resulting in considerable economic gains for the local people. Thus the reduced upwelling during an El Niño event can lead to decreased fish stocks, declines in the

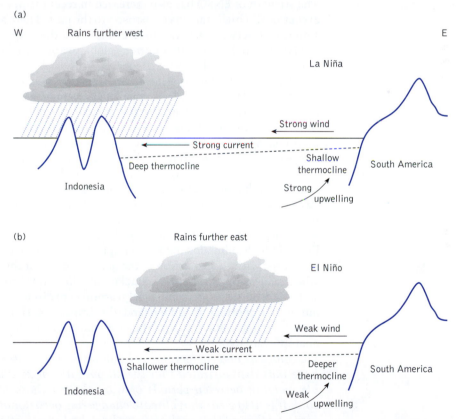

Fig. 12.26 The pattern of oceanic currents and rainfall in the southern Pacific Ocean showing two contrasting situations. In (a), sometimes called La Niña, there is a strong airflow from east to west and hence a strong movement of surface water from continental South America toward Indonesia. This creates an upwelling of cold, nutrient-rich water along the South American coast and leads to high rainfall over Indonesia. In (b), called El Niño, the easterly airflow and surface ocean currents are weak, resulting in a poor upwelling of cold water along the South American coast and a shortage of rainfall in Indonesia.

populations of piscivorous mammals, such as the fur seals, and economic hardship for local people. The lighter east winds result in the rain-bearing clouds failing to reach Indonesia, which then experiences drought. Drought in Indonesia can be very severe during periods of a pronounced ENSO cycle, as in 1982–1983, when forest fires devastated large areas of Indonesia, and again in 1997–1998. The effects of El Niño are widespread throughout the world. The 1982–1983 event was associated with droughts in Australia, Central America, and East Africa, together with floods in Florida and the Caribbean. The alternative state, when east winds are stronger and upwelling is more pronounced, is known by the feminine term "La Niña." Exactly what causes ENSO remains unclear, but its global significance has stimulated a great deal of research into the cycle.

Corals in the oceans grow by depositing layers of calcium carbonate in annual bands, and these can be analyzed isotopically to provide a record of past ocean temperature, which is affected by ENSO. Studies of the history of ENSO [70] reveal that its strength has varied considerably over the past thousand years, being particularly prominent in the mid-17th century. The strength of ENSO has also increased in recent times but is not in an exceptionally high state in comparison to the past. There appears to be no simple link between general global climate and the frequency or strength of ENSO, which means that current global warming is unlikely to be a strong influence on the system.

A similar cycle takes place in the Indian Ocean to the west of the Indonesian islands, called the **Indian Ocean Dipole**. Work on corals from this region [71] indicates that the two systems are not linked, as was once thought, but that the Indian Ocean circulation is dominated by the Asian monsoon system, independently of ENSO.

Biogeographers need to understand the underlying patterns of climate change, including short-term change, in order to predict the future changes in the distribution patterns of organisms on the planet. We are clearly in one of the warm interludes, or interglacials, that have regularly interrupted the generally cold conditions of the Pleistocene. In comparison with past interglacials, we are well advanced into the warm phase and likely to be heading for the next cold event. All climatic evidence supports this, with the exception of the actual records of the past 200 years, when rising temperatures have bucked the downward trend. The causes and consequences of current global warming are discussed in Chapter 14, but it is important to understand the impact of short-term climatic cycles, such as ENSO, so that the other contributory factors can be adequately evaluated.

Geological and meteorological studies of the past 2 million years have thus demonstrated that the climate has been extremely unstable, varying between periods of warmth and those of considerable cold. It is also clear that climate change has occasionally taken place very rapidly and that such changes have had strong effects on the biogeography of the planet. It is against the background of these fluctuating conditions that our own species enters the story, bringing new pressures to bear on the global environment. This is the theme of the next chapter.

1 The overall downward trend in global temperature observed in the Cenozoic finally resulted in the Quaternary Ice Age of the last 2 million years.

2 Within the Quaternary, glacials have alternated with interglacials in a cyclic pattern, but generally the cold phases have dominated.

3 Biomes have altered their distribution patterns in response to changes, and communities of plants and animals have broken up and reconstituted in new assemblages. Some species have become extinct in the process.

4 Many areas that are now regarded as biodiversity hotspots were greatly altered during cold episodes, including the rainforests of equatorial regions, which are likely to have been reduced in area, changed in their composition, and fragmented.

5 Many factors have contributed to the development of the glacial/interglacial cycle, primarily external, astronomic factors, but also including internal changes, such as ocean/atmospheric circulation patterns, feedbacks with living ecosystems, and perhaps volcanism.

6 Climate change is not necessarily a smooth, steady process; it can alter abruptly and vary significantly over a matter of decades, as in the case of the cold Younger Dryas stadial at the opening of the current warm episode. Critical in such change is the failure of the oceanic conveyor to maintain its global heat transfer.

7 Pollen analysis of lake and peat sediments has permitted detailed reconstruction of the rates of movement of major tree genera and the directions they took during the warming of the climate in the current interglacial.

8 Climatic changes over the past 10,000 years (the Holocene) are well recorded in lake sediments, ice sheets, marine deposits, and in human historical records. There appears to have been a climatic optimum in the first half of the Holocene, after which conditions have become cooler. Global sea-level changes during these times have altered and created barriers to dispersal for species in the course of their spread.

9 Atmosphere/ocean interactions are responsible for some short-term climatic alterations and cycles, such as the El Niño Southern Oscillation (ENSO), which is centered on the southern Pacific Ocean but which has climatic impacts throughout the world.

Summary

Further Reading

Beerling DJ, Chaloner WG, Woodward FI, eds. Vegetation–climate–atmosphere interactions: past, present and future. *Phil Trans R Soc* London B 1998; 353: 1–171.

Delcourt HR, Delcourt PA. *Quaternary Ecology: A Palaeoecological Perspective.* London: Chapman & Hall, 1991.

Lowe JJ, Walker MC. *Reconstructing Quaternary Environments.* London: Longman, 1996.

Moore PD, Chaloner B, Stott P. *Global Environmental Change.* Oxford: Blackwell Science, 1996.

National Research Council *Abrupt Climate Change: Inevitable Surprises.* Washington DC: National Academy Press, 2002.

Pielou EC. *After the Ice Age: The Return of Life to Glaciated North America.* Chicago: University of Chicago Press, 1991.

Tallis JH. *Plant Community History.* London: Chapman & Hall, 1991.

Willis, KJ, Bennett, KD, Walker D, eds. The evolutionary legacy of the Ice Ages. *Phil Trans R Soc London* B 2004; 359: 155–303.

References

1 Edgar KM, Wilson PA, Sexton PF, Suganuma Y. No extreme bipolar glaciation during the main Eocene calcite compensation shift. *Nature* 2007; 448: 908–916.

2 Zanazzi A, Kohn MJ, MacFadden BJ, Terry DO. Large temperature drop across the Eocene–Oligocene transition in central North America. *Nature* 2007; 445: 639–642.

3 Eldrett JS, Harding IC, Wilson PA, Butler E, Roberts AP. Continental ice in Greenland during the Eocene and Oligocene. *Nature* 2007; 446: 176–179.

4 West RG. *Pleistocene Geology and Biology*, 2nd ed. London: Longman, 1977.

5 Jansen E, Sjoholm J. Reconstruction of glaciation over the past 6 Myr from ice-borne deposits in the Norwegian Sea. *Nature* 1991; 349: 600–603.

6 Moore PD, Chaloner B, Stott P. *Global Environmental Change*. Oxford: Blackwell Science, 1996.

7 Winograd IJ, Szabo BJ, Coplen TB, Riggs AC. A 250,000-year climatic record from Great Basin vein calcite: implications for Milankovitch theory. *Science* 1988; 242: 1275–1280.

8 Emiliani C. Quaternary paleotemperatures and the duration of high temperature intervals. *Science* 1972; 178: 398–401.

9 Zachos JC, Dickens GR, Zeebe E. An early Cenozoic perspective on greenhouse warming and carbon-cycle dynamics. *Nature* 2008; 451: 279–285.

10 Moore PD, Webb JA, Collinson ME. *Pollen Analysis*, 2nd ed. Oxford: Blackwell Scientific Publications, 1991.

11 Stevenson AC, Moore PD. Pollen analysis of an interglacial deposit at West Angle, Dyfed, Wales. *New Phytologist* 1982; 90: 327–337.

12 Lowe JJ, Walker MJC. *Reconstructing Quaternary Environments*, 2nd ed. London: Longman,1996.

13 Kerr RA. Second clock supports orbital pacing of the ice ages. *Science* 1997; 276: 680–681.

14 Lopez de Heredia U, Carrion JS, Jimenez P, Collada C, Gil L. Molecular and palaeoecological evidence for multiple glacial refugia for evergreen oaks on the Iberian Peninsula. *J Biogeogr* 2007; 34: 1505–1517.

15 Leroy SAG, Arpe K. Glacial refugia for summer-green trees in Europe and south-west Asia as proposed by ECHAM3 time-slice atmospheric model simulations. *J Biogeogr* 2007; 34: 2115–2128.

16 FAUNMAP Working Group. Spatial response of mammals to Late Quaternary environmental fluctuations. *Science* 1996; 272: 1601–1606.

17 Svenning J-C, Skov F. Ice age legacies in the geographical distribution of tree species richness in Europe. *Global Ecology and Biogeography* 2007; 16: 234–245.

18 Dansgaard W, Johnsen SJ, Clausen HB, Dahl-Jensen D, Gundestrup NS, Hammer CU, Hvidberg CS, Steffensen JP, Sveinbjornsdottir AE, Jouzel J, Bond G. Evidence for general instability of past climates from a 250-kyr ice-core record. *Nature* 1993; 364: 218–220.

19 Schmidt MW, Vautravers MJ, Spero HJ. Rapid subtropical North Atlantic salinity oscillations across Dansgaard-Oeschger cycles. *Nature* 2006; 561–564.

20 Whitlock C, Bartlein PJ. Vegetation and climate change in northwest America during the past 125 kyr. *Nature* 1997; 388: 57–61.

21 Kershaw AP. A long continuous pollen sequence from north-eastern Australia. *Nature* 1974; 251: 222–223.

22 Prideaux GJ, Long JA, Ayliffe LK, Hellstrom JC, Pillans B, Boles WE, Hutchinson MN, Roberts RG, Cupper ML, Arnold LJ, Devine PD, Warburton NM. An arid-adapted middle Pleistocene vertebrate fauna from south-central Australia. *Nature* 2007; 445: 422–425.

23 Jolly D, Taylor D, Marchant R, Hamilton A, Bonnefille R, Buchet G, Riollet G. Vegetation dynamics in central Africa since 18,000 yr BP: Pollen records from the interlacustrine highlands of Burundi, Rwanda and western Uganda. *J Biogeogr* 1997; 24: 495–512.

24 Tallis JH. *Plant Community History.* London: Chapman & Hall, 1991.

25 Colinvaux PA, De Oliveira PE, Bush MB. Amazonian and Neotropical plant communities on glacial time-scales. The failure of the aridity hypothesis. *Quaternary Sci Revs* 2000; 19: 141–169.

26 Bonaccorso E, Koch I, Peterson AT. Pleistocene fragmentation of Amazon species ranges. *Diversity Distrib* 2006; 12: 157–164.

27 Wells G. Observing Earth's environment from space. In: Friday L, Laskey R, eds. *The Fragile Environment.* Cambridge: Cambridge University Press, 1989.

28 Brown JH. The desert pupfish. *Sci Am* 1971; 225 (11): 104–110.

29 Ganopolski A, Rahmstorf S, Petoukhov V, Claussen M. Stimulation of modern and glacial climates with a coupled global model of intermediate complexity. *Nature* 1998; 391: 351–356.

30 Hays JD, Imbrie J, Shackleton NJ. Variations in the Earth's orbit: pacemaker of the ice ages. *Science* 1976; 194: 1121–1132.

31 Petit JR et al. Climate and atmospheric history of the past 420,000 years from the Vostok ice core, Antarctica. *Nature* 1999; 399, 429–436.

32 Sigman DM, Boyle EA. Glacial/interglacial variations in atmospheric carbon dioxide. *Nature* 2000; 407: 859–869.

33 Galbraith ED, Jaccard SL, Pedersen TF, Sigman DM, Haug GH, Cook M, Southon JR, Francois R. Carbon dioxide release from the North Pacific abyss during the last deglaciation. *Nature* 2007; 449: 890–893.

34 Woodward FI. Stomatal numbers are sensitive to increases in CO2 from pre-industrial levels. *Nature* 1987; 327: 617–618.

35 McElwain JC. Do fossil plants signal palaeoatmospheric CO2 concentration in the geological past? *Phil Trans R Soc Lond B* 1998; 353: 83–96.

36 Broecker WS, Denton GH. What drives glacial cycles? *Sci Am* 1990; 262 (1): 43–51.

37 Schmidt MW, Vautravers MJ, Spero HJ. Rapid subtropical North Atlantic salinity oscillations across Dansgaard-Oeschger cycles. *Nature* 2006; 443: 561–564.

38 Bray JR. Volcanic triggering of glaciation. *Nature* 1976; 260: 414–415.

39 Houghton J. *Global Warming: The Complete Briefing*, 3rd ed. Cambridge: Cambridge University Press, 2004.

40 Benson L, Burdett J, Lund S, Kashgarian M, Mensing S. Nearly synchronous climate change in the Northern Hemisphere during the last glacial termination. *Nature* 1997; 388: 263–265.

41 Moreno PI, Jacobson GL, Lowell TV, Denton GH. Interhemispheric climate links revealed by late-glacial cooling in southern Chile. *Nature* 2001; 409: 804–808.

42 Broecker WS, Kennett JP, Flower BP, Teller JT, Trumbo S, Bonani G, Wolfli W. Routing of meltwater from the Laurentide

Ice Sheet during the Younger Dryas cold episode. *Nature* 1989; 341: 318–321.

43 Fairbanks RG. A 17,000-year glacio-eustatic sea level record: influence of glacial melting rates on the Younger Dryas event and deep-ocean circulation. *Nature* 1989; 342: 637–642.

44 Smith JE, Risk MJ, Schwarcz HP, McConnaughey TA. Rapid climate change in the North Atlantic during the Younger Dryas recorded by deep-sea corals. *Nature* 1997; 386: 818–820.

45 McManus JF, Francois R, Gherardi J-M, Keigwin LD, Brown-Leger S. Collapse and rapid resumption of Atlantic meridional circulation linked to deglacial climate changes. *Nature* 2004; 428: 834–837.

46 Clark PU, Pisias NG, Stocker TF, Weaver AJ. The role of the thermohaline circulation in abrupt climate change. *Nature* 2002; 415: 863–869.

47 Camill P, Umbanhower CE, Teed R, Geiss CE, Aldinger J, Dvorak L, Kenning J, Limmer J, Walkup K. Late-glacial and Holocene climatic effects on fire and vegetation dynamics at the prairie-forest ecotone in south-central Minnesota. *Journal of Ecology* 2003; 91, 822–836.

48 Bennett KD. Postglacial population expansion of forest trees in Norfolk, U.K. *Nature* 1983; 303: 164–167.

49 Jacobson GL, Webb T, Grimm EC. Patterns and rates of change during the deglaciation of eastern North America. In: Ruddiman WF, Wright HE, eds. *The Geology of North America*, Vol. K-3, pp. 277–288. New York: Geological Society of America, 1987.

50 Huntley B, Birks HJB. *An Atlas of Past and Present Pollen Maps for Europe: 0–13,000 Years Ago.* Cambridge: Cambridge University Press, 1983.

51 Schefuss E, Schouten S, Schneider RR. Climatic controls on central African hydrology during the past 20,000 years. *Nature* 2005; 437: 1003–1006.

52 Waller MP, Street-Perrott FA, Wang H. Holocene vegetation history of the Sahel: pollen, sedimentological and geochemical data from Jikariya Lake, north-eastern Nigeria. *J Biogeogr* 2007; 34: 1575–1590.

53 Rosignol-Strick M, Nesteroff W, Olive P, Vergnaud-Grazzini C. After the deluge: Mediterranean stagnation and sapropel formation. *Nature* 1982; 295: 105–110.

54 Sancetta C. The mystery of the sapropels. *Nature* 1999; 398: 27–28.

55 Lezine AM. Late Quaternary vegetation and climate in the Sahel. *Quaternary Res* 1989; 32: 317–334.

56 Singh G, Joshi RD, Chopra SK, Singh AB. Late Quaternary history of vegetation and climate of the Rajasthan Desert, India. *Philosophical Transactions of the Royal Society of London* 1974; B267, 467–501.

57 Gibbard P. Europe cut adrift. *Nature* 2007; 448: 259–260.

58 Moore PD. The origin of blanket mire, revisited. In: Chambers FM, ed. *Climate Change and Human Impact on the Landscape*, pp. 217–224. London: Chapman & Hall, 1993.

59 Bond WJ, van Wilgen BW. *Fire and Plants*. London: Chapman & Hall, 1996.

60 Delcourt PA, Delcourt HR. *Prehistoric Native Americans and Ecological Change.* Cambridge: Cambridge University Press, 2004.

61 Moore PD. Down to the woods yesterday. *Nature* 2005; 433, 588–589.

62 Hailer F et al. Phylogeography of the white-tailed eagle, a generalist with large dispersal capacity. *J Biogeogr* 2007; 34: 1193–1206.

63 De Toledo MB, Bush MB. A mid-Holocene environmental change in Amazonian savannas. *J Biogeogr* 2007; 34: 1313–1326.

64 Barber K, Maddy D, Rose N, Stevenson AC, Stoneman RE, Thompson R. Replicated proxy-climate signals over the last 2000 yr from two distant UK peat bogs: new evidence for regional palaeoclimate teleconnections. *Quaternary Sci Revs* 2000; 19, 481–487.

65 Blackford, JJ. Palaeoclimatic records from peat bogs. *Trends in Ecology and Evolution* 2000; 15, 193–198.

66 Payette S, Filion L, Delwaide A, Begin C. Reconstruction of tree-line vegetation response to long-term climate change. *Nature* 1989; 341, 429–432.

67 Laird KR, Fritz SC, Maasch KA, Cumming BF. Greater drought intensity and frequency before AD 1200 in the Northern Great Plains, USA. *Nature* 1996; 384: 552–554.

68 Moberg A, Sonechkin DM, Holmgren K, Datsenko NM, Karlen W. Highly variable Northern Hemisphere temperatures reconstructed from low- and high-resolution proxy data. *Nature* 2005; 433: 613–617.

69 Houghton J. *Global Warming*, 3rd ed. Cambridge: Cambridge University Press, 2004.

70 Cobb KM, Charles CD, Cheng H, Edwards RL. El Niño/Southern Oscillation and tropical Pacific climate during the last millennium. *Nature* 2003; 424: 271–276.

71 Abram NK, Gagan MK, Liu Z, Hantoro WS, McCulloch MT, Suwargadi BW. Seasonal characteristics of the Indian Ocean Dipole during the Holocene epoch. *Nature* 2007; 445: 299–302.

The Advent of Humanity

The Pleistocene was a time of climatic instability, which had considerable impact on the distribution patterns of organisms over the face of the Earth. It was a time of extinctions, but it was also a time of diversification for some types of organisms. There has been much debate concerning whether speciation became faster or slower during the Quaternary Ice Age, and the general conclusion is that extinction rates within the Pleistocene exceeded speciation rates [1]. But for mammals it was a time of extensive evolution, and most living species of mammal evolved during Quaternary times, influenced by unstable Quaternary environments [2]. Among these evolving groups of mammals were the primates, and these were about to generate a species that would have an even greater impact on the biogeography of the Earth than the Ice Ages had achieved. That species was our own, Homo sapiens. *Our arrival on the planet heralded the start of a global transformation.*

The Emergence of Humans

The fossil history of humans is still very incomplete, but each year brings new fossil material to light, helping to fill in the gaps and providing a more detailed picture of how anatomically modern humans emerged. The primates of the New World and the Old World became separated from each other some 40 million years ago, and it is the Old World branch that is ancestral to humans. The separation of the human ancestral branch (the **hominins**) from the great apes (jointly known as **hominids**) is thought to have taken place around 7 million years ago (see Fig. 13.1) [3]. A major source of evidence on which this estimate is based is the genetic similarity between humans and chimpanzees; almost 99% of human genetic makeup is shared with the chimpanzee, so their evolutionary divergence must have been relatively recent in geological terms. Palaeontological research into this relationship has been hampered by a lack of fossil material of chimpanzees: the earliest chimpanzee fossil, found in East Africa, dates back only 0.5 million years [4]. In 2002, Michel Brunet and his fellow researchers [5] discovered six specimens of fossil bones (a cranium and some lower jaws) that had hominin similarities. They were assigned to a new genus on the human line of evolution and called *Sahelanthropus tchadensis*. Until this time most finds associated with early human evolution had been

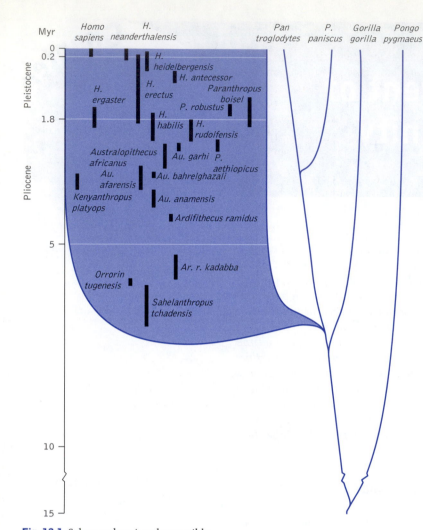

Fig. 13.1 Scheme showing the possible relationships between the various hominid lines of evolution. The divergence of the hominins (blue shading) from the great apes took place sometime in the region of 7 million years ago. From Carroll [3].

discovered in East Africa, but this set of fossils came from the African state of Chad, in the Sahel region, south of the Sahara Desert. The fossil fauna associated with these hominin finds suggests a date of late Miocene, around 6–7 million years ago. So, if the genetic estimate of the separation from the ape line of evolution is correct, *Sahelanthropus* could be one of the very first organisms on the human line of development, as opposed to the great ape line.

Trying to establish the biogeography of the early hominids is extremely difficult because of the lack of fossil data. One approach has been to study the distribution and spread patterns of mammal groups, such as the hyaenids (hyenas) and the proboscideans (mammoths and elephants) that are often associated with the hominids [6]. These share a common set of patterns involving speciation in Africa in the early Miocene, expansion into Europe, Asia, and North America during the middle Miocene, followed by a movement back into Africa. It can be argued that hominids (encompassing the ancestors of apes and humans) followed similar changing patterns of distribution in the Miocene.

There is still controversy about whether a single line of development emerged from the common ancestor with the chimpanzee, or whether a rather untidy series of developments interbred and concluded in the development of the human line. Whichever is the case, the hominin line, diverging from the apes, involved more bipedalism, a larger brain, and greater manual dexterity. While the chimpanzee line remained in the lower canopy of the trees as a forest animal, the ancestors of humans took to the savannah woodlands and grasslands. The discovery of *Sahelanthropus* fossils in the Sahel region suggests that the hominin evolutionary process in the late Miocene was taking place over a wide area and was not localized in East Africa. The skull of *Sahelanthropus* is remarkable in that it has the features of a chimpanzee when viewed from behind, but the front view looks very like the genus *Australopithecus*, a genus of hominins that was to become very important some 3 million years later. Its discovery has given strong support to those who prefer to think of human evolution as stemming from a single line, for it has precisely the combination of characters one might expect of such a stem.

The fossil remains of the **australopithecines** (the name given to members of the *Australopithecus* genus) that succeeded *Sahelanthropus* are widely recorded from eastern and southern Africa; the earliest finds were in Tanzania and Ethiopia and dated from around 4 million years ago. Among the fossils of this age is the partial skeleton that has come to be known as "Lucy," and that has supplied a great deal of anatomical information about the early hominins. These fossils have been assigned the specific name *Australopithecus afarensis*, and it is believed that they walked upright on their hind legs (bipedalism). This conclusion is supported by the extraordinary discovery of fossil human footprints in volcanic ash from Tanzania [7]. The habitat in which these organisms lived was open woodland and savanna, far from the dense tropical forests, but very little is known of their precise way of life and the ecological niche they occupied in the ecosystem. Walking on two legs, they would have been able to hold tools with their free hands, much in the way that a chimpanzee is capable of doing, but there is no evidence that they were able to make extensive modifications to natural tools. In fact, analysis of the bone structure of Lucy and other fossils of *A. afarensis* indicate that they did not have the appropriate hand structure for toolmaking [8]. They probably contented themselves with using the sticks and stones that they found. Whatever its ecological role in the ecosystem, we must regard this primitive hominin as one more species in the complex food web of the ecosystem that it occupied. There is no reason to believe that it was more influential than any other species. Molecular studies are helping to sort out the diet and have the ecological role of the australopithecines (see Concept Box 11.1).

Australopithecine Diet

Concept Box 13.1

It is difficult to be sure about the diet of the australopithecines. It is possible that they were largely vegetarian but also took occasional animal prey, just like the chimpanzee. One attempt to reconstruct the australopithecine diet has used the enamel from their fossil teeth. Tooth enamel is extremely hard-wearing and can survive intact over millions of years. The chemistry of tooth enamel may reflect diet, especially in the isotopes of carbon that are contained within it. The technique is based on the fact that the two photosynthetic systems operating in plants, C_3 and C_4 (plus CAM), accumulate carbon from the atmosphere as a result of the activity of two different enzymes (see Chapter 2). The two enzymes have different capacities for discriminating between the two isotopes of carbon, ^{13}C and ^{12}C, with ^{13}C being enriched by the C_4/CAM system. So the organic products of photosynthesis differ depending on which type of plant has constructed them. The different proportions of carbon isotopes are also transmitted to the animals that consume them, so analysis of plant or animal organic matter can help determine their photosynthetic origins. When tooth enamel of australopithecine fossils was analyzed [9], it was found to be rich in ^{13}C, suggesting that the australopithecines had a diet that was rich either in C_4 plant species (such as the tropical grasses, including their roots or seeds) or in the meat of herbivores that consumed C_4 plant species. These herbivores may have included large mammals, such as the ancestors of giraffes, gazelles, and bovids, together with insects that fed on grasses. It is important to remember, however, that the australopithecines were small in stature, only about the size of chimpanzees (100–150 cm (3–5 ft) tall and weighing 30–50 kg (70–110 lbs)), so their ability to prey on very large mammals must have been limited. Further evidence, including the lack of wear on the tooth enamel and the lack of any tools that would be needed for grinding fibrous grasses, suggests that the consumption of grazing animals is the most likely explanation for the carbon isotope ratios in australopithecines. So the evidence implies that these hominins lived in open, savanna environments and also suggests that animals formed an important part of their diet.

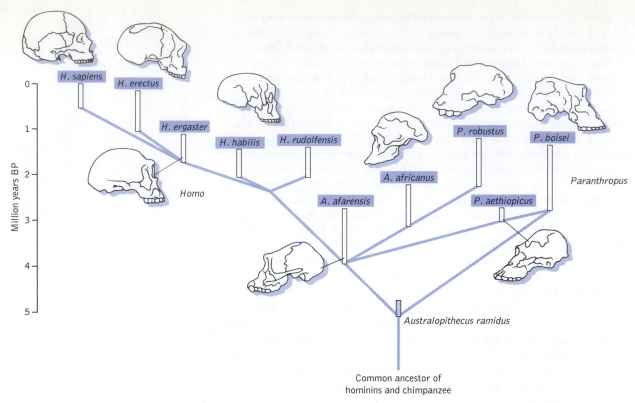

Fig. 13.2 A scheme showing the possible interrelationships between hominins over the past 5 million years. The diversification of hominins over the past 2 million years is apparent. Based on the scheme of Bernard Wood [13].

By 2 million years ago, evolutionary developments had taken place in the hominin line, and at least three species of hominins were present in East Africa (Fig. 13.2): two of these were species of *Australopithecus*, but the third belonged to our own genus and has been called *Homo habilis*. This species had a larger brain than its relatives, and there is evidence that it was able to modify stone tools [10], thus showing a marked cultural change that was to continue with remarkable consequences along the *Homo* line of development. Surprisingly, following the development of a toolmaking culture, no real change in technology occurred for the next million years. Knowledge of the diet and the ecological role of *Homo habilis* is still fragmentary, but it is believed that the meat content of the diet increased.

Although bipedalism has advantages, such as freeing the hands for other tasks and elevating the head above the ground vegetation, it is not associated with great speed. Many of the larger mammalian quadrupeds present at the time of these early species of *Homo* would have been considerably faster, and this group included both prey animals and predators. One possible attribute associated with bipedal striding locomotion, however, is the capacity to maintain long-distance running [11]. Although no sprinter, the early human line of evolution provided the right body form for endurance running.

Sometime around 1.9 million years ago a new hominin appeared on the scene, *Homo ergaster*, closely followed by *H. erectus*, and there is every reason to believe that these new species were the direct descendants of *Homo habilis*. Footprints dated to 1.5 million years ago and believed to belong to *Homo ergaster/erectus* have been found at Ileret, Kenya, and analysis of these impressions suggests that the feet of this species were essentially the same as modern humans [12]. Two aspects of *Homo erectus* are of biogeographical significance: its cultural developments and its distributional changes. On the cultural side we find much more sophisticated stone tools, including structures that could be termed hand axes, and even more significantly there is evidence that some populations of *Homo erectus* used fire. This latter development provided the species with the capacity to modify its environment more profoundly than any other cultural development to date. Although fire would have been used as a means of food preparation, its potential as an aid in hunting must surely have been appreciated by this intelligent species, and the consequences must have been felt in terms of the new pressures placed on the flora and fauna of its surroundings. The distribution of the species is also of interest. Although *Homo habilis* is believed to have spread from Asia into the Eurasian region of modern Georgia [14], *Homo erectus* is the first species of our genus to be found far beyond Africa, and by 1.8 million years ago it had spread into eastern Asia [15]. Some excavations in Java indicate that *Homo erectus* may have survived in Southeast Asia as late as the last Ice Age (50,000 years ago), in which case it would have overlapped with our own species in the area [16]. By 700,000 years ago the species had reached what are now the British Isles, leaving samples of worked flints in the organic deposits of the Cromerian interglacial [17] (see Fig. 12.6).

The fossil record in Africa, Europe and western Asia shows how, over a period of about 750,000 years, *H. ergaster* and *H. erectus* gradually evolved into the common ancestor of both our own species and a similar species that once lived alongside us, Neanderthal Man, *Homo neanderthalensis*. The British palaeontologist Chris Stringer [18] suggests that this common ancestor is the hominin that has been named *Homo heidelbergensis*, the remains of which have been found in Europe. There were no sharp breaks between the various successive species in the history of human evolution from *Australopithecus* to *H. habilis* to *H. erectus* to *H. heidelbergensis*, and then to *H. neanderthalensis*

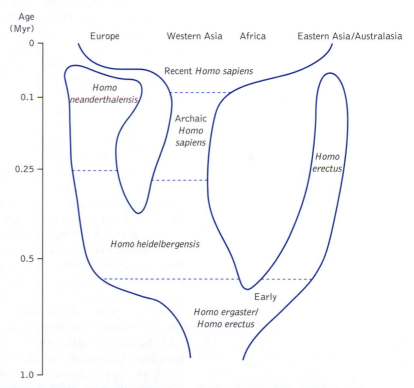

Fig. 13.3 An interpretation of the relationships and groupings of Pleistocene hominins in time and space. The timescale is not linear. Based on the proposals of Stringer [18].

and *H. sapiens*. These "species" should be regarded as stages that palaeontologists find convenient to recognize in what was really a gradual process of evolutionary change. A scheme representing their proposed relationship is shown in Fig. 13.3.

A unique insight into the way of life of these ancestors of our species has been provided by the discovery of hunting spears, buried in compressed peat deposits in north Germany [19]. They are dated to 400,000 years ago, long before any known fossil remains of *Homo sapiens*. It suggests that the ancestors of modern humankind who occupied the northern regions of Europe were big-game hunters and gives weight to the argument that the hunting and butchering of animals using tools may extend far back into human ancestry.

Our own species, *Homo sapiens* (the "Wise Man"), is thus a very recent arrival on the planet. Many ancient fossil fragments of bones that have been assigned to our species have been found, dating between 260,000 and 130,000 years ago [20]. Unfortunately, most of these fragments have not been well dated, so it is impossible to put a clear date on our origins. Perhaps the best dates come from Ethiopian fossils that can definitely be assigned to *Homo sapiens*, and they fall in the region of 160,000 to 154,000 years ago [21]. These remains provide support for the growing evidence, based on the wide scatter of fossil remains, that our species first arose in Africa. Outside that continent, the earliest reliable dates come from Israel, with a date of 115,000 years. So it is likely that the human population of Africa began to expand and spread into other parts of the world at around that time.

One problem that is still not fully resolved is how closely we are related to a second hominin species, *Homo neanderthalensis*, which also existed at this time. Undoubtedly, we shared a common ancestor, probably *Homo heidelbergensis*, but the question is, how recently? Fossil remains from Spain [22] have been described that could be interpreted as a common ancestor, but this does not fit well with the evidence for African origins of the main hominin lines. Genetic evidence from the DNA preserved in Neanderthal fossils [23] indicates that this species diverged from our own somewhere between 690,000 and 550,000 years ago. But the picture may have been confused if there has been genetic interchange by interbreeding between the two in the last half million years. Certainly, by 400,000 years ago the Neanderthal line of evolution was apparent in the fossil record, and archaeological work has shown that both *Homo neanderthalensis* and *H. sapiens* were present and widespread in Europe and Asia Minor 40,000 to 35,000 years ago [24]. Analysis of DNA from bones dating 38,000 years ago from Uzbekistan in Central Asia have shown Neanderthal affinities, so the species may well have spread extensively through Asia [25]. Neanderthals disappeared from the record about 28,000 years ago (though some would claim survival to 24,000 years ago in Gibraltar [26]). How the two species interacted is a matter of very active debate. It seems likely that the Neanderthals spread through Europe in advance of *Homo sapiens*, so our species invaded territory already occupied by them. Direct competition with humans resulting in the extinction of the Neanderthals seems a possible explanation for their disappearance; genetic evidence from DNA analysis suggests that there may have been some interbreeding, but probably very little. There is no morphological evidence that any such interbreeding left a mark on human anatomy. It is also possible that climate played a part in their demise, but there are problems with precise dating using radiocarbon at this point in prehistory, so correlations are fraught with difficulty. If the Neanderthals survived until about 24,000 years ago, then their disappearance coincides with a major expansion in global ice volume leading to the maximum extent of the last glaciation, which may have placed an additional strain on Neanderthal survival [27]. One fact is clear, however: only *Homo sapiens* remained in Europe at the beginning of the Holocene.

Archaeologists have long believed that *Homo sapiens* was the only member of the genus present on Earth during the Holocene, but a discovery on the island of Flores, Indonesia, in September 2003 [28] threw this idea into doubt. During the excavation of cave sediments, the skeleton of a very small adult hominin, only about a meter tall (3 ft 4 ins) was unearthed that dated back only 18,000 years and was believed to be a new species, *Homo floresiensis*. Speculation has run rife as to the ancestry of this diminutive creature, some claiming that it is the direct descendant of *Homo erectus* and is an endemic island dwarf, while others suggest that it could be a pygmy form of our own species, but the cranial structure does not encourage this latter view. Additional fossil bones of other members of the population were found in the cave in 2004, so the original discovery is not just a single aberrant individual [29]. These discoveries have provided opportunities to examine in detail the anatomical features of the population. Work on the foot structure suggests that these small people were indeed bipedal but they had exceptionally long feet, and their foot structure is in some ways more ape-like than human. This raises the possibility that this hominin may even not be a direct descendant of *Homo erectus*, but may be derived from some other line of primate development [30]. The hunt continues for further samples, but the debate about human ancestry in this part of the world remains very active.

Even before the last glaciation had begun, *Homo sapiens* was spreading out of Africa into the Middle East and also penetrating further south into the African continent (see Fig. 13.4). By the middle of the glaciation, our ancestors had reached the Asian interior, north of the Caspian Sea, and had also spread across the Tibetan Plateau into Southeast Asia. Australia became populated by about 50,000 years ago, and humans had also spread into Europe by 50,000 years ago and reached the extremities of Europe and eastern Asia by the time the glaciation was at its peak, around 20,000 years ago. From eastern Asia the peopling of the Americas began.

When the last glaciation was at its height, a very large volume of the world's water was locked up in the ice of the expanded ice caps and glaciers (see Chapter 12), which meant that the sea level of the world's oceans was considerably lower, perhaps by around 100 m. Regions that are now beneath the oceans were then exposed as landmasses, and a substantial area of the Bering Straits between what is now Siberia and Alaska formed a land bridge, linking the two continents. This is the most likely route by which

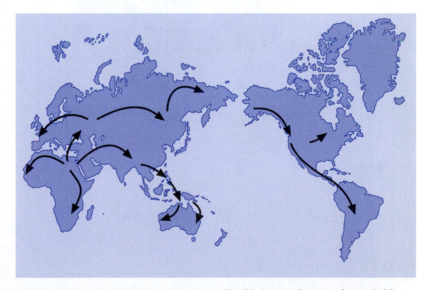

Fig. 13.4 Map showing the probable routes taken during the spread of *Homo sapiens* out of Africa over the past 100,000 years.

North America was colonized. Fossil plant material dating from about 24,000 years ago, the time of the maximum glacial extent, has been found in Yukon and has revealed much about the nature of the environment on this land bridge. Grasses and prairie sage (*Artemisia frigida*) were abundant [31], so the tundra steppe vegetation would have supported herds of large herbivores, including woolly mammoth, horses, and bison. The hunting peoples of eastern Asia probably followed these herds across into the New World. The exact date of arrival is still disputed, but indirect evidence from the geography of human languages suggests that the invasion must have taken place before the major advance of the last glaciation at about 22,000 years ago. The linguistic research of R. A. Rogers [32] has revealed three distinct groups of Native American languages, and these are centered on the three ice-free refugial areas of North America during the height of the last glaciation (Fig. 13.5). It seems likely that human populations were isolated in these three areas during the glacial maximum and subsequently spread to other regions. The extinct language of Beothuk, once spoken in Newfoundland, could belong to another population isolated in that eastern refugium. An alternative explanation is that there have been three separate invasions of North America from Asia, thus accounting for the three language groups [33].

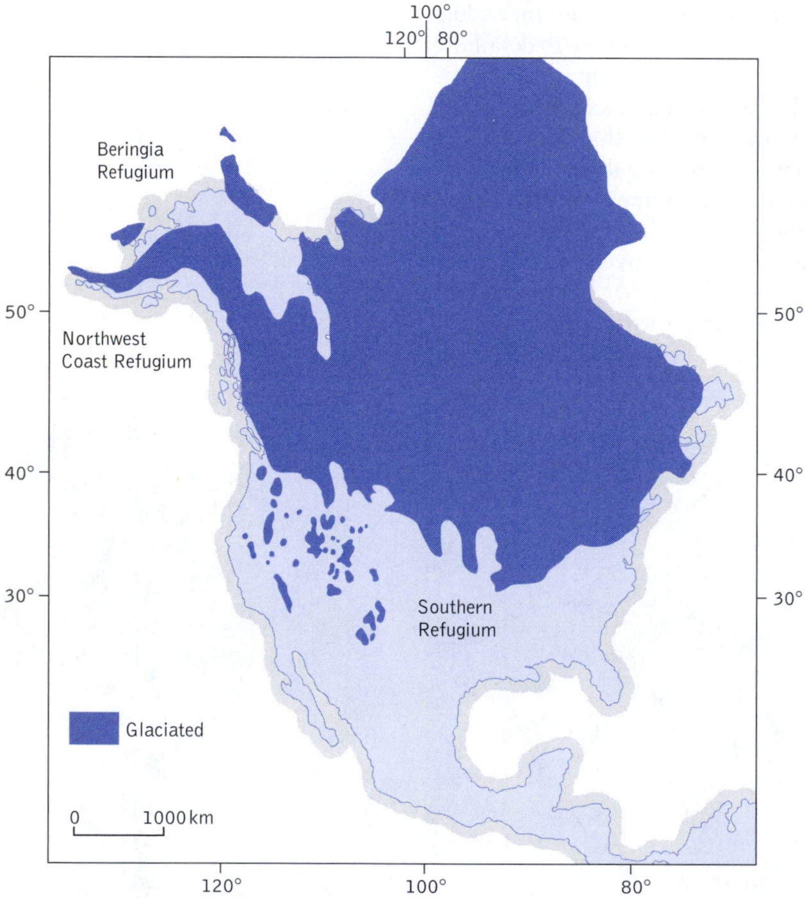

A complication in all theories of the peopling of the Americas is the existence of early human occupation in southern Chile, far from the proposed Beringian point of entry [34]. This settlement at Monte Verde dates from around 12,500 years ago, so it is possible that an early wave of immigrant peoples from eastern Asia made their way as far as the southern part of South America.

Information based on the study of ancient skulls in the Americas is also leading to new ideas being generated concerning the peopling of the continent. Fossil human skulls with an age greater than 10,000 years are very rare in both North and South America, but a collection of 33 skulls has been found in Baja California, Central America, which complicates the story [35]. The structure of these skulls resembles that of southern Asian peoples rather than those from the northeast. This raises a whole new series of questions concerning the mode of migration, the number of invasive episodes, and the degree of isolation between the early human settlements in the Americas. The invasion from Beringia at the end of the

Fig. 13.5 Maximum extent of the last (Wisconsin) glacial in North America, showing the three ice-free areas (refugia), which correspond to the language groups of Native Americans. From Rogers et al. [32].

last glaciation is likely to account for much of the human occupation of the Americas, but some believe that other arrivals survived as separate groups, as in the case of the Baja California settlement that may have been isolated by the aridity of surrounding regions. Genetic analysis of people from a wide range of Native American races, however, has indicated a very close similarity [36]. This suggests that there was just one founding population of immigrants, who have subsequently diverged into communities as widely separated as the Aleuts of Alaska and the Yanomami people of Brazil. Evidence will continue to be accumulated, but the current body of research points to a single origin for the native peoples of America.

Modern Humans and the Megafaunal Extinctions

The spread of the human species during the last glaciation was accompanied by an alteration in the fauna of North America and of Europe, namely, the extinction of many species of large mammals, called the **megafauna**. It was long assumed that these extinctions were the result of the climatic changes, but the American anthropologist Paul Martin [37] suggested that humans may have been the culprits. Martin pointed out that most of the animals that became extinct were large herbivorous mammals or flightless birds, weighing over 50 kg body weight—in other words, precisely the fauna humans might have been expected to hunt. He also observed that similar extinctions had taken place in other, more southern areas than North America, and suggested that the timing of these extinctions varied: in each case the time corresponded with the evolution, or arrival, of a race of humans with relatively advanced hunting techniques. In Africa, for example, where *Homo sapiens* probably evolved, the extinctions of many large herbivores apparently took place before those in the Northern Hemisphere. However, it has proved difficult to date these precisely, and even more difficult to correlate them with any changes in the hominin cultures of that continent. The same is true of the extinctions that took place in South America.

It is in North America that the record of extinctions has been studied in the greatest detail. Martin suggests that 35 genera of large mammal (55 species) became extinct in North America at the end of the last (Wisconsin) glaciation, over twice as many as had taken place during all the earlier glaciations, and this at a time when the climate was already becoming warmer. This combination of features seems to support the idea that some agent other than climate was responsible, and it is reasonable to suspect the hunting activities of humans. But the American anthropologist J. E. Grayson [38] has shown that there was a similar rise in the level of extinctions of North American birds (ranging from blackbirds to eagles) at that same time. Since it is unlikely that early humans were responsible for the extinction of these birds, this observation throws doubt on the whole hypothesis of the dominant role played by humans in Pleistocene extinctions in general.

On the other hand, the fact that so many North American species became extinct at the same time (12,000–11,000 years ago) as the arrival of hunting peoples, whereas in Europe the extinctions were spread over a longer period, provides a strong body of circumstantial evidence to support Paul Martin's claim that humans are to blame [39]. The debate continues, and

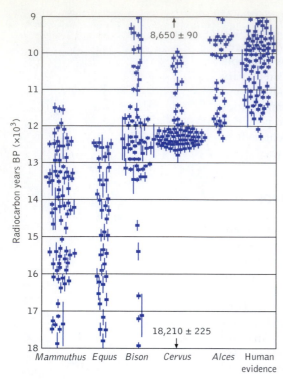

Fig. 13.6 Radiocarbon dates of fossil mammal bones from North America showing the loss of mammoth and horse around the time of the Pleistocene/Holocene transition. The expansion of bison, moose, and wapiti populations is also shown, as are dates for human occupations, mainly based on charcoal from hearths. From Guthrie [40].

the extinction of so many different species of mammals may not have been caused by one single factor, but an increasing number of studies show that the time of extinction can be precisely correlated with the arrival or intensification of settlement by human populations. R. Dale Guthrie of the University of Alaska has obtained radiocarbon dates for the late Pleistocene and early Holocene remains of many large mammal fossils from Alaska and the Yukon Territory [40], (see Fig. 13.6). Although the horse (*Equus ferus*) and the mammoth (*Mammuthus primigenius*) became extinct at about the time that humans were settling the area, other large mammals, such as wapiti or elk (*Cervus canadensis*) and bison (*Bison priscus*, which later evolved into *B. bison*) had begun to expand their populations prior to human invasion. The implication is that environmental changes, involving both climate and vegetation, had already created new conditions under which the balance of competition between the large herbivores had changed. It is always possible, of course, that human predation added to pressures on the declining species, so humans cannot be totally absolved from blame as a result of this work, but the situation is clearly complex.

In Europe, the mammoth survived into the Younger Dryas stadial, but had vanished by the opening of the Holocene. The same was true of the giant Irish "elk" (*Megaloceros giganteus*), which stood two meters (6 ft 6 in) high at its shoulder and bore antlers with a 3.5-m (11 ft 6 in) span. Both species survived much longer in Asia; the giant elk was present in the Ural Mountains until 7,000 radiocarbon years ago (7,700 solar years ago) and the mammoth survived in western Siberia until just 3,700 radiocarbon years ago (4,000 solar years ago) [41]. Again, this suggests complex interactions between climate, vegetation, competition, and human predation.

In Australia, where human populations had maintained their hunting activities in the midlatitude regions through the global cooling of the last glacial episode, megafaunal extinction came earlier than in Europe and America. All 19 species of marsupials that exceeded 100 kg in weight and 85% of all animal species greater than 44 kg became extinct during the Late Pleistocene (see Concept Box 13.2). Some of these, such as the ostrich-sized flightless bird *Genyornis newtoni*, have been studied in detail. By radiocarbon dating 700 of the fossil eggshells of *Genyornis*, it has proved possible to trace its decline from widespread and common 100,000 years ago to a sudden disappearance about 50,000 years ago, corresponding to the arrival of humans [42]. The circumstantial evidence linking human predation to megafaunal extinction is thus strong in this case.

The case for human involvement in the megafaunal extinction process is complicated by the fact that it took place at a time of rapid climatic and environmental change, as explained in Chapter 12. It has been proposed that some form of planetary impact took place at around 12,900 years ago that could have affected the climate and the megafauna [43]. The evidence is based on the discovery of tiny diamonds in sediments marking the commencement of the Younger Dryas cold stadial. These roughly spherical nanodiamonds (less than 300 nm in diameter) have been found throughout

One intriguing observation concerning the megafaunal extinction at the close of the last glaciation could prove helpful in interpreting the event. Fossil evidence suggests that several of the species heading for extinction underwent a period of size reduction prior to their final disappearance. In North America, for example, about 70% of the large mammals became extinct about 13,000 to 11,000 years ago, as determined by radiocarbon dating. Of these, the horses comprise an important group, and these have been studied intensively. Dale Guthrie of the University of Alaska has examined fossil metacarpal bones of two horse species in Alaska at the close of the last glaciation [45] and has found a 14% decline in the size of the bones between 20,000 and 12,000 years ago. He considers this finding to be due to the climatic changes taking place at that time leading to a decline in the availability of forage for the grazers. It is possible, therefore, that the climate change could have finally resulted in the extinction of these species. A similar decline in size has been found in the kangaroos of Australia around this time.

But is the size decline in herbivorous mammals necessarily proof of the climatic cause of megafaunal extinction? One could look at the data in another way. Human populations of the time may well have concentrated in their hunting on the larger individual animals, and this could have resulted in an evolutionary selection in favor of smaller individuals. Modern trophy hunting is known to have an impact on the overall size of a population of prey animals, as in the case of bighorn sheep [46]. So human hunting in the late Pleistocene could have selected against large individuals. A further consideration is the rate of reproduction among the prey species. On the whole, it was the slower-breeding animals that became extinct at this time, again suggesting that they were unable to recover from constant harvesting by humans [47].

North America, from Canada to Arizona, and in Germany. The layer is reminiscent of the iridium layer at the KT (end of Cretaceous) boundary also associated with mass extinction. But the presence of the diamonds does not fully confirm the impact of a shower of comet fragments, and certainly cannot explain all aspects of the megafaunal extinction data [44].

Plant Domestication and Agriculture

The success of *Homo sapiens* can be explained in many different ways, including high brain capacity, manual dexterity, toolmaking, and social organization. Another feature that led to the survival of this species through times of fluctuating climate was adaptability. When prey animals became scarce, humans readily turned to alternative sources of food. By about 164,000 years ago, some human groups in South Africa had resorted to marine habitats for their main food supply, harvesting shellfish and other intertidal organisms [48]. Hunting as a means of subsistence was supplemented by gathering available resources, which included both animal and plant products.

By the end of the last cold period, people living in southwest Asia were experimenting with a new technique for enhancing their food supplies. In the fertile region of Palestine and Syria grew a number of annual grasses with edible seeds, the ancestors of our wheat and barley. The humans occupying these regions at the time were hunters and gatherers of the Upper Palaeolithic [49], feeding on a rich variety of animals, including gazelles, birds, rodents, fish, and molluscs. Some of their settlements have been discovered in Israel dating from 19,500 radiocarbon years ago (about 23,000 calendar years ago), and their hearths contain residues of charred dough made from the seeds of wild barley, wheat, and other grass species. Recognizing the value of these plants, successive generations of people

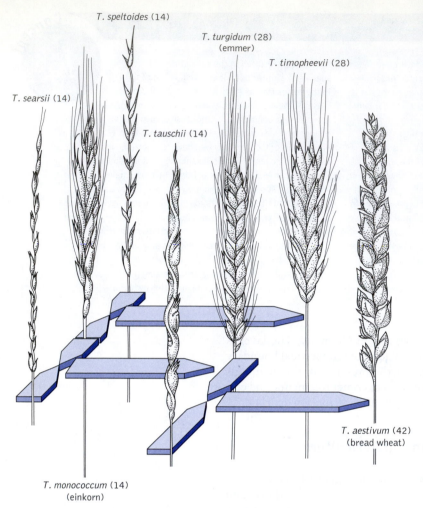

T. speltoides (14)

T. turgidum (28)
(emmer)

T. timopheevii (28)

T. searsii (14)

T. tauschii (14)

T. aestivum (42)
(bread wheat)

T. monococcum (14)
(einkorn)

Fig. 13.7 The evolution of modern bread wheat. This reconstruction is tentative, but represents the probable course of crossings among the wild wheat species that led to the early domesticated forms of the genus *Triticum*, and the subsequent crossings of domesticated wheats with wild species and chromosome doubling that led to bread wheat. Figures in parentheses after names represent chromosome numbers.

must have encouraged their growth of such useful plants by removing shade-casting trees and shrubs, and disturbing soils in which their seeds germinated more effectively. It was then a simple step to retain some of the seeds from one season to the next, and the selection of those strains with the highest investment in the edible grains.

Tracing the ancestry of our modern species of cereals is difficult, but Fig. 13.7 represents a possible scheme for the evolution of modern wheat, based on a study of the chromosome numbers of the various wild and cultivated species. The original wild wheat species had a total of 14 chromosomes (seven pairs) in each cell. *Triticum monococcum* (einkorn) was the first wild wheat to be extensively used as a crop plant. It probably hybridized with other wild species, but the hybrids would have proved infertile because the chromosomes could not pair up prior to the formation of gametes. Faulty cell division in one of these hybrids, however, could solve this problem because once the chromosome number had doubled (**polyploidy**, see p. 186) chromosome pairing could take place and the species would become fertile. Fertile polyploid hybrid species formed in this way included another important crop species, emmer (*T. turgidum*, sometimes called *T. dicoccoides*) with 28 chromosomes. This evolutionary development probably took place naturally, without any human intervention, because emmer wheat was one of the plants found in association with Upper Palaeolithic gatherers [45]. But the bread wheat of modern times (*T. aestivum*) has 42 chromosomes, and this probably arose as a result of polyploidy following the hybridization of emmer with another wild species, *T. tauschii*. This species comes from Iran, and it probably interbred with emmer as a result of the transport of that wheat species to Iran by migrating human populations. Thus, early agricultural peoples began not only to modify their environment but also to manipulate the genetics of their domestic species. The genetic modification of food-producing domesticated species is a process as old as agriculture itself.

More sophisticated studies have used not simply chromosome number but DNA fingerprinting of wheat seeds. Manfred Heun, from the Agricultural University of Norway, and his colleagues [50] have analyzed 338 samples of the most primitive wheat species, einkorn, which still grows wild in the Near East. Their aim was to determine precisely where wheat domestication first took place. They made two important assumptions as a basis of their work; first, that the DNA constitution of the species had not altered substantially since domestication took place, and second, that they could determine how closely populations of the plant were related to one another by reference to marker sections of the DNA sequence. They were able to locate wild einkorn populations in a region of the Karacadag Mountains in southeastern Turkey and close to the Euphrates River that provide genetic evidence of being the progenitors of domesticated einkorn. The application of molecular techniques in the study of domestication and of biogeographical origins of many plants and animals is becoming increasingly important in solving some important and some long-unanswered biogeographical and anthropological questions.

Molecular techniques have also been used to investigate whether plant domestication was a rapid process involving a limited number of ancestral forms, or a protracted series of trial-and-error domestications complicated by a constant genetic influx from wild species [51]. This process of mixing would have been constantly modified as the farmers selected for specific traits and brought in new genetic material to improve their crops [52]. As a result of the complete analysis of the genome of baker's yeast (*Saccharomyces cerevisiae*) in 1996, it has been appreciated that even the domestication of this fungus involved a complex series of mixing wild strains and selecting the required characteristics [53].

Humans also grew and cultivated other wild plants in the Middle East (Fig. 13.8), among them barley, rye, oat, flax, alfalfa, plum, and carrot.

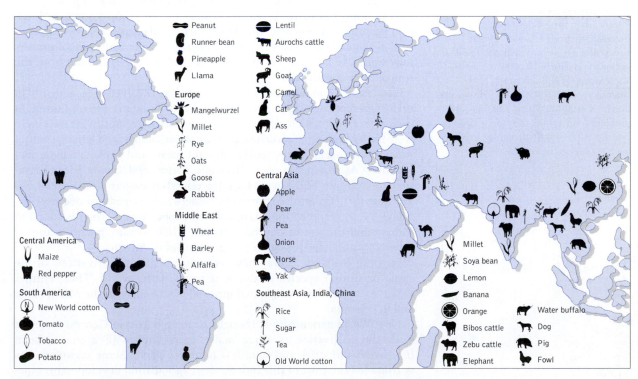

Fig. 13.8 Areas where different animals and plants were probably first domesticated.

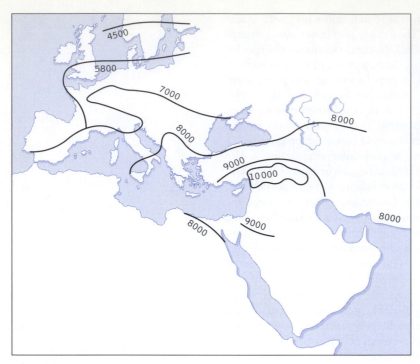

Fig. 13.9 Map of Europe showing the spread of agriculture from the area of the Fertile Crescent. Dates are in radiocarbon years before present. Radiocarbon timescales diverge increasingly from calendar or solar timescales as one moves into the past. A radiocarbon date of 10,000 years ago is approximately equivalent to 11,500 calendar years ago (9500 BC). The pattern is greatly simplified here, and there are problems concerning the precise time and direction of agricultural spread in such regions as the Balkans [55].

Farther west, in the Mediterranean Basin, yet more native plants were domesticated, including pea, lentil, bean, and mangelwurzel. The idea of domestication and organized agriculture resulted in rapid changes in the diet of early Neolithic people [54], and generated a steady spread of agricultural techniques from the Near East across the continent of Europe during the Holocene (Fig. 13.9), accompanied by the transport of domesticated plants from their native regions. It is still disputed whether this process involved the movement of peoples or the diffusion of a new technique [55]. If agriculture had proved an efficient means of stabilizing food supplies and avoiding some of the chance catastrophes of hunting and gathering, then it could have led to population expansion and the need to move to new areas. Some archaeologists have investigated the association between agricultural spread and the assumption of dominance by certain language groups. In Europe and Asia, for example, 144 of its languages belong to the group known as Indo-European, and the reason for this dominance could be its spread with agriculture in the early Holocene [56]. The languages of the early agriculturalists may well have become "steamroller" tongues as they spread out from the sources of domestication.

The idea of plant domestication seems to have evolved independently in many different parts of the globe (Fig. 13.10) and at many different times. In each area, appropriate local species were exploited: in southwest Asia there were millet, soybean, radish, tea, peach, apricot, orange, and lemon; Central Asia had spinach, onion, garlic, almond, pear, and apple; in India and Southeast Asia there were rice, sugarcane, cotton, and banana. Rice cultivation, for example, began in the coastal wetlands of eastern China over 7,500 years ago [57]; settlers used fire to clear alder swamps ready for rice growing. Maize, New World cotton, sisal, and red pepper were originally found in Mexico and the rest of Central America, while tomato, potato, common tobacco, peanut, and pineapple first grew in South America.

In some cases there may have been independent cultivations of the same or similar species in different parts of the world. Thus, emmer wheat may well have originated quite independently in the Middle East and in Ethiopia.

New World agriculture is thought to have begun in Central America with the cultivation of three major crops, maize (*Zea mays*), bean (*Phaseolus vulgaris*), and squash (*Cucurbita pepo*). Some argument has surrounded the time of this independent agricultural development, especially in relation to agricultural origins in other parts of the world.

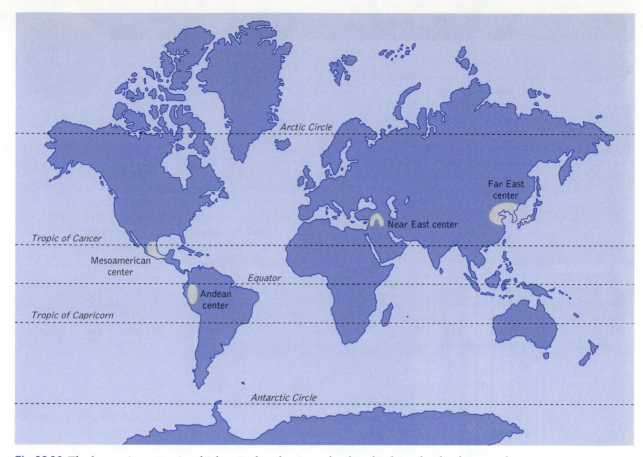

Fig. 13.10 The four main centers in which agricultural systems developed independently of one another.

Radiocarbon dates of squash seeds from caves in Oaxaca, Mexico, place their cultivation at about 10,000 years ago, so an early origin for New World agriculture is now well established [58]. The origins of maize have long been debated, as explained in Concept Box 13.3.

Animal Domestication

The domestication of certain animals may have preceded that of plants. There is some evidence, for instance, that early cultures had domesticated the wolf or, in some North African communities, the jackal. It is likely that such animals were of considerable use in driving and tracking game and hunting down the wounded prey, but determining when the dog was domesticated has proved very difficult using conventional archaeological techniques. Bones of dog/wolf associated with human settlements (found as far back as 400,000 years) could, after all, mean that people ate the animal rather than domesticated it. Joint burials, which provide a reasonable indication of domestication, are known that date from the early Holocene in the Near East, but little is known of earlier

One of the most puzzling problems in the study of plant domestication has been the origin of maize. Most research workers into this subject now agree that the most likely ancestor is the Mexican annual grass, teosinte. Indeed, both maize and teosinte are now regarded as subspecies of *Zea mays*, but structurally they are very different, especially in their flower and fruit structure. Teosinte has its fruits arranged in just two rows along the axis of the inflorescence, and each seed is surrounded by a persistent woody case that is shed with the seed. Its value as a food plant is therefore limited because it is difficult to separate the nutritious seed from its unpalatable case. Corn, or maize, has many rows of fruits in its inflorescence and, most important, the seeds are easily separated from their cases. The real problem is that there are no intermediate types between teosinte and maize, and neither are there any archaeological records of wild teosinte grains being collected and used for food. Very often, maize appears quite suddenly in the archaeological record of the diet of early human populations, and it may then become the dominant food resource. Its arrival at a site can be traced by isotopic changes in fossil human bone collagen. Maize is a C_4 plant (see Chapter 3), and one of the characteristics of these plants is that the ratio of ^{13}C to ^{12}C in the sugars produced by photosynthesis differs from that found in C_3 plants. This difference is retained in animals feeding on the plant, so it is possible to trace the importance of C_4 plants in an animal's diet. In this case it can be used to record the adoption of maize cultivation during the history of a human population.

The massive structural transformation involved in the evolution of maize from teosinte, however, may not have required major genetic change. The loss of the persistent fruitcase, which is the most critical limitation on the use of the plant for food production, involves a change in a single gene [59]. This change probably arose as a random mutation but was effectively selected out by the early farmers who recognized the value of this particular variety. This discovery not only goes a long way toward solving the mystery of maize evolution, it also demonstrates that relatively small genetic changes can result in considerable alterations in an organism's form, and hence its value as a domesticated species.

associations. Molecular studies may again give the best clue. Carles Vila, from the University of California Los Angeles, and his colleagues [60] have analyzed mitochondrial DNA samples from 162 wolves and 140 dogs, representing 67 different breeds. Their studies support the idea that the dog evolved from the wolf, but the differences between the two groups suggest that the evolutionary separation (presumably associated with domestication and isolation of dogs from wolves) took place about 100,000 years ago. This would place dog domestication at least as far back as the last interglacial. As in the case of plant domestication, however, the combined processes of selection and backcrossing with wild races has probably confused the genetic record.

It is likely, however, that many of the other animals that became associated with humans, such as sheep and goats, were domesticated much later, during the Early Neolithic Period and soon after the first cultivation of plants. These were initially herded for their meat and hides, but would have also been a source of milk, once tame enough to handle. The first traces of domesticated sheep come from Palestine around 8,000 radiocarbon years ago. These may have originated from one of the three European and Asiatic sheep, or may have resulted from interbreeding among these species. The Soay sheep that has survived in the Outer Hebrides in Scotland almost certainly originated from the moufflon, either the European *Ovis musimon* or the Asiatic *O. orientalis*. Domestication of these animals may have resulted from the adoption of young animals orphaned as a consequence of hunting activity. In Israel there is a marked shift in diet between 10,000 and 8,000 years ago, gazelle and deer being replaced by goat and sheep. Almost certainly this was a consequence of domestication.

The aurochs, *Bos primigenius*, a species of wild cow, was a frequent inhabitant of the mixed deciduous woodland that was spreading north over Europe during the postglacial period. The bones of this animal have been found in many of the sites where fossil remains of these forests have been preserved, such as in buried peats and submerged areas. The aurochs is now extinct, but it has long been supposed that European cattle resulted from the domestication of this wild bovid. The analysis of DNA in modern cattle, however, has led to a very different conclusion [61]. The domestic cattle of Europe (which were subsequently introduced into North America) have DNA that is very distinct from that of fossil aurochs bones and is more closely related to *Bos taurus* of the Middle East. The cattle of Africa and India, on the other hand, seem to have arisen from different stock, suggesting that cattle domestication has taken place in several different locations and was based on different bovid species.

Humans modified the genetic constitution of their domesticated animals by selecting for certain qualities, such as placid behavior, as well as high meat or milk production. But the close association of people with animals also led to genetic modifications in the human species, as in the case of milk digestion. It has proved difficult to establish the earliest date of human use of milk from domesticated mammals, but molecular and isotopic evidence can be used to detect the presence of those fatty acids associated with milk fat. It is likely that the milking of cattle was being practiced in the Near East about 9,000 years ago [62]. Milk is a product of mammals that is associated with the nutrition of the very young, and contains the carbohydrate lactose as a major constituent. Lactose is digested in young mammals by the enzyme lactase, but the gene controlling the production of this enzyme is normally switched off in humans as they are weaned. European races are exceptional in that lactase production continues through adulthood, so they can still digest milk. The same is true of many East African peoples, but not of those in West Africa. The work of Sarah Tishkoff of the University of Maryland on the genetics of African peoples has indicated that the mutation leading to lactase persistence in adulthood took place relatively recently, probably since 7,000 years ago [63]. It seems likely, therefore, that the increasing use of domesticated cattle in East Africa during the Holocene led to selection for the maintenance of the activity of the lactase gene. Here is a case of domesticated animals exerting an evolutionary selective pressure on their human masters. The geographical regions from which many types of domestic animals originate are shown in Fig. 13.8.

Diversification of *Homo sapiens*

As humans spread around the world and developed their own cultures and food resources, they continued to diversify in response to the environmental pressures placed on them. Some of the modifications they developed resulted from their choice of food, as in the case of the persistence of lactase production in populations that consumed milk. There was evidently a strong survival advantage for any individuals in pastoral populations who were able to digest lactose even into adulthood. But some of the diversity found among human races is less easy to explain in terms of natural selection, as in the case of blood groups. The blood group Rh, for example, is found in 30% of all Caucasians and yet is rare in the peoples of the Far East and among Native Americans. The AB blood group is totally absent in Native Americans, yet is common among Caucasians. As yet, it has proved impossible to explain these differences in terms of their selective value under different environmental circumstances. Some blood groups are associated with higher or lower incidence of particular diseases. For example, individuals with blood group O are more likely to suffer from stomach ulcers, and those with group A have a higher incidence of stomach cancer. But these are conditions normally found in older people, so one could argue that they are unlikely to affect the survival of a population and so will not have a strongly deleterious effect in evolutionary terms. On the other hand, the survival of a postbreeding, senior cohort in a population could have other advantages, such as assistance with rearing the young, or the transmission of tribal wisdom in complex societies, so the health of the old could still have evolutionary advantages for the population as a whole.

One of the most visually obvious sources of diversity among humans is the color of their skin. There is a great range of skin colors and shades within the human species, and there is an evident geographical correlation between skin types. On the whole, the peoples of the equatorial regions have darker skins than those of the high latitudes, and the most obvious explanation for this is protection from the intense light of the tropics, especially the harmful **ultraviolet radiation** (UVR). UVR can cause skin cancer and is also responsible for the destruction of certain B vitamins, such as folic acid, in the skin. The intensity of UVR is greater near the equator because light from the Sun passes along a shorter path through the atmosphere, as explained in Chapter 3, while in higher latitudes light passes at a shallow angle through a greater depth of atmosphere, where more energy is reflected or absorbed. But there are complications, such as the effect of increasing altitude, when UVR increases, so the relationship is not a simple latitudinal one. Recent research into skin color in humans has focused on the correlation between intensity of pigmentation and the local UVR level. The results show that between 70 and 77% of the variation in skin can be accounted for by UVR [64]. This is quite a high level of correlation and seems to account for the predominance of dark skin types in the tropics. But there is also a positive advantage for weaker skin pigmentation in the peoples of higher latitudes because a certain degree of UVR penetration is needed for the production of vitamin D. Vitamin D is required for calcium metabolism and bone growth, and it can be manufactured in the skin from biochemical precursors when exposed to UVR, so lack of UVR exposure can lead to vitamin D deficiency. There is, therefore, a clear advantage for the people of the high latitudes, such as the Nordic races, to have a pale skin color that allows the penetration of ultraviolet radiation. In the course of evolution, different races have developed the most efficient level of skin pigmentation to assure the optimal level of vitamin D production while protecting tissues from excessive UVR.

In excessive amounts, vitamin D can be toxic, so a dark skin prevents this from occurring. Vitamin D balance may also explain one of the most obvious exceptions to the general latitudinal variation in skin color, namely, the relatively dark coloration of the Inuit people of the far north.

These tundra-dwelling people feed mainly on fish, walruses, seals, and polar bears, and the livers of these animals contain very high levels of vitamin D. The Inuit avoid eating large quantities of liver from their prey, but this source of vitamin D means that there is no need for weak skin pigmentation in order to boost its production. Another apparent exception to the UVR correlation with skin color is the Bantu people of southern Africa, who have a darker skin than might be expected for the southern temperate latitudes they now occupy. But these people migrated south into these regions only within the last 2,000 years, so once again the lack of correlation with expectation can be explained.

The recent movements of people around the world have obviously complicated any studies of human biogeography and adaptation, but there are occasions where such movements have themselves produced new evolutionary developments. One good example is the blood condition called sickle-cell anemia, which, as its name suggests, causes anemia and other malfunctions of the blood system. One would expect evolution to select strongly against this condition, but in West Africa the gene for sickle-cell anemia is strongly prevalent, occurring in over 20% of the population. The reason for this retention of a potentially harmful gene in the population is that the gene also provides the carrier with a high degree of resistance to the malarial parasite. As in the case of skin color, there is a trade-off between positive and negative effects. When people were forcibly transported from West Africa to North America as slaves, they encountered conditions where malaria was less common and thus operated less strongly as a selective factor in population survival. Under these circumstances, the gene for sickle-cell anemia became distinctly less advantageous, and its incidence among African Americans in North America has now dropped to below 5%. Farther south in Central America, where malaria remains a greater risk, the gene is still found in 20% of the people of West African origin.

The Biogeography of Human Parasitic Diseases

When our species first evolved, as hunter-gatherers on the plains of Africa, the people would, like any other species, have been subject to a variety of diseases, some caused by viruses and bacteria, others by parasitic organisms. Some of these diseases may have infected the ancestors of early humans, such as *Homo habilis* and *Homo erectus*. So our early African ancestors were probably already liable to such widespread infections as roundworms (*Ascaris*), hookworms (*Necator*), and amoebic dysentery (*Entamoeba histolytica*). All of these have infective stages that are passed out with the feces of the infected individual and then wait in the soil or water until it is ingested by the next individual. As hunter-gatherers on the plains of Africa, however, our ancestors probably suffered less from such diseases than do the people who live there today in more sedentary settlements. Their habit of continually moving on from temporary campsites would have ensured that they did not remain for long near their own feces, which might otherwise have acted as infective agents for the eggs or larvae of parasites. Furthermore, the members of each small, independent group would have been closely related to one another, and therefore all would have had a similar amount of immunity to any viral or bacterial infections. So any such infection might rapidly lead to the death of most of the individuals in a particular group, but any survivors would be immune to future infections. As a result, our ancestors would not have suffered from epidemics that spread from group to group.

The life cycle of most parasites involves not only the final, **definitive host** (such as a human being), but also an **intermediate host**, or **vector**, within whose body the parasite multiplies and is transformed into a stage that can infect a new definitive host. Flying, blood-sucking insects are particularly well-suited to the role of intermediate host. They were probably quick to take advantage of the thin skin and reduced covering of body hair of this new species of hominid, even if it was as yet only present in low-population densities. The best-known of these diseases today is malaria, which is transmitted by the mosquito *Anopheles* and is caused by a protozoan (*Plasmodium*) that lives in the bloodstream of humans; a related species of *Plasmodium* infects monkeys. The mosquito *Aedes* similarly transmits the virus that causes yellow fever and, along with other genera of mosquito, transmits the infective stage of the nematode worms (*Brugia* and *Wuchereria*) that cause elephantiasis. But mosquitoes are not the only culprits in such insect-transmitted diseases. Sandflies (*Phlebotomus*) carry the infective stage of the African disease known as leishmania, the blackfly *Simulium* carries the infective stage of the nematode *Onchocerca* that causes river blindness, and the tsetse fly *Glossina* carries the infective protozoan *Trypanosoma* that is responsible for sleeping sickness. Finally, the vectors of the disease known as bilharzia or schistosomiasis are snails that live in the streams, rivers, and lakes, the infective stage boring into humans when they enter the water.

Studies of the distribution patterns of parasitic diseases have shown a strong latitudinal gradient in the frequency of the diseases associated with protozoan parasites [65], with higher concentrations of such diseases in the tropics. The distribution pattern of any parasitic disease that requires an intermediate host is naturally limited by the environmental needs of both the final host and the vector. Thus the year-round warmth of the tropics and subtropics provides a genial environment for all of them, and so it is not surprising that such diseases are prevalent in the low latitudes. What is less clear is why diseases caused by viruses and helminths (flatworms) do not display such latitudinal correlations. As our ancestors spread northward from Africa, they were thus accompanied by most of the above diseases. Only sleeping sickness did not succeed in spreading to Asia, apparently because the tsetse fly is restricted to Africa and the Arabian Peninsula. The rest are all prevalent in southern Asia, including the Indian subcontinent, while leishmania is also found in southern Europe, and malaria has occurred as far north as southern England. (It is also possible that all these diseases were already present in Eurasia, having been taken there earlier when *Homo erectus*, the ancestor of *Homo sapiens*, spread to that continent.)

In time, people changed their way of life from nomadic hunter-gathering into more permanent settlements, surrounded by the animals and plants that they had domesticated. But their new closeness to animals also brought with it a greater variety of disease exposure. The tapeworm *Taenia* finds its intermediate host in cattle and pigs, while the human diseases smallpox, tuberculosis, and measles are all closely related to similar diseases of cattle. Similarly, the nematode worm *Trichinella*, which encysts in muscle cells, infects humans when they eat inadequately cooked pork (which may be why pigs are considered unfit for human consumption in the Middle East). At the same time, the irrigation systems that early farmers constructed in the Fertile Crescent of the Middle East may, by placing water courses permanently near their villages, have made it more likely that they would be infected by diseases transmitted by mosquitoes (whose larvae live in water). At the same time, their homes and storehouses would have provided shelter and food for rats, from which they could have caught typhus. Finally, the higher population densities that accompanied all these changes would have made these humans more vulnerable to epidemics of disease. So there were certainly drawbacks, as well as advantages, to the development of domestication.

Just as humans gradually modified their domesticated animals to fit with their new environments, so the diseases and parasites the animals carried evolved to exploit the new opportunities offered by close human contact [66]. Many diseases of animals, whether prey species or domesticates, cannot be transferred to human beings. The characteristics of the disease organisms are so closely tuned to the nature of their host that they cannot pass through the barriers to enter and infect a different species. For example, most malarial parasites are species specific, so humans cannot be infected by any other than the human form. Other diseases, however, can be transmitted from animals to humans, such as rabies or bird influenza. Often such diseases are caused by agents that are relatively inefficient in terms of persistence, either because they prove fatal to the new host or because they are unlikely or unable to be transmitted between individuals of the new host. Thus any epidemic among humans is unlikely. In the next evolutionary stage, transmission from human to human becomes possible, as in the case of the ebola virus and dengue fever. With human-to-human transmission, the disease becomes much more serious because outbreaks in human populations can persist. In later evolutionary developments, the agent may become increasingly adapted to its new human host, eventually being transmissible only between human beings, as in the case of the HIV-1 virus, which probably originated in wild primate populations but is now a specifically human disease.

One biogeographic problem with diseases that have become exclusively human is tracing their origins. A very large number of such diseases seem to have originated in the Old World, but tracing the precise source can be difficult. In the case of the bacterium *Helicobacter pylori*, which is present in approximately 50% of all human stomachs, genetic analysis has provided an effective means of tracing its biogeographical history [67]. It is an unwelcome agent in the stomach because it can cause peptic ulcers and may even be a causative agent of stomach cancer. An extensive survey of the genetics of the bacterium shows that its genetic diversity decreases with distance from East Africa, suggesting that this was the original center of infection and evolutionary development. The current widespread distribution pattern of this organism suggests that early human populations became infected prior to their spread into other parts of the world.

When humans first migrated through the colder lands of northern Asia and across the Bering Straits into the New World, they left behind their domesticated animals and, with them, their associated diseases. Perhaps the only parasite that is found naturally in both the Old World and the New World is the hookworm *Ankylostoma*. But in South America, a New World version of leishmania evolved independently of that of the Old World, caused by a different species of the parasite and carried by a different fly. South America is also the home of Chagas' disease, caused by a species of *Trypanosoma* related to that which causes sleeping sickness in Africa; it is present in South American mummies dating from 2000 B.C. (Interestingly, the disease is endemic in the marsupial opossum of South America, but the animal is hardly affected by the infection, perhaps because this ancient inhabitant of the continent has, unlike humans, had many millions of years in which to adapt to the infective relationship and mitigate its effects.)

Another factor that reduced the prevalence of disease in the pre-Columbian peoples of North America was that they did not succeed in domesticating the large mammals of that continent in the same way that early humans domesticated those of Eurasia. The reasons for this are discussed by the biologist Jared Diamond of the University of California Los Angeles [68], who surveyed the reasons the Western version of civilization came to conquer or dominate the rest of the world. Biogeography, at a worldwide scale, was important. Eurasia is the world's largest landmass, extending widely from west to east, and so has a great variety of temperate, subtropical, and tropical environments. As a result, a greater number of large mammals, possible candidates for domestication, evolved there than in any other continent. Furthermore, by chance, many of those of Eurasia could be domesticated (e.g., cattle, sheep, goat, pig and horse), while this was not true of any of the large mammals of North America, and is true only of the llama of South America. (Even today, none of the other large mammals of North America or Africa have been domesticated, and none exist in Australia.) But the absence of domesticated animals in the New World also meant that their peoples were not exposed to the rise of the diseases that had accompanied domestication in the Old World.

This advantage for the humans of the New World was accompanied by a corresponding disadvantage, for these peoples had no immunological defenses when they were confronted by the expanding peoples of the Old World. So the diseases that Eurasians had caught from their domesticated animals (smallpox, measles, influenza, and typhus) ravaged the pre-Columbian Native Americans of the New World, killing 95% of those of North America and 50% of the Aztecs of Mexico and of the Incas of Peru. Smallpox, in particular, similarly devastated the peoples of southern Africa, Australia, and the Pacific islands.

On the other hand, because the Europeans who colonized the rest of the world came from higher latitudes, where tropical diseases were rare or absent, these diseases for many centuries acted as a barrier to large-scale European colonization of tropical Africa. What we now call the Ivory Coast was once known as the White Man's Grave. So geography, climate, and mammalian evolution have all played important roles in controlling the variety and incidence of the diseases that plague our species.

Inevitably, the incidence of some of these diseases has been affected by the increasing environmental impact of humanity and its activities. For example, human operations such as mining, deforestation, and road-building have led to increases in the predominance of *Plasmodium falciparum* (the more virulent parasite species that causes cerebral malaria), at the expense of *P. vivax*, which causes a less serious type of malaria. This has also been aided by the evolution of drug-resistant strains of *P. falciparum* and by the construction of dams and large-scale irrigation projects, which increase the area of water in which the mosquitoes can breed, as well as placing the water close to areas in which people live. The breakdown of control measures in the highlands of East Africa and Madagascar after the 1950s also led to an increase in malaria in these regions. In Madagascar this was aided by climatic changes involving an increase in temperature during the period of maximum mosquito abundance in December–January.

All these human activities, however, have also increased the size and widespread distribution of human populations available as hosts to *Leishmania* and *Trypanosoma cruzi*. Increasing aridity in parts of southern Africa has caused the movement of tsetse flies and *Simulium* blackflies into new areas, leading to an increase in sleeping sickness and river blindness. On the other hand, the loss of forest in some parts of Africa has led to the loss of these vectors and a consequent reduction in these illnesses [69]. Environmental degradation, as well as human population pressures, resource exhaustion, and disease, can thus combine to cause the collapse of human societies [70]. When societies do collapse, the environment often recovers remarkably rapidly, as is evidenced by the effects of the Black Death pandemic in Europe in 1347–1352. When this plague spread into Europe from Asia, it resulted in 30 to 60% mortality among the human population. Analysis of contemporaneous pollen profiles from lakes has shown that arable farming was abandoned and pastoral activity was greatly reduced, resulting in the regrowth of many forest areas that had previously been cleared for agriculture [71].

The Environmental Impact of Early Human Cultures

Environmental modification was an essential consequence of domestication and subsequent human spread. As the postglacial climatic optimum passed and conditions in the temperate regions became generally cooler and wetter, the agricultural concept continued to spread into higher latitudes. The incentive to modify the environment to make it more suitable for enhanced productivity of domestic animals and plants became a major driving force for the expanding human populations. Temperate forests are unsuitable for the growth of domesticated plants, most of which have a southern origin and a high demand for light. Similarly, domestic animals such as sheep and goats are not at their most efficient in a woodland habitat, preferring open grassland conditions. Cattle and pigs, on the other hand, can be herded within forest, but even they can be managed more efficiently in a habitat that is more open. In preagricultural times, the Mesolithic people of northern Europe discovered that the opening of forest and burning to retain open glades provided a higher productivity of red deer (*Cervus elaphus*). Like some of the

Native American tribes of North America, who became closely dependent on the bison, the European people of the Middle Stone Age were often reliant on the red deer (known in North America as the wapiti).

The intensification of forest clearance with the coming of agriculture in northern Europe is very apparent from pollen diagrams, where the pollen of open habitat species (such as grasses, plantains, and heathers) rises and the proportion of tree pollen falls. The precise pattern of forest clearance and the development of heathland, grassland, moorland, and blanket bog as consequences of this activity varies from one area to another, depending on the local conditions and the pattern of human settlement. By 2,000 years ago the impact was severe through much of central and western Europe, although the forests of the far north had been little influenced by humankind at that time. Some of the most severe deforestation, judging from the pollen record, had taken place in the northwest of Europe, including the British Isles. Perhaps it was in this region that the forest was least able to recover from human impact, and the maintenance of heavy grazing kept the area relatively open.

In North America, the hunting and foraging strategies of the Native Americans produced a patchwork of seasonal settlements and camps that involved opening the forest. The use of fire led to the development of sharp divisions between habitats, such as prairie grasslands and woodland. Many groups subsisted on the gathering of nuts from trees, and it is likely that they managed habitats, clearing unwanted species and opening the canopy to enhance nut production [72]. Agriculture in the temperate zone in pre-European times was confined largely to the growing of maize and other open-habitat, weedy species, including purslane (*Portulaca oleracea*). This involved the clearance of small areas of forest, and the effect of these clearings can be detected in pollen diagrams [73]. On abandonment, these clearings seemed to recover, and there is little evidence for large-scale forest destruction of the European type. However, some changes in the composition of the North American forests may well have resulted from the activities of agricultural peoples. The burning and cutting of forests is often associated with a loss in certain species, such as sugar maple and beech, and an increase in the abundance of fire-resistant pines and oaks, together with a general increase in the frequency of birch. Intensive clearance in the eastern United States and Canada was usually delayed until the arrival of European settlers in the 18th and 19th centuries. It is often marked in the pollen diagrams by a marked rise in the ragweed (*Ambrosia*).

There is still considerable debate about the impact of aboriginal peoples on the development of Australian vegetation. It is likely that the first invaders were mainly associated with the savanna regions, and undoubtedly they used fire as a means of managing and hunting game. Like the Native Americans, this practice would have resulted in sharp boundaries between savanna and forest habitats. Fire-resistant species, such as the eucalypts, would have been favored by the use of fire [74].

Environmental change has accelerated in the last few centuries, and the face of the Earth is changing ever faster. Figure 13.11 illustrates the rate of destruction of primary forest in the United States over the past 400 years [75]. The destruction that our temperate forests have experienced during these recent centuries is now being repeated in the tropical forests.

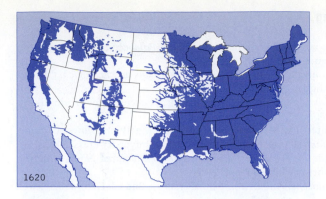

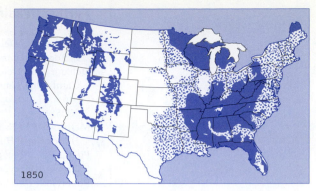

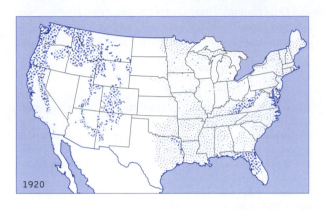

Fig. 13.11 The extent of virgin forest in what is now the United States of America in 1620, 1850, and 1920, showing the progressive destruction of American old-growth forests [75].

The arrival of human beings on the planet created a whole new set of conditions for the species already present. Some were granted new opportunities for extending their ranges as they adapted to the new circumstances, but many experienced new stresses either because of direct destruction by humans or because their habitats were increasingly modified as human populations expanded. We examine the consequences of recent and continuing environmental change in Chapter 14.

Summary

1 The human line of evolution separated from the great apes around 7 million years ago and led to the development of many types of hominin (members of the human line of evolution rather than that of the great apes).

2 Africa is the likely location for the origin of the genus *Homo* and of our own species, *Homo sapiens*.

3 Of the various types of hominin, only *Homo sapiens* survived beyond the last glaciation, although the discovery of *H. floresiensis* fossils could indicate the local persistence of a separate line of evolution.

4 The extinction of many large vertebrate animals at the end of the last glacial had not occurred in former climatic cycles, and circumstantial evidence suggests that humans were involved in the demise of many of them.

5 The domestication of animals and plants provided new opportunities for human food production and hence population expansion. It took place in a number of different locations around the world.

6 Many parasites and disease organisms evolved alongside humans and adopted this primate as their host. Other parasites and diseases of animals were favored by domestication, and some of these adapted to enable them to move from domestic animals to humans.

7 Human spread, population increase, and the need for agricultural development has meant that humans have increasingly manipulated and modified their environment.

Further Reading

Delcourt HR, Delcourt PA. *Quaternary Ecology: A Paleoecological Perspective.* London: Chapman & Hall, 1991.

Delcourt HR, Delcourt PA. *Prehistoric Native Americans and Ecological Change.* Cambridge: Cambridge University Press, 2004.

Diamond J. *Collapse: How Societies Choose to Fail or Succeed.* New York: Viking, 2005.

McIntosh RJ, Tainter, JA, McIntosh SK. *The Way the Wind Blows: Climate, History, and Human Action.* New York: Columbia University Press, 2000.

Pielou EC. *After the Ice Age: The Return of Life to Glaciated North America.* Chicago: University of Chicago Press, 1991.

Roberts N. *The Holocene: An Environmental History*, 2nd ed. Oxford: Blackwell Publishers, 1998.

Russell EWB. *People and Land through Time: Linking Ecology and History.* New Haven, CT: Yale University Press, 1997.

References

1 Willis K, Niklas KJ. The role of Quaternary environmental change in plant macroevolution: the exception or the rule? *Phil Trans R Soc Lond* 2004; 359: 159–172.

2 Lister AM. The impact of Quaternary Ice Ages on mammalian evolution. *Phil Trans R Soc Lond* 2004; 359: 221–241.

3 Carroll SB. Genetics and the making of *Homo sapiens*. *Nature* 2003; 422: 849–857.

4 McBrearty S, Jablonski NG. First fossil chimpanzee. *Nature* 2005; 437: 105–108.

5 Brunet M, et al. A new hominid from the Upper Miocene of Chad, Central Africa. *Nature* 2002; 418: 145–155.

6 Folinsbee, KE, Brooks, DR. Miocene hominoid biogeography: pulses of dispersal and differentiation. *J Biogeogr* 2007; 34: 383–397.

7 Hay RL, Leakey MD. The fossil footprints of Laetoli. *Sci Am* 1982; 246 (2): 38–45.

8 Kerr RA. Tracing the identity of the first toolmakers. *Science* 1997; 276: 32–33.

9 Sponheimer M, Lee-Thorp JA. Isotopic evidence for the diet of an early hominid, *Australopithecus africanus*. *Science* 1999; 283: 368–370.

10 Semaw S, Renne, Pl. Harris JWK, Feibel CS, Bernor RL, Fesseha N, Mowbray K. 2.5-million-year-old stone tools from Gona, Ethiopia. *Nature* 1997; 385: 333–336.

11 Bramble DM, Lieberman DE. Endurance running and the evolution of *Homo*. *Nature* 2004; 432: 345–352.

12 Bennett MR et al. Early hominin foot morphology based on 1.5-million-year-old footprints from Ileret, Kenya. *Science* 2009; 323, 1197–1201.

13 Wood B. The oldest hominid yet. *Nature* 1994; 371: 280–281.

14 Lordkipanidze, D et al. Postcranial evidence from early *Homo* from Dmanisi, Georgia. *Nature* 2007; 449: 305–310.

15 Ciochon RL, Bettis EA. Asian *Homo erectus* converges in time. *Nature* 2009; 458: 153–154.

16 Swisher CC, Rink WJ, Anton SC, Schwarcz HP, Curtis GH, Suprijo A, Widiasmoro. Latest *Homo erectus* of Java: potential contemporaneity with *Homo sapiens* in Southeast Asia. *Science* 1996; 274: 1870–1874.

17 Parfitt SA, et al. The earliest record of human activity in northern Europe. *Nature* 2005; 438: 1008–1012.

18 Stringer C. Modern human origins: progress and prospects. *Phil Trans R Soc Lond* 2002;357: 563–579.

19 Thieme H. Lower Palaeolithic hunting spears from Germany. *Nature* 1997; 385, 807–810.

20 Stringer C. Out of Ethiopia. *Nature* 2003; 423, 692–695.

21 White, TD, Asfaw B, DeGusta D, Gilbert H, Richards GD, Suwa G, Howell FC. Pleistocene *Homo sapiens* from Middle Awash, Ethiopia. *Nature* 2003; 423: 742–747.

22 Bermudez de Castro JM, Arsuaga JLO, Carbonell E, Rosas A, Martinez I, Mosquera M. A hominid from the Lower Pleistocene of Atapuerca, Spain: possible ancestor to Neanderthals and modern humans. *Science* 1997; 276: 1392–1395.

23 Carroll SB. Genetics and the making of *Homo sapiens*. *Nature* 2003; 422: 849–857.

24 Mellars P. Neanderthals and the modern human colonization of Europe. *Nature* 2004; 432: 461–465.

25 Krause J et al. Neanderthals in central Asia and Siberia. *Nature* 2007; 449: 902–904.

26 Delson E, Harvati K. Return of the last Neanderthal. *Nature* 2006; 443: 762–763.

27 Tzedakis PC, Hughen KA, Harvati K. Placing late Neanderthals in a climatic context. *Nature* 2007; 449: 206–208.

28 Brown P, Sutikna T., Morwood MJ, Soejono RP, Jatmiko, Saptomo EW, Due RA. A new small-bodied hominin from the Late Pleistocene of Flores, Indonesia. *Nature* 2004; 431; 1055–1061.

29 Morwood MJ, Brown P, Jatmiko, Sutikna T, Saptomo EW, Westaway KE, Due RA, Roberts RG, Maeda T, Wasisto S, Djubiantono T. Further evidence for small-bodied hominins from the Late Pleistocene of Flores, Indonesia. *Nature* 2005; 437: 1012–1017.

30 Jungers WL, Harcourt-Smith WEH, Wunderlich RE, Tocheri MW, Larson SG, Sutikna T, Due RA, Morwood MJ. The foot of *Homo floresiensis*. *Nature* 2009; 459, 81–84.

31 Zazula GD, Froese DG, Schweger CE, Mathewes RW, Beaudoin AB, Telka AM, Harington CR, Westgate JA. Ice-age steppe vegetation in east Beringia. *Nature* 2003; 423: 603.

32 Rogers RA, Rogers LA, Hoffmann RS, Martin LD. Native American biological diversity and the biogeographic influence of Ice Age refugia. *J Biogeogr* 1991; 18: 623–630.

33 Gibbons A. The peopling of the Americas. *Science* 1996; 274: 31–32.

34 Meltzer D. Monte Verde and the Pleistocene peopling of the Americas. *Science* 1997; 276: 754–755.

35 Gonzalez-Jose R, Gonzales-Martin A, Hernandez M, Pucciarelli HM, Sardi M, Rosales A., Van der Molen S. Craniometric evidence for Palaeoamerican survival in Baja California. *Nature* 2003; 425: 62–65.

36 Stoneking M. From the evolutionary past. *Nature* 2001; 409: 821–822.

37 Martin PS, Wright HE *Jr. Pleistocene Extinctions: The Search for a Cause*. New Haven, CT: Yale University Press, 1967.

38 Grayson JE. Pleistocene avifaunas and the overkill hypothesis. *Science* 1977; 195: 691–693.

39 Stuart AJ. Mammalian extinctions in the Late Pleistocene of Northern Eurasia and North America. *Biol Rev* 1991; 66: 453–562.

40 Guthrie RD. New carbon dates link climatic change with human colonization and Pleistocene extinctions. *Nature* 2006; 441: 207–209.

41 Stuart AJ, Kosintsev PA, Higham TFG, Lister AM. Pleistocene to Holocene extinction dynamics in giant deer and woolly mammoth. *Nature* 2004; 431: 684–689.

42 Miller GH, Magee JW, Johnson BJ, Fogel ML, Spooner NA, McCulloch, MT, Ayliffe LE. Pleistocene extinction of *Genyornis newtoni*: Human impact on Australian megafauna. *Science* 1999; 283: 205–208.

43 Kennett DJ, Kennett JP, West A, Mercer C, Que Hee SS, Brement L, Bunch TE, Sellers M, Wolbach WS. Nanodiamonds in the Younger Dryas boundary sediment layer. *Science* 2009; 323, 94.

44 Kerr RA. Did the mammoth slayer leave a diamond calling card? *Science* 2009; 323, 26.

45 Guthrie RD. Rapid body size decline in Alaskan Pleistocene horses before extinction. *Nature* 2003; 426: 169–171.

46 Coltman DW, O'Donoghue P, Jorgenson JT, Hogg JT, Strobeck C., Festa-Bianchet M. Undesirable evolutionary consequences of trophy hunting. *Nature* 2003; 426: 655–657.

47 Cardillo M, Lister A. Death in the slow lane. *Nature* 2002; 419: 440–441.

48 Marean CW, et al. Early human use of marine resources and pigment in South Africa during the Middle Pleistocene. *Nature* 2007; 449: 905–908.

49 Piperno DR, Weiss E, Holst I, Nadel D. Processing of wild cereal grains in the Upper Palaeolithic revealed by starch grain analysis. *Nature* 2004; 430: 670–673.

50 Heun M, Schafer-Pregl R, Klawan D, Castgna R, Accerbi M, Borghi B, Salamini F. Site of einkorn wheat domestication identified by DNA fingerprinting. *Science* 1997; 278: 1312–1314.

51 Allaby RG. The rise of plant domestication: life in the slow lane. *Biologist* 2008; 55 (2), 94–99.

52 Purugganan MD, Fuller DQ. The nature of selection during plant domestication. *Nature* 2009; 457, 843–848.

53 Liti G. et al. Populations genomics of domestic and wild yeasts. *Nature* 2009; 458, 337–341.

54 Richards MP, Schulting RJ, Hedges, REM. Sharp shift in diet at the onset of Neolithic. *Nature* 2003; 425: 366.

55 Willis KJ, Bennett KD. The Neolithic transition—fact or fiction? Palaeoecological evidence from the Balkans. *Holocene* 1994; 4: 326–330.

56 Diamond J. The language steamrollers. *Nature* 1997; 389: 544–546.

57 Zong Y, Chen Z, Innes JB, Chen C, Wang Z, Wang H. Fire and flood management of coastal swamp enabled first rice paddy cultivation in east China. *Nature* 2007; 449: 459–462.

58 Smith B. The initial domestication of Cucurbita pepo in the Americas 10,000 years ago. *Science* 1997; 276: 932–934.

59 Wang H, Nussbaum-Wagler T, Li B, Zhao Q, Vigouroux Y, Faller M, Bomblies K, Lukens L,Doebley JF. The origin of the naked grains of maize. *Nature* 2005; 436: 714–716.

60 Vila C, Savlainen P, Maldonado JE, Amorim IR, Rice JE, Honeycutt RL, Crandall KA, Lundeberg J, Wayne RK. Multiple and ancient origins of the domestic dog. *Science* 1997; 276: 1687–1689.

61 Troy CS, MacHugh DE, Bailey JF, Magee DA, Loftus RT, Cunningham P, Chamberlain AT, Sykes BC, Bradley DG. Genetic evidence for Near-Eastern origins of European cattle. *Nature* 2001; 410, 1088–1091.

62 Evershed RP et al. Earliest date for milk use in the Near East and southeastern Europe linked to cattle herding. *Nature* 2008; 455, 528–531.

63 Check E. How Africa learned to love the cow. *Nature* 2006; 444: 994–996.

64 Diamond J. Geography and skin colour. *Nature* 2005; 435: 283–284.

65 Nunn CL, Altizer SM, Sechrest W, Cunningham AA. Latitudinal gradients of parasite species richness in primates. *Diversity Distrib* 2005; 11: 249–256.

66 Wolfe ND, Dunavan CP, Diamond J. Origins of major human infectious diseases. *Nature* 2007; 447: 279–283.

67 Linz B, et al. An African origin for the intimate association between humans and *Helicobacter pylori*. *Nature* 2007; 445: 915–918.

68 Diamond J. *Guns, Germs, and Steel: The Fates of Human Societies*. New York: W. W. Norton, 1997.

69 Molyneux DH. Common themes in changing vector-borne disease scenarios. *Trans R Soc Trop Med Hygiene* 2003; 97: 129–132.

70 Diamond J. Collapse: *How Societies Choose to Fail or Succeed*. New York: Viking Penguin, 2005.

71 Yeloff D, van Geel B. Abandonment of farmland and vegetation succession following the Eurasian plague pandemic of AD 1347-52. *J Biogeogr* 2007; 34: 575–582.

72 Delcourt PA, Delcourt HR. *Prehistoric Native Americans and Ecological Change*. Cambridge: Cambridge University Press, 2004.

73 Andrews JH. Human disturbance of North American forests and grasslands: the fossil record. In: Huntley B, Webb III T, eds. *Vegetation History*, pp. 673–697. Dordrecht: Kluwer, 1988.

74 Bowman DMJS. *Australian Rainforests: Islands of Green in a Land of Fire*. Cambridge: Cambridge University Press, 2000.

75 Williams M. *Americans and Their Forests*. Cambridge: Cambridge University Press, 1989.

Into the Future

The evolutionary developments among the primates that eventually led to the emergence of human beings proved to have profound biogeographical implications. Many organisms had already displayed the capacity to modify their environment, so-called ecological engineers, such as beavers. Some organisms, such as ants and termites, had even developed techniques of exploiting other species in a manner that could be regarded as a form of agriculture. But humans proved to be in a different league. Their high level of intelligence, combined with manual dexterity and social organization, placed them in a unique position for global dominance, which they soon managed to achieve. It is no exaggeration to say that within the course of the current interglacial, some 12,000 years, they have transformed the planet. In this chapter we examine some of these transformations and take a speculative look at possible future biogeographical trends in response to continuing global changes.

One consequence of human success in diverting the energy resources of the planet into the provision of a reliable food base has been rapid population expansion. The degree of control over parasites and diseases has assisted in this process, but the seemingly inevitable aggression within the human society, which has resulted in some highly destructive wars in recent centuries, has operated in the opposite direction. With rising populations the demand for food production has increased dramatically, leading to more land, once regarded as marginal or useless, being transformed for agriculture. The outcome has been the destruction and fragmentation of natural habitats and the instigation of a cascade of extinctions leading to a global loss of biodiversity. The development of industry has generated greater quantities of waste materials, including the carbon dioxide returned to the atmosphere after millions of years of incarceration in the form of fossil fuels. Such waste products in the atmosphere are now recognized as a major cause of current changes in global climate.

The current situation is thus one of rapid change. Species boundaries have changed as a result of local extinction, interrupted migration corridors, habitat destruction, inadvertent and deliberate introduction, and changes in climatic patterns that have altered the capacity of species to survive within their past ranges. Change is not a new experience for the great assemblage of living organisms on Earth; the past 2 million years have been characterized by constant change. Even rapid climate change is nothing new, and the consequences of such past changes are apparent in many distribution patterns of modern organisms (see Chapter 12). But the current rapid change in climate has been accompanied by considerable alterations in the landscape as a result of human activity. Plants and animals now face a combination of severe challenges. It is the responsibility of biogeographers to foresee and predict the consequences and outcomes of these changes. It is also incumbent upon the science to find ways of ameliorating some of the harmful possibilities, such as a catastrophic loss of global biodiversity. Understanding biogeographical principles should provide a means of developing informed conservation strategies.

Human Population

The future of the planet and its living inhabitants is intimately tied to the future of our own species. Our population will ultimately be controlled by the limited resources of the planet, but the exploitation of those resources will place great stress on all other organisms. When the climate began to ameliorate at the end of the last glaciation, there were probably fewer than 20 million people on the planet, and perhaps fewer than 2 million. Agriculture and settlement provided a more reliable food base, and populations began to expand. By the beginning of the Christian Era, about 8,000 years later, the world population was likely around a quarter of a billion (250 million) [1]. By 1650, the human population had doubled from this figure to reach half a billion, and the first billion was achieved around 1830. From this point on, the explosion really began, the second billion mark being passed in 1930. By 1974, the population had doubled to 4 billion, and the turn of the millennium saw over 6 billion people on the Earth. Many people alive today are able to claim that the population of the world has doubled within their lifetime. Improved agriculture, pest control and, most important of all, medical advances involving greater longevity and reduced infant mortality, have all contributed to this population explosion. Growth in population will undoubtedly continue, especially in the developing nations, where the beneficial effects of medical developments and agriculture will greatly reduce mortality both in the very young and the old. Infant survival into maturity is of particular demographic significance because this will increase general fecundity in the population. Thus by the year 2020 it is expected that 80% of the world's population will live in the developing nations [2].

In the developed nations, general life expectancy will also continue to rise as mortality in later life declines with improved medical techniques and care. The expectancy has already risen in some developed countries, such as the United Kingdom, from about 50 years in 1850 to around 78 in 2008. There is a limit to how much further this survival curve can extend because of the innate biological changes associated with aging, and the improved survival of older people does not have the same general effect on population fecundity seen with improved survival among the young. But an expansion in the number of older people still presents demands on the world's resources in terms of living space needs and the requirement for food, water, and energy. Increasingly scarce, and therefore expensive, resources in the developed world, coupled with high lifestyle expectations, is resulting in deliberate delays in child-bearing by the young, so this factor leads to a slowing of the population growth rate.

Human population growth cannot continue indefinitely. As in the case of all animal, plant, and microbe populations, it will eventually become limited by resource depletion, or by disease and parasitism, or by some social factor, such as war. The rate of population growth has already begun to decline, so the peak of the explosion has passed (somewhere around 1965–1970), but the growth rate is still fast and must decline much further if the population is to stabilize. One of the major resources of the planet that will soon be exploited to its limits is the production of fixed energy by photosynthesis, which is the ultimate source of our food supply. Back in 1986, it was calculated that humans consume, feed to their domestic animals, use for other purposes, or divert around 40% of the Earth's primary production [3]. Included in this figure is the production lost because we have built cities where productive plants once grew so that the land has been wasted by human activities. This figure demonstrates that resources of primary production are very limited and that a further doubling of population would put a great strain on the food production capacity of the planet.

Apart from food, human beings have a high demand for energy, which creates a whole range of environmental problems to be discussed later, but one of the proposed solutions is the use of renewable energy derived from sunlight through the mediation of green plants or microbes, so-called bioenergy. Energy plantations for the supply of bioenergy will consume yet more land resources that will thus be lost to agriculture or to biodiversity conservation. The effects of this problem began to be felt in 2008. The high price of oil meant that alternative biofuels also became more valuable and farmers derived economic benefit by growing fuel crops rather than food. Continuation of this trend can only lead to food shortages and rising food prices worldwide. Biofuels, therefore, are unlikely to provide a complete alternative to fossil fuels.

Part of the demand we place on the planet results from our dependence on domesticated animals for the proportion of our food supply. Just as our populations have expanded rapidly, so have the populations of our domesticated animals. The population of chickens, for example, doubled between 1981 and 1991 to reach 17 billion. There are now more than three chickens for every human on Earth. Our larger herbivores, such as cattle, sheep, and goats, are estimated to total 4.3 billion. Although we outnumber these domestic stock animals, they have a larger biomass than the entire human population of the planet. Since they all consume primary production (some of it, like grain, perfectly suitable for human consumption), it is clear that our domesticates add to the pressures that we place on the resources of the planet.

Water is also a vital commodity that is in short supply. Freshwater is needed for our personal use, for the irrigation of crops, for domestic animals, and for industry. By 1996 we were already consuming over half of the Earth's terrestrial freshwater runoff [4], and demand has continued to grow. More than a billion people already lack access to clean drinking water, and an estimated 2.4 billion have inadequate sanitation [5]. By 2025, more than half of the world's nations will be suffering from water shortages, rising to 75% of nations by 2050 [6]. Already many areas of the

West and Midwest of the United States are removing more water from the ground each year than arrives by rainfall. This situation is clearly not sustainable in the long term, especially if changes in climate lead to less precipitation in these regions and if the projected population growth of 40–50% over the period 2000 to 2020 is realized. The oceans, of course, contain an abundance of water, but desalinization demands energy, and this is a resource that is also in limited supply.

Some elements may prove limiting to human development and population growth, such as phosphorus, which is present at relatively low concentrations in most rocks, yet is needed in relatively large quantities by living organisms, including humans. The recovery of phosphorus from fecal wastes, especially from colonies of seabirds, has long been an important industry, and in future more general recycling of such elements that have passed through human populations may prove both feasible and economically viable. As in most recycling issues, economics lies close to the heart of ultimate solutions. The recycling of materials will take place when it proves to be an economically attractive alternative to replacement.

Given the limitations that face humanity, the pressing question is, how many people can the Earth support? All ecosystems have a carrying capacity for any particular species, and any species that exceeds this value is bound to decline before it can stabilize. Has humanity already exceeded the Earth's carrying capacity for our species, or do we yet have some room for expansion? What are the future biogeographic prospects for humanity? Very much depends on the nature of our expectations of life. If everyone on Earth were to demand the quality of life and the abundance of resources available to the developed nations in North America and Europe, then it is probable that the Earth could not sustain the present population of humans, let alone any future expansion. If everyone became vegetarian and reduced their daily calorific intake and their demand for an excess of water, then our present population could be supported and could even expand somewhat. Because of the diversity of factors involved, estimates of the human population occupying the Earth in 2050 vary between about 8 and 12 billion. Most demographers agree that a doubling of the present population must be close to or even beyond the limit of the Earth's carrying capacity [7] unless considerable advances are made in the technology of resource usage.

Apart from resource depletion, human populations are restrained by social factors, such as birth control, where a planned limitation of population growth can give rise to higher individual quality of life. War, involving mass population extermination, has been a constant element in history, and the prospect of global conflict on scales even greater than those seen in the 20th century, though unthinkable, can never be discounted. The impact of disease on human mortality has been greatly reduced in the past two centuries, which has contributed to population expansion, especially because of improved infant survival, but high population density and high mobility even between continents presents disease organisms with new opportunities for evolution and spread.

The success of medical science has itself led to evolutionary and biogeographic developments in the microbial world. The extensive use of antibiotics to eliminate harmful pathogens in the environment inevitably disturbs the entire ecological balance of microbial communities. In the human gut, for example, various benign strains of the bacterium *Escherichia coli* are vital for digestive processes. Oral antibiotics upset the competitive interactions between these strains and other, less benign microbial occupants of the gut. This can lead to strong selective pressures for antibiotic-resistant strains, among which some may prove pathogenic. The strains of *E. coli* are given identity numbers, and a strain called 0157:H7 first emerged in the United States in the 1970s and proved harmful, particularly to the very young and the old, who developed fatal renal illness when they encountered this food-borne microbe. Literally hundreds of such emergent diseases have been reported over the past 60 years and, significantly, they are most common in the United States, United Kingdom, Australia, Japan, and Germany. It is possible that this collection of emergent diseases is more common in the more medically advanced countries simply because they are more often recognized and recorded there, but it is more likely that medical applications, particularly the profligate use of antibiotics, are simply generating new diseases [8].

Despite the unpredictable possibilities of wars and pandemics arresting human population growth, the general consensus of opinion among population scientists is that human population expansion is likely to continue for some time. On the basis of this assumption, biogeographers can anticipate certain consequences for the living world, including increasing habitat loss as land is turned to agriculture and urban development, increasing pollution leading to global problems in the oceans and the atmosphere, and inevitable repercussions on global biodiversity.

Pollution of the Planet

The size, growth rate, widespread distribution, and demands placed on global resources by the human population inevitably impact the entire biosphere of the Earth. Some of these impacts are chronic and widespread, such as the chemical changes created by pollution of the atmosphere and the hydrosphere (including the oceans). Many of the pollution problems that have resulted from human activity have their origin in a deliberate enhancement or speeding up of natural element cycles. An excellent example is the nitrogen cycle. For as long as agriculture has existed, farmers have appreciated the value of fertilization, often in the form of applying animal dung to fields. Nitrogen is an essential component of amino acids, and therefore proteins, so all organisms demand it. Most green plants take up their nitrogen from the soil in the form of nitrates, and decomposing matter, including fecal materials, are rich sources of the element. Industrial societies have supplemented traditional management of the natural nitrogen cycle by tapping the vast nitrogen resources of the atmosphere (where N_2 gas takes up 81% by volume). This process requires large energy investment to attain the temperatures and pressures required, so it relies heavily on fossil fuel energy. It is estimated that the total amount of nitrogen fixed by industrial processes is equal to the entire natural fixation of the world's land surface [9]. This acceleration of natural nitrogen fixation encourages faster plant growth and hence agricultural production. But this also has its negative side, for much of the nitrate fertilizer applied to fields and intended to enhance crop growth is lost. Nitrates are soluble in water and are not well retained in soils, so the application of excess fertilizer often results in the enrichment of drainage waters, resulting in the enrichment, or **eutrophication**, of streams, rivers, lakes, and even oceans. Resulting algal productivity and inevitable death and decomposition lead to oxygen depletion by microbial activity, thus affecting the quality of freshwater habitats and decreasing their biodiversity.

There are natural processes that act in opposition to this enrichment, such as the denitrifying bacteria in soils and waterlogged sediments that use nitrates as an electron source in respiration rather than oxygen. They thus reduce nitrates (NO_3^-) to nitrogen gas (N_2) and return it to the atmosphere. But even here there is an environmental snag because the oxidation process is sometimes left incomplete and the bacteria may generate nitrous oxide (N_2O) instead of molecular nitrogen [10]. This is a potent greenhouse gas that contributes to atmospheric changes and consequent global warming. Biogeographers and ecologists are becoming increasingly aware of the complexity of natural element cycles and the dangers inherent in their modification.

Some of the nitrates spread on the land by farmers find their way into the atmosphere, and this is supplemented by the nitrogenous compounds produced by the combustion of organic materials, including fossil fuels. This is a form of atmospheric eutrophication, and it is likely to be causing global impacts. The atmospheric deposition of biologically active nitrogen compounds (such as nitrates) over the ecosystems and biomes of the world is generally stimulating to plant growth. The temperate and boreal forests of the Northern Hemisphere are picking up much of the aerial deposition of nitrates, and evidence suggests that their rate of growth is being stimulated as a result [11]. The same is also likely to be true of the tropical forests as they recover from clearance [12]. Forest clearance often involves burning, which generates oxides of nitrogen that are lost to the atmosphere, but precipitation returns some of this to the soil and contributes to tree growth and recovery. Such responses to human-induced pollution are to be welcomed for many reasons. Quite apart from the healing of damaged habitats, the regrowth of forests takes carbon dioxide out of the atmosphere and so helps to lower the levels of greenhouse gases. But the process operates only for as long as nitrogen limits tree growth. In many areas, especially the tropics, phosphorus becomes limiting to growth, as it is rapidly absorbed by the growing trees, and then the rate of biomass increase in the ecosystem declines. Changes in the nitrogen balance of ecosystems will also result in alterations in competitive interactions between plant species, favoring some and proving harmful to others. Sensitive species and those lacking the capacity for robust growth under the eutrophic conditions will suffer, and biodiversity is lost.

Increasing the nitrate load of the atmosphere thus does not always result in a favorable outcome. Ecosystems that have nitrogen-poor soils have developed a distinctive flora and fauna that survive simply because of their high levels of specialization. As a consequence of the isolation of the Arctic regions from major centers of industrial activity, the actual quantities of nitrogen deposited from the air are relatively small, between 0.1 and 1.0 g m^{-2}. But the natural nitrogen cycle in these regions is also very slow, and the input of nitrogen to tundra ecosystems from all other sources is of the same order as the aerial deposition. Fertilization from the atmosphere thus doubles the rate of nitrogen input, encouraging the growth of more competitive, nitrogen-demanding plants and reducing the survival of the more specialist species [13]. In the subarctic zone of Sweden, the aerial input of nitrogen compounds has been shown to disrupt the plant communities, stimulating the growth of robust grasses such as *Calamagrostis lapponica*, and leading to a tall grass canopy that overtops and shades out the characteristic dwarf shrub community of the tundra [14]. In the high Arctic, where nitrogen-demanding grasses are not present, the mountain avens (*Dryas octopetala*, see Fig. 12.1) is initially favored by the aerial inputs and grows faster [15], but continued enhanced supply of nitrates eventually leads to winter injury in the plant. Small alterations in the chemical composition of the atmosphere can thus disrupt the natural balance of sensitive ecosystems and lead to unforeseen changes.

Oxides of sulphur are also produced as waste products in the combustion of organic materials. These too have accumulated in the atmosphere, leading to biogeographical consequences. Sulphur has long been known as a potent fungicide, copper sulphate being the basis of the Bordeaux Mixture used in the treatment of fungal pathogens of domestic plants. An increase in sulphur in the atmosphere and the rainfall therefore impacts on microbial populations in soils and on plant surfaces. One of the incidental effects of sulphur dioxide pollution in urban areas in the middle of the last century was a decrease in the abundance of leaf spot and mildew diseases of roses. Subsequent reductions in sulphur emissions, a consequence of clean air legislation, resulted in a return of these familiar fungal garden blights in temperate zone cities.

The effects of sulphur on decomposition processes in natural ecosystems are not easy to observe, but lichens, which are formed by a combination of algal and fungal components, are more easily visible and they have reacted to raised sulphur levels in ways that are apparent. Epiphytic species (those growing on the twigs and branches of trees and shrubs) have proved particularly sensitive, and surveys of epiphytic lichen distributions have provided reliable information relating to atmospheric sulphur levels in regions where no chemical data are currently available. Lichen biodiversity has been shown to correlate closely with the incidence of lung cancer in a studied area in the Veneto region of Italy [16], but interpreting correlations of this type is fraught with difficulties. Much research remains to be done to determine precisely how this kind of linkage can be explained. An understanding of the physical demands and limits of particular species of plants and animals, however, can enable us to use them as **bioindicators** of environmental conditions.

Injury by sulphur has also been recorded from plants other than lichens. The impact of sulphur on the bog mosses (*Sphagnum* species) has been severe in heavily polluted wetland areas, leading to the disruption of vegetation and consequent erosion of peatlands. The southern part of the Pennine Mountains in Britain, a peatland region closely adjacent to areas of high industrial activity, is a particularly severe example of pollution-induced erosion, with large expanses of exposed eroding peat surfaces, which have a limited capacity for recolonization and "healing" in the absence of sulphur-sensitive bog mosses.

The development of urban environments, associated with increasing use of internal combustion engines to provide transport, has considerably altered the atmosphere close to the ground. The action of sunlight on some products of fossil fuel combustion can result in the generation of the highly reactive gas ozone. Ozone is a valuable component of the stratosphere encircling the Earth (between 40 and 60 km above the ground), forming a protective layer that filters out ultraviolet radiation, but when generated close to the Earth's surface it can prove harmful to both plants and animals. It is an extremely reactive oxidizing agent, and consequently it can irritate the linings of the respiratory systems of animals. Humans suffering from asthma and bronchitis are particularly susceptible to its effects. In plants, damage from ozone exposure is visible, and when first described in California it was referred to as "smog damage." The photosynthetic pigment chlorophyll is damaged by ozone exposure, and leaves become scarred by lesions or may even die, as is the case with white clover (*Trifolium repens*). Although leaves lost in this way are often replaced quite quickly, chronic exposure leads to a loss in the overall productivity of the plant; some forest studies suggest that the productivity loss could be as great as 10% [17]. Crop plants are also affected, as are many wild plants, although some species and some varieties are more resistant than others. Ozone pollution is becoming an increasing threat in many cities throughout the world, even in the cooler temperate areas where climatic changes (as well as cleaner air resulting from a reduction in particulates) are permitting the penetration of higher intensities of sunlight to ground level. In the long term, chronic exposure will influence patterns of competition and selection of genotypes in plant communities.

Chemical Overload

The acceleration of element cycles by human activities has also involved elements normally found in low concentrations in the natural world, but now becoming locally abundant. Some of these are toxic to living organisms, such as some of the heavy metals, including lead, copper, mercury, zinc, nickel, and cadmium. These elements are widely used in industrial processes and have often been released into the environment in waste products. In the Santa Barbara Basin, Southern California, for example, the concentration of mercury in sediments deposited 1,500 years ago was 0.04 $\mu g\ g^{-1}$. By the late 20th century this concentration had risen by over four times as a result of industrial discharge [9]. This metal has undoubtedly moved through food webs in the course of history, as has been demonstrated in Virginia, where blood samples of birds that have fed upon aquatic insects show elevated levels of mercury.

Industrial development has resulted in an ever increasing demand for metals. Metal mining was an important aspect of life in Roman times, when it is estimated that 15,000 tons a year of copper were extracted each year from the Roman Empire, together with 100,000 tons of lead and 10,000 tons of zinc [18]. While many species of plants and animals are damaged by these pollutants, some species and some ecosystems seem capable of accumulating heavy metals and may provide a means of environmental cleansing and rehabilitation. Reed beds and aquatic macrophytes are being used for filtering water contaminated with mine waste before releasing it back into the environment. Mine spoil heaps, which are slow to revegetate following abandonment, may be reclaimed by planting pollution-tolerant species and varieties, such as white pine (*Pinus strobus*) in the **ecological rehabilitation** of coal mine spoil [19]. As our knowledge of ecosystem function and tolerance increases, it will become increasingly important to apply such information to the repair and rehabilitation of damaged landscapes.

Mining continues to expand. Between 1970 and 2000, the total extraction of construction and industrial minerals around the world doubled from 4 to 8 billion tons, and the trend continues. New industries have led to new demands for minerals, such as the expansion in a demand for columbite-tantalite that is needed in the electronics industry for construction of cell phone capacitors. The development of mining for such minerals in the Republic of Congo has resulted in the disturbance and destruction of much of the Okapi Reserve and the Kahuzi-Biega National Park, and a consequent loss of over 80% of the population of lowland gorillas in the region. Mine waste disposal is often more damaging than the extraction, especially in the case of those minerals and metals with very low concentrations in the rocks from which they are extracted. Iron ore, for example, contains around 40% of the metal, and aluminium ore about 20%. But copper ores have less than 1% copper, and gold may even be as low as 0.00033%, so mining these metals generates proportionately very large quantities of waste. Canadian mines produce 60 times as much waste material as all of the Canadian cities combined. More efficient mining technology may help reduce the waste disposal problem, but in the long term an increased dependence on recycling elements must emerge [20].

Apart from the acceleration of natural cycles and the artificial accumulation of certain elements, such as metals, human ingenuity has also created and disseminated a number of chemical compounds that are designed to have toxic consequences for target plants and animals, the **pesticides**. The ecological and biogeographical consequences of the use of such compounds as DDT and the organophosphorus group of chemicals have been appreciated since the 1950s, but the problems associated with the chemical suppression of organisms considered pests by human societies have not gone away. New pesticides have replaced some of the older ones, but these can still cause unforeseen catastrophes, sometimes resulting in the abrupt decline of nontarget species. A recent example is the population collapse of several species of Asian vultures as a result of the widespread use of a nonsteroidal anti-inflammatory drug, dicofenac, in cattle (used for both the European cow, *Bos taurus*, and the Indian hump-backed cow, *B. indicus*). The drug becomes concentrated in fat deposits, as well as the intestine, kidney, and liver of the cattle and, if the cow dies, these organs are often consumed by local vulture flocks, including the Indian white-backed vulture (*Gyps bengalensis*), the long-billed vulture (*G. indicus*), and the slender-billed vulture (*G. tenuirostris*). Carcasses of animals that have received drug treatment within a week of death are highly toxic to the vultures, usually resulting in 10% mortality among feeding birds [21]. The outcome has been a population collapse among Indian vultures, and there are now fears for related African species, such as the African white-backed vulture (*G. africanus*), as the drug comes into use on that continent. Conservationists are recommending the withdrawal of dicofenac from veterinary use and are hopeful that this will be put into effect before these valuable bird species become totally extinct. It has often proved to be the case in ecological systems that the removal of one key species can cause cascades of additional extinctions as associated species are affected. It is possible that the large carrion-feeding vultures will prove to be keystone species within their ecological food webs. Their loss will result in considerable changes in the structure of food chains based on carrion.

It is fortunate for the vultures that the likely source of their decline has been identified quite rapidly and can be tackled. Many declines of species are more difficult to understand, resulting in delays in taking any remedial activity. The general and worldwide decline in amphibians is an example of a complex combination of pressures that is proving difficult to unravel. Loss of wetland habitats has undoubtedly played a part in many highly populated countries, but the decline in amphibians is not confined to such areas. Pollution of waterways with pesticides may have reduced food availability and contaminated the invertebrate life that remains. One compound, atrazine, the most commonly used herbicide in North America, has devastating effects on the gonads of male leopard frogs (*Rana pipiens*), effectively turning males into females. The compound induces the production of an enzyme that converts the male androgen hormones into the female estrogen hormones. Crops are often sprayed with atrazine in the early spring as a weed suppressant, which is precisely the time when amphibians begin their breeding and are most sensitive to hormonal disturbance. The compound is widely found in North American waters, even in rainwater in some locations, so it is not surprising that the frogs are in decline [22]. This must generate concern not only for the future welfare of frogs but of humans as well, and the amphibian decline may prove an early warning of worse problems to come.

The Cascade Mountains of western North America have some of the cleanest water in the continent, but even there such amphibians as the western toad (*Bufo borealis*) is experiencing a dramatic decline [23]. The cause here is not pollution, but a complicated interaction of weather conditions and parasitic fungi. The immediate cause of death is the infection of toad embryos while still embedded in their spawn and is caused by a parasitic fungus, *Saprolegnia ferax*. The increasing frequency of this parasitism is linked to the rise in ultraviolet (UV-B) radiation experienced by the toad eggs as they incubate in shallow water ponds. There has been some dispute about whether the UV-B increase is itself a consequence of the destruction of ozone in the stratosphere by human-produced chemical compounds, or whether it is a consequence of ponds becoming dryer and therefore shallower in the spring. UV-B is strongly absorbed by water, five times less radiation penetrating to a depth of 80 cm than is available just 5 cm below the surface. The spring dryness and shallow ponds, in turn, are the result of shifting climatic patterns, associated with the El Niño Southern Oscillation cycle in the Pacific Ocean. It is possible, therefore, that climatic changes ultimately underlie the decline in this and other amphibians, though computer modelling and projections suggest that amphibians overall should benefit from climate warming because of their ability to occupy more northerly regions [24].

These examples demonstrate how human activities can have biogeographical consequences on both local and global scales. Species of plant and animal can act as bioindicators providing early warning systems of inherent chemical dangers in the environment. Sometimes the messages they convey can be interpreted relatively easily, as in the case of the Indian vultures, but in other cases the sheer complexity of ecological interactions renders their message opaque.

Albedo is an index of reflectivity. It has 1.0 as its highest value, meaning that all incoming radiation is reflected; fresh snow, for example, has an albedo of 0.95. All vegetation types absorb more radiant energy than snow, but the albedo varies according to the nature of its surface. A uniform deciduous forest canopy has an albedo of about 0.2, whereas a coniferous forest canopy has a value of about 0.15 because its darker color absorbs more of the incident energy. When a uniform canopy is broken and fragmented by disturbance, such as clearing areas for agriculture or grazing, the complex pattern that results absorbs more solar energy than the original uniform surface. The clearings in a forest, for example, can act as sheltered sun traps in which the energy is reflected between layers and more becomes absorbed (see Fig 14.1). A broken forest canopy may have an albedo of about 0.1. Once the entire forest has been cleared and replaced by a crop such as wheat, or by pasture grassland, however, the uniform canopy more than regains the reflectivity of the original forest and has an albedo of about 0.25. The replacement of grassland by bare soil or sand as a result of overgrazing raises the albedo to between 0.35 and 0.45. So, when we consider the agricultural activities of humans on a continental or global scale, their impact on the energy budget of the Earth is profound.

Landscape Transformation

Humans have been altering the face of the planet by their activities for many thousands of years, as examined in Chapter 13. By clearing natural vegetation and creating suitable conditions for their crop plants and their grazing animals, people gradually fragmented the natural landscape and created a mosaic of seminatural and highly managed patches. The development of cities, linked by road and rail systems, has further impacted on natural habitats and created a disruptive network through the landscape. Some of these act as barriers to the movement of organisms, while others form corridors along which species can spread. Changing the vegetation has also altered the radiation balance of the Earth's surface, because different types of vegetation differ in the degree to which they reflect incoming solar radiation [25]. The reflectivity of a surface is measured by its **albedo** (see Concept Box 14.1).

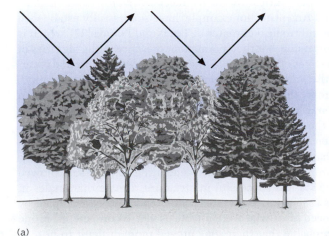

(a)

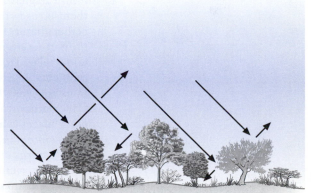

(b)

Fig. 14.1 The albedo (reflectivity) of a forest is reduced when it is partially cleared and the canopy becomes broken. Radiant energy is trapped within the clearings instead of being reflected by the uniform canopy. Landscape modification thus creates a new complexity of microclimates.

Agriculture has increased the reflectivity of land surfaces (see Box 14.1 and Fig. 14.1), and an increasing movement of people into cities in recent centuries has further enhanced our impact on global albedo. Concrete surfaces have an albedo of around 0.2, which is not very different from a uniform vegetation cover, but tarred roads absorb much energy, having an albedo of less than 0.1. Overall, a city absorbs more heat than an equivalent area of vegetation. In addition to its greater absorbance of solar energy, cities also generate heat. Cities are energy sinks into which energy-rich materials, such as fossil fuels, are imported and are then combusted in machinery, automobiles, and power stations, to supply the work that supports the activities of the city. The conversion of the chemical energy of these materials into the kinetic energy of motion and work is an inefficient process leading to the release of heat into the environment. Additional heat, used for the space heating of homes and offices, also leaks into the surrounding atmosphere, resulting in the city microclimate often being 4°C or more above the surrounding regions. As urbanization has proceeded, the sum of city microclimates, especially in highly populated parts of the world, such as North America, China, India, and Europe, have modified climate on a continental scale. In particular, they have contributed to a general trend in reduced diurnal temperature change. The radiation of heat from cities at night has led to smaller differences between daytime and night-time temperatures. How much cities have contributed to this trend has proved difficult to determine, but recent models [26] suggest that around half of this trend can be accounted for by urban expansion, and the overall contribution of cities to global warming could be as high as 0.27°C per century.

The raised temperature of cities has created biogeographic islands that have proved attractive to organisms from lower latitudes. Various types of parakeets, for example, are successfully invading higher latitude cities and managing to establish permanent breeding populations. This group of birds is helped, of course, by the fact that they are often kept as pets and subsequently escape into the wild, so the problem of dispersal to a new site is overcome for them. But once present they are becoming an increasing feature of many northern cities. Monk parakeets (*Myiopsitta monachus*) from Paraguay and Argentina, for example, have successfully established themselves in the subtropical climates of Texas and Florida [27], which is perhaps hardly surprising, but their permanent presence in the big cities of the northeastern states of the United States is more remarkable and is certainly related to the warmer winter microclimate generated in urban environments. In Europe, the ring-necked parakeet (*Psittacula krameri*) from India now has a large breeding population as far north as London.

In addition to their microclimatic peculiarities, cities are also usually noisy places when compared with the open countryside, and noise can affect animal behavior patterns. In Alberta, for example, studies of ovenbirds (*Seiurus aurocapilla*) close to noise-generating compressors showed that males were 15% less successful in attracting a mate than males in quieter locations [28]. As a result, such sites tended to be occupied by young inexperienced birds who were unable as yet to compete for prime territories. It is probable that their song proved more difficult to hear or was less impressive to females in noisy sites. The influence of noise on singing patterns has been studied in urban great tit (*Parus major*) populations in the Netherlands [29]. Males in noisy territories sing at significantly higher pitch in order to be heard above the general background of low-frequency sound. City habitats are thus producing new selective pressures on those species that seek to exploit them and are leading to new evolutionary developments.

The establishment of new patches of landscape, together with the fragmentation of the natural vegetation by human activity, presents new biogeographical challenges to plants and animals, similar to those encountered by dwellers on island archipelagos in the oceans. Island biogeographic principles (see Chapter 8) can therefore be applied to the study of fragmented landscapes, though this must be conducted with great caution [30]. As previously discussed, large areas of habitat are likely to contain greater numbers of species than small areas, but this varies with habitat type and with the precise species involved.

Habitat fragmentation has had its most severe effects on large organisms with slow generation times that have limited capacity for spreading to new habitats or crossing alien ones. The most sensitive species are those that require large areas of habitat, free from the disturbance created and possible enhanced predation effects found at habitat edges. An example is the forest-dwelling tiger of Sumatra [31]. In a study of the 36,000 km^2 (14,000 sq mls) Kerinci Seblat region of Sumatra, where tropical forest fragmentation has taken place, it was found that larger patches of forest contained the highest populations of tigers and that there was a strong correlation between tiger occurrence and the distance from roads. Tigers evidently avoid regions of human disturbance, and this makes it difficult for them to maintain contact between fragmented populations. Isolation of small populations not only renders them liable to local extinction by catastrophe, such as disease or famine, but also reduces their capacity to maintain genetic variability, which is so essential for continued evolution and adaptation. From the conservation point of view, therefore, such fragmented populations (termed **metapopulations** by ecologists) require assistance in ensuring continued genetic movement between the subpopulations. The most obvious way to ensure such genetic transfer is to encourage migration along maintained corridors of habitat. Conservationists refer to the presence of corridor networks of this type as the landscape **connectivity**.

There is experimental evidence that small subpopulations in a metapopulation of a species are both genetically impoverished and more liable to local extinction [32]. Studies of the Glanville fritillary butterfly (*Melitaea cinxia*) in Finland, for example, have shown that egg hatching, larval survival, and adult longevity were all higher in larger subpopulations and that these features correlated with higher genetic diversity. There is likely to be less mating between closely related individuals and therefore less accumulation of unfavorable genetic characters. Links providing the opportunity for genetic exchange between subpopulations are thus valuable attributes for a metapopulation.

Maintaining connectivity in highly fragmented or heavily modified landscapes may prove extremely difficult, especially if the object of conservation is a large and sensitive animal such as a tiger or a cougar, which requires wide corridors if it is to move through them without undue stress. In the case of such potentially dangerous animals, there is also likely to be considerable opposition from local residents, fearful for their own safety and that of domestic stock. Roads, especially large and busy highways, present very considerable barriers to the movement of large mammals, and the establishment of connectivity in a landscape rich in highways can be both difficult and expensive for planners. Building wide, vegetated underpasses or overpasses requires considerable financial investment, and genetic outcomes need to be monitored to ensure that they prove effective. In Southern California, for example, the busy Ventura Freeway separates the Santa Monica Mountains from the fragmented habitats to the north. This wide 10–12 lane highway carries about 150,000 vehicles per day. In a study of bobcats (*Lynx rufus*) and coyotes (*Canis latrans*) in the region, it was found that 11.5% of the bobcats and 4.5% of the coyotes tracked over a seven-year study period crossed the road at some time [33]. This sounds like a very reasonable degree of mobility, but analysis of the consequences of such movements in the molecular genetics of populations on either side of the highway showed very much smaller degrees of genetic mixing. Many of the animals plucking up courage to cross the road were young seeking to establish new territories, but they presumably met with territorial opposition on arrival at the other side and were repulsed without donating any genetic material to the resident population. Such results will be useful in generating computer models to assess the effectiveness of different degrees of connectivity in landscapes and establishing the most cost-effective solution to the maintenance of genetic movements among isolated subpopulations of a species [34]. In a situation where climatic changes are expected over the coming years, the movement of species into new areas of potential occupation will become increasingly important to ensure that plants and animals can adjust their ranges to the new set of climatic circumstances. Landscape connectivity, therefore, is likely to become an increasingly important area for biogeographical research.

Apart from connectivity, landscapes can also vary in their texture. Landscape texture incorporates the vertical structural complexity with the degree of horizontal variability. Thus a uniform, simple texture is created by a landscape with little vertical complexity and low lateral variation, while a complex landscape texture combines structural complexity with varied spatial patterns. This approach to landscape description appears to be meaningful in biogeographic terms. In a study of Australian birds [35], it was found that large-bodied birds were associated with landscapes of a simple texture (grazed open grassland), while complex textures in the landscape (tall, wet eucalypt forests) favored small-bodied birds. The concept needs to be tested in other localities with additional groups of organisms, but the model may prove useful in landscape design and biological conservation.

The Changing Climate

Meteorological records from around the world span back over the past century or more, and therefore trends in global climate can now be documented without recourse to proxy methods such as those described in previous chapters. We do not need to look to oxygen isotope profiles or to indirect records of climate in fossil animals and plants when we have plentiful direct sets of data. This means that meteorologists are currently in a position to determine whether global climate is changing and at what rate such alteration is taking place. It is also possible to project these changes into the future on the basis of past experience and by making certain assumptions about future trends [36].

Figure 14.2 presents the record of global temperature since the mid-19th century and shows a clear over-

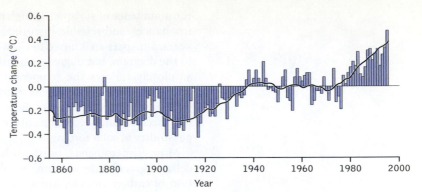

Fig. 14.2 Mean annual global combined sea surface and air temperatures for the past 130 years relative to the average for the period 1951–1980.

all upward trend especially during the 20th century. There is no doubt that the Earth has become warmer (by about 0.6°C) over the past century, and it is possible to resolve this increase in warmth into three main stages: first between 1900 and 1940, followed by a fairly stable plateau for around 30 years, and then a further rise since about 1970. Projecting into the future demands that we understand the underlying causes of the rise so far. Perhaps it is part of a natural cycle, in which case we need to understand the detailed form of the cycle in order to predict its next move. Or perhaps we are observing a climate change in which human activity is playing a significant part, such as by contributing heat-absorbing gases to the atmosphere and creating a thermal blanket or **greenhouse effect**. Meteorologists have been cautious about adopting the second of these explanations until all alternatives have been adequately explored.

If one considers the climate history of the Earth over the past million years, as reflected in the oxygen isotope curves of ocean sediments and ice cores (see page 372), then the pattern of climate change during a normal interglacial is one of rapid initial temperature rise to a peak within about 5,000 years, followed by a gradual, if erratic, fall in temperature over the next 100,000 years. If, as seems reasonable, we are currently experiencing the latest in a series of interglacials, then one can predict that we are past the expected thermal optimum and should be on our gradual way toward the next glaciation. The present global warming appears to be a denial of expectation, although warm intervals have been known to interrupt the temperature declines in past cycles.

The warming we are now seeing is relatively recent in origin and was preceded by the Little Ice Age in which global cooling was very evident. It is possible that this reversal is a consequence of a shorter-term climate cycle than that of glacial and interglacial scale. If we look at changes in solar irradiance [37], which represents a possible explanation for climatic changes on this

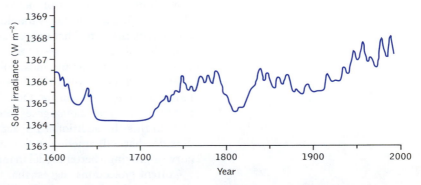

Fig. 14.3 Changes in solar irradiance over the past 400 years based on historical records of sunspot number.

kind of timescale (Fig. 14.3), we find a pattern that generally follows the observed changes in global temperature. Solar irradiance is dependent on the abundance of sunspots (a high number of sunspots leading to lowered irradiance), and reliable historical records of these go back almost 400 years. Sunspots are known to show a regular 11-year cycle; this is evident in the diagram, but of greater significance is the long-term variation that so closely follows the temperature curve. The precise mechanism by which solar irradiance might influence world climate remains obscure; possibly ozone production in the upper atmosphere is involved. Whatever the mechanism, predicting future climate must take into account the possibility of solar forcing.

Most climatologists, on the other hand, now believe that the much-debated greenhouse warming is taking place as a result of human injection of carbon dioxide, ozone, methane, chlorofluorocarbons, oxides of nitrogen, and other infrared-absorbing gases into the atmosphere. Water vapor is a much neglected greenhouse gas, however, so all modeling of atmospheric and climatic changes needs to take into account changes in humidity. Current global levels of atmospheric humidity are on the rise [38], which will also affect future patterns of cloud formation and precipitation. Analysis of precipitation trends suggests that observed alterations in patterns, such as the poleward deflection of rainfall patterns from subtropical into temperate regions, is itself a consequence of human modification of the atmosphere [39]. The picture has been complicated by human generation of other waste gases, such as sulphur dioxide, that are released into the atmosphere and that have the opposite effect of the greenhouse gases and lead to some degree of cooling. Future predictions need to take into account the whole range of gaseous pollutants that we generate and how their output may increase or decrease in the future. An international group of scientists examining the evidence for climate change and the human role in this process (the Intergovernmental Panel on Climate Change, IPCC), after much careful debate and analysis, came to the conclusion that "the balance of evidence suggests that there is a discernible human influence on global climate." The question of whether human activity is causing climate change now seems to be settled. More relevant are the questions concerning what will happen next (a meteorological question), what can be done about it (a political and economic question), and what are its consequences (a biogeographical question).

The prediction of the biogeographical consequences of climate change depends on the forecasts of meteorologists about future climates. Given all the physical and human variables involved, it is not surprising that the future remains somewhat unpredictable. It is possible, however, to set up a number of scenarios based on certain assumptions and then to check which of these is proving most robust as time and climate change proceeds. Figure 14.4 gives an example of two scenarios extrapolating global temperature into the future. One is based simply on the continued projected input of greenhouse gases into the atmosphere, and the lower curve adds to this the additional impact of other pollutants, such as sulphur dioxide, that will ameliorate the temperature rise. At present, the lower curve is proving a better model in matching the observed temperature rise.

Current projections suggest that the high latitudes will warm up faster than most other regions, which will impact strongly on tundra habitats. The arid regions of the world will also be sensitive to temperature rise as these are already in a delicate state of balance as far as natural and agricultural ecosystems are concerned [40, 41]. An important question that will affect much of the world's climate is whether global temperature change will influence the frequency and intensity of the El Niño phenomenon in the equatorial Pacific. Recent years have shown how this periodic buildup of warm water in the eastern Pacific can have widespread repercussions, sometimes with catastrophic consequences from Indonesia to California and Peru, and perhaps even more extensively. The 1997–1998 El Niño was very carefully studied and modeled, but the future of these events is still uncertain [42].

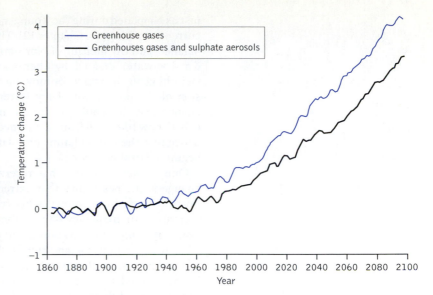

Fig. 14.4 Calculated global surface air temperature modeled to take into account the production of greenhouse gases alone and the combination of greenhouse gases and sulphate aerosols. Observed values of temperature follow the lower curve, which should therefore be regarded as a more reliable forecast of future developments.

Global warming, especially if it is concentrated in the higher latitudes, may be expected to cause the melting and breakup of the ice sheets around the Antarctic Ice Cap. Observations suggest that this process has indeed begun and is accelerating. Over the last 50 years there has been considerable ice-shelf retreat around the Antarctic Peninsula. Workers from the British Antarctic Survey team have come to the conclusion that a monthly average temperature of −2.5°C is the threshold for melting; above that temperature the melt is quite rapid. The number of summer days in which melting occurs has been increasing at the rate of about one extra day per year since the late 1970s, and consequential ice-shelf collapse has been rapid, as evidenced by satellite photographs. An obvious possible result of such melting is a rise in global sea level, especially if sea temperatures rise with a resultant expansion in volume. Many models have been developed to predict just how great a rise might be expected and how rapid the sea-level rise is likely to be. The problem is that this estimate is dependent on models of global temperature rise, which, as we have seen, are themselves still open to question. An average figure derived from numerous models suggests a global sea-level rise of 38.7 cm by the year 2100. Thermal expansion would account for most of this (28.8 cm); mountain glaciers and ice caps might account for 10.6 cm; the Greenland ice sheet could contribute a further 2.4 cm; but the Antarctic may actually accumulate more ice as a result of increased precipitation in the region, thus decreasing ocean volume. Many uncertainties remain, however, both about the behavior of the two major ice sheets and the contribution of mountain glaciers [43]. Rising sea level, however, will give rise to many biogeographical problems, especially in low-lying islands and coastal regions of the world.

Another serious potential problem associated with the oceans is the possibility of disruption in the oceanic thermohaline circulation (the conveyor belt) that transports heat around the world and that depends for its circulatory mechanism on subtle changes in seawater salinity and density. Changes in global temperature and their resulting inputs of additional freshwater from melting ice could disrupt these movements of water masses, just as happened during the Younger Dryas event in the early stages of the current warming (see Chapter 12). The North Atlantic is a key area in this balance, and the melting of Northern Hemisphere glaciers and the lower density water created by thermal expansion could prove critical in switching off the conveyor movement and leading to a southward extension of the polar front of icy waters from the Arctic. There is even a chance that this could precipitate another ice age. Oceanographers still debate how likely such an occurrence would be [44], but there is some evidence that the thermohaline circulation of the North Atlantic is already beginning to slow down [45].

One further issue that has received attention in relation to climate change is the possibility that extreme weather events, such as storms, hurricanes, and exceptionally heavy rainfall leading to flooding may be increasing in frequency. As in the case of all environmental change issues, it is important to know about patterns in the past if we are to assess present-day conditions and trends. One approach has been the study of lake sediments, checking for times in the past when high loads of eroded sediments entered the basin. An analysis of a number of lake basins in the northeastern United States has revealed a distinct pattern of erosion episodes during the Holocene [46]. There have been four peaks of increased storm frequency in the last 12,000 years (12 kyr), at 11.9, 9.1, 5.8, and 2.6 kyr. The stormy episodes have tended to last about a thousand years each. There is a regularity to these figures; they suggest that storm frequency increases roughly every 3,000 years in the North Atlantic region and that we are approaching the next peak. This observation complicates the interpretation of the current apparent increase in extreme weather events. We cannot assume that such a development is an outcome of human activities or of general global warming.

There are many uncertainties concerning the future climate of the planet, but one thing is certain: there will be changes and these changes are likely to be considerable and may prove rapid. It is also inevitable that such changes will have extensive biogeographical consequences.

Biogeographical Consequences of Global Change

Predicting the future climate of the Earth is extremely important for agriculturalists, economists, and conservationists, and much research is currently being conducted into improving the global models on which such predictions are based. Estimates of future conditions constantly change as models become more sophisticated, but the general consensus of opinion suggests that increasing carbon dioxide in the atmosphere will continue to lead to thermal insulation of the planet, resulting in a global temperature rise of approximately 4°C (figures range between 2° and 6°C) during the present century [2]. Such changes will undoubtedly alter the range distributions of individual species of organisms, modify the species compositions of ecosystems, and change the positions of global biomes.

Biomes are defined by the life forms of their constituent components and are in the main climatically determined in their distribution, subject to human modification. The French ecologist Wilfried Thuiller has calculated that a change in global temperature of 1°C causes a change in the biogeographical position of the world's biomes of about 160 km (100 mi) [47]. A rise in global temperature of 4°C should therefore produce a general poleward movement of the biomes by about 500 km (310 mi), assuming that soils and other factors do not prove limiting to such a shift. In the case of altitudinal zonation, this would entail an upward shift of about 500 m (1,600 ft) in the zone boundaries. Past changes in climate are now well documented, as are past changes in vegetation in many parts of the world. Thus it is possible to use the past as a key to the present [48]. In the Swiss Alps, for example, the rapid rise in temperature between 11,600 and 11,500 years ago (perhaps as great as 3–4°C in a century) caused rapid changes in the forest composition and the timberline [49]. Larch invaded alpine tundra, and lower down the mountains pine began to replace larch. The timberline may have moved upward by 800 m (2,600 ft) in less than 300 years. Therefore, rapid changes in climate and vegetation are not entirely new to the planet, but some conditions have changed during the last 10,000 years that complicate the issue. The impact of humans on the landscape has been considerable by creating new assemblages of plants and animals through agricultural development and the application of pesticides, and by fragmenting the remaining natural ecosystems so that the movement of organisms has often been interrupted. In consequence, it has become difficult to determine whether changes in species ranges or ecosystem boundaries are truly a reflection of climate change or of some other cause [50].

One must also differentiate between shifts in biome boundaries as a result of climate change, and changes in the distribution pattern of individual species. Species change in their pattern in individual ways and do not all follow the movements of the biome in concert. Species, on the other hand, rarely move from one biome to another, and this has some important implications for any biome that contracts in its area as a result of climate change. The tundra biome, for example, whether in the polar regions or in alpine locations, is likely to contract as climate warms, but its component species are unlikely to change biome and so are in particular danger of extinction [51].

It is easy to assume that the responses of ecosystems to climate change will be gradual and relatively slow, but this may not be the case. It is possible for an ecosystem to reach a critical threshold and then to undergo rapid change [52]. This may result from some stochastic catastrophe that finally triggers a shift to a new state. For example, an outbreak of fire in a boreal forest, or a population eruption of an insect pest, such as the mountain pine beetle (*Dendroctonus ponderosae*) can cause massive changes in the ecosystem [53], and recovery may take decades or may even fail to occur if the ecosystem is close to its climatic limit. Ecosystem shifts such as forest devastation can raise the further problem of positive feedback as they lead to greater inputs of carbon into the atmosphere and enhance the climate warming process. As winters become warmer, outbreaks of defoliating insects will increase, as demonstrated by the European sawfly (*Neodiprion sertifer*) and the geometrid moth *Epirrita autumnata*, which infest boreal pine and birch forest, respectively. These insects overwinter as eggs, so warmer winters result in better egg survival and a higher likelihood of a devastating outbreak the following summer [54]. Thus the positive feedback spiral develops.

But ecosystems are composed of individual species of plants, animals, and microbes, and the impact of climate change on an ecosystem is likely to be seen in the changing composition of these species. As described in Chapter 12, communities of species do not usually migrate as units in response to climate change. Different species react in an individualistic manner and become assembled into new combinations as communities disintegrate and reform. Predicting the outcome of climate change on the living world must therefore concentrate on an analysis of individual species' requirements and reactions. This is not an easy task, especially because the prediction of future physical conditions is itself so tentative.

The simplest approach is to begin with a single species, perhaps even a single population of a species, and experiment on its responses to artificially induced changes. This type of work is essentially an exercise in applied physiology, but is a starting point for understanding biogeographical response processes. A simple experiment, for example, would be to grow a particular plant species under conditions of double the normal carbon dioxide levels and compare its growth with those achieved at current levels. This type of experiment has been carried out for many plant species, both C3 species and C4 species (see Chapter 2). Take a common C3 weed species such as *Chenopodium album* (fat hen) [55]. In a growth experiment under present levels of carbon dioxide in the atmosphere (360 p.p.m.), this plant was able to increase its weight by 0.22 g per day for every gram of plant tissue present over the course of its growth period. This type of measure is called relative growth rate and is a convenient way of measuring growth in terms that allow comparison between different-sized plants. When carbon dioxide is raised from 360 p.p.m. to 700 p.p.m., the relative growth rate of fat hen becomes 1.08 g per day for every gram of plant tissue present. This is a fairly typical figure for annual C3 plants and is explained by the fact that photosynthesis is limited by how fast the plant can accumulate carbon through its pores (stomata). In general, elevated carbon dioxide in the atmosphere means that a plant can fix carbon faster and hence grow more rapidly, as long as other factors, such as water supply and light, are not limiting. C4 species, however, are more efficient at collecting carbon dioxide at low atmospheric concentration, so they generally do not respond as strongly to raised levels. This means that we might hazard a prediction that C3 species will gain advantages over C4 species in a greenhouse world and extend their ranges into those currently occupied by C4 species. Perhaps the balance line of the two groups of species in the Great Plains region (see Fig. 2.22) will shift southward. On the other hand, C4 species tend to perform better than C3 at higher temperatures, so this may compensate to some extent. There is fossil evidence to suggest that changes in the proportions of C3 and C4 plants have occurred in the past. The isotopic composition of fossil herbivores in South America, for example, shows that they fed largely on C3 plants up to 7 mya, when conditions became warmer, increasingly seasonal, and arid. From then on C4 species formed a higher proportion of their diet, suggesting a shift in the proportions of grasses.

Some of our most important crop plants, especially in the temperate zone, are C3 and C4 grasses, including wheat, barley, and maize. Is it possible that the productivity of these plants will increase with more carbon dioxide and higher temperatures? Higher temperature will probably favor the C4 species, like maize and sugarcane, but the higher carbon dioxide, as we have seen, will be particularly advantageous to the C3 species, such as wheat and barley. At present, maize can be grown in Britain only as a silage crop for the production of cattle food. Grain maturation is poor in the relatively cool moist summers of the area because ripening requires 850 degree-days above a base temperature of 10°C (number of days above 10°C multiplied by the number of days experiencing these temperatures). If the average temperature were to rise by 3°C, only the extreme north of Scotland would be unsuitable for growing grain maize. In the United States, an increase in temperature of 3°C and an increase in precipitation of 8 cm (3 in.) would cause a shift in the grain belt to the northeast, around the Great Lakes.

Similar experimental information relating to temperature optima could also lead to certain predictions about future species ranges. Such is the likely case, for example, with the beech tree (*Fagus grandifolia*) in North America [56] (Fig. 14.5). Two possible scenarios are considered, but both show a northward shift in distribution pattern in response to higher temperature. At the other end of the process, the retreat of species from areas where they have formerly succeeded is already apparent in some locations. The red spruce (*Picea rubens*), for example, is declining over much of its range in the eastern United States and is believed to have been

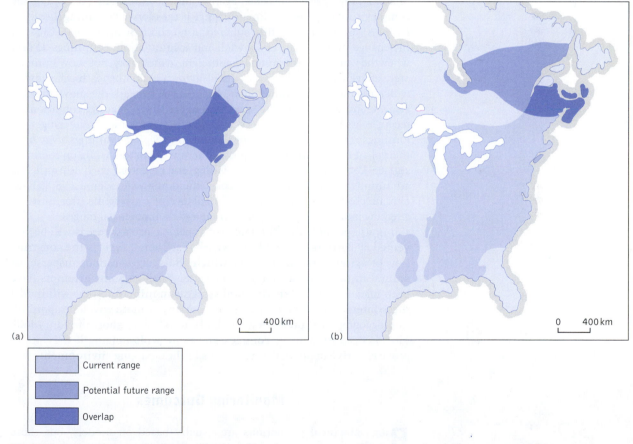

Fig. 14.5 Projected change in the distribution of beech (*Fagus grandifolia*) in North America under two greenhouse scenarios: (a) less severe and (b) more severe. From Roberts [56].

in decline since the early part of the 20th century. It has been proposed [57] that this change is a response to a general increase in both mean annual and summer temperatures.

As the oceans become warmer, cold-requiring organisms may be forced deeper, just as alpine species may be pushed to higher altitudes on land. In a study of the North Sea, the winter bottom temperature has risen by 1.6°C over the past 25 years, and, as a consequence, the fish communities have begun to occupy deeper waters. On average the depth requirements of fish have changed by 3.6 m (12 ft) [58]. Such depth changes among fish are more consistent than any observed latitudinal changes, probably because of changing currents in the oceans.

The argument that observed range changes in organisms is a response to climate change may sometimes be too simple. It assumes that plants and animals inhabit "climate envelopes," and that when the climate shifts the organism's distribution will alter accordingly. The real world is likely to be more complex because we have to take into account the responses of other species. Some species may prove either more or less competitive under the new set of conditions, thus modifying the expected response. A species may also be affected by the reaction to climate change of those other species on which it is dependent or with which it is associated, such as food plants, pollinators, prey species, dispersal agents, and parasites. An example of the complexity of response to climate change is the Norway lemming (*Lemmus lemmus*), which has recently suffered in northern Scandinavia because of the change in the nature of snowfall and accumulation. Lemmings do best when the warmth of the ground is sufficient to melt a layer of snow immediately above the ground surface, thus leaving a region beneath the snow in which the rodents can move about. In this zone beneath the snow, they are also well protected from the attentions of predators. Warmer conditions in recent years have resulted in a sequence of snow melt and refreezing, thus forming ice immediately over the ground and keeping lemmings away from their plant food sources and exposing them to predators while on the snow surface. Lemming populations are falling, and predators, such as Arctic fox, are turning to alternative prey species, including willow grouse [59]. Thus climate change can upset the entire balance of ecosystems, this, in turn, affecting individual species populations and their potential ranges.

It is also possible that the spread of a species into new areas will be prevented by barriers of some kind, whether artificial, such as the construction of agricultural land across which the species cannot disperse, or natural in the form of mountain or water barriers. Some attempts have been made to set up experimental systems in miniature that will model these interactions, and they confirm that any climate envelope approach to biogeographical predictions is likely to fall well short of reality [60]. Models need to be more complex and more wide ranging if they are to prove effective in predicting responses to the changing environment.

Monitoring Outcomes

Biogeographical predictions are usually based on models, but these models require raw information if they are to be more than sheer speculation. That information must be based on the records of changing distribution patterns of species on the ground, especially alterations in a species' range [61]. There have been numerous surveys of range changes in organisms, including plants, insects, birds, and mammals, over recent years, and in general there has been a very clear and detectable series of range alterations in response to warming climatic conditions that follow expectations based on known climatic tolerance limits for those species [62]. One problem, however, is that we cannot always be sure about the precise causes in such range changes. Is it simply a response to climate or to some other unseen factor (such as human impact in one form or another)? An example of a range change that seems closely correlated with climate is Edith's checkerspot butterfly (*Euphydryas editha*) from the western coastal area of North America. Camille Parmesan, from the University of California Santa Barbara [63], has studied the range and the fortunes of this butterfly and has found distinct patterns of range alteration even in the absence of any marked human impact. Given the climatic range of the species as a whole, one would expect that, as temperature increases, populations in the South and at low altitude would be more likely to become extinct than those in the North or at high altitude. This is precisely what has been found. Populations in Mexico are four times as likely to have become extinct than populations in Canada.

In the British Isles there have been several studies of butterfly distribution, but that area's high density of human population and intensive land use and habitat fragmentation make the data more difficult to interpret. In the case of the speckled wood butterfly (*Pararge aegeria*), for example, the observed northward extension of its range over the past 60 years may have been limited in part by the availability of suitable habitat [64]. This butterfly requires lightly shaded woodland, a habitat that is in short supply in, for example, the Highland mountain regions of Scotland. Its climatic potential range may not, therefore, become fully occupied until (if given the opportunity) forest habitats extend to higher latitudes and altitudes, and this may not take place if current sheep and deer-grazing intensities of the Highlands are maintained. The ability of a species to occupy its full potential climatic range will thus depend on its degree of specificity as regards habitat and food. The high brown fritillary (*Argynnis adippe*) is a specialist whose larvae depend on just two species of violets for their food, preferably under a cover of bracken. The comma butterfly (*Polygonia c-album*), on the other hand, is a generalist whose larvae are prepared to accept nettles, hops, elms, and willows as food plants. Thus, it is much more likely than the high brown fritillary to be able to take advantage of climatic warming and move northward in Europe, despite their similar climatic requirements [65].

A more complicated situation arises in the case of migratory organisms. A bird that breeds in the Northern Hemisphere but spends its non-breeding period in the Southern Hemisphere, for example, may be sensitive to environmental change in either of its residential areas or on the route in between. An example is the whitethroat (*Sylvia communis*), a small insectivorous warbler that breeds in Europe and spends its winters in West Africa, having crossed the Sahara Desert on its way. Regular recording of its abundance in breeding sites by the British Trust for Ornithology showed a collapse in populations in the late 1960s that continued until the

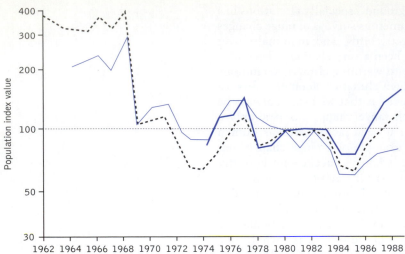

mid-1980s, after which it began to recover (Fig. 14.6). Perhaps its most hazardous activity is migrating through the Sahara on the way to and from its wintering ground, and examination of the rainfall records of the Sahel region of the southern Sahara (Fig. 14.7) shows a marked decline in rainfall through the late 1960s and some recovery in the late 1980s. The population has effectively stabilized at this level since that time. Although it is dangerous to argue from correlations, this does provide quite strong circumstantial evidence that the population decline was related to the climatic changes resulting in greater stress for the bird during its migrations.

Climatic warming has resulted in changing **phenology** among plants and animals. The term *phenology* covers such activities as growth, flowering, and fruiting in plants, and breeding cycles in animals. Warm conditions in early spring encourages an early start in plant growth and an early commencement in breeding among insects. As a consequence, the optimum timing of breeding for insectivorous birds has also moved to an earlier position in the year. In the case of the great tit (*Parus major*), for example, the time of breeding in southern Britain is now a full 14 days earlier than it was in the 1960s. The birds have responded to this in an intelligent manner and now hatch their young two weeks earlier to take advantage of the early flush in food availability. This adaptation has taken place quickly because it involves only behavioral change that is not genetically controlled. In the case of migratory birds, however, any adjustment to a changed climate involves more complex alterations in physiology and genetically controlled behavior. An example of this problem is afforded by the pied flycatcher (*Ficedula hypoleuca*), a small insectivorous bird of western European woodlands that spends its winters in West Africa [66]. Populations of the bird in the Netherlands have declined by 90% in the past 20 years, and its decline seems to be caused by its mistiming of the breeding season. The arrival of the migrants comes too late for the peak in caterpillar abundance, so breeding has a very low success rate. Because the migratory timing is genetically determined, it will take time for the selective pressures to operate on its modification.

Evolutionary adjustment can occur quite quickly, however, and in an unexpected manner, as exemplified by the blackcap (*Sylvia atricapilla*). This insectivorous migrant also spends its summers in western Europe but does not always migrate as far south as Africa, sometimes spending its winters in the Mediterranean area. In the past 50 years it has increasingly been found in southern Britain during the winter, and this was widely assumed to be a consequence of the warmer, relatively frost-free conditions associated with the warming climate. But this has proved to be only part of the story. Marking birds with rings or bands around their legs has enabled ornithologists to trace the origin of the blackcaps overwintering in Britain, and they have found that they come from continental Europe, mainly Germany [67]. It is probable that the behavior pattern began with what is normally a harmful change in migratory behavior, namely, a reverse migration in which the bird's orientation is confused and it flies in precisely the wrong direction. German birds arrived in Britain by mistake, but found that they were able to survive the winter in the newly improved conditions. In the spring they migrate back to Germany, but here they are at a great advantage: they arrive before the birds that have spent the winter further south and so are better able to occupy the best

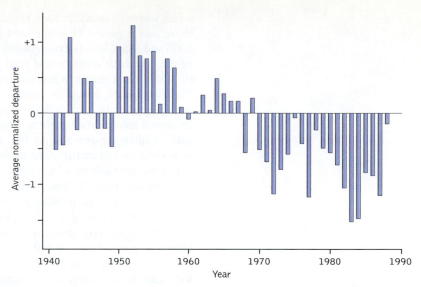

Fig. 14.7 Rainfall trends in the Sahel region of the southern Sahara expressed as a departure from the long-term mean (represented as 0). Note the general trend toward drought conditions, particularly in the late 1960s, resulting from the failure of monsoon rains to penetrate this region. Compare the pattern of this record with that of the whitethroat population in Fig. 14.6.

breeding sites and obtain an early supply of insect food. The population of blackcaps spending the winter in Britain is now increasing, demonstrating the selective advantages of this particular adaptation.

Climate may not always be the cause of such population changes in migratory organisms, however. Many North American migrant birds, for example, have declined in recent years. The golden-winged warbler is down by 46% over the past 25 years; the wood thrush is 40% down, and the orchard oriole has seen a 29% reduction. These changes have been related to an alteration in habitat management in their wintering grounds in Mexico, Central America, and the Caribbean. In the past, the traditional agriculture of these regions has been growing coffee under a canopy of trees, but high-yielding coffee varieties demand more open conditions, and the varied habitat formerly available has been much modified. Thus, the loss of migrants in this case may have more to do with human land use in their wintering areas than with climatic change, but it is also possible that breeding mistiming and other environmental impacts have played a part in the general decline. Conservationists need to ask biogeographical questions, such as precisely where the migrants from different populations spend their winter and what food supplies and other habitat requirements are available to them. In this respect, molecular studies have proved extremely valuable. Small samples of blood from the birds can provide information on stable isotopes of carbon that reveals much of their recent gastronomic history and is a simple and nondestructive way of answering many of these questions [68].

Even when observational data are sound and reliable and when range changes are clearly demonstrated from historical records, it is still difficult to be certain that such changes are induced by climate change. The identification of cause is often based on statistical correlation of events. One example is the recent advance of white spruce (*Picea glauca*) into the tundra regions around the eastern shore of Hudson Bay in Canada [69]. The annual growth rings of trees enables botanists to determine their age and the times in the past when they grew most rapidly, so a survey of the tree population of the timberline region provides clear documentation of when waves of invasion took place. Invasion began around AD 1660, but there was a period of suppressed growth and no recruitment between 1800 and 1850. Stunted trees began to take on a more robust growth form in the late 19th century and seed production increased. At the present time there is a vigorous recruitment of seedlings into the shrub tundra. The conclusion, based on correlations with what is known of past climate, is that the spruce has reacted to times of climatic warmth by increased growth and higher seed production. There is always a danger in such studies, however, that the argument can become circular. Tree growth rates and recruitment at timberlines has often been used as evidence on the basis of which climatic reconstructions have been carried out. One must then be cautious in reversing the argument and using that climatic data to explain plant population changes.

The simplest way of avoiding circular argument is to use climatic data that are not derived from proxy biological sources. Ideally, meteorological records can be used, but even in the absence of these, alternative sources of information may be available. The Sonoran Desert of Arizona, for example, does not have a long historical record of meteorological data, so an analysis of the population dynamics of cacti, such as the saguaro (*Carnegiea gigantea*), in relation to past climate presents many difficulties. Ecologists Taly Dawn Drezner and Robert C. Balling [70] have approached the problem by using the historical records of volcanic eruptions over the past century. Volcanic eruptions inject large and widespread volumes of dust into the stratosphere, which reduces the penetration of solar rays; the incidence, intensity, and duration of such events in recent times are well documented by climatologists. Studies on the growth rates of the cacti enabled the researchers to establish an age/height relationship and hence to examine the population age structure of the species in the Sonoran Desert. They found a strong positive correlation between the intensity of stratospheric dust and recruitment of saguaro cacti into the population. Cooler summer temperatures following volcanic eruptions would have reduced the likelihood of seedling desiccation and death.

The presence of foreign materials in the atmosphere, such as volcanic dust, can thus create climatic changes quite apart from the overall warming that results from the accumulation of greenhouse gases. These foreign materials include pollutants resulting from human activities. Aerosol pollution of this type reflects some solar radiation and needs to be included in climate models used for predictions. In a study of the western Amazonian rainforests, for example, the drought of 2005 has been linked to decreasing levels of aerosol pollution in the atmosphere [71]. It is ironic that lower atmospheric pollution levels can induce global environmental problems.

When we attempt to predict the impact of such global changes on whole communities of plants and animals, however, the most promising opportunities lie with the compilation of species data and combining them together. The lessons of palaeoenvironmental studies are that species operate as independent entities in response to environmental change, so if we want to predict, say, the impact of global change on the organisms and biodiversity of the Amazon tropical forest, then the most obvious method is to work from the bottom up and begin with the individual species. Biogeographers have, in fact, attempted this daunting task [72] by selecting

69 angiosperm species covering a range of life forms and functional types for which there was good information regarding range and ecological requirements. When subjected to modeled conditions of climatic change, these species reacted in various ways. The more generalist species were least affected in distributional terms, whereas those with narrow niches and strict habitat requirements were most severely affected. Those with slower reproductive processes or poorer dispersal potential showed little reaction simply because the set time frame of a century was inadequate to allow them to move and equilibrate to the new conditions. In conservation terms, the conclusion was relatively obvious: large areas of forest need to be conserved if biodiversity is to be retained in the long term.

Future research in this area needs not only to improve the model systems on which predictions are based, but also to improve knowledge in such areas as species taxonomy, distribution, phenology, physiology, and population dynamics. This is a very tall order given how little has been achieved so far even in the basic taxonomic field for many organisms, but since ecosystems assemble themselves from available species, it is the study of individual species that will prove critical for the effective operation of predictive models.

Population Declines and Extinctions

A decline in the number of individuals of any species can result in a decrease in its genetic diversity, especially if the decline involves the loss of whole populations in part of the species' range. Declining genetic diversity, in its turn, can lead to a lack of adaptability in the species, which may lead to a failure of that organism to cope with changing environmental conditions and can eventually culminate in extinction. The loss of a species, and even the loss of a particular genotype, contributes to a decline in biodiversity (see Chapter 4). It is now widely appreciated that the impact of humans on the Earth is resulting in rapid contraction in the range of many species, and this heralds a forthcoming rash of extinctions.

One of the problems that biogeographers face when trying to document these trends is the lack of information about so many species. As described in Chapter 4, we have not even recognized and described most species of invertebrate animals, let alone monitored their distribution ranges. So any study of range changes and extinction patterns must be confined to the groups of species that are reasonably well known, such as flowering plants, birds, and mammals. In one of the most thorough analyses of range changes, Gerardo Ceballos and Paul Ehrlich [73] selected a sample of mammals with well-known ranges and calculated their range contraction from historical information. They chose 173 species from all continents (apart from Antarctica), representing almost 4% of the 4,650 known mammals. Not surprisingly, their general finding was that loss of mammals was greatest where human populations were highest, especially in Southeast Asia and India, where local losses of 75 to 100% of mammal species occurred over 60% of the area. A combination of habitat loss and hunting to extinction seem to underlie most of the range contraction that was observed. In Africa, there was a slightly different trend; the highest losses were not associated with high human population density.

The greatest losses were in the Saharan region, where human population is relatively low, but even so the hunting intensity is high, especially for gazelles and larger herbivores. Competition with domestic herbivores for scarce forage is also important in these regions. The great apes are declining catastrophically as a consequence of bushmeat hunting in the equatorial forests [74], exacerbated by the spread of the deadly ebola virus. The overall conclusion reached by Ceballos and Ehrlich is that the 173 species analyzed have experienced a combined loss of 50% of their historic ranges, mainly as a direct consequence of human activities.

Even where direct human activity is low, such as in the Arctic, the dangers associated with climate change threaten many species, especially large mammals. The polar bear (*Ursus maritimus*) currently has a world population of about 25,000 [75]. Its hunting grounds for its main prey, seals, consists largely of the sea ice in the Arctic Ocean, and this is declining rapidly as the polar regions become warmer. The United States Geological Survey estimates that 42% of the pack ice will have been lost by the middle of this century, possibly resulting in a decline in the polar bear population by about two-thirds. The sea ice is already breaking up three weeks earlier than it did 30 years ago, and the body mass of bears is declining dramatically. Large mammals are thus in a particularly vulnerable position in our changing world, but the attention given to mammals far exceeds that expended on many other organisms. Global surveys indicate that birds, reptiles, amphibians, fish, invertebrates, and plants are all facing similar threats of widespread extinction [76].

Direct human predation on wild animals is perhaps most severely concentrated on fish. Throughout human history wild fish have comprised an important source of food, and this has been maintained even in cultures highly dependent on domesticated stock and arable agriculture. Fish generally cope with high intensities of predation by producing large numbers of young that replace those harvested, but the increasing efficiency of techniques for detecting and removing fish from the oceans has led to harvesting exceeding replacement in many species. The Atlantic cod (*Gadus morrhua*) is an example of a species that has been overharvested. The Grand Banks off Newfoundland supported an annual harvest of 800,000 tons of fish in 1968, but by 1992 it was closed because of the total depletion of stock. A similar process is now taking place in the North Sea off Europe. As the large individuals become scarcer, increasingly small fish are taken until there are no fish reaching sufficient maturity to breed and the stock collapses. Industrial fishing can reduce the biomass of fish by 80% within 15 years of exploitation of a stock [77]. Climatic fluctuations have also contributed to the decline in some species, such as salmon, anchovies, and sardines in the Pacific [78]. One of the main problems in fish exploitation is that most of the commercially important sea fish are predatory, so that we are harvesting high in the food chain (Fig. 14.8). This is energetically very inefficient—the equivalent of eating foxes rather than rabbits.

One consequence of population declines and range contraction as a result of local population extinction is that the overall genetic variation within a species can be reduced, leaving it ill equipped to cope with future environmental pressures. Species that have experienced population crashes, reducing numbers to just a few individuals, are said to have passed through a **genetic bottleneck**, and this can often be identified by examining the degree

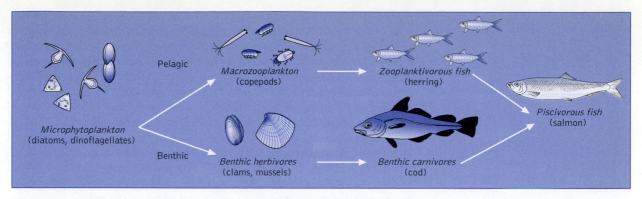

Fig. 14.8 Food web of the salmon, a marine predatory fish. Harvesting an animal so high in the food chain is energetically inefficient.

of genetic variation left in the recovering population. The European natter-jack toad (*Bufo calamita*) has been tested in this way, using molecular methods for examining the residual genetic variability in individual toads [79]. Local populations that are known to have been reduced to very low levels in the past did indeed show signs of genetic impoverishment that could prove disadvantageous in the future evolution of the species. In the Californian sea lion, for example, an increasing degree of sickness and pathogen susceptibility in wild populations has been linked to inbreeding resulting from population reduction and loss of genetic diversity [80]. Species may be saved from extinction by such remedial programs as captive breeding from a limited number of individuals, but the resulting gene pool may be severely depleted and the species weakened as a consequence. The experience of captive breeding of endangered birds, such as the peregrine falcon, and its subsequent release in the midwestern United States, confirms that breeding stock should come from a wide geographical and genetic base. Seven subspecies of the falcon from four continents were used in the breeding program, and the success of the released birds has confirmed the value of maintaining a diverse genetic stock [81]. The genetic constitution of a species has often taken many millions of years to become assembled and to be sorted by the process of natural selection to provide a high degree of fitness to the organism.

An important question for biogeographers is how to recognize species that are at risk of extinction and then how to prevent their loss. Are there particular characters or factors associated with extinction risk? Several research projects have looked for correlations; as yet there are no clear indications, but some factors seem to indicate risk in certain groups of animals and plants. Size has already been mentioned in relation to mammals. There is both historical and current evidence that large mammals have been more subject to decline and extinction than small ones. Perhaps small mammals, such as voles, mice, and moles, are able to escape extreme conditions (such as frost and fire) by burrowing underground and thus surviving to breed again. Being small means that they can maintain higher population levels per unit ground area, and their reproductive rates are also generally faster, so recovery from population collapse can quickly be established. Large mammals are thus in need of particular conservation attention.

Do the same criteria apply to plants? A study of tropical angiosperms suggests not [82]. Size does not seem to be associated with extinction risk in this group, but certain types of specialism may prove fatal. Epiphytes, plants dependent on mammal pollinators, and plants associated with deep forest environments are all at greater risk. Some of these also have restricted geographical distribution; this can leave an organism in a vulnerable position, as has been shown by a study of frogs at a global level [83]. Species with limited ranges were clearly at greater risk of extinction. A complication in this study, however, was the correlation between larger body size and poorer fecundity with restricted range. Is it the low breeding success or the restricted range that leaves the animal more liable to extinction? Correlations always have to be interpreted with great care before precise conclusions about cause and effect can be made.

The loss of a species is essentially irreparable, and even the loss of a distinctive population with its own genetic peculiarities is inevitably going to impoverish not only the species, but also the biodiversity of the planet. But ecologists and biogeographers must argue their case for biodiversity conservation in the light of growing human populations and hence scarcity of resources, especially land, water, food, and energy. Balancing the needs of people against the requirement to retain biodiversity is one of the greatest dilemmas facing humankind.

Conservation Biogeography

It is the responsibility of biogeographers not only to predict likely changes to the biota of the planet, but also to recommend ways to reduce the rate of species loss. The discipline has progressed considerably over the past few decades and is now much closer to understanding the multitude of factors that determine plant and animal distribution patterns and the ways in which species interact to form communities and ecosystems. Historical studies have enabled us to observe the responses of living things to climatic and other environmental changes in the past, and ecological and physiological research is providing a clearer picture of how organisms react individually to change. But even these detailed and extensive types of study cannot provide all that is needed for future prediction and remediation. The current circumstances of the Earth are quite unlike anything that has been previously experienced, so historical studies have limited application. The abundance of industrial humanity on the planet has created a new global landscape and with it new biogeographical problems.

Reduction in the physical and chemical pollution of atmosphere, waters, and soils is a subject that has relevance not only in biogeography but also in areas of human health and agriculture, so it will receive continued global attention despite the constant constraints of economics and politics. The human modification of landscape, however, is a topic where biogeographers have specific interest and expertise and can apply these to remedial activities [84]. Whatever action is taken now to reduce pollution and its effects on climate change, there is no doubt that climatic conditions will continue to change over the coming century. Plants and animals will need to adjust to these changes either by genetic adaptation or by changes in their ranges, or both.

Genetic adaptation, involving either physiology, phenology, or behavior, can be quite rapid, as in the case of the blackcap mentioned earlier in this chapter. Much will depend, however, on the degree of genetic variability present within populations of each organism and how effectively new genetic combinations can spread between populations. In the case of plants in the Arctic, for example, there is a high level of genetic variability within the populations of many species. Many, such as the purple saxifrage (*Saxifraga oppositifolia*), are present in a range of different growth forms, or, as in the case of the mountain avens (*Dryas octopetala*), have several physiological ecotypes with differing drought resistance [85]. Such genetically diverse species have a high likelihood of survival through the coming time of warmth. Bob Crawford of the University of St. Andrews, Scotland, has made an extensive study of plants at the margins of habitats and the edges of biomes, and concludes that in these areas of constant instability there may be a high degree of residual genetic variability and hence the capacity to cope with climatic and other changes [86].

Wild organisms that are hunted or harvested by human populations for food or other purposes are particularly at risk of genetic depletion. In the Himalayas of northern India, *Nardostachys grandiflora* (Valerianaceae) is a perennial herb that is widely exploited for its medicinal properties and also as a source of incense [87]. It grows in clumps, extending vegetatively by means of rhizomes, and it is these organs that are harvested. It is strongly dependent on vegetative spread for maintaining its populations, so harvesting can lead to local extinction, especially in rocky habitats where its flowering and seed production is very low. It is more resilient in meadow locations. The loss of local populations in the rocky parts of its range, coupled with its low fecundity in such situations, inevitably leads to reduced genetic variability. Conservation regimes must concentrate on directing harvesters to the more tolerant meadow populations and protecting the more sensitive highland sites. Exploited species of this kind clearly need careful management and protection if they are to avoid extinction.

The survival of pockets of genetic variability is not enough for the avoidance of extinction, however, unless those genetic features that improve fitness can be shared between individuals and populations. This requires genetic connectivity, which itself is dependent on habitat connectivity coupled with an efficient dispersal capacity. The most obvious way of ensuring these features is by the active conservation of large areas of a habitat in which there are no barriers to gene flow. This must always be the biogeographer's ideal, but the demand on land for other purposes, such as agriculture, road development, or urban growth, means that the ideal is not always available. In conditions where fragmentation is inevitable, fragments should be kept as large as possible and connectivity should be assisted by the establishment of corridors, as discussed earlier in this chapter. In this way the flow of genes can be encouraged, and species can be assisted in their adaptation to new conditions.

Habitat connectivity in the future world is required not only for gene flow, but for the physical spread and reassembly of plants and animals, and hence alterations in range in response to climatic and other environmental changes. The effectiveness of corridors and other aids to dispersal will vary from one species to another, and the design of a landscape, ideally, needs to take into account the dispersal capacity of the target species

[88]. Some species, such as flying insects and birds, have high dispersal capacities, but others, even apparently mobile mammals, may find that a fragmented landscape presents too many barriers for movement. In the case of hedgehogs, for example, even roads can limit movements and hence the exchange of genes and the extension of range [89]. Mobility and dispersal capacity, coupled with adaptability and opportunism, will undoubtedly prove the most valuable of all assets for survival into the future. There are so many uncertainties, both in the prediction of future conditions and in the understanding of the adaptability of organisms, that models predicting the future biogeographic patterns of the planet are generally inconclusive [90]. But some of the more pessimistic conclusions have claimed that the rates of spread organisms will need to ensure that survival may far exceed their known capacity [91].

Innumerable species may require intensive human assistance to ensure survival, and some will survive only if taken into captive care, either in zoos or botanical gardens. In the case of plants, there is the option of long-term survival in a dormant state, especially as seeds. Britain's Kew Gardens has established a seed bank project on the Arctic island of Spitzbergen in the Svalbard archipelago. In an underground vault within the permafrost, the seeds of thousands of plants are being accommodated as a final resort in long-term conservation [92]. Domestic species and their varieties take precedence (including 100,000 varieties of rice), but the project will eventually also contain over 24,000 wild plant species, stored for centuries at a temperature of $-18°C$. It is possible that this approach may be extended in future to cover endangered livestock breeds and wild mammals by storing frozen eggs, sperm, or embryos.

Although informed conservation requires much more information about the individual species, including their physiology, climatic requirements, breeding strategy, and interactions with other species, the current speed of global change requires urgent action even before such scientific data can be accumulated. The designation of representative areas of natural and semi-natural habitats as protected and managed reserves is essential if a cascade of extinctions is to be avoided. Regions regarded as biodiversity hotspots, in particular but not exclusively the tropical forests, must form a focal point for global conservation effort. At a time when all nations and most politicians are becoming concerned about global climate change, it is salutary to note that the tropical forests are still acting as a sink for atmospheric carbon and are therefore helping to reduce the accumulation of greenhouse gases. For a long while it has been assumed that mature forests are likely to be carbon neutral, having achieved the maximum biomass possible within the climatic conditions prevailing. It is reasonable to suppose that in regions of abandoned farmland, now reverting to forest, as in some areas of the former Soviet Union, there will be an accumulation of carbon as the biomass rapidly grows [93]. This is demonstrably the case, but, more surprisingly, studies in the African tropical forests [94] and the temperate old-growth forests of the Northern Hemisphere are also proving these ecosystems to be carbon sinks [95]. Thus the conservation of large areas of such habitats will assist not only in the maintenance of global biodiversity, but also in a program of climate control.

As we have seen in this book, the study of biogeography involves a consideration of the past (historical biogeography) and the present requirements of species (ecological biogeography). In combination, these approaches provide us with a better understanding of why plants and animals are distributed in their present patterns. They also present us with a basis for predicting the biogeographical outcome of future change, together, we hope, with the ability and the wisdom to modify the future and to conserve this planet's rich biodiversity.

The future undoubtedly holds changes for the Earth's biosphere, only some of which are predictable. Perhaps it is the unpredictable ones that are most to be feared. The changes in climate and in the physical and chemical conditions of our planet will result in a modification of its biogeography, inevitably involving the extinction of many species. The study of the requirements and the interactions of plants and animals, including their past distribution patterns and evolution, provides an important means of understanding the complexity of nature and of our own role within it. The story of our planet has been one of constant change and adjustment, and the biosphere has proved remarkably resilient to all the changes so far experienced. It may seem complacent to have faith in the ability of the natural world to cope with any forthcoming stresses, including those directed at it by our own species, but the probability is that this is the case. The most important question facing humanity, however, is whether we are equally resilient, and whether we are equally able to adapt socially and technologically to changing conditions fast enough to permit the long-term survival of Homo sapiens, *especially at current population levels. Nature, in some form, will undoubtedly survive the next few centuries, but our own future may prove more questionable and may well involve population contraction. Such a future raises considerable social and political issues, but the biogeographer will also have a part to play in the continuing history of our species as well as that of the other species with which we share the planet.*

Summary

1 Taking the past and the present as the key to the future, it should be possible to turn biogeography into a predictive science and to make projections concerning the future of the Earth. The current situation, however, is unlike any experienced in the past because of the extensive impact of humans on the planet, so predictions based on history may prove unreliable.

2 Predictions about future climate and nutrient cycling changes, which will vary both in response to external forces, such as solar irradiance, and as a consequence of less predictable aspects of human activity, remain vague, which restricts the scope of biogeographical projections.

3 Ecosystems are built from individual species, so the species needs to be a focus for the development of predictive models.

4 Knowledge of the physiological requirements of individual species provides a basis for prediction in biogeography, but it does not take into account the reactions of other species, nor the problems of dispersal to new sites.

5 Monitoring changes in species ranges is critical for the establishment of more reliable predictive models.

6 Population decline and genetic impoverishment are often a prelude to local or widespread extinction. Conservation biogeography needs to concentrate on maintaining connectivity in landscapes that encourages the survival of populations and their genetic communication.

Further Reading

Committee on Abrupt Climate Change. *Abrupt Climate Change: Inevitable Surprises*. Washington, DC: National Academy Press, 2002.

Cowie J. *Climate Change: Biological and Human Aspects*. Cambridge: Cambridge University Press, 2007.

Crawford RMM. *Plants at the Margin: Ecological Limits and Climate Change.* Cambridge: Cambridge University Press, 2008.

Cuff DJ, Goudie AS. *The Oxford Companion to Global Change.* Oxford: Oxford University Press, 2009.

Houghton J. *Global Warming: The Complete Briefing.* 3rd ed. Cambridge: Cambridge University Press, 2004.

Oldfield F. *Environmental Change: Key Issues and Alternative Approaches.* Cambridge: Cambridge University Press, 2005.

Spellerberg IF. *Monitoring Ecological Change.* 2nd ed. Cambridge: Cambridge University Press, 2005.

References

1 Cohen, JE. Population, economics, environment and culture: an introduction to human carrying capacity. *Journal of Applied Ecology* 1997: 34, 1325–1333.

2 Houghton J. *Global Warming: The Complete Briefing.* 3rd ed. Cambridge: Cambridge University Press, 2004.

3 Vitousek PM, Ehrlich PR, Ehrlich AH, Matson PA. Human appropriation of the products of photosynthesis. *Bioscience* 1986: 36, 368–373.

4 Postel SL, Daily GC, Ehrlich PR. Human appropriation of renewable fresh water. *Science* 1996: 271, 785–787.

5 Gleick PH. Soft water paths. *Nature* 2002: 418, 373.

6 Hightower M, Pierce SA. The energy challenge. *Nature* 2008: 452, 285–286.

7 Lutz W, Sanderson W, Scherbov S. Doubling of world population unlikely. *Nature* 1997: 387, 803–805.

8 Woolhouse EJ. Emerging diseases go global. *Nature* 2008: 451, 898–899.

9 Moore PD, Chaloner B, Stott P. *Global Environmental Change.* Oxford: Blackwell Science, 1996.

10 Mulholland PJ, et al. Stream denitrification across biomes and its response to anthropogenic nitrate loading. *Nature* 2008: 452, 202–205.

11 Magnani F, et al. The human footprint in the carbon cycle of temperate and boreal forests. *Nature* 2007: 447, 848–850.

12 Davidson EA, et al. Recuperation of nitrogen cycling in Amazonian forests following agricultural abandonment. *Nature* 2007: 995–998.

13 Robinson CH, Wookey PA, Callaghan TV, Press MC. Plant community responses to simulated environmental change at a high arctic polar semi-desert. *Ecology* 1998: 79, 856–866.

14 Lee JA. Unintentional experiments with terrestrial ecosystems: ecological effects of sulphur and nitrogen pollutants. *J Ecol* 1998; 86: 1–12.

15 Wookey PA, Robinson CH, Parsons AJ, Walker JM, Press MC, Callaghan, TV, Lee JA. Experimental constraints on the growth, photosynthesis and reproductive development of *Dryas octopetala* to simulated environmental change in a high arctic polar semi-desert. *Oecologia* 1995; 102: 478–489.

16 Cislaghi C, Nimis PL. Lichens, air pollution and lung cancer. *Nature* 1997; 387: 463–464.

17 Chappelka AH, Samuelson LJ. Ambient ozone effects on forest trees of the eastern United States: a review. *New Phytologist* 1998: 139, 91–108.

18 Nriagu, JO. A history of global metal pollution. *Science* 1996: 272, 223–224.

19 Holl, KD. Long-term vegetation recovery on reclaimed coal surface mines in the eastern U.S.A. *J Applied Ecology* 2002: 39, 960–970.

20 Sampat, P. Scrapping mining dependence. In: Starke L, ed. *State of the World 2003: Progress towards a Sustainable Society*, pp. 110–129. London: Earthscan, 2003.

21 Green RE, Taggart MA, Devojit D, Pain DJ, Kumar CS, Cunningham AA, Cuthbert R. Collapse of Asian vulture populations: risk of mortality from residues of the veterinary drug diclofenac in cacasses of treated cattle. *J Applied Ecology* 2006: 43, 949–956.

22 Hayes T, Haston K, Tsui M, Hoang A, Haeffele C, Vonk A. Feminization of male frogs in the wild. *Nature* 2002: 419, 895–896.

23 Kiesecker JM, Blaustein AR, Belden LK. Complex causes of amphibian population declines. *Nature* 2001: 410, 681–684.

24 Araujo MB, Tuiller W, Pearson RG. Climate warming and the decline of amphibians and reptiles in Europe. *J Biogeography* 2006: 33, 1712–1728.

25 Guyot G. *Physics of the Environment and Climate.* Chichester: Wiley, 1998.

26 Kalnay E, Cai M. Impact of urbanization and land-use change on climate. *Nature* 2003: 423, 528–531.

27 Baughman M., ed. *Reference Atlas to the Birds of North America.* Washington, DC: National Geographic Society, 2003.

28 Habib L, Bayne EM, Boutin S. Chronic industrial noise affect pairing success and age structure of ovenbirds Seiurus aurocapilla. *J Applied Ecology* 2007: 44, 176–184.

29 Slabbekoorn H, Peet M. Birds sing at a higher pitch in urban noise. *Nature* 2003: 424, 267.

30 Whittaker RJ, Fernández-Pacios JM. *Island Biogeography: Ecology, Evolution and Conservation.* 2nd ed. Oxford: Oxford University Press, 2007.

31 Linkie M, Chapron G, Martyr DJ, Holden J, Leader-Williams N. Assessing the viability of tiger subpopulations

in a fragmented landscape. *J Applied Ecology* 2006: 43, 576–586.

32 Saccheri I, Kuussaari M, Kankare M, Vikman P, Forelius W, Hanski I. Inbreeding and extinction in a butterfly metapopulation. *Nature* 1998: 392, 491–494.

33 Strasburg JL. Roads and genetic connectivity. *Nature* 2006: 440, 875–876.

34 Epps CW, Wehausen JD, Bleich VC, Torres SG, Brashares JS. Optimizing dispersal and corridor models using landscape genetics. *J Applied Ecology* 2007: 44, 714–724.

35 Fischer J, Lindenmayer DB, Montague-Drake R. The role of landscape texture in conservation biogeography: a case study on birds in south-eastern Australia. *Diversity and Distributions* 2008: 14, 38–46.

36 Cowie J. *Climate Change: Biological and Human Aspects*. Cambridge: Cambridge University Press, 2007.

37 Lean J, Beer L, Bradley RS. Reconstruction of solar irradiance since A.D. 1600: implications for climate change. *Geophys Res Lett* 1995.

38 Willett KM, Gillett NP, Jones PD, Thorne PW. Attribution of observed surface humidity changes to human influence. *Nature* 2007: 449, 710–712.

39 Zhang X, Zwiers FW, Hegerl GC, Lambert FH, Gillett NP, Solomon S, Stott PA, Nozawa T. Detection of human influence on twentieth-century precipitation trends. *Nature* 2007: 448, 461–465.

40 Hopkin M. Climate takes aim. *Nature* 2007: 446, 706–707.

41 Lioubimtseva E. Climate change in arid environments: revisiting the past to understand the future. *Progress in Physical Geography* 2004: 502–530.

42 Webster PJ, Palmer TN. The past and the future of El Niño. *Nature* 1997; 390: 562–564.

43 Raper SCB, Braithwaite RJ. Low sea level rise projections from mountain glaciers and icecaps under global warming. *Nature* 2006: 439, 311–313.

44 Clark PU, Pisias N, Stocker TF, Weaver AJ. The role of the thermohaline circulation in abrupt climate change. *Nature* 2002: 415, 863–869.

45 Quadfasel D. The Atlantic heat conveyor slows. *Nature* 2005: 438, 565–566.

46 Noren AJ, Bierman PR, Steig EJ, Lini A, Southon J. Millennia-scale storminess variability in the northeastern United States during the Holocene epoch. *Nature* 2002: 419, 821–824.

47 Tuiller W. Climate change and the ecologist. *Nature* 2007: 448, 550–552.

48 Moore PD. Back to the future: biogeographical responses to climate change. *Progress in Phys. Geogr.* 2003: 27, 122–129.

49 Tinner W, Kaltenrieder P. Rapid responses of high-mountain vegetation to early Holocene environmental changes in the Swiss Alps. *Journal of Ecology* 2005: 93, 936–947.

50 Rosenzweig C. et al. Attributing physical and biological impacts to anthropogenic climate change. *Nature* 2008: 453, 353–357.

51 Crisp MD, et al. Phylogenetic biome conservatism on a global scale. *Nature* 2009: 458, 754–756.

52 Scheffer M, Carpenter S, Foley JA, Folkes C, Walker B. Catastrophic shifts in ecosystems. *Nature* 2001: 413, 591–596.

53 Kurz WA, Dymond CC, Stinson G, Rampley GJ, Neilson ET, Carroll AL, Ebata T, Sanfranyik L. Mountain pine beetle and forest carbon feedback to climate change. *Nature* 2008: 452, 987–990.

54 Neuvonen S, Niemela P, Virtanen T. Climatic change and insect outbreaks in boreal forests: the role of winter temperatures. *Ecological Bulletins* 1999: 47, 63–67.

55 Bunce JA. Variation in growth stimulation by elevated carbon dioxide in seedlings of some C3 crop and weed species. *Global Change Biol* 1997: 3, 61–66.

56 Roberts L. How fast can trees migrate? *Science* 1989: 243, 735–737.

57 Hamburg SP, Cogbill CV. Historical decline of red spruce populations and climatic warming. *Nature* 1988: 331, 428–431.

58 Dulvy NK, Rogers SI, Jennings S, Stelzenmuller V, Dye SR, Skjoldal HR. Climate change and deepening of the North Sea fish assemblage: a biotic indicator of warming seas. *J Applied Ecology* 2008: 45, 1029–1039.

59 Coulson T, Malo A. Case of the absent lemmings. *Nature* 2008: 456, 43–44.

60 Davis AJ, Jenkinson LS, Lawton JH, Shorrocks B, Wood S. Making mistakes when predicting shifts in species range in response to global warming. *Nature* 1998: 391, 783–786.

61 Spellerberg IF. *Monitoring Ecological Change*. Cambridge: Cambridge University Press, 2005.

62 Walther G-R, Post E, Convey P, Mensel A, Parmesan C, Beebee TJC, Fromentin J-M, Hoegh-Guldberg O, Bairlein F. Ecological responses to recent climate change. *Nature* 2002: 416, 389–395.

63 Parmesan C. Climate and species' range. *Nature* 1996: 382, 765–766.

64 Hill JK, Thomas CD, Huntley B. Climate and habitat availability determine 20th century changes in a butterfly's range margin. *Proc. R. Soc. Lond.* B 1999: 266, 1197–1206.

65 Hill JK, Fox R. Climate change and British butterfly distributions. *Biologist* 2003: 50, 106–110.

66 Both C, Bouwhuis S, Lessellls CM, Visser ME. Climate change and population declines in a long-distance migratory bird. *Nature* 2006: 441, 81–83.

67 Berthold P, Helbig AJ, Mohr G, Querner U. Rapid microevolution of migratory behaviour in a wild bird species. *Nature* 1992: 360, 668–670.

68 Levy S. Atomic detectives. *Nature* 2006: 442, 504–506.

69 Caccianiga M, Payette S. Recent advance of white spruce (*Picea glauca*) in the coastal tundra of the eastern shore of Hudson Bay (Quebec, Canada). *J Biogeography* 2006: 33, 2120–2135.

70 Drezner TD, Balling RC. Regeneration cycles of the keystone species *Carnegiea gigantea* are linked to worldwide volcanism. *J Vegetation Science* 2008: 19, 587–596.

71 Cox PM, Harris PP, Huntingford C, Betts RA, Collins M, Jones CD, Jupp TE, Marento JA, Nobre CA. Increasing risk of

Amazonian drought due to decreasing aerosol pollution. *Nature* 2008: 453, 212–214.

72 Miles L, Grainger A, Phillips O. The impact of global climate change on tropical forest biodiversity in Amazonia. *Global Ecology and Biogeography* 2004: 13, 553–565.

73 Ceballos, G. & Ehrlich, P.R. Mammal population losses and the extinction crisis. *Science* 2002: 296, 904–907.

74 Walsh, PD, et al. Catastrophic ape decline in western equatorial Africa. *Nature* 2003: 422, 611–614.

75 Courtland R. Polar bear numbers set to fall. *Nature* 2008: 453, 432–433.

76 Mackay R. *The Atlas of Endangered Species*. London: Earthscan, 2002.

77 Myers RA, Worm B. Rapid worldwide depletion of predatory fish communities. *Nature* 2003: 423, 280–282.

78 Finney BP, Gregory-Evans I, Douglas MSV, Smol JP. Fisheries productivity in the northeastern Pacific Ocean over the past 2,200 years. *Nature* 2002: 729–733.

79 Beebee T, Rowe G. Application of genetic bottleneck testing to the investigation of amphibian declines: a case study with Natterjack toads. *Conservation Biology* 2001: 15, 266–270.

80 Acevedo-Whitehouse K, Gulland F, Greig D, Amos W. Disease susceptibility in California sea lions. *Nature* 2003: 422, 35.

81 Tordoff HB, Redig PT. Role of genetic background in the success of reintroduced peregrine falcons. *Conservation Biology* 2001: 15, 528–532.

82 Sodhi N. et al. Correlates of extinction proneness in tropical angiosperms. *Diversity and Distributions* 2008: 14, 1–10.

83 Cooper N, Bielby J, Thomas GH, Purvis A. Macroecology and extinction risk correlates of frogs. *Global Ecology and Biogeography* 2008: 17, 211–221.

84 Schwartz MW, ed. *Conservation in Highly Fragmented Landscapes*. New York: Chapman and Hall, 1997.

85 Crawford RMM, Abbott RJ. Pre-adaptation of arctic plants to climate change. *Botanica Acta* 1994: 107, 271–278.

86 Crawford RMM. *Plants at the Margin: Ecological Limits and Climate Change*. Cambridge: Cambridge University Press, 2008.

87 Ghimire SK, Gimenez O, Pradel R, McKey D, Aumeeruddy-Thomas Y. Demographic variation and population viability in a threatened Himalayan medicinal and aromatic herb *Nardostachys grandiflora*: matrix modelling of harvesting effects in two contrasting habitats. *Journal of Applied Ecology* 2008: 45, 41–51.

88 Bullock JM, Kenward RE, Hails RS, eds. *Dispersal Ecology*. Oxford: Blackwell Publishing, 2002.

89 Rondinini C, Doncaster CP. Roads as barriers to movement of hedgehogs. *Functional Ecology* 2002: 16, 504–509.

90 Higgins SI, et al. Forecasting plant migration rates: managing uncertainty for risk assessment. *J of Ecology* 2003: 91, 341–347.

91 Malcolm JR, Markham A, Neilson RP, Garaci M. Estimated migration rates under scenarios of global climate change. *J of Biogeography* 2002: 29; 835–849.

92 Hopkin M. Frozen futures. *Nature* 2008: 452, 404–405.

93 Henebry GM. Carbon in idle croplands. *Nature* 2009: 457, 1089–1090.

94 Lewis SL, et al. Increasing carbon storage in intact African tropical forests. *Nature* 2009: 457, 1003–1006.

95 Luyssaert S, Schulze E-D, Boerner A, Knohl A, Hessenmoeller D, Law BE, Ciais P, Grace J. Old-growth forests as global carbon sinks. *Nature* 2008: 456, 213–215.

Glossary

Abyssal plain The deep ocean floor, which lies between the continental shelves (q.v.).

Abyssal zone The marine life zone above the abyssal plain (q.v.).

Adaptive radiation The evolutionary radiation of a group, based on a novel set of characteristics, that allow it to adapt to a wide range of ways of life.

Albedo An index of the extent to which incoming radiation is reflected, rather than absorbed.

Allele One of sometimes several versions of a gene (q.v.), located at a single position on the chromosome.

Allopatric speciation The evolution of a new species in isolation from its parent species.

Alpha diversity The biodiversity (usually measured as species richness) in a local area.

Amphitropical distribution See *bipolar distribution*.

Angiosperms Flowering plants.

Antitropical distribution See *bipolar distribution*.

Apomorphic The derived state of a characteristic.

Arborescent Tree-like.

Area biogeography An approach to biogeography that begins with the identification of areas of endemism.

Area of endemicity Area within which one or more taxa are found exclusively. Also known as an area of endemism.

Arid Extremely dry, with an annual precipitation of less than 10 cm.

Assembly rules The hypothesis that some species are only found on islands where another particular species is absent, or in islands containing a larger total assemblage of species. More generally in ecology, the principles that determine the aggregation of various species to make up a community.

Australasia The Australian continent, plus the islands on the Australian continental plate—New Guinea, Tasmania, Timor, and New Caledonia, and other smaller islands.

Barrier An obstruction to the passage of organisms.

Bathyal zone The marine life zone above the continental slope (q.v.); also known as the archibenthal zone.

Bathypelagic zone Zone of complete darkness in the sea, from below the mesopelagic zone (q.v.) to a depth of 6,000 m (20,000 ft).

Benthic Organisms that live on the seafloor.

Beta diversity The rate or amount of change in species composition between local areas. Also known as turnover or differentiation diversity, beta diversity translates the number of species in local areas (alpha diversity) into the number of species in larger regions (gamma diversity). Low beta diversity means similar sets of species between patches or local areas, such that the gamma diversity will not be much higher than the alpha diversity.

Biodiversity A census of the diversity of species in different parts of the planet and in the whole of the planet. It may include the genetic variation within species.

Biogeography Study of the distribution, and of the patterns of distribution, of living organisms at all levels, ranging from genes to whole organisms and biomes, and of the evolution of these.

Biological species concept See *species*.

Biome A large-scale ecosystem, such as desert or tundra, found in different parts of the world and characterized by a similar life forms of animals and plants.

Biota The total of all organisms that inhabit a given region, that is, its fauna plus its flora.

Bipolar distribution Distribution in which related organisms are found in temperate or polar environments on either side of the equatorial region, but not in the equatorial region itself. Also known as antitropical or amphitropical distribution.

Bloom Rapid, seasonal increase in the mass of marine phytoplankton.

Boreal Found throughout the Northern Hemisphere; usually referring to plant distributions in the cool temperate zones of the north.

Boreotropical Confined to the tropics of the Northern Hemisphere.

Bottom water A mass of cold, dense, saline water that sinks down to the ocean floor along the eastern side of Greenland and near the Antarctic Peninsula.

Brooks parsimony analysis or **BPA** A technique of pattern-based historical biogeographical analysis (q.v.).

Buffon's Law The observation that similar environments, in different parts of the world, contain different groupings of organisms.

C$_3$ plants Those plants in which the first product of photosynthesis is a sugar containing three carbon atoms; this is most common form of photosynthesis.

C$_4$ plants Those plants in which the first product of photosynthesis is a sugar containing four carbon atoms; this is most advantageous in high light intensity and temperature.

Carrying capacity The number of species, or the biomass, or the highest population level of a species, that can be supported by a given area of a particular habitat or environment.

Chromosome The thread-like body within the nucleus of a cell that carries its DNA (q.v.).

Circumboreal A pattern of distribution that stretches around the northern regions.

Clade A group of taxa that includes their common ancestor and all its descendants.

Cladistic biogeography The analysis of patterns of endemism using cladistics (q.v.). Also known as vicariance biogeography.

Cladistics The system of analysis of the evolutionary relationships of taxa that views this as a series of dichotomously branching events, using the presence of shared derived characters to identify the location of each branch.

Cladogram A diagram portraying the results of a cladistic analysis.

Clements approach An interpretation of plant communities that suggests that they behave as integrated units, rather like an individual organism. See also *Gleason approach*.

Climatic envelope The sum of all the climatic variables that limit the distribution of a species or a biome (q.v.).

Community An assemblage of different species that live together in a particular habitat and interact with one another.

Competitive exclusion A situation in which the presence of one species prevents the presence of another.

Continental rise The wedge of sediments, derived from the erosion of the continents, that lies at the foot of the continental shelf and extends onto the edge of the abyssal plain (q.v.).

Continental shelf That portion of the lower-lying parts of a continent that are covered by shallow sea.

Continental slope That region in the sea, below the edge of the continental shelf, at which the seafloor descends steeply until it reaches the floor of the deep ocean that lies between the continental shelves.

Convergence Lines in the oceans along which water that is more dense and saline sinks below adjacent water that is lighter and fresher.

Corridor A pathway that permits the passage of most organisms.

Cosmopolitan group One that is widely distributed throughout the world.

Crown group The earliest common ancestor of a clade (q.v.), plus all of its descendants, which therefore have all of the characteristics found in all of those descendants. Contrast with *stem group* (q.v.).

Darwin's finches Finches that colonized the Galápagos Islands off the coast of western South America, where they underwent a radiation into a variety of forms.

De Geer route A former connection between northeastern Greenland and northwestern Europe.

Deciduous Perennial plants that shed their leaves during a cold or dry season.

Deoxyribonucleic acid A complex molecule, able to reduplicate itself, that lies at the heart of the genetic system; also known as DNA.

Diploid The genetic condition of most body cells, which have pairs of each chromosome, one having originated from each parent.

Disharmonic biota An island biota that lacks some of the normal components because these have been unable to reach, or survive in, the island.

Disjunct A pattern of distribution in which the areas occupied by a given organism are discontinuous, separated from one another.

Dispersal biogeography An approach to historical biogeography based on the assumption that related taxa that occupy ranges that are separate from one another arrived in them by crossing preexisting barriers.

Dispersal-vicariance analysis (DIVA) A type of parsimony-based tree fitting (q.v.), designed to cope with a reticulate pattern of relationship (q.v.), in which no general area cladogram is assumed.

DNA See *deoxyribonucleic acid*.

Duplication The presence of two related species in an area, when it is uncertain whether this arose by dispersal or vicariance.

East Pacific Barrier The barrier to the dispersal of the short-lived larvae of organisms that live on the continental shelves, formed by the wide, deep, almost island-free expanse of the East Pacific Ocean.

Ecological biogeography The study of aspects of biogeographical phenomena that focuses on the interactions between the organisms and their environment.

Ecological species concept See *species*.

El Niño Southern Oscillation or **ENSO** The cycle of changes in climate through much of the world that is caused by variations in wind and current intensity in the southern Pacific Ocean.

Endemic An organism that is found only in a particular region.

Energy flow The process by which solar energy is fixed by plants and then passes in turn to herbivores, carnivores, detritus feeders and decomposers.

Energy hypothesis The explanation of latitudinal gradients of species diversity as the result of variation in the amount of energy that is captured by vegetation.

Epicontinental sea Shallow sea that covers the lower-lying parts of the continents.

Epipelagic zone The upper, warmer layers of the sea, containing a high concentration of living organisms; up to 200 m (650 ft) deep.

Epiphytes Plants that use other plants for support.

Equatorial submergence The phenomenon whereby the level of the transition between the cold waters of the bathyal marine life zone (q.v.) and the warmer waters above it is deeper at lower latitudes.

Euphotic zone The upper few tens of meters depth of the sea, in which there is enough sunlight for photosynthesis to take place.

Eurytopic An organism that has wide ecological tolerance.

Eustatic A change in sea level caused by a change in the volume of water in the sea.

Evapotranspiration The sum total of water that evaporates directly from the surface of the ground, plus that lost by the uptake and loss of water by plants.

Event-based methods Methods of cladistic biogeographical analysis that specify what event (vicariance, duplication, dispersal, or extinction) took place at each point of branching of a biological area cladogram.

Evolutionary relicts A pattern of distribution in which a formerly more dominant organism now inhabits only the scattered remains of a formerly continuous area, due to competition with another species or group.

Extant group One that is still alive today.

Facilitation A situation in which the presence of one species aids the addition of another to the community.

Family A group of genera (q.v.).

Filter An ecological barrier that prevents the passage of certain categories of organism.

Gamma diversity The total species richness in a region comprising a number of smaller patches or habitats.

Gene A region of the DNA (q.v.) that is responsible for one or several characteristics of the organism.

General area cladogram or **GAC** See *pattern-based methods*.

General dynamic model The hypothesis that volcanic oceanic islands show a progression in which there is a linkage between area, altitude, erosion, habitat diversity, number of species and the proportion of them that are endemic to the island.

Generalized track See *track*.

Genotype The total of all the genes of an organism, making up its total genetic inheritance.

Genus, plural **genera** A group of species (q.v.) that are closely related to each other.

Geological area cladogram A cladogram (q.v.) that shows the sequence of separation of the various continents.

Gleason approach An individualistic interpretation of plant communities that emphasizes the varied ecological requirements of its component species. See also *Clements approach*.

Gondwana The supercontinent that, in past geological time, was formed of the southern continents (South America, Antarctica, Africa, and Australia) plus India, before these split apart.

Greenhouse gas A gas that contributes to the **greenhouse effect**, in which clouds and carbon dioxide absorb the heat reradiated by the Earth, so causing an increase in its temperature.

Gyre The huge mass of horizontally rotating water that fills a large part of an ocean basin.

Habitat The general type of environment within which an organism lives, for example, forest or marsh.

Hadal zone The marine life zone at a depth of over 6 km (3.5 mi).

Hadopelagic zone Deepest part of the ocean, within the submarine trenches (q.v.).

Halocline The level within the sea at which there is a rapid change in the salinity of the water.

Haploid The genetic condition of a sperm or ovum, which has only one of each pair of chromosomes.

Hermatypic corals The reef-forming corals in which there is a symbiotic relationship between the polyps of the coral and the algae, known as zooxanthellae, that live within them.

Historical biogeography The study of aspects of biogeographical phenomena that focuses on the origins and subsequent history of lineages and taxa.

Holarctic Found throughout the Northern Hemisphere; usually referring to mammals.

Hominids The group including humans and great apes.

Hominins The evolutionary lineage that branched away from that leading to the great apes, and that includes humans and their ancestors.

Hotspot A location, deep within the Earth, from which a plume of hot material rises to the surface. Where this occurs within an ocean, it leads to the formation of a volcano, which either reaches the surface as an island or remains as a submerged "seamount" or "guyot." The term is also used of a region of the Earth with unusually high biodiversity.

Hybrid The result of a mating between two different species or divergently adapted genotypes.

Hydrological cycle The movement of water from oceans via water vapor to precipitation and back to the ocean via streams and rivers.

Hydrothermal vents Points in the midoceanic spreading ridges where cold seawater penetrates into the rocks surrounding the area where hot lava is emerging, and reacts chemically with them so that minerals are precipitated out.

Incidence The pattern of occurrence of a species on islands and the factors that affect this pattern.

Indo-Pacific plant kingdom The region including India, Southeast Asia, and the islands of the Pacific Ocean, which contains a characteristic flowering plant flora.

Interglacial A period of time, between glacial events, that was warm enough for temperate vegetation to establish itself.

Interstadial A period of warmth, between glacial events, that was too short or too cool for temperate vegetation to establish itself.

Island arc The line of islands that form along a mid-oceanic region where old ocean crust is disappearing into the Earth.

Isolating mechanism Genetic systems that prevent mating between two different species, or that lead to any offspring having reduced fertility.

Isostatic A change in sea level relative to land level caused by a change in the level of the continental surface.

Keystone species A species that has an important influence on many other species in the ecosystem.

K-selected species Species that are slower to reproduce than r-selected species (q.v.), but are more able to sustain their population when this is close to the carrying capacity (q.v.). Characteristic of later colonists in a succession.

Laurasia The supercontinent that, in past geological time, was formed of the northern continents (North America and Eurasia), before these split apart.

Lessepsian exchange A link between marine faunas, through which they can exchange organisms; named after the link between the Mediterranean Sea and the Red Sea formed by the Suez Canal.

Life form A type of plant characterized by an assemblage of structural and physiological features that adapt it to life in a particular type of environment.

Limiting factor One that is responsible for limiting the pattern of distribution of an organism.

Macroecology The study of the assembly and structure of biotas that focuses on their general, large-scale patterns and mechanisms.

Marsupial In this, one of the two major groups of living mammals, the young leave the uterus at a very early stage and complete their development in the mother's pouch, in contrast with those of the placentals (q.v.).

Mediterranean climate One with hot, dry summers and cool, wet winters.

Megafauna The large terrestrial vertebrate component of a fauna.

Megatherm A plant that prefers temperatures above 20°C.

Mesopelagic zone The zone of reduced light intensity in the sea, lying below the pycnocline (q.v.), and which extends down to a depth of about 1,000 m (3,300 ft).

Mesotherm A plant that prefers temperatures between 13°C and 20°C.

Metabolic theory The explanation of latitudinal gradients of species diversity as being the result of variation in the metabolic activity of organisms.

Microclimate The physical conditions of temperature, intensity of light, humidity, and so on, that are found in a particular small-scale environment.

Microhabitat The fine-scale environment within which an organism lives (e.g., forest floor).

Microtherm A plant that prefers temperatures below 13°C.

Mid-Continental Seaway A shallow seaway that at one time ran across North America from the Arctic Ocean to the Gulf of Mexico.

Mid-domain effect An explanation for gradients of diversity that suggests that it is due merely to variations in the range of individual species.

Milankovich Cycles An explanation of the sequence of alternating warm and cold climatic episodes, over the last 2 million years, as being the result of variations in the Earth's orbit and in the tilt of its axis.

Model-based methods Methods of cladistic biogeographic analysis based on stochastic models, in which the biological processes involved are quantified and the principle of parsimony is not used in selecting the most likely explanation. Also known as parametric methods.

Monophyletic group One in which all its members are descended from a single common ancestor.

Natural selection The process whereby, due to differential survival and reproduction, the more advantageous characteristics persist into the next generation, while the less advantageous gradually disappear.

Nekton Organisms that swim in the waters of the sea.

Neritic The shallow-sea realm.

Nested clade analysis A type of phylogeographic biogeography (q.v.) in which the various states of a given molecule in the taxa involved are arranged into a series of groups differing in only one mutational alteration, and that is continued until they are all included in a single clade.

Neutral theory of biodiversity The proposal that assemblages of species are merely a collection of randomly selected species.

Niche The set of physical and biological conditions and resources within which an organism is found, and the role that a species plays within the community.

Null hypothesis A statistical technique that estimates how much similarity there would be between the results of the action of two sets of phenomena, assuming that there is no causal relationship between them. This can then be compared with the actual degree of similarity in order to find out whether or not this is the result of chance.

Nutrient cycling The process by which nutrients pass from organism to organism, and finally into the soil, from which they can be reused by plants.

Ocean baselines See *track*.

Oceanic circulation The worldwide pattern by which seawater is warmed in tropical latitudes and then circulated to higher latitudes before being returned to the tropics. Also known as the oceanic conveyor belt or thermohaline circulation.

Oceanic conveyor belt See *oceanic circulation*.

Oriental zoogeographic region India plus Southeast Asia, a region that contains a characteristic mammalian fauna.

Oroboreal flora Flora found only in the mountainous areas of Asia and North America.

Palaeomagnetism A technique that uses the presence of magnetized particles in rocks to deduce the movements of the rocks through time, and therefore of the landmasses in which they lay.

Palmae Province The Cretaceous equatorial region, containing a characteristic megathermal (q.v.) forest.

Panbiogeography An approach to historical biogeography based on the identification of generalized tracks (q.v.) and that relies on vicariance (q.v.) rather than dispersal (q.v.).

Pangaea The supercontinent that, in past geological times, was formed of all today's continents before these split apart.

Panthalassa The single, worldwide ocean that existed when all the continents were united in Pangaea (q.v.).

Parametric methods See *model-based methods*.

Parsimony A principle of analysis in which the explanation involves the minimum number of assumptions; also known as economy of hypothesis.

Parsimony-based tree fitting A type of event-based biogeographical analysis (q.v.) that uses the principle of parsimony (q.v.) in deciding on the most likely explanation.

Pattern-based methods Methods of cladistic biogeographical analysis that commence with an attempt to find a common pattern of relationships, known as a general area cladogram or GAC, which shows the physical history of the relationships between the areas of endemism that are involved.

Pelagic Organisms that swim or float in the sea.

Phenotype The total characteristics of an organism, resulting from the action of its genes.

Phylogenetic biogeography The analysis of patterns of distribution of organisms using groups whose interrelationships are analyzed using cladistics (q.v.).

Phylogeography A type of phylogenetic biogeography (q.v.) in which the interrelationships of the taxa are based on data from their DNA.

Phytodetritus The remains of phytoplankton, forming an important constituent of the muds and oozes that cover the seafloor.

Picoplankton Tiny, single-celled planktonic organisms.

Placental In this, one of the two major groups of living mammals, the whole period of development of the young takes place in the uterus, in contrast with the situation in the marsupials (q.v.).

Plankton Minute organisms that float in the waters of the sea.

Plant formation A large-scale ecosystem, such as desert or forest, found in different parts of the world, and characterized by a similar set of life forms of plants. (If the fauna also is included, the result is known as a biome.)

Plate tectonics The explanation of the history of the continents and oceans as resulting from the movements of the tectonic plates (q.v.).

Plesiomorphic The original ancestral or primitive state of a characteristic.

Polyploidy The doubling or multiplication of the whole set of chromosomes within the cells of an organism.

Polytypic species A species that contains many races or subspecies.

Productivity The amount of plant material that accumulates in a given area in a given time.

Pycnocline The level within the sea at which there is a rapid change in the density of the water.

Race A genetically or morphologically distinct set of populations of a species, confined to a particular area.

Range The geographical area within which an organism is found.

Rapoport's Rule The observation that organisms found in high latitudes tend to have broader geographical and altitudinal ranges and ecological tolerances than those found in lower latitudes.

Refugium, plural **refugia** A location in which some organisms have been able to survive a period of unfavorable conditions.

Relict An organism that now has a more limited distribution than it once had. In the case of a **habitat relict** or **climatic relict**, this is because of climatic change; in the case of a **glacial relict**, the organism has been left behind, in areas of cold climate, by the northward retreat of the Ice Age climates.

Rescue effect When local extinction of a species is prevented by the immigration of individuals of that species from elsewhere.

Reticulate pattern A network-like pattern of relationship between areas of endemism, caused by their having had more than one type of relationship with one another over time.

r-selected species Species with a high potential rate of population increase. Characteristic of early colonists of a succession. See also *K-selected species*.

Sclerophyll Having tough, thick, evergreen leaves. Also known as scleromorph.

Seafloor spreading See *spreading ridges*.

Seamount A submerged volcano that formed above a hotspot (q.v.); also known as a guyot.

Seismic waves Shock waves caused by earthquakes.

Shelf break The edge of the continental shelf, at which point the seafloor starts to descend more steeply, forming the continental slope, until it reaches the deep ocean floor or abyssal plain.

Shelf zone The marine life zone above the continental shelf (q.v.).

SLOSS debate The argument as to the relative advantages of Single Large, or Several Small, nature reserves in retaining species and reducing extinctions.

Species The fundamental unit of the taxonomic system, which can be defined in a variety of ways. Using the **biological species concept**, the species is a group of natural populations whose members can all breed together to produce offspring that are fully fertile, but that in the wild do not do so with other such groups. Using the **ecological species concept**, the species is a group of natural populations whose members all possess a set of characteristics (morphological, behavioral, phys-

iological, etc.) that adapt it to a particular ecological niche.

Species-energy theory The hypothesis that the number of species on an oceanic island is controlled by the amount of energy that falls upon it.

Spreading ridges The worldwide system of chains of submarine volcanic mountains, where new seafloor is being formed as the regions on either side move apart, a process known as seafloor spreading.

Stem group The ancestors of a crown group (q.v.), which may have had some of the characteristics of the crown group (q.v.).

Stenotopic An organism that has limited ecological tolerance.

Stochastic A result or process produced by chance, that is, not caused by the action of a regulatory force.

Subduction The process by which old ocean floor is drawn back into the Earth at the system of oceanic trenches (q.v.).

Subspecies A genetically or morphologically distinct set of populations of a species, confined to a particular area.

Succession A regular change in a community over time. When this begins with bare ground, it is sometimes called primary succession, as distinct from secondary succession, which refers to changes after the collapse or destruction of an existing community.

Sweepstakes route A dispersal route through which it is extremely difficult to pass, so that organisms can only do so by a chance combination of favorable circumstances, or by special adaptations to facilitate its passage.

Sympatric speciation The evolution of a new species within the same area as that of its parent species.

Taxon, plural **taxa** Any unit in the system of naming and classifying organisms (e.g., **species**).

Taxon cycle The hypothesis that the distribution and ranges of individual species in island communities go through a cycle of expansion and contraction.

Taxon-area cladogram A cladogram (q.v.) in which the name of each taxon has been replaced by that of the area in which it is found.

Taxonomy The study of the naming of organisms and their placement into a hierarchical system of classification.

Tectonic plates The assemblage of regions of the Earth's surface, containing ocean floor with or without continents, and which move across the face of the Earth, fusing or subdividing.

Temporal separation A situation in which two species occupy similar niches in the environment but at different times of the day.

Terminal Eocene Event The marked climatic cooling that took place at the end of the Eocene Epoch, involving a decrease in mean annual temperature and an increase in the mean annual temperature range.

Terrane A small area of originally marine or volcanic rocks, which have become scraped off against the edge of a continent as the ocean floor that bore them has become subducted below it, and which therefore totally differ from the other rocks that now surround them.

Tethys Ocean An ocean that once separated the southern continents from the northern continents, now represented only by the Mediterranean Sea.

Theory of island biogeography or **TIB** A theory that suggests that the changing, and interrelated, rates of colonization and extinction of organisms on oceanic islands eventually lead to an equilibrium between these two processes. The rate of replacement of species, or turnover rate, then becomes approximately constant, as does the number of species on the island. The theory also suggests that there is a strong correlation between the area of the island and the number of species that it contains at equilibrium.

Thermocline The level within the sea at which there is a rapid change in the temperature of the water.

Thermohaline circulation The worldwide pattern by which seawater is warmed in tropical latitudes and then circulated to higher latitudes, before being returned to the tropics. Also known as the oceanic conveyor belt or oceanic circulation.

Thulean route A former land connection between Greenland and Europe across the area now occupied by Iceland.

TIB See the *Theory of Island Biogeography*.

Till A clay-rich deposit, containing quantities of unsorted rounded and scarred boulders and pebbles, left behind during the melting and retreat of a glacier.

Track A line that connects the separate ranges of a set of related taxa, used in the theory of panbiogeography (q.v.). Where a number of unrelated sets of taxa show identical tracks, this is known as a generalized track; where these run across ocean basins, they are known as ocean baselines.

Trenches The system of deep submarine canyons where old ocean floor is consumed, disappearing downward into the Earth.

Turgai Sea A shallow seaway that once separated Europe from Asia; also known as the Obik Sea.

Turnover rate. The rate of replacement of species on oceanic islands. See the *Theory of Island Biogeography*.

Ungulate Hoofed.

Vicariance biogeography An approach to historical biogeography based on the assumption that related taxa, occupying ranges that are separate from one another, arrived in them before the appearance of the barriers that now separate them, rather than by dispersal across the barriers after they had formed.

Vicariant speciation A process of speciation that arises after an organism has become isolated as a result of vicariance (q.v.).

Wallace's Line The north-south line, running through Wallacea (q.v.), that separates the predominantly Asian fauna to the west from the predominantly Australian fauna to the east.

Wallacea The region, containing many islands, that lies between the continental shelves of Southeast Asia and Australia.

Zonation A regular spatial sequence of replacement of species caused by a similar sequence of change in physical or chemical conditions.

Index

Page numbers in **bold** refer to major concepts, and page numbers followed by t, f, and b refer to tables, figures, and boxes, respectively.